Springer Series in Chemical Physics

Edited by Vitalii I. Goldanskii

Springer Series in Chemical Physics

Editors: Vitalii I. Goldanskii Fritz P. Schäfer J. Peter Toennies

Managing Editor: H. K.V. Lotsch

40 **High-Resolution Spectroscopy of Transient Molecules**
By E. Hirota

41 **High Resolution Spectral Atlas of Nitrogen Dioxide 559–597 nm**
By K. Uehara and H. Sasada

42 **Antennas and Reaction Centers of Photosynthetic Bacteria**
Structure, Interactions, and Dynamics
Editor: M. E. Michel-Beyerle

43 **The Atom-Atom Potential Method.**
Applications to Organic Molecular Solids
By A. J. Pertsin and A. I. Kitaigorodsky

44 **Secondary Ion Mass Spectrometry SIMS V**
Editors: A. Benninghoven, R. J.Colton, D. S. Simons, and H. W. Werner

45 **Thermotropic Liquid Crystals, Fundamentals**
By G. Vertogen and W. H. de Jeu

46 **Ultrafast Phenomena V**
Editors: G. R. Fleming and A. E. Siegman

47 **Complex Chemical Reaction Systems**
Mathematical Modelling and Simulation
Editors: J. Warnatz and W. Jäger

48 **Ultrafast Phenomena VI**
Editors: T. Yajima, K. Yoshihara, C. B. Harris, and S. Shionoya

49 **Vibronic Interactions in Molecules and Crystals**
By I. B. Bersuker and V. Z. Polinger

50 **Molecular and Laser Spectroscopy**
By Zu-Geng Wang and Hui-Rong Xia

51 **Space-Time Organization in Macromolecular Fluids**
Editors: F. Tanaka, M. Doi, and T. Ohta

52 **Multiple-Photon Laser Chemistry**
By. R. V. Ambartzumian, C. D. Cantrell, and A. Puretzky

53 **Ultrafast Phenomena VII**
Editors: C. B. Harris, E. P. Ippen, G. A. Mourou, and A. H. Zewail

54 **Physics of Ion Impact Phenomena**
Editor: D. Mathur

55 **Ultrafast Phenomena VIII**
Editors: J.-L. Martin, A. Migus, G. A. Mourou, and A. H. Zewail

56 **Clusters of Atoms and Molecules.**
Salvation and Chemistry of Free Clusters, and Embedded, Supported and Compressed Clusters
Editor: H. Haberland

57 **Radiationless Transitions in Polyatomic Molecules**
By E. S. Medvedev and V. I. Osherov

58 **Positron Annihilation in Chemistry**
By O. E. Mogensen

59 **Soot Formation in Combustion**
Mechanisms and Models
Editor: H. Bockhorn

60 **Ultrafast Phenomena IX**
Editors: P. Barbara, W. H. Knox, G. A. Mourou, and A. H. Zewail

Volume 1–39 are listed on the back inside cover

E. S. Medvedev V. I. Osherov

Radiationless Transitions in Polyatomic Molecules

With 31 Figures

Springer-Verlag
Berlin Heidelberg New York
London Paris Tokyo
Hong Kong Barcelona
Budapest

Dr. Emile S. Medvedev
Professor Dr. Vladimir I. Osherov

The Institute of Chemical Physics in Chernogolovka
Russian Academy of Sciences. 142432 Chernogolovka, Russia

Title of the original Russian edition:
Teoriya Besyslutchatel'nykh Perekhodov v Mnogoatomnykh Molekulakh
© Nauka, Moscow 1983

ISBN-13: 978-3-642-85111-7 **e-ISBN-13: 978-3-642-85109-4**
DOI: 10.1007/978-3-642-85109-4

Library of Congress Cataloging-in-Publication Data. Medvedev, Ė. S.(Ėmil ' Samuilovich) [Teoriia bezyzluchatel 'nykh perekhodov v mnogoatomnykh molekulakh. English] Radiationless transitions in polyatomic molecules / E. S. Medvedev, V. I. Osherov. p. cm. – (Springer series in chemical physics; v. 57) Includes bibliographical references and index. ISBN 0-387-57769-6 1. Molecular theory. 2. Radiationless transitions. I. Osherov, V. I. (Vladimir Iosifovich) II. Title. III. Series. QD461.M4413 1995 539'.6–dc20 94-19897

Softcover reprint of the hardcover 1st edition 1995

Typesetting: Macmillan India Ltd., Bangalore 25
SPIN: 10069991 54/3140 - 5 4 3 2 1 0 - Printed on acid-free paper

Preface

Radiationless transitions comprise an important class of physical phenomena occurring in the excited states of molecules. They affect the lifetimes of the excited states and govern primary photochemical and photophysical processes.

Much effort has been devoted to the understanding of radiationless transitions. Still, owing to recent advances, the field continues to attract attention. The demand for a book on the theory of these processes naturally arises in attempting to comprehend a large body of literature, as the famous review article by K. F. Freed and the book by R. Englman do not encompass some issues of current interest.

Our intent is to highlight the underlying physical principles and methods in such a way that the book both in its content and its presentation is instructive for a wide audience.

The basic ideas are treated in simple mathematical terms intelligible to experimentalists and to readers unfamiliar with the field. Complicated theoretical methods are always expounded from first principles, so that a knowledge of quantum mechanics and mathematics at the graduate-student level will enable the reader to easily follow the derivations. Experts will find efficient methods of calculating the transition rates, as well as new applications of quasiclassical methods and fresh treatments of standard problems.

Details of measurements are not discussed, and experimental data are only invoked to illustrate the theory.

The authors are much indebted to Professors E. E. Nikitin, M. D. Frank-Kamenetskiĭ and Dr. M. Ya. Ovchinnikova for reading the manuscript and helpful criticism. Our particular gratitude is due to Dr. V. G. Ushakov for his valuable assistance in solving the problem of Sect. 5.7. We are also very grateful to Dr. E. B. Sveshnikova for the help in writing Sect. 6.7. One of us (ESM) expresses gratitude to Professor D. W. Pratt for his hospitality during a joint research at the University of Pittsburgh, and to the U.S. National Research Council and the International Science Foundation for financial support.

Chernogolovka, September 1994

Emile S. Medvedev
Vladimir I. Osherov

Preface

[illegible] transitions constitute an important class of physical phenomena [illegible] excited states of molecules. They affect the lifetimes of the ex[illegible] states [illegible] governing many photochemical and photophysical processes.

[illegible] the [illegible] to the understanding of radiationless transi[illegible] [illegible] the field continues to attract attention. [illegible] book on the theory of these processes naturally arises in [illegible] and a large body of literature; as the famous review article [illegible] and the book [illegible] do not encompass some issues of [illegible].

[illegible] is to highlight the underlying physical principles and methods [illegible] the book [illegible] its content and its presentation is instruc[illegible] [illegible] audience.

[illegible] ideas are treated [illegible] simple mathematical terms intelligible to [illegible] and to readers [illegible] familiar with the field. Complicated theore[illegible] [illegible] derived from first principles, so that a knowledge [illegible] the graduate-student level will [illegible] easily follow the derivations. Experts will find efficient [illegible] transition rates, as well as new applications of [illegible] solutions of standard problems.

[illegible] discussed, and experimental data are only [illegible] to the [illegible].

[illegible] to Professors [illegible] for reading the manuscript and [illegible] is due to Dr. V.G. Ushakov for his [illegible] of Sect. 5.[illegible]. We are also very [illegible] for the help in writing Sect. 6.7. One of us [illegible] for his hospitality during [illegible] at the University of Pittsburgh, and to the U.S. National [illegible] Science Foundation for financial sup[illegible]

[illegible]

Emile S. Medvedev
Vladimir I. Osherov

Preface

Radiationless transitions comprise an important class of physical phenomena occurring in the excited states of molecules. They affect the lifetimes of the excited states and govern primary photochemical and photophysical processes.

Much effort has been devoted to the understanding of radiationless transitions. Still, owing to recent advances, the field continues to attract attention. The demand for a book on the theory of these processes naturally arises in attempting to comprehend a large body of literature, as the famous review article by K. F. Freed and the book by R. Englman do not encompass some issues of current interest.

Our intent is to highlight the underlying physical principles and methods in such a way that the book both in its content and its presentation is instructive for a wide audience.

The basic ideas are treated in simple mathematical terms intelligible to experimentalists and to readers unfamiliar with the field. Complicated theoretical methods are always expounded from first principles, so that a knowledge of quantum mechanics and mathematics at the graduate-student level will enable the reader to easily follow the derivations. Experts will find efficient methods of calculating the transition rates, as well as new applications of quasiclassical methods and fresh treatments of standard problems.

Details of measurements are not discussed, and experimental data are only invoked to illustrate the theory.

The authors are much indebted to Professors E. E. Nikitin, M. D. Frank-Kamenetskiĭ and Dr. M. Ya. Ovchinnikova for reading the manuscript and helpful criticism. Our particular gratitude is due to Dr. V. G. Ushakov for his valuable assistance in solving the problem of Sect. 5.7. We are also very grateful to Dr. E. B. Sveshnikova for the help in writing Sect. 6.7. One of us (ESM) expresses gratitude to Professor D. W. Pratt for his hospitality during a joint research at the University of Pittsburgh, and to the U.S. National Research Council and the International Science Foundation for financial support.

Chernogolovka, September 1994

Emile S. Medvedev
Vladimir I. Osherov

Contents

1. Introduction

The transfer of electronic excitation energy to atomic vibrations with no emission of photons is one of the principal mechanisms for electronic relaxation in excited states of molecules. The rate of this process in real systems varies very widely, ranging from 0.1 to more than 10^{12} s^{-1}, and frequently dominates the lifetime, a most important physical characteristic of the excited state. Radiationless transitions often populate metastable excited electronic and vibrational states, which are, as a rule, hard to prepare by direct optical excitation because of low oscillator strengths. The radiationless population of the metastable states gives rise to a variety of interesting physical phenomena, e.g., phosphorescence, delayed fluorescence, triplet–triplet absorption, etc.

The physical basis and mathematical tools of the theory of radiationless transitions have been addressed in [1.1–38]. Yet some fundamental points, which are of great importance to the understanding of the nature of radiationless transitions, as well as the physical hypotheses underlying current methods of calculating the rate constants for these processes, have up until now not been elucidated. Another motivation for writing this book was the need of a discussion of the main principles and methods of the theory that would be complete enough and still be intelligible to those who are not experts in the field. We have also considered the solutions to some theoretical problems previously given by a number of authors and now widely recognized as unsatisfactory.

In many respects the content of this book differs from the works cited above. The book opens with a qualitative discussion of electronic relaxation (Chap. 2) based on primitive models using only a bare minimum of mathematics. The detailed account of early theories in a modern context may sound archaic. However, one should bear in mind that the sophisticated mathematical treatments developed later, outlined in Chap. 3, contributed nothing to the main physical concepts formulated semiintuitively; as often happens, these mathematical methods just formed a solid foundation for the methods in current use and enabled one to specify the phenomenological parameters entering the theory. In spite of involving numerous and sometimes very drastic simplifications, the earlier theories correctly explained the nature of irreversibility of radiationless transitions in isolated large molecules and connected electronic relaxation with vibrational relaxation in the final electronic state.

It is especially worth mentioning the Bixon–Jortner theory. This theory resorts to such a highly idealized model of molecular energy levels and couplings that it is usually referred to only to illustrate the origin of the irreversible behavior of isolated molecules, whereas its transition rate expression is not considered adequate for the description of real molecules [1.1, 3]. An alternative point of view developed in Chap. 2 is that, in the so-called large-molecule statistical limit, where the transition rate is independent of the molecule–medium coupling and falls exponentially with the electronic energy gap, the dominant contribution to the transition rate comes from the final states that do satisfy the basic Bixon–Jortner assumptions (we call them effective states). Thus, the Bixon–Jortner rate expression is adequate for real systems under the statistical limit provided that only the effective states are taken into account.

The rationale behind the above point of view is given in Chap. 6, where a way is shown by which the effective states can be found in reality and their parameters calculated from a given model of potential surfaces. The results obtained are then used to develop a new method for calculating the radiationless transition rate in the statistical limit. The conventional saddle-point method is also described in the same chapter and the two methods are compared. Their areas of application do not overlap completely, so that new results can be obtained by the effective-states method (Sect. 6.4).

In practice, however, the Metz method of summing up the Franck–Condon factors as well as the semiempirical Sveshnikova–Ermolaev method are used to calculate the transition rate in the statistical limit. They are presented in Sects. 6.6, 7.

The effective-states method also enables the necessary statistical-limit condition to be formulated correctly, which requires a high enough density of the effective states. This essentially differs from the Freed–Jortner criterion which demands a high total density of *all* the final states, and which is considered here the criterion for irreversibility (Sect. 2.4.2). The molecule obeying the Freed–Jortner irreversibility criterion behaves irreversibly even under conditions of complete isolation, though it can still be categorized as either the statistical limit or the intermediate case, depending on the density of the effective states (Sect. 2.11 and Chap. 7).

Applying the effective-states method requires a simple explicit dependence of the Franck–Condon factor on vibrational quantum numbers. The necessary formulas can be obtained by the Landau method [1.39, 40] for calculating the probabilities of classically forbidden (tunnel) transitions, invoking analytical continuations of quasiclassical wave-function asymptotes into the complex plane. The Landau method, described in Chap. 5, allows an extension of the Franck–Condon principle to radiationless transitions and states that major contributions to the rate of the tunnel radiationless transition (or, which is the same, to the Franck–Condon factor) come from electronic term crossings in the classically forbidden nuclear coordinate region and/or even in the complex plane. This sheds a new light on the Condon approximation and on the problem of selection rules for radiationless transitions (Chap. 4).

It should be emphasized that Chap. 5 treats problems rarely met in the traditional approaches to radiationless transitions. One reason may lie in the complexity of the corresponding analysis. The detailed derivations presented are intended to help introduce the reader to this new and interesting research area, which is still waiting for further development. At first glance some issues discussed, including the collision model, the overtone transitions and the strongly coupled two-state quasiclassical energy levels, may seem to be beyond the scope of this book. However, their close relevance to the problems of radiationless transitions becomes evident as soon as the above generalized Franck–Condon principle is invoked. For instance, higher overtone intensities were shown to provide new information on the ground-state potential energy, important in calculating the radiationless transition rate to this state.

The concept of effective states also provides a natural way to define the intermediate case of large molecules (Sects. 2.11, 3.5), which is characterized by a low density of the effective states and, simultaneously, by a high total density of all the final states. The latter condition leads to a level broadening of the effective states due to irreversible intramolecular relaxations (vibrational and/or electronic to a third state) while the former demands that the broadening does not cover the energy spacings between the effective states.

In Sects. 2.5–7 dynamical aspects of the intermediate case and the statistical limit are discussed in terms of the mono- and biexponential decay kinetics as well as the oscillatory behavior of the fluorescence emission. In Chap. 7 the method of correlation functions for calculating the intermediate case radiationless transition rate is described. In the intermediate case the electronic relaxation has much to do with the processes of intramolecular vibrational redistribution and vibrational energy relaxation, so the actual problem is to derive kinetic equations for the density matrix similar to those in the theory of magnetic resonance.

Some significant aspects of the theory have not been given proper attention. This especially pertains to such an important matter as the comparison between theory and experiment. Notwithstanding that an abundance of both experimental data and theoretical analyses are now available, only a limited number of experimental findings have so far received a credible quantitative interpretation. One reason is that the set of data on excited electronic states is, as a rule, insufficiently complete to allow the calculation of the radiationless transition rate. Moreover, the process is often complicated by the presence of an intermediate electronic state [1.28, 29, 33, 41–45].

The theory of radiationless transitions in molecules and in optical centers of solids essentially stems from the same fundamentals [1.2, 15, 16, 19, 20, 24, 25], but for the sake of clarity mainly polyatomic molecules are treated in this monograph. The discussion is confined to the radiationless transitions from low-lying vibrational sublevels of the initial electronic state, in which case the transition involves tunneling; at higher energies of the initial vibrational excitation, additional channels for the electronic relaxation via excited electronic states are opened, which significantly complicates the analysis of such processes. Only in the broad terms discussed in Chap. 4 are the nature and magnitude of the electronic interactions responsible for the radiationless transitions [1.23, 32, 46–53].

The reader is not expected to have any prior acquaintance with the theory of radiationless transitions. Some sections call for serious mathematical knowledge, though not going beyond university courses. This is particularly true for Chap. 5, but it may safely be omitted on a first reading (except for Sects. 5.3.3, 5.4).

Literature on radiationless transitions is quite extensive and is far from being covered with the references given, though an attempt has been made to list reviews and monographs as completely as possible.

2. Qualitative Theory of Radiationless Transitions

The lifetimes of the majority of the excited electronic states of polyatomic molecules are very short and therefore are not easy to measure experimentally. At present the low-lying states with lifetimes in excess of 10^{-9} s are mostly studied in great detail. Indirect information on extremely short lifetimes may in some cases be drawn from absorption linewidths [2.1–3] and emission quantum yields [2.4–12]. The advent of picosecond and subpicosecond techniques [2.3, 13–17] greatly expands the manifold of the electronic states under investigation.

The experimental methods to produce electronically excited molecules include direct photoexcitation, chemical reactions, and intra- and intermolecular electronic excitation energy transfer. In many practically important cases the excitation process is fast enough to ensure that the subsequent excited state evolution relates only to the properties of the system under study and not to the details of the excitation process. When the latter condition is not obeyed, excitation and emission should be regarded as a unique compound process [2.6, 7, 18].

This chapter is devoted to a qualitative description of the instantaneously excited electronic state dynamics in the framework of simply treatable models amenable to clear physical interpretations, and providing adequate representations of the basic physical ideas. The chapter starts with a discussion of the simplest conceivable luminescence behavior governed by a simple kinetic (balance) equation, and with definitions of basic parameters to characterize the excited states. Experimental observations are briefly described, and then a more extensive treatment of the fluorescence decay, including the nonkinetic behavior, is elaborated in frames of solvable models.

2.1 Balance Equation

The conventional way to study the properties of the excited electronic states is to measure temporal, spectral and integral features of the luminescence under the condition of *instantaneous excitation*, i.e., excited by a light pulse with the *pulse duration* τ_{pulse} much shorter than the *observed emission decay time* τ_{obs} (which is just the full lifetime of the excited state),

$$\tau_{\text{pulse}} \ll \tau_{\text{obs}} . \tag{2.1.1}$$

Let $P(t)$ stand for the ratio of the number of excited molecules to their total number at time t, and τ_{rad}^{-1} for the rate of spontaneous emission associated with a transition to the ground electronic state; τ_{rad} is the *radiative lifetime* of the excited state. The *luminescence intensity*, i.e., the average photon emission rate per molecule, is

$$I(t) = \tau_{\text{rad}}^{-1} P(t) . \tag{2.1.2}$$

At present additional powerful methods are elaborated to prepare the excited state (e.g., electron-impact excitation [2.19]) and to directly measure the $P(t)$ (chemical timing [2.20], photoionization [2.21–23]), but for clarity we will refer to the pure photoexcitation and photoemission methods.

An important luminescence characteristic is the *quantum yield* φ defined as the ratio of the total number of photons emitted at $t > 0$ to that absorbed during the excitation process at $t = 0$; under one-photon excitation, the number of photons absorbed per molecule is obviously equal to $P(0)$. Thus

$$\varphi = \frac{1}{P(0)} \int_0^\infty I(t)\, dt . \tag{2.1.3}$$

The simplest luminescence kinetics is described by the *balance equation*

$$\dot{P}(t) = -(\tau_{\text{rad}}^{-1} + k)\, P(t) , \tag{2.1.4}$$

where k usually designates the *electronic radiationless transition rate* to the ground state or a lower-lying excited state. In general, however, it is the total nonradiative decay rate of the fluorescent state; specifically, it may include the rate of vibrational energy redistribution within the same electronic state providing that such a process stops light emitting in a given spectral range.

According to (2.1.2, 4) the luminescence intensity decays exponentially,

$$I(t) = I(0)\, e^{-t/\tau_{\text{obs}}}, \tag{2.1.5}$$

with the observed lifetime

$$\tau_{\text{obs}} = \frac{\tau_{\text{rad}}}{1 + k\tau_{\text{rad}}} \tag{2.1.6}$$

and the quantum yield

$$\varphi = \frac{\tau_{\text{obs}}}{\tau_{\text{rad}}} = \frac{1}{1 + k\tau_{\text{rad}}} . \tag{2.1.7}$$

Thus, if the luminescence decay follows the simple exponential law (2.1.5), then (2.1.6, 7) allow the determination of the luminescent state parameters τ_{rad} and k via the measured values of τ_{obs} and φ. When the oscillator strength of a given transition is known from the absorption spectrum, τ_{rad} can be calculated independently by the Einstein relations. Discrepancies between the τ_{rad} values thus found and the values directly obtained from emission experiments (e.g., in acetone [2.24] and other molecules [2.25]) may signal the breakdown of the above simple decay scheme based on the balance equation (see Sect. 2.2). The τ_{rad} values estimated from the absorption spectra are also employed for the determination of k from (2.1.6) in the cases where no data on the quantum yield are available and only τ_{obs} is measured.

When condition (2.1.1) is violated, $I(t)$ departs from (2.1.5), now depending on the shape of the exciting light pulse. Let $F(t)$ denote the *excitation function*, i.e., the relative number of molecules excited per unit of time. It is proportional to the source intensity at time t and is only nonvanishing within a time interval of τ_{pulse} around $t = 0$. Adding $F(t)$ to the right-hand side of (2.1.4) and integrating the new equation gives the luminescence intensity (2.1.2) in the form

$$I(t) = \tau_{rad}^{-1} \int_{-\infty}^{t} e^{-(t-t')/\tau_{obs}} F(t')dt' , \tag{2.1.8}$$

where τ_{obs} is again given by (2.1.6). Thus the luminescence intensity is proportional to the convolution of the monoexponential decay (2.1.5) with the excitation function.

Now the quantum yield definition (2.1.3) needs a refinement. We define the quantum yield as the ratio of the total number of emitted to the total number of absorbed photons

$$\varphi = \int_{-\infty}^{+\infty} I(t)dt \Big/ \int_{-\infty}^{+\infty} F(t)dt . \tag{2.1.9}$$

Substituting (2.1.8) and changing the integration order (or, equivalently, directly integrating (2.1.4) with $F(t)$ added) gives the same expression (2.1.7), which proves thus to be independent of the shape and duration of the exciting pulse. Now, however, as distinct from the decay (2.1.5) at the short-time excitation, the τ_{obs} values cannot be directly estimated by the luminescence decay curve (2.1.8), and deconvolution procedures should be exploited.

Under condition (2.1.1) of prompt excitation at time $t = 0$, the exponential in the integrand of (2.1.8) is a slowly varying function of time as compared to the excitation function. Taking it off the integration sign at $t' = 0$ (for $t > 0$) and integrating to infinity leads, as expected, to (2.1.5) and transforms (2.1.9) back to (2.1.3) since the integral of $F(t)$ is just the total number of the photons absorbed per molecule, i.e., $P(0)$. In this limit the excitation function obviously takes the form [1.3]

$$F(t) = P(0)\,\delta(t) , \tag{2.1.10}$$

where $\delta(t)$ is the Dirac delta function.

In the opposite limit of stationary excitation, $\tau_{pulse} \gg \tau_{obs}$, taking the slowly varying function $F(t)$ off the integration sign in (2.1.8) at $t' = t$ (or, equivalently, setting $\dot{P}(t) = 0$ in equation (2.1.4) with $F(t)$ added) one obtains

$$I(t) = \varphi F(t)\,, \tag{2.1.11}$$

i.e., the emission follows the excitation pulse. If $F(t)$ (and $I(t)$ in parallel) changes at the front and back portions of the pulse only, being flat at times in-between, then measuring the stationary absorption ($F(t) \equiv F_{\text{stat}}$) and emission ($I(t) \equiv I_{\text{stat}}$) intensities at $0 < t < \tau_{\text{pulse}}$ by collecting the transmitted and emitted light, respectively, one can find the quantum yield by the formula

$$\varphi = I_{\text{stat}}/F_{\text{stat}}\,. \tag{2.1.12}$$

The main features of (2.1.4) are that it introduces the radiative and nonradiative decays, balancing them with the production of the excited molecules (introduced by either an initial condition on $P(t)$ or the addition of an excitation function to its right-hand side), and the total decay rate is a sum of the partial ones. In a more general case the build-up of the excited molecules in the luminescent state may include contributions from other excited states (Sect. 7.2).

The balance equation operates with state populations since only diagonal elements of the density matrix enter it. The most general reason for the lack of coherences, i.e., off-diagonal elements, is the existence of an efficient mechanism for a loss of the phase memory leading to a fast relaxation of the coherence (if it was created initially), and hence to the lack of a backward phase-correlated effect of other (dark) states on the bright state Φ_s.

The experimental data referred to in the next section both confirm the simplest decay pattern pictured by the balance equation and reveal numerous violations of the predicted exponential decay, indicating the breakdown of the balance equation and necessitating the analysis, to be performed later in this chapter, of the conditions for its validity as well as of the new decay patterns having emerged.

2.2 Experimental Observations and Empirical Rules

Convincing experimental evidence for the radiationless transitions was obtained in photoluminescence studies of organic molecules in solid and liquid phases as well as in dense and rarefied gases. A new powerful method to produce perfectly isolated rotationally cooled molecules in supersonic free jets has become available recently [2.26]. Electronic states of the organic molecules are specified by the total spin [2.27], and they are arranged on the energy scale by the *Jabłoński* diagram shown in Fig. 2.1 [1.36, 2.28]. The ground state is the singlet S_0. There are manifolds of singlet ($S_1 \ldots S_n$) and triplet ($T_1 \ldots T_n$) excited states, the T_1 state normally being intermediate between S_0 and S_1. Higher spin states are disregarded as little information on them is available [2.28, 29].

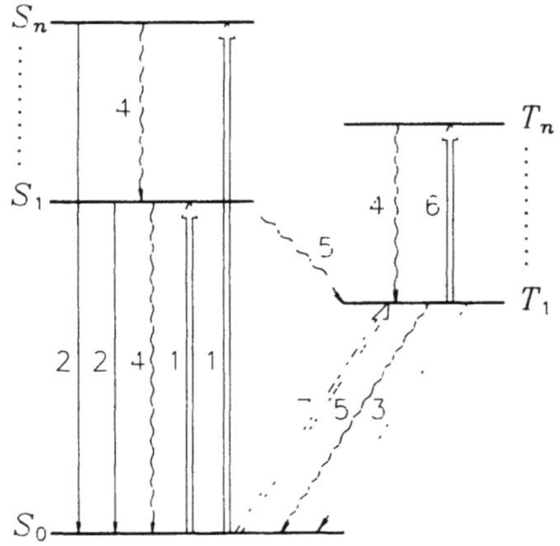

Fig. 2.1. Electronic energy levels of organic molecules (the Jabłoński diagram) and the observed radiative (→) and nonradiative (⇢) transitions: (*1*) excitation (singlet–singlet absorption); (*2*) fluorescence; (*3*) phosphorescence; (*4*) internal conversion; (*5*) intersystem crossing; (*6*) triplet–triplet absorption; (*7*) singlet–triplet absorption

The molecules photoexcited by visible or UV radiation display two kinds of luminescence: Fluorescence generated by singlet–singlet radiative transitions with lifetimes of $10^{-8} - 10^{-6}$ s and phosphorescence due to spin-forbidden triplet–singlet transitions lasting for $10^{-4} - 10^{1}$ s. The occurrence of the radiationless transitions is evidenced by the following experimental findings.

Firstly, the photoexcitation of the organic molecules in the $S_1 \leftarrow S_0$ absorption region brings about the $S_1 \rightarrow S_0$ fluorescence with a quantum yield lower than unity [2.30]. According to (2.1.7) this attests to the existence of a radiationless deactivation process in the S_1 state of which the rate is comparable to that of spontaneous emission. A further observation is that the $S_1 \leftarrow S_0$ excitation in both condensed and dense gas phases often gives rise to the $T_1 \rightarrow S_0$ phosphorescence and the $T_n \leftarrow T_1$ absorption [2.31], and this permits to qualify the above deactivation processes as the radiationless singlet–triplet $S_1 \dashrightarrow T_1$ transition called *intersystem crossing*[1] [1.37].

Secondly, the vast majority of the organic molecules obey *Kasha*'s rule [1.37], which reads that, independent of the excitation process (particularly, of the exciting wavelength in the case of the photoexcitation), the emission always originates from the lowest excited electronic state of a given multiplicity. In other words, the fluorescence and phosphorescence are always emitted from S_1 and T_1, respectively, (exceptions are azulene [1.28, 2.38, 39] and its derivatives [2.40], where the fluorescence comes from S_2, as well as pyrene [2.41, 42],

---

[1] This term originates from the initial wrong thinking that the existence of actual intersections of the potential surfaces in the classically accessible region is a necessary condition for the radiationless transition [2.32–34], which is thus viewed as an "intramolecular collision of the second kind" [2.32]. However, it was shown in [2.35–37] that the radiationless transitions in polyatomic molecules, as a rule, involve tunneling due to small relative displacements and distortions of the potential surfaces (see [1.18]). The current meaning of this term is "the radiationless transition between states of different multiplicity". For the role of surface intersections in tunnel transitions and for the collision model see Sects. 5.4.2, 5.7.

acetone [2.43], and some other compounds [2.44] where S_1 and S_2 are fluorescing simultaneously). Kasha's rule is accounted for in a natural way by the occurrence of fast radiationless transitions between the excited states of a given multiplicity (*internal conversion* [2.32]), of which the rate steeply increases upon decreasing the energy gap separating adjacent electronic states. Since the singlet energy spacings between the excited states are significantly smaller than those between S_1 and S_0, the suppression of the luminescence from higher states S_2, S_3, etc. due to internal conversion from S_n to S_1 is much more efficient than from S_1 due to the conversion to S_0 (this argument does not hold for azulene since its S_1 state is approximately half way between S_0 and S_2). Experimental data imply that the internal conversion within the excited singlet [1.28, 2.4–13, 45–49] and triplet [1.28] manifolds are, as a rule, much faster than $10^{10}\ s^{-1}$ (see, however, [2.12]).

Kasha's rule may equally be formulated as the following empirical trend in the absorption spectra of large polyatomic molecules. Only the lowest excited state of a given multiplicity displays sharp features in the absorption spectrum whereas the spectra of higher states are diffuse [1.11, 37, 2.4–12, 48].

Experimental data show that the dominating channel for the physical intramolecular degradation of the emission from the first excited singlet state is provided by the intersystem crossing, i.e., by the transition into the lower triplet state (*Ermolaev–Sveshnikova*'s rule [1.26, 27]). Though the $S_1 \rightsquigarrow S_0$ internal conversion may sometimes well compete with the $S_1 \rightsquigarrow T_1$ intersystem crossing (e.g., in propynal [2.50], phenol [2.51] and other molecules [2.6, 52]), especially upon increasing the initial vibrational energy in S_1 [1.3, 2.53–64].

The initial amount of the S_1 vibrational energy depends, in particular, on the wavelength of the exciting radiation, being larger the larger the photon energy (the latter should not, of course, exceed the S_2 excitation origin). The fast vibrational relaxation occurring in condensed phases establishes the Boltzmann distribution over the S_1 vibrational states. As this distribution only depends on the medium temperature and not on the initial vibrational energy, *Vavilov*'s law holds in condensed phases: The fluorescence spectrum and the quantum yield in polyatomic molecules are independent of the wavelength of the exciting radiation over the whole absorption spectrum [2.65]. Vavilov's law does not apply to vapors where the fluorescence quantum yield drops with increasing energy of the photon absorbed[2] [2.65–67]. In this case the relaxation rate is not high enough so that the emission occurs from states with over-thermal vibrational energy. Therefore, the violation of Vavilov's law is explained by the increase of the radiationless transition rate with the initial vibrational energy. There are molecules that do not show any emission in low-pressure vapors at all but start emitting upon addition of an inert gas that accelerates the vibrational relaxation

[2] Except for anthracene which obeys Vavilov's law in the gas phase, too. Some anthracene derivatives even display an increase in the gas phase fluorescence yield with the exciting photon energy [2.68, 69]. Polyatomic molecules may follow Vavilov's law in very dense vapors as well.

in the initial electronic state (the so-called *fluorescence stabilization*) [1.35, 36, 2.34, 66, 70, 71]. On the other hand, no fluorescence enhancement is seen if the radiationless transition is insensitive to the vibrational energy (as in perylene [2.72]).

Numerous studies have revealed that the electronic relaxation processes in a variety of molecular systems are intramolecular in origin [1.3, 4, 17, 33]. For instance, the luminescence lifetimes and quantum yields for organic molecules are often known to be nearly insensitive to the molecular environment in the condensed phase. Moreover, Kasha's rule holds true [1.18, 2.34, 73, 74] and the quantum yield remains lower than unity [2.75] even in the gas phase at very low pressures when the time interval between successive collisions is significantly longer than the excited-state lifetime.

An important observation is that the magnitude of the energy gap ΔE between two electronic states involved in the radiationless transition strongly affects the very nature of the process itself. Thus, at $\Delta E > 10000\,\mathrm{cm}^{-1}$ the so-called statistical limit of large molecules is usually reached, and the luminescence kinetics is described by the balance equation of Sect. 2.1. The *energy gap law* and *Siebrand*'s rule, or the *normal deuterium isotope effect*, reflect the following properties inherent to most organic molecules in the statistical limit: The radiationless transition rate falls off exponentially with the increase of ΔE and diminishes upon deuteration, i.e., upon substitution of deuterium for hydrogen (Sect. 2.11). *El-Sayed*'s rule [1.23, 29–34, 51] states that the rate of intersystem crossing between different electronic configurations (i.e., $n\pi^* \leftrightarrow \pi\pi^*$) is two to four orders of magnitude higher as that between similar configurations (i.e., $n\pi^* \rightarrow n\pi^*$ or $\pi\pi^* \rightarrow \pi\pi^*$).

On the other hand, a number of polyatomic systems under collisionless conditions in low-pressure vapors or supersonic jets display more complicated fluorescence kinetics than that described in Sect. 2.1. The biexponential decay behavior was first found in pyrazine [2.76] and then in pyrimidine and its methyl derivatives [2.77–81], *s*-triazine [2.82], quinoxaline [2.83, 84], biacetyl [2.85–88], glyoxal [2.89], methylglyoxal [2.86, 89, 90], acetaldehyde [2.91], acetone [2.43, 92, 93], and other organic [2.42, 94–106] and inorganic [2.107] polyatomic molecules, as well as in cations [2.108, 109] and triatomic molecules [2.110–113]. Another deviation from the simple exponential decay manifests as chaotic or quasi-periodic quantum beats of the luminescence intensity of these molecules with or without external magnetic field [2.87, 88, 99–105, 111–134].

The above phenomena are manifestations of the coupling between the electronic states (mostly between S_1 and T_1) separated by an energy gap $\Delta E < 5000\,\mathrm{cm}^{-1}$. They correspond to the intermediate case of large molecules or even to the small molecule limit (Sect. 2.11). The balance equation is no longer valid, and the excited state dynamics has to be described with detailed use of quantum mechanics (Sects. 2.3–7).

The interest toward the nonstatistical-limit molecules was greatly enhanced when molecular rotations were shown to significantly influence the nonexponential fluorescence decay [2.135–138]. Owing to deep rotational cooling (up to a few Kelvin) reached in supersonic jets, the rotational effects are now investigated in great detail [2.82, 87, 88, 114–134, 138–158], as well as polarization properties [2.151–160], magnetic field effects [2.102–105, 113, 129, 132, 133, 151–153, 156–159], the influence of the exciting pulse duration [2.83, 131, 161], of the excitation wavelength, and of the gas pressure. The benzene phosphorescence-decay acceleration by three orders of magnitude in going from the solid to the gaseous phase is attributed to the rotational effect [2.162]; a similar effect was observed in pyrazine [2.19]. A useful means to investigate the radiationless transitions is to observe the effects produced by a formation of van der Waals complexes [2.163, 164]. The experimental results obtained are summarized in review articles [1.3, 2.117, 123, 165–168].

Additional experimental facts, which may seem at first glance to be out of the scope of this book, are worth to be mentioned. These are fluorescence quantum beats and biexponential decays in larger molecules such as anthracene [2.169–174] and others [2.174–176] caused by *vibrational* coupling within the S_1 state. Obviously, the vibrational energy redistribution rather than the electronic energy relaxation operates in this case. However, the approaches to solve the two problems and the results obtained are very similar. Thus, the nonstatistical limit Intramolecular Vibrational energy Redistribution ("restricted IVR" in *Zewail*'s terminology [2.172]) is responsible for the nonexponential fluorescence decay of anthracene.

2.3 Molecular Energy Level Model

This section describes a simple energy level model which serves as a basis for the theory of radiationless transitions in polyatomic molecules [1.41, 2.36, 37, 177–184].

Specification of the molecular states rests on the separation of the electronic and nuclear motions under the adiabatic (Born–Oppenheimer) approximation when the total wave function is represented as the product (Sect. 4.1)

$$\Psi_{rm}(x,q) = \varphi_r(x,q)\,\chi_{rm}(q)\,, \tag{2.3.1}$$

where x and q are sets of electronic and nuclear coordinates, respectively; $\varphi_r(x,q)$ is the electron wave function at fixed values of the nuclear coordinates (it is an eigenfunction of the total molecular Hamiltonian minus the nuclear kinetic energy) and $\chi_{rm}(q)$ is the wave function for the nuclear motion in an effective potential, which depends on the distribution of the electronic density $|\varphi_r(x,q)|^2$ for nuclei at rest. The approximate wave function (2.3.1) is specified by the electronic and vibrational quantum numbers (r and m, respectively), and the corresponding energy is the sum of the electronic and vibrational energies.

The total molecular Hamiltonian H is not diagonal in the basis of the electronic–vibrational wave functions (2.3.1). We express it in the form

$$H = H_0 + V, \tag{2.3.2}$$

where H_0 is a zeroth-order Hamiltonian diagonal in basis (2.3.1) and V is a perturbation. Given the set of functions (2.3.1), the Hamiltonian decomposition in (2.3.2) is performed in a unique way provided that the diagonal matrix elements of V are bound to vanish. However, it is important to bear in mind that the very choice of functions (2.3.1) is not unique. Indeed, in addition to the nuclear kinetic energy, some other small terms (e.g., spin–orbit coupling) may well be omitted in solving the (electronic) equation for $\varphi_r(x, q)$, thus resulting in a different basis set of functions. Then the terms dropped in the electronic equation are to be included in the perturbation on equal footing with the operator breaking down the adiabaticity of nuclear motion. The specific choice of decomposition (2.3.2) is not arbitrary, since it must stem from the physical essence of the problem under consideration. In particular, a situation is conceivable when the decomposition is not justified at all so that exact eigenfunctions and eigenenergies of the total Hamiltonian,

$$H\Phi_n = E_n\Phi_n, \tag{2.3.3}$$

must be dealt with (Sect. 2.4.1).

The zeroth-order Hamiltonian as a function of the electronic variables is diagonal in the basis of real-valued functions $\varphi_r(x, q)$,

$$\int \varphi_r(x, q) H_0 \varphi_r(x, q)\, dx = H_r(q)\, \delta_{rr'}, \tag{2.3.4}$$

since all the components of the total Hamiltonian mixing different adiabatic electronic states are included in V. The diagonal matrix element $H_r(q)$ is still an operator acting on the nuclear variables and representing the Hamiltonian for the nuclear motion in the r-th electronic state. Thus, despite the presence of the strong electron–nucleus interaction, the adiabatic approximation allows one to introduce a notion of electronic states and of a vibrational (or *vibronic*) Hamiltonian governing the nuclear motion in a given electronic state.

So the energy spectrum of a polyatomic molecule consists of a sequence of electronic energy levels (as in Fig. 2.1 for organic molecules), and each level has its own set of vibrational sublevels, whose energies and wave functions depend on the electronic state (the latter circumstance is often stressed by using the term *vibronic states*). The simplified diagram of the unperturbed levels of the zeroth-order Hamiltonian, which is widely used in the theory, is shown in Fig. 2.2. It is assumed that there exists only one state Φ_0 (it is a vibrational sublevel of the ground electronic state) connected by an allowed dipole transition with one of

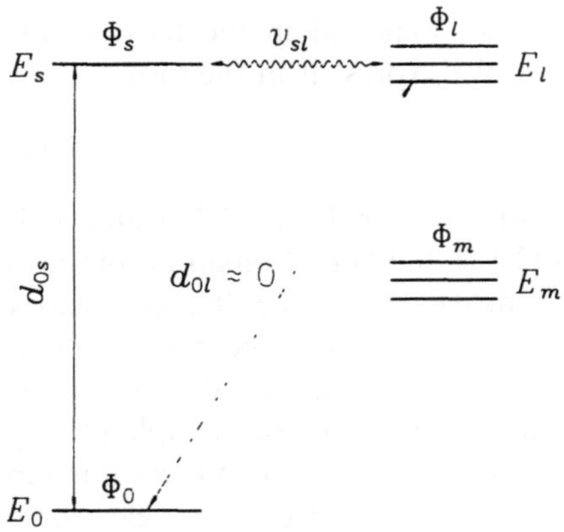

Fig. 2.2. A simplified scheme for the electron-vibrational (vibronic) molecular levels. E_0: ground state; E_s: excited electronic state connected with E_0 by a dipole transition; E_l: vibrationally excited sublevels of the final electronic state isoenergetic to E_s; E_m: vibrationally relaxed sublevels of the final electronic state

the lowest sublevels E_s of an excited electronic state. The state Φ_s is called optically *bright* or *doorway state*. In addition, there are a large number of vibrational sublevels E_l and E_m belonging to a different excited state (or to the ground state), the levels E_l being nearly isoenergetic with E_s and the levels E_m representing vibrationally relaxed states. For example, if E_s is associated with the first excited singlet state S_1 of an organic molecule (see Fig. 2.1), then the levels E_l and E_m belong to the ground state S_0 or to the first excited triplet T_1 in the cases of the $S_1 \rightsquigarrow S_0$ internal conversion and the $S_1 \rightsquigarrow T_1$ intersystem crossing, respectively. The states Φ_l are supposed to be optically *dark states*, i.e., no optical transitions from or to these states are observed, and they are impossible to be photoexcited directly from the ground state as the transition dipole moment is negligibly small, $d_{0l} \approx 0$. It is further assumed that the interaction operator V only has the following nonzero matrix elements:

$$v_{sl} = \langle \Phi_s | V | \Phi_l \rangle \ . \tag{2.3.5}$$

Under the instantaneous excitation condition (2.1.1) one may speak of an *initially prepared state* meaning the state created by a prompt radiation pulse from a flash lamp or a laser source. The theory aims at describing the initially prepared excited state and its subsequent time evolution as well as the calculation of the radiationless transition rate constant in the case when the simple kinetic description of Sect. 2.1 is valid.

Particular attention will be given in this book to the situation widely met in real practice when it is the bright state Φ_s that is prepared initially, and the radiationless transition, if any, occurs from Φ_s to the manifold of nearby isoenergetic Φ_l states. In order to formulate the problem correctly, one has first to choose the zeroth-order Hamiltonian H_0 and the corresponding perturbation V in an appropriate way, and then to find out the conditions to be fulfilled to prepare the initial state Φ_s. As long as radiationless transitions are concerned, additional conditions for irreversibility are to be established since we mean in this book by the *radiationless transition* just an irreversible process, also called *electronic relaxation*, that can be introduced consistently only in the context of the balance equation of Sect. 2.1.

2.4 Physical Nature of Radiationless Transitions

The following peculiarities of the radiationless transitions have to be borne in mind in order to understand the nature of these processes.

First, a radiationless transition is a transition between *nonstationary* states of a polyatomic system. Indeed, if a stationary state has been prepared during the excitation process, the system is residing in this state infinitely long unless it interacts with an electromagnetic field, and is otherwise emitting spontaneously with the lifetime $\tau_{obs} = \tau_{rad}$ and the quantum yield $\varphi = 1$. Therefore, a primary condition for the radiationless transition to exist is that the excitation process produces a nonstationary excited state.

The second essential feature of the radiationless transitions is their irreversibility. Just this property allows one to introduce a notion of the transition rate, i.e., the transition probability per unit of time, and to invoke the balance equation (2.1.4). Consequently, the observation of the exponential luminescence decay (2.1.5) in the presence of a radiationless transition is a straightforward experimental proof of the irreversibility of the transition.

It should be stressed once more that, in general, it makes sense to speak of a radiationless transition if and only if the electron–vibration energy transfer is irreversible. Therefore, the term "*reversible* radiationless transition", sometimes met in the literature, has not any physical significance (for *reverse* transitions see [2.185, 186] and Sect. 7.2). If the physical conditions are such that there is no mechanism for irreversibility (for instance, in isolated diatomic molecules), the intramolecular couplings between various unperturbed states manifest quite differently: Depending on what initial state is prepared by the excitation, these couplings may lead to nonexponential decays and quantum beats of the luminescence intensity, to line perturbations in the absorption spectra and to other phenomena having nothing to do with radiationless transitions. If we consent to reckon the above phenomena to radiationless transitions then we must adopt that the radiationless transitions are also present in isolated small molecules and atoms, and their primary meaning as a property of large polyatomic systems disappears.

2.4.1 The Nature of the Initial State

The nature of the initially prepared state depends on both the properties of the system and the spectral and temporal characteristics of the excitation process [2.187–191]. For simplicity, photoexcitation of an isolated molecule by a weak field is considered, and the interactions other than the electric dipole ones and the quantum properties of the radiation are neglected.

Let the molecules interact with a monochromatic radiation pulse of mean frequency $\bar{\omega}$, spectral bandwidth $\Delta\omega$ and duration τ_{pulse}. The bandwidth has several components.

First, the radiation field from a real thermal source is a superposition of primary fields generated by elementary radiators (e.g., atoms or molecules). Phases of the primary fields are random and their angular frequencies ω fill up the interval

$$|\omega - \bar{\omega}| < \Delta\omega_{\text{incoh}} , \tag{2.4.1}$$

where $\Delta\omega_{\text{incoh}}$ is the incoherent bandwidth of the pulse; it is also present in laser radiation. The frequency spread in (2.4.1) is due to unequal external conditions for different elementary radiators. For instance, the radiating species in a solid source interact with various local environments that differently modify the radiation frequency; frequency shifts in vapors are on account of the Doppler effect and collisions.

Second, according to the uncertainty principle there is an extrinsic coherent broadening due to a finite pulse duration

$$\Delta\omega_\tau \equiv 2\pi\Delta\nu_\tau = 1/\tau_{\text{pulse}} , \tag{2.4.2}$$

where $\Delta\nu_\tau$ is the extrinsic coherent linewidth on the cyclic frequency scale. Thus, expressing τ_{pulse} in ns and $\Delta\nu_\tau$ in MHz one obtains $\Delta\nu_\tau \simeq 160/\tau_{\text{pulse}}$. Finally, the emission of the elementary radiators also is of finite duration, which is the intrinsic coherence time τ_{coh}; the corresponding intrinsic coherent bandwidth $\Delta\omega_{\text{coh}}$ is expressed in terms of τ_{coh} by the same relation (2.4.2).

The value of $\Delta\omega_{\text{coh}}$ is negligibly small since $\tau_{\text{coh}} \gg \tau_{\text{pulse}}$ under typical experimental conditions. Consider for example a common thermal source such as the mercury arc lamp. Here, mercury atoms excited by an electric discharge are the elementary radiators, and τ_{coh} simply coincides with their excited state lifetime, which is on the order of 10^{-8} s for dipole-allowed transitions. The τ_{coh} of laser sources is still longer and may be as high as 10^{-2} s since lasers operate at forbidden transitions. On the other hand, usually τ_{pulse} is on a nanosecond or even subnanosecond time scale since the observed fluorescence lifetime in most polyatomic molecules falls within the interval $\tau_{\text{obs}} = 10^{-6}$–$10^{-8}$ s and condition (2.1.1) has to be satisfied for the excitation and emission processes being separated in time. Then the excitation process "prepares" an excited molecular state that next evolves with time due to both intramolecular couplings and the interaction of the molecule with its environment and with the zero-point vibrations of the electromagnetic field (the latter causing spontaneous emission).

The above arguments show that τ_{coh} is practically infinite and the intrinsic coherent emission bandwidth for the elementary radiators may be neglected so that the total bandwidth of the exciting pulse is written as a sum of the incoherent and extrinsic coherent contributions,

$$\Delta\omega = \Delta\omega_{\text{incoh}} + \Delta\omega_\tau . \tag{2.4.3}$$

Since the emission phases for different elementary radiators are random, the excitation process is additive in the sense that the density matrix for the excited molecules is a sum of terms each describing the effect of a single elementary radiator. In other words, the molecule excitation process initiated by a real pulse can be treated as a set of independent elementary excitation events each corresponding to the interaction of the molecule with a monochromatic field of frequency ω during a finite time interval $0 \leq t \leq \tau \equiv \tau_{\text{pulse}}$; the bandwidth of the elementary excitation pulse is simply $\Delta\omega_\tau$ from (2.4.2).

Now let us ascertain what kind of excited molecular state is created by the elementary excitation act. The solution to this problem, the weak-field limit assumed, is given by the perturbation method. Let $W(t)$ stand for the interaction energy of the molecular dipole moment with the electromagnetic field. We transform the Schrödinger equation

$$i\hbar\dot{\Psi}(t) = [H + W(t)]\,\Psi(t) \tag{2.4.4}$$

into the interaction representation

$$\Psi(t) = e^{-iHt/\hbar}\Phi(t)\,. \tag{2.4.5}$$

Treating $W(t)$ as a perturbation, one obtains for $\Phi(t)$ at $t = \tau$:

$$\Phi(\tau) = \left[1 - \frac{i}{\hbar}\int_0^\tau e^{iHt/\hbar}W(t)e^{-iHt/\hbar}dt\right]\Phi_0 \equiv \Phi_0 + \Phi_{\text{ex}}\,, \tag{2.4.6}$$

where Φ_0 is the ground-state wave function at energy E_0, and Φ_{ex} the wave function of the excited molecule at the time of pulse termination.

Let Φ_n in (2.3.3) refer to the excited molecular states, and let the perturbation vanish when averaged over the ground state:

$$\langle\Phi_0|\,W(t)|\,\Phi_0\rangle = 0\,. \tag{2.4.7}$$

Then Φ_{ex} in (2.4.6) can be expanded as

$$\Phi_{\text{ex}} = \sum_n a_n\Phi_n\,, \tag{2.4.8}$$

$$a_n = -\frac{i}{\hbar}\int_0^\tau e^{i(E_n - E_0)t/\hbar}\,W_{n0}(t)dt\,, \tag{2.4.9}$$

$$W_{n0}(t) = \langle\Phi_n|\,W(t)|\,\Phi_0\rangle\,. \tag{2.4.10}$$

As follows from (2.4.9), only such excited states contribute to the sum in (2.4.8) that are allowed with regard to the dipole transitions from the ground state, i.e., the matrix element (2.4.10) does not vanish.

Coefficients a_n contain exhaustive information on the nature of the excited state emerging at time $t = \tau$ when the pulse stops acting. Moreover, the

Schrödinger wave function of excited molecules, $\Psi_{ex}(t)$, at any time $t > \tau$ is also expressed in terms of these coefficients. Thus, considering that $W(t > \tau) \equiv 0$ and $\Phi(t \geq \tau) \equiv \Phi(\tau)$ one derives from (2.4.5, 6, 8)

$$\Psi_{ex}(t \geq \tau) = e^{-iHt/\hbar}\Phi_{ex} = \sum_n a_n e^{-iE_n t/\hbar}\Phi_n . \qquad (2.4.11)$$

The derivation being quite general, this equation describes all types of time evolution of excited molecules also including those which can be qualified as radiationless transitions in the sense explained at the beginning of Sect. 2.4.

Turning to the analysis of the wave function (2.4.11), we recall that $W_{n0}(t)$ is a matrix element of a perturbation oscillating at frequency ω. Therefore, the a_n's in (2.4.9) will be markedly nonvanishing only under the condition:

$$|E_n - E_0 - \hbar\omega| < \hbar\Delta\omega_\tau , \qquad (2.4.12)$$

i.e., only the energy levels are involved of which the detunings are smaller than the quantum uncertainty in the excitation frequency.

Consider first the most trivial case when condition (2.4.12) holds for a single level number n_0 and, consequently, only one coefficient a_{n_0} does not vanish. This is possible if the density of excited levels, ϱ_n, near the energy $E_0 + \hbar\omega$ is quite low and if the quantum uncertainty of the photon energy is smaller than the mean level separation,

$$\hbar\Delta\omega_\tau \ll 1/\varrho_n . \qquad (2.4.13)$$

Such a situation is met, e.g., in isolated diatomic molecules whose excited level spacings in the optical and UV spectral ranges are high enough. Then it is clear that the *excitation process prepares the stationary state* n_0, and (2.4.11) takes the form

$$\Psi_{ex}(t \geq \tau) = a_{n_0} e^{-iE_{n_0}t/\hbar}\Phi_{n_0} . \qquad (2.4.14)$$

Obviously, the probability to find the system in the n_0-th stationary state is independent of time, and therefore *no radiationless transitions occur in this case.*

Exactly this same state is produced when the molecule is excited by a real source provided that the total bandwidth (2.4.3) of the pulse is small enough,

$$\hbar\Delta\omega \ll 1/\varrho_n , \qquad (2.4.15)$$

and the mean frequency $\bar{\omega}$ is chosen according to the condition

$$|E_{n_0} - E_0 - \hbar\bar{\omega}| < \hbar\Delta\omega . \qquad (2.4.16)$$

Now we assume that inequality (2.4.15) is broken down for a real source because of a high value of $\Delta\omega_{\text{incoh}}$,

$$\Delta\omega_{\text{incoh}} \gg 1/\varrho_n , \qquad (2.4.17)$$

but condition (2.4.13) is still fulfilled. Then, as was established above, a definite stationary state n_0 is prepared in each elementary excitation act, but n_0 is not the same for all molecules since the molecules are permitted, by (2.4.17), to absorb photons of varying frequency within the limits of $\Delta\omega_{\text{incoh}}$ around $\bar{\omega}$. As a result, there arises a statistical ensemble of molecules in the stationary states n_0 whose energies are restricted by (2.4.16); the populations of these states are equal to $|a_{n_0}|^2$ and are independent of time. Thus, the above conclusion that no radiationless transitions are observed under condition (2.4.13) persists when the actual properties of the exciting radiation are properly taken into account.

Since the density of the molecular excited states in a given spectral range ϱ_n steeply increases with the number of nuclear degrees of freedom, another interesting limit arises under the condition

$$\hbar\Delta\omega_\tau \gg 1/\varrho_n \tag{2.4.18}$$

when the *excitation of a single stationary state becomes impossible*, even potentially. Indeed, inequality (2.4.18) implies that condition (2.4.12) is fulfilled for a large number of levels $\hbar\Delta\omega_\tau\varrho_n$, and therefore many terms now contribute to the sums in (2.4.8, 11). It is seen that in this case a *nonstationary state is formed* at the moment of pulse termination, and the subsequent evolution in time of this state is given by (2.4.11). The same situation may also occur in small molecules provided that two or more levels E_n are very close in energy [in which case the mean separation between these nearly degenerate levels should be substituted for $1/\varrho_n$ on the right-hand side of (2.4.18)].

In order to clarify the nature of the initial nonstationary state, we write down the total Hamiltonian as a sum of terms

$$H = H_1 + H_2 + H_3 + \cdots \tag{2.4.19}$$

satisfying the symbolic inequalities $H_1 \gg H_2 \gg H_3 \gg \cdots$, which should be read as

$$\Delta E_1 \gg \Delta E_2 \gg \Delta E_3 \gg \cdots, \tag{2.4.20}$$

where ΔE_1 denotes the mean separation between the H_1 levels (in general degenerate) to which the dipole transitions from the ground state are allowed, ΔE_2 the mean splitting of the degenerate H_1 levels caused by H_2, etc. It should be stressed that the above Hamiltonian expansion is not an absolute one, but instead refers to a certain very narrow domain of the molecular energy spectrum near $E_0 + \hbar\omega$. Thus, in the optical and UV spectral ranges of interest here, H_1 is the electronic Hamiltonian with the nuclei fixed, H_2 the vibronic Hamiltonian in a given electronic state, and H_3 may involve rotational, nonadiabatic and spin–orbit contributions.

Comparing the successive terms in (2.4.20) with $\hbar\Delta\omega_\tau$ one may group the individual contributions to (2.4.19) so as to write the total Hamiltonian in such a form (2.3.2) that this symbolic inequality holds:

$$H_0 \gg \hbar\Delta\omega_\tau \gg V'. \tag{2.4.21}$$

It means that the spacings between the unperturbed degenerate levels chosen, to which the dipole transitions from the ground state are allowed, are large as compared to the coherent bandwidth of the elementary radiator $\hbar \Delta \omega_\tau$, while the small splittings of the unperturbed levels caused by the perturbation V fall within this bandwidth. In other words, any given degenerate unperturbed level can be excited singly, but its sublevels cannot, all being coherently excited in every elementary excitation event.

Considering (2.4.2, 21) one has

$$\tau \equiv \tau_{\text{pulse}} \ll \tau_V = \hbar / V , \qquad (2.4.22)$$

where τ_V is a characteristic time for quantum beats between components of the degenerate unperturbed state split by perturbation V. According to (2.4.22) the quantity $Vt/\hbar$ appearing in the exponents of (2.4.6), after substitution of (2.3.2), is small at $t < \tau$, which, to some extent, favors its neglect when calculating the wave function of the excited state since the coefficients a_n in (2.4.8) will change insignificantly. However, when a large number of terms contribute to the sum (2.4.8), the pulse duration must be confined by a more restrictive condition of a small variation in the total sum as a whole, not only in its individual entries.[3] This is assumed from now on.

Evidently, the intramolecular perturbation V does not play any role in the excitation process because the pulse duration is such short that the perturbation has not enough time to affect the wave function appreciably, and may be regarded as being "turned off" during excitation. Hence the wave function of the excited molecules can be approximated with the expression

$$\Phi_{\text{ex}} = \left[-\frac{\mathrm{i}}{\hbar} \int_0^\tau \mathrm{e}^{\mathrm{i} H_0 t/\hbar} W(t) \mathrm{e}^{-\mathrm{i} H_0 t/\hbar} dt \right] \Phi_0 , \qquad (2.4.23)$$

which differs from (2.4.6) only by substitution of H_0 for H. It should be emphasized that under the assumptions made both equations (2.4.6, 23) describe exactly the same initially prepared excited state.

Let $\Phi_m^{(0)}$ and $E_m^{(0)}$ be the eigenfunctions and eigenenergies of the excited states of H_0 (Φ_0 is an eigenfunction for both H and H_0 since the matrix elements of V for transitions between the excited states of interest and the ground state are assumed to vanish). Expanding Φ_{ex} of (2.4.23) in the functions $\Phi_m^{(0)}$, we obtain

$$\Phi_{\text{ex}} = \sum_m a_m^{(0)} \Phi_m^{(0)} , \qquad (2.4.24)$$

$$a_m^{(0)} = -\frac{\mathrm{i}}{\hbar} \int_0^\tau e^{\mathrm{i}(E_m^{(0)} - E_0)t/\hbar} W_{m0}(t) dt ,$$
$$W_{m0}(t) = \langle \Phi_m^{(0)} | W(t) | \Phi_0 \rangle . \qquad (2.4.25)$$

---

[3] For example, if the probability to find the molecule in the excited state decays exponentially as in (2.1.2, 5), then condition (2.1.1) should be satisfied in addition to (2.4.22).

Equations (2.4.8, 24) are the expansions of one and the same wave function Φ_{ex} into two different complete sets of functions Φ_n and $\Phi_m^{(0)}$.

Just as previously, the main contribution to the sum (2.4.24) comes from levels $E_m^{(0)}$ that satisfy the condition

$$|E_m^{(0)} - E_0 - \hbar\omega| < \hbar\Delta\omega_\tau \tag{2.4.26}$$

similar to (2.4.12). However, due to the left-hand inequality (2.4.21), in the present case the frequency ω of the exciting radiation can freely be chosen to satisfy condition (2.4.26) for *only one* (degenerate) level $E_m^{(0)}$. Then

$$\Phi_{ex} = \sum_{m_0} a_{m_0}^{(0)} \Phi_{m_0}^{(0)}, \tag{2.4.27}$$

where the summation is taken over all states with a given energy $E_{m_0}^{(0)}$. Since (2.4.27) is an eigenfunction of H_0 at the energy $E_{m_0}^{(0)}$ we arrive at the following important conclusion: Under the said limitations *the initially excited nonstationary state of the system is a stationary state of this very same system with the perturbation V "turned off"*. However, the fact that the initial state is nonstationary becomes apparent only when the perturbation gradually is "turned on" at times $t \gg \tau$.

The selection rules for the matrix elements of the external perturbation $W(t)$ are usually such that only one state contributes to the sum (2.4.27), as occurs in the example below. Then the principal feature of the wave functions $\Phi_m^{(0)}$ is formulated thus: These functions describe such nonstationary states of the system that can be prepared by excitation with a small enough pulse duration, provided that appropriate selection rules are obeyed. While the two full sets of functions, Φ_n and $\Phi_m^{(0)}$, are entirely equivalent from the mathematical point of view, physically it is clear that the second set must be employed to describe the system, because the problem of radiationless transition from a given initial state to all final states of the same energy only makes sense for the states $\Phi_m^{(0)}$, which, unlike Φ_n, can be prepared in a real physical experiment.

These arguments imply that the wave function of the excited molecules at any time t is naturally expressed in terms of an expansion in $\Phi_m^{(0)}$ rather than in Φ_n (2.4.11). If the m_0-th excited state of the system was initially prepared at $t = 0$, then

$$\Psi_{ex}(t \geq 0) = e^{-iHt/\hbar} \Phi_{m_0}^{(0)} \equiv \sum_m c_{mm_0}(t) \Phi_m^{(0)}, \tag{2.4.28}$$

where

$$c_{mm_0}(t) = \langle \Phi_m^{(0)} | e^{-iHt/\hbar} | \Phi_{m_0}^{(0)} \rangle \tag{2.4.29}$$

are the matrix elements of the evolution operator taken with basis set $\Phi_m^{(0)}$ under the initial condition $c_{mm_0}(0) = \delta_{mm_0}$, where δ_{mm_0} is Kronecker's symbol. Equation (2.4.28) differs from (2.4.11) by a shift in the zero-time reference point and by a common normalization factor of the wave function. Other than this, the two expressions are identical within the approximations involved, and they are expansions of one and the same wave function in two different basis sets. It is worthwhile to note that the above arguments are equally valid for both large molecules with a high density of states fulfilling condition (2.4.18), small molecules, and even atoms with nearly degenerate levels provided that inequality (2.4.18) holds for the level splitting instead of $1/\varrho_n$. Therefore, the subsequent evolution of the initially prepared state according to (2.4.28) may result in either the radiationless transitions or the quantum beats of the fluorescence intensity such as those seen in atomic spectroscopy [2.192].

The restrictions on pulse duration can now be established more precisely. In the case of a large molecule the initially prepared state Φ_{ex} changes for a characteristic time τ_{obs} of (2.1.6); therefore, the said restrictions involve (2.1.1) along with (2.4.22). On the other hand, for the small molecules condition (2.4.22) alone is sufficient since τ_V is the only parameter governing the evolution rate of Φ_{ex}.

As an illustration, let us discuss the simplified energy level model drawn in Fig. 2.2 and explained in detail in Sect. 2.3. The Hamiltonian decomposition (2.3.2) employed there first needs to be considered from the viewpoint of conditions (2.4.21). The experimental excitation pulse duration typically varies within the limits of 10^{-8}–10^{-10} s; correspondingly[4],

$$\hbar \Delta\omega_\tau \simeq 5\times(10^{-4}-10^{-2})\,\mathrm{cm}^{-1}\,. \qquad (2.4.30)$$

The states Φ_0 and Φ_s in Fig. 2.2 are the lowest vibrational sublevels of the ground and the first excited electronic states, respectively, connected to each other through a dipole-allowed transition. The next states are separated from them by the energy on the order of a vibrational quantum so that the left-hand inequality (2.4.21) is fulfilled with a large reserve. The vibrational frequency differences between the ground and the excited electronic states largely exceeding the value in (2.4.30) (they are on the order of 10–100 cm^{-1}), the exciting frequency $\bar{\omega}$ may always be chosen to fall in resonance with a certain pair of levels; the incoherent excitation of the adjacent levels may also be avoided at small enough values of $\Delta\omega_{incoh}$. These arguments show that the states Φ_0 and Φ_s may only be considered with no regard to other vibrational sublevels[5].

The dark states Φ_l in Fig. 2.2 are excited vibrational sublevels of a certain (ground or excited) electronic state, which are isoenergetic with the bright state Φ_s. The direct photoexcitation of these states is supposed to be impossible, e.g., because of the spin-selection rule as far as the triplet state is concerned (in fact the singlet–triplet absorption is observed in benzene [2.193] and other [2.194–196] molecules). In order to check the fulfilment of the right-hand inequality (2.4.21), the matrix elements (2.3.5) of the perturbation V, to which we

[4] The energy unit used is $1\,\mathrm{cm}^{-1} = 2\times10^{-23}$ J (the energy of a photon with the wavelength 1 cm); then $\hbar = 1/2\pi c = 5.3\,\mathrm{cm}^{-1}$ ps/rad.

[5] Single vibronic levels are still manifolds of rotational levels. This may also be neglected if rotation has no effect on the excited state dynamics (see, however, the end of this section as well as Sect. 4.4.3).

assign both nonadiabatic and small electronic (spin–orbit) interactions, have to be estimated.

The matrix elements strongly depend on the purely electronic energy difference ΔE between the states Φ_s and Φ_l since they are proportional to the overlap integral for the corresponding vibrational functions, the integral squared being the Franck–Condon factor (see Chap. 4). The adiabatic potentials for the nuclear motion in the organic polyatomic molecules change only slightly between the lower electronic states, which results in a very steep (exponential) fall-off in the Franck–Condon factor upon increasing ΔE (see Sects. 2.11 and 4.8). In parallel, the density of states ϱ_l increases faster the greater the number of nuclear degrees of freedom N is. Then inequalities (2.4.18, 22) will be satisfied simultaneously with ΔE and N high enough (a slight difference between ϱ_l and ϱ_n is disregarded). Such a situation really takes place in polyatomic molecules, and if the excitation time is short enough to meet the requirements (2.1.1) and (2.4.22), *the bright state Φ_s is prepared initially*. For instance, in organic molecules the states Φ_0 and Φ_s belong to the singlet manifolds S_0 and S_1, respectively, and Φ_l to S_0 or T_1; then the sublevel Φ_s of the S_1 singlet state is prepared initially under the said conditions.

This statement, being always taken as an axiom in interpretations of experimental data on fluorescence of organic molecules, but sometimes being a point of confusion for theorists, now turns out to be trivial in the light of the previous considerations. Indeed, if the excitation time τ is such short that the intramolecular coupling between Φ_s and Φ_l is "turned off" within the interval $0 \leq t \leq \tau$, the molecule subject to the action of the exciting radiation can only pass from the ground state to the excited state Φ_s since, by assumption, the $\Phi_l \leftarrow \Phi_0$ oscillator strengths vanish. For the level diagram (Fig. 2.2) where a single bright state Φ_s enters, this result can be established in a formal way immediately from (2.4.8–10) by setting $\tau \to 0$ and $W(t) = A\delta(t)$ [instantaneous excitation (2.1.10)] with A being proportional to the molecular dipole moment. Then

$$\Phi_{ex} = -\frac{i}{\hbar}\sum_n \langle \Phi_n | A | \Phi_0 \rangle \Phi_n \,. \tag{2.4.31}$$

Here use will be made of the completeness of the sets of functions $\{\Phi_n\}$ and $\{\Phi_s, \Phi_l\}$, which reads

$$\sum_n \Phi_n^* \Phi_n = \Phi_s^* \Phi_s + \sum_l \Phi_l^* \Phi_l = 1 \,. \tag{2.4.32}$$

Since the transition matrix element of the operator A is nonzero for the state Φ_s alone and vanishes for the Φ_l's, (2.4.31) after substitution of (2.4.32) becomes

$$\Phi_{ex} = -\frac{i}{\hbar}\langle \Phi_s | A | \Phi_0 \rangle \Phi_s \,, \tag{2.4.33}$$

explicitly expressing the fact that Φ_s is indeed the initially prepared state.

Using $\{\Phi_s, \Phi_l\}$ as a complete basis set of eigenfunctions $\Phi_m^{(0)}$ of the unperturbed Hamiltonian, expansion (2.4.28) is rewritten in the form

$$\Psi(t) = c_{ss}(t)\Phi_s + \sum_l c_{ls}(t)\Phi_l \,, \tag{2.4.34}$$

where $c_{ss}(t)$ and $c_{ls}(t)$ are given by (2.4.29), e.g.,

$$c_{ls}(t) = \langle \Phi_l | \mathrm{e}^{-\mathrm{i}Ht/\hbar} | \Phi_s \rangle \,. \tag{2.4.35}$$

It follows that the probability to find the system in the initially prepared state (differing from $P(t)$ in Sect. 2.1 by a normalization factor),

$$P_{ss}(t) = |c_{ss}(t)|^2 \,, \tag{2.4.36}$$

is not stationary, as Φ_s is not an eigenfunction of the molecular Hamiltonian H. As was stated at the beginning of this section, the radiationless transition from the initial state Φ_s to all final states Φ_l may only be spoken of under the condition that the process is irreversible. In the simplest, but most important case of the single bright state Φ_s and the instantaneous excitation (2.1.1) under consideration this means that the theory of Sect. 2.1 applies, giving

$$P_{ss}(t) = \mathrm{e}^{-k_s t} \,, \tag{2.4.37}$$

where k_s is the radiationless transition rate constant, i.e., the transition probability per unit of time. The criterion for irreversibility of the electronic transitions in isolated polyatomic molecules is discussed in the following subsection.

Excitation with a real source having a large $\Delta\omega_{\mathrm{incoh}}$ but a small $\Delta\omega_\tau$, the latter satisfying conditions (2.4.21), prepares a statistical mixture of the nonstationary states Φ_s (e.g., rotational sublevels of a single vibronic level) with energies within $\Delta\omega_{\mathrm{incoh}}$, and with populations $|a_s|^2$ proportional to the oscillator strengths of the corresponding transitions. If the irreversibility criterion is satisfied, the populations decay in time, each component decaying independently of the others because the excitation of different states is incoherent under the present conditions. Consequently, the total probability to detect the system in an emitting excited state will be proportional to $\sum_s |a_s|^2 \exp(-k_s t - \tau_{\mathrm{rad}(s)}^{-1} t)$, where the rate constant $\tau_{\mathrm{rad}(s)}^{-1}$ accounts for the finite probability of spontaneous emission from Φ_s. It follows that the appreciably differing k_s's and/or $\tau_{\mathrm{rad}(s)}^{-1}$'s result in a nonexponential fluorescence decay for long times, whereas at short times the decay behavior can be approximated with an exponential of which the decay rate is an average of the partial ones over the initial distribution of the populations,

$$k = \sum_s |a_s|^2 k_s \Big/ \sum_s |a_s|^2 \,. \tag{2.4.38}$$

Possible ways of averaging the observables under these conditions are discussed in [2.197, 198].

2.4.2 Freed–Jortner Irreversibility Criterion

Early experimental studies on electronic relaxation in molecules were only carried out in condensed phases and dense vapors, so the irreversibility of radiationless transitions was first incorrectly thought of as always and solely being a straightforward consequence of the vibrational relaxation toward the thermal equilibrium due to interaction with the medium [2.199]. However, the very first fluorescence studies on aromatic hydrocarbons in low-pressure vapors showed that the quantum yield is still appreciably below unity even when the molecules are nearly isolated from their environment, i.e., when their fluorescence lifetime is essentially shorter than the time between successive collisions. Thus *Kistiakowsky* and *Parmenter* [2.75] found $\varphi = 0.34$ for benzene at a pressure of 0.01 Torr. Further experiments were performed at still lower pressures and with a number of other species [1.3, 10, 28, 30], and presently the existence of the radiationless transitions in isolated large molecules may be regarded as to be evidenced conclusively.

If a molecule interacts with a medium, an electronic–vibrational (vibronic) transition is irreversible since the continuous energy spectrum of the system "molecule + medium" allows the transfer of either all or a part of the intramolecular electronic energy to the medium. As for an isolated large polyatomic molecule, the energy spectrum of which is discrete, the principal issue arises with regard to the mechanism for irreversibility of the vibronic transition.

The physical origin of the irreversible behavior of isolated large molecules was discussed by *Freed* [1.1]. He invoked the analogy with a closed macroscopic system of particles obeying the classical equations of motion. The Poincaré recurrence theorem [2.200] states that such a system returns arbitrarily close to its initial state after a large enough period of time, T. The irreversible behavior of the system is nevertheless observed due to the fact that the Poincaré period T is excessively long, so that the phenomena due to the quasiperiodicity of motion are not seen in practice.

The same principle of "practical irreversibility" is applied to the problem of radiationless transitions in isolated polyatomic molecules. Let the system be prepared at an initial point of time in the starting vibronic state Φ_s coupled with a great number of vibronic states Φ_l belonging to another molecular electronic state. Equations (2.4.34–36) imply that the probability to detect the system in the state Φ_s at times not very large is below unity. Here again the question arises about the length of the "theoretical recurrence" time T (similar to the Poincaré period) after which the probability is restored close to unity, the latter possibility being obviously assured by the discreteness of the vibronic energy spectrum. The recurrence time is to be compared with a time limit placed upon the experiment that depends on either the maximum duration of the real measurement (usually on the order of the excited-state lifetime) or the largest period of time during which the molecule is isolated, i.e., it does not undergo collisions with other molecules, with the walls of the vessel, etc. The expectation is that the condition

$$T \gg \tau_{max} \tag{2.4.39}$$

will be valid for a molecule large enough to have a high density of final states ϱ_l, isoenergetic with the initial state. This is just the irreversibility criterion by *Freed*. If it is not fulfilled, radiationless transitions in the isolated molecule are impossible and the quantum beats occur among the unperturbed states already mentioned at the beginning of this section.

In order to give a quantitative significance to the Freed criterion, the parameter T should be expressed in terms of known molecular parameters. *Bixon* and *Jortner* [1.41] introduced a specific model that provided the following exact expression for T:

$$T = 2\pi\hbar\varrho_l \, . \tag{2.4.40}$$

Later it was demonstrated [1.3, 2.201] that this estimate is valid for some more general molecular models as well. In general, that recurrence time for the quasiclassical states of a system at a given energy E can be shown [2.200] to be proportional to the volume of the phase-space domain bounded by two isoenergetic surfaces at E and $E + dE$, the volume in turn being proportional to the density of states at energy E [1.40].

Substituting (2.4.40) into (2.4.39) gives *the Freed–Jortner irreversibility criterion* in the form [1.10, 2.202]

$$2\pi\hbar\varrho_l \gg \tau_{\max} \, . \tag{2.4.41}$$

This criterion can be used to evaluate the possibility of the existence of radiationless transitions in particular molecules (see, e.g., [1.10, 41, 2.178]) since there are sufficiently reliable methods to calculate the density of states [2.203].

Inequality (2.4.41) is often referred to as a condition for the statistical limit of large molecules [1.1, 2.204] since the molecules fulfilling this condition behave irreversibly, such as statistically large systems, even without environment. Conventionally, however, the statistical limit is understood in a narrower sense. Namely, it is additionally supposed that the balance equation of Sect. 2.1 is valid (as we shall see in Sects. 2.5–7, it is not always assured by practical irreversibility alone) and that the transition rate constant entering it is independent of the vibrational molecule–medium coupling and decreases exponentially with increasing electronic energy gap. For this reason we call (2.4.41) the irreversibility criterion. The complete set of requirements implied by the statistical limit are summarized in Sect. 2.11 following the discussion of this problem in the next sections.

Thus, the irreversibility of electronic transitions in isolated polyatomic molecules is explained by a long time for the theoretical recurrence; for this reason the recurrences cannot be observed in a real experiment. As for the decay behavior of the probability to find the system in the initial state $P_{ss}(t)$, of which the principle of practical irreversibility, as it stands, says nothing, it will be exponential (2.4.37) under conditions involving both the Freed–Jortner criterion and some additional limitations. This is demonstrated in Sects. 2.5–7 within the Bixon–Jortner model in the strong coupling limit and within the Trifonov-

Shekhtman model in the weak coupling limit; a more general treatment will be given in Chap. 3.

We note that introducing τ_{max} implies the assignment of an effective width $\hbar\ \tau_{max}$ to the energy levels of the isolated molecule. On the other hand, if there is a surrounding medium the levels are broadened by $\hbar/\tau_{VR}$, where τ_{VR} is the vibrational relaxation time. In both cases the broadenings change the discrete molecular spectrum into a continuous one, which is a common reason of irreversibility in these cases. However, there are also two essential distinctions between them.

1) According to (2.4.41) the level width $\hbar/\tau_{max}$ for the isolated molecule has to be large as compared to the average level spacing ϱ_l^{-1}, while with the medium present the corresponding width $\hbar/\tau_{VR}$ may take on both large and small values. Because of this difference (which can be qualified as the difference between dynamical and statistical broadenings) the intramolecular electronic relaxation without medium is only possible in large polyatomic molecules having large ϱ_l's, whereas in the medium this limitation is relaxed, viz., the electronic relaxation (while now not always being intramolecular) occurs also in the molecules with small ϱ_l's.

2) From the definition of τ_{max} the width $\hbar/\tau_{max}$ cannot exceed $\hbar/\tau_{obs}$, while $\hbar\ \tau_{VR}$ can. Moreover, inequality $\tau_{VR} \ll \tau_{obs}$ must be fulfilled for the balance equation to be valid. As mentioned before, the practical irreversibility alone does not lead to the true statistical behavior embodied in the balance equation since the latter demands an either inter- or intramolecular relaxation process in the final electronic state faster than the observable decay.

The practical irreversibility introduces a new feature in the level diagram of Fig. 2.2. The basic energy levels E_s and E_l involved in the intramolecular interaction under consideration are free of limitations regarding their characteristic parameters so that criterion (2.4.41) may or may not be fulfilled for these levels. However, a large molecule is supposed to have very dense manifolds of other levels interacting with E_s and E_l and satisfying the above criterion. Then, E_s and E_l acquire finite widths and hence become quasistationary levels.

2.5 General Description of Luminescence Kinetics: Intermediate Case and Statistical Limit

The level diagram in Fig. 2.2 and the excitation process are characterized by several parameters: the interaction matrix element v_{sl}, the mean level spacing ϱ_l^{-1}, the level half-widths γ_s and γ_l, the detuning

$$\Delta = E_s - E_{l=0} \tag{2.5.1}$$

($E_{l=0}$ is the level nearest to E_s), and the duration of the exciting monochromatic pulse τ_{pulse}. An important parameter is also

$$k_s = \frac{2\pi}{\hbar} v_{sl}^2 \varrho_l \,. \tag{2.5.2}$$

Depending on the situation, k_s may either have the meaning of a transition rate constant or be merely considered as a formal quantity.

Relative values of these parameters affect both the characteristics of the excited state prepared initially and its subsequent evolution. In view of the large number of possible situations it is appropriate to furnish, as a guide, a classification and a general description of some limiting cases [2.205] to be studied further in Sects. 2.6, 7, where proofs will be given for the results presented in the following.

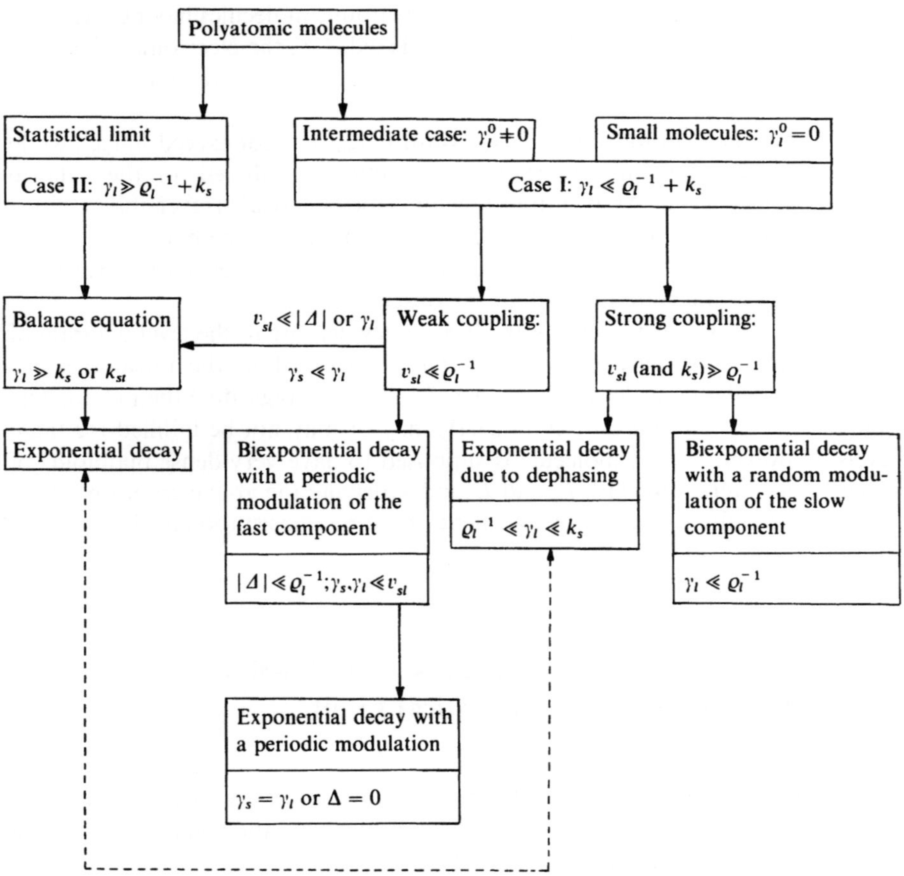

Fig. 2.3. Basic types of the time evolution of the fluorescence intensity from the initially prepared state Φ_s manifesting at various relations between the parameters γ_s, γ_l, ϱ_l^{-1}, Δ, and v_{sl} [or k_s from (2.5.2)]. Constants 2π and $\hbar$ are omitted in the relations given in the figure. These relations and cases I, II and III (the latter is not shown) are discussed in Sects. 2.6, 7. The exponential decays connected by the dashed line are of different physical nature

To elucidate the presentation and to give a clear view of the results of the analysis, we depict in Fig. 2.3 a scheme capturing the most interesting cases. Within this scheme the following discussion proceeds from top to bottom.

The evolution of the many-level system of Fig. 2.2 is governed by a complicated set of coupled equations for populations and "coherences" i.e., for diagonal and off-diagonal elements of the density matrix. There exist, however, cases where coherence effects are either absent at all or clearly separated in time so that the situation becomes simpler. It depends on the specific relations between the parameters, of which the most important is the density of dark states, ϱ_l. Following *Robinson* [1.11, 2.179], *Jortner* [1.8, 9] and *Bixon* and *Jortner* [1.41, 2.206], one distinguishes the small molecule limit, the intermediate case and the statistical limit of large molecules depending on the value of ϱ_l (Sect. 2.11).

If the exciting pulse has a long duration and a narrow spectral bandwidth, an exact molecular eigenstate of (2.3.3) can be prepared in the small-molecule limit and sometimes in the intermediate case (never in the statistical limit). Then, there remains only one balance equation for the population of that state. Another example of a simple description relates to a short pulse with a large coherent bandwidth, where the initial state is a coherent superposition (2.4.8) of the molecular eigenstates that approximately coincides with the bright state, Φ_s (Sect. 2.4.1). If the rate of phase relaxation is much faster than that of population relaxation, the luminescence decays in two stages: the fast stage of loss of coherence followed by the slow stage corresponding to the population decay of the molecular eigenstates, the latter again being described by the balance equation.

In the above two examples the dynamics of the system is connected with the mixed states Φ_n. Therefore, the concept of a radiationless transition from an initial state Φ_s to final states Φ_l loses its significance. In contrast to this, the third example pertains to the statistical limit that arises at a very fast relaxation in the final electronic state (see Fig. 2.3). All molecular states Φ_n merge in a united quasistationary level $\tilde{E}_s$ the population of which obeys the balance equation, where the rate constant for the $\Phi_s \rightarrow \Phi_l$ radiationless transition is equal to k_s.

Turning now to a more detailed discussion of the luminescence decay kinetics, we introduce important definitions of the strong,

$$v_{sl} \text{ (and } k_s) \gg \varrho_l^{-1} , \tag{2.5.3}$$

and weak,

$$v_{sl} \ll \varrho_l^{-1} , \tag{2.5.4}$$

coupling limits. From a formal point of view the difference between them is that, in the first case, a large number of final levels E_l,

$$N = 2\pi^2 v_{sl}^2 \varrho_l^2 \gg 1 , \tag{2.5.5}$$

(Sect. 2.6.1) contribute to the excited-state dynamics, while essentially one level $E_{l=0}$ contributes in the second case (though this may be different at a large level broadening, see Sect. 2.5.4). In physical terms these cases are distinguished by manifestations of the effects of coherence, intra- and interstate relaxations and detuning.

2.5.1 Strong Coupling

First consider a situation where the initially prepared state is Φ_s, which is a coherent superposition (2.4.8) of N molecular states Φ_n. Here coherence means the property of the generally complex expansion coefficients a_n to have such strictly correlated phases that the sum on the right-hand side of (2.4.8) exactly reduces to Φ_s. The subsequent evolution changes the Schrödinger wave function $\Psi_{ex}(t \geq 0)$ according to (2.4.11), and therefore, at $t > 0$ the wave function is not reduced to a unique term proportional to Φ_s but instead contains also contributions from Φ_l due to the fact that the coefficients a_n are multiplied by additional phase factors $\exp(-\mathrm{i}E_n t/\hbar)$. In other words, a loss of the initial coherence occurs in the sense that the phase differences for the expansion coefficients in (2.4.11) are increasing with t and thus become more and more random.

The loss of the initial coherence, or dephasing, results in an exponential decay (2.4.37) of excited molecules and, as a consequence, of the luminescence with a characteristic time

$$\tau_{\mathrm{deph}(s)} = 1/k_s \ll T\,, \tag{2.5.6}$$

in which case k_s plays the role of the rate constant for the dephasing (provisionally it is assumed that the relaxation and radiative decays of the state Φ_s are negligible since the respective lifetimes, $\tau_{\mathrm{rel}(s)}$ and $\tau_{\mathrm{rad}(s)}$, are much longer than $\tau_{\mathrm{deph}(s)}$). Therefore in *the first* (*fast*) stage of evolution, a transition from the initial bright state Φ_s to *mixed* (bright + dark) states Φ_n occurs with the rate constant equal to k_s, and the process itself is intramolecular in nature. The main condition for the state Φ_s to be prepared initially is (2.1.1), which now takes the form $\tau_{\mathrm{pulse}} \ll \tau_{\mathrm{deph}(s)}$ [the second condition (2.4.22) derives from it and from (2.5.3)].

The ultimate result of the dephasing is that the initial population of the state Φ_s is "spread" over all N molecular states Φ_n strongly coupled with Φ_s, i.e., over these states for which the coefficients a_n in the expansion (2.4.8) are markedly nonvanishing. The average population of each of them will thus be on the order of $1/N$. Furthermore, the oscillator strengths for the dark states Φ_l vanish, whereas that for the bright state Φ_s is again distributed in equal portions among N states Φ_n. Hence the emission intensity from each molecular state after dephasing will be on the order of $1/N^2$, and the total intensity (which is now simply a sum of the partial intensities) on the order of $1/N$. **At very large** N [$\gg N_0$, where $1/N_0$ is the lowest detectable intensity, see definition (2.6.55)] the emission intensity after dephasing practically disappears, and the emission terminates. If $\tau_{\max} \gg \tau_{\mathrm{deph}(s)}$ is chosen, the whole decay process is described by one exponential (2.1.5) with $\tau_{\mathrm{obs}} = \tau_{\mathrm{deph}(s)}$.

This result remains valid when account is taken of the various relaxation processes in the bright and dark states, viz., dissociation, internal conversion and intersystem crossing to an outside (say, the ground) electronic state and vibrational relaxation. These relaxations with characteristic times τ_{rel} are introduced phenomenologically through the Lorentzian level broadening. We shall assume that no radiative damping is present in the dark states, and that they all have equal halfwidths at half maximum:

$$\gamma_l = \gamma_{rel(l)} = \hbar \tau_{rel(l)}^{-1}/2 \equiv \text{const} . \tag{2.5.7}$$

The halfwidth of the bright state is

$$\gamma_s = \gamma_{rel(s)} + \gamma_{rad(s)} = \hbar(\tau_{rel(s)}^{-1} + \tau_{rad(s)}^{-1})/2 . \tag{2.5.8}$$

The latter should not be too large, otherwise it is difficult to prepare Φ_s since condition (2.1.1) cannot be satisfied at a too low $\tau_{obs(s)}$ (which is still lower than $\hbar/2\gamma_s$ due to the electronic relaxation from Φ_s to Φ_l). Consider first the influence of γ_l alone.

At very large N when the emission terminates completely just during the first stage of evolution of the excited state, the relaxations affect neither the decay law nor the decay rate irrespective of the value of γ_l. This can be explained qualitatively as follows. For

$$\gamma_l \ll \hbar k_s/2 \tag{2.5.9}$$

the lifetime $\tau_{rel(l)}$ is long as compared with $\tau_{deph(s)}$ so that the luminescence terminates before the relaxation effects can emerge. On the other hand, for the opposit limit of

$$\gamma_l \gg \hbar k_s/2 , \tag{2.5.10}$$

the molecular levels E_n acquire large halfwidths γ_n and therefore merge in a united quasistationary level $\tilde{E}_s$ of which the halfwidth is equal to $\hbar k_s/2$. The luminescence again decays exponentially, and just with the same rate constant k_s as was the case for $\gamma_l = 0$; but here, k_s is no longer the dephasing rate since no Φ_n states exist now. Instead, k_s is the rate constant for the true $\Phi_s \rightsquigarrow \Phi_l$ radiationless transition.

The distinction between dephasing and true radiationless transition, both having the same rate constant k_s, becomes explicit when damping (2.5.8) is taken into account. Let for instance

$$\gamma_s < \gamma_l \tag{2.5.11}$$

hold. In this case γ_s must be added to the right-hand sides of inequalities (2.5.9, 10). Then, under condition (2.5.10), the halfwidth of the united quasi-stationary level turns out to be $(\hbar k_s/2) + \gamma_s$, and hence it decays with the rate constant

$$\tau_{obs(s)}^{-1} = k_s + (2\gamma_s/\hbar) = k_s + \tau_{rel(s)}^{-1} + \tau_{rad(s)}^{-1} , \tag{2.5.12}$$

i.e., the phenomenological relations (2.1.5–7) are valid. The parameters are $\tau_{obs} = \tau_{obs(s)}$, $\tau_{rad} = \tau_{rad(s)}$, $k = k_s + \tau_{rel(s)}^{-1}$, and k_s is the usual radiationless transition rate. On the other hand, in the limit of (2.5.9), the decay is generally nonexponential because the dephasing process is modulated by an additional decay of the E_n levels, which all have different halfwidths γ_n and therefore decay at different rates. Though, in the present case of (2.5.5, 11), these rate differences are not large because $\gamma_n \approx \gamma_l + \Delta\gamma_n$, where $\Delta\gamma_n$ is small. If they are neglected and γ_n is replaced by a mean value

$$\bar{\gamma}_n = \gamma_l + \frac{1}{N}\gamma_s , \tag{2.5.13}$$

the exponential decay law (2.1.5) is restored. However, instead of the commonplace phenomenological relation (2.5.12), the following observed decay rate of the fast component of the emission is obtained (see [2.76a] (4) and [2.76b], Table 1):

$$\begin{aligned} \tau_{fast}^{-1} &\equiv k_s + (2\bar{\gamma}_n/\hbar) \\ &= k_s + \tau_{rel(l)}^{-1} + \frac{1}{N}(\tau_{rel(s)}^{-1} + \tau_{rad(s)}^{-1}) . \end{aligned} \tag{2.5.14}$$

The peculiarity of this expression as compared to (2.5.12) is, firstly, that the lengthened radiative lifetime of the molecular states

$$\tau_{rad(n)} = N\tau_{rad(s)} \tag{2.5.15}$$

enters instead of the radiative lifetime of the emitting state Φ_s. Secondly, the contribution from the bright-state relaxation is diminished by a factor of N, but, in turn, a term is added due to the dark-state relaxation. All these effects are explained by the fact that the decay observed is that of a set of quasistationary levels $\tilde{E}_n$ rather than that of one united level $\tilde{E}_s$. *Van der Werf* and *Kommandeur* [2.86] neglected this point and used (2.5.12) for τ_{fast}^{-1} instead of (2.5.14).

Owing to (2.5.9, 11) the additional τ^{-1}-terms in (2.5.14) are small as compared to k_s. Therefore, one obtains $\tau_{obs(s)}\,(\equiv\tau_{fast}) \simeq \tau_{deph(s)}$ so that the quantum yield of luminescence at this first (fast) stage is low according to the left-hand side equality in definition (2.1.7):

$$\varphi_{fast} = \tau_{fast}/\tau_{rad(s)} \ll 1 , \tag{2.5.16}$$

while no limitations are imposed on the relative values of different terms in (2.5.12) so that the corresponding quantum yield can take on arbitrary values.

Equation (2.5.16) needs a comment because of the existence of two radiative lifetimes, $\tau_{rad(s)}$ and $\tau_{rad(n)}$. From (2.1.2, 6) it is seen that τ_{rad} in (2.1.7) means the radiative lifetime of the state prepared by the initial excitation (which, from the Einstein relation, is inversely proportional to the oscillator strength of the corresponding absorption), and τ_{obs} the decay time of this state.

Note that the departure from the exponential behavior becomes pronounced at $\gamma_l \ll \gamma_s \ll \hbar k_s/2$ when the second term in (2.5.13), which is valid in this case, may be comparable with the first term. A final note is that the above considerations apply not only at very large N, as was stated initially, but also at a **moderate relaxation rate** to the dark states consistent with (2.5.9),

$$\varrho_l^{-1} \ll \gamma_l \ll \hbar k_s/2 , \tag{2.5.17}$$

since the relaxation at such a rate suppresses the emission on the second stage.

In the case of not very large N (i.e., $1 \ll N \ll N_0$) and of very slow relaxation,

$$\gamma_l \ll \varrho_l^{-1} , \tag{2.5.18}$$

the emission does not cease after dephasing is completed, an *the second (slow) stage* sets in where relaxation plays a crucial part. The termination of dephasing implies that the system has reached such a state which is an incoherent superposition of the molecular states Φ_n, and the emission from this state is a sum of those from the states Φ_n. As before, the deviation from the exponential behavior is insignificant, and the decay rate of the slow component of the luminescence coincides with the average observed decay rate of the states Φ_n being equal to

$$\tau_{\text{slow}}^{-1} \equiv \tau_{\text{obs}(n)}^{-1} \equiv 2\bar{\gamma}_n/\hbar = \tau_{\text{rel}(l)}^{-1} + \frac{1}{N}\tau_{\text{rel}(s)}^{-1} + \tau_{\text{rad}(n)}^{-1} \tag{2.5.19}$$

according to (2.5.13, 15). Here, the second and third terms are small due to (2.5.5, 11), so

$$\tau_{\text{obs}(n)} \approx \tau_{\text{rel}(l)} . \tag{2.5.20}$$

In order to detect the second stage emission, two obvious conditions should be fulfilled: 1) the intensity on the order of $1/N$ is observable experimentally; 2) the characteristic lifetimes satisfy the inequalities

$$\tau_{\text{deph}(s)} \ll \tau_{\text{obs}(n)} < \tau_{\max} . \tag{2.5.21}$$

The left inequality stems from relation (2.5.20) and from inequality (2.5.9), the latter having been written for the case of (2.5.11); in the opposite case of $\gamma_s > \gamma_l$, inequality (2.5.21) again derives from (2.5.9), where γ_l is now replaced by γ_s.

Taking the ratio of (2.5.14) to (2.5.19) and considering (2.5.21) one obtains

$$\frac{\tau_{\text{slow}}}{\tau_{\text{fast}}} = \frac{\hbar k_s}{2\bar{\gamma}_n} + 1 \approx \frac{\hbar k_s/2}{\gamma_l + \gamma_s/N} \gg 1 . \tag{2.5.22}$$

When the two stages of luminescence are observed in experiment, the intensity of the emission from the initially prepared state Φ_s may be written formally as a sum of the fast and slow exponentials (the coherent and incoherent components) [2.76]:

$$I_s(t) = \tau_{\text{rad}(s)}^{-1}\left[\exp(-t/\tau_{\text{fast}}) + \frac{1}{N}\exp(-t/\tau_{\text{slow}})\right] , \tag{2.5.23}$$

where τ_{fast} and τ_{slow} are given by (2.5.14, 19). In fact, of course, only one component exists at every point of time. Indeed, the fast component alone is present during the time interval of $0 \le t \le \tau_{fast}$ because of a large difference between τ_{fast} and τ_{slow} (the constant $\sim 1/N$ is neglected), and only the slow component at $\tau_{fast} \ll t < \tau_{max}$. To simultaneously observe both components within a common long-enough time interval would demand the relation $\tau_{fast} \lesssim \tau_{slow}$, in which case, however, the incoherent component would be negligible due to (2.5.5).

According to (2.1.2, 3), the total quantum yield of emission can be written as a sum of contributions from the fast and slow components,

$$\varphi = \varphi_{fast} + \varphi_{slow} , \tag{2.5.24}$$

where φ_{fast} is given by (2.5.16) and

$$\varphi_{slow} = \frac{\tau_{slow}}{N\tau_{rad(s)}} = \frac{\tau_{obs(n)}}{\tau_{rad(n)}} \tag{2.5.25}$$

[see (2.5.15)]. The latter equality is consistent with the fact that an incoherent superposition of the states Φ_n, each having a population of $1/N$, is prepared at the moment dephasing has terminated. Since $\tau_{obs(n)}$ and $\tau_{rad(n)}$ are independent of n, this ensemble of N states acts as a single state with unit population. Then it follows from (2.1.7) that φ_{slow} is the observed-to-radiative lifetime ratio for this state.

Substituting (2.5.19) into (2.5.25) gives

$$\varphi_{slow} = \frac{\gamma_{rad(s)}}{\gamma_{rad(s)} + \gamma_{rel(s)} + N\gamma_{rel(l)}} \approx \frac{\gamma_{rad(s)}}{N\gamma_l} \ll 1 . \tag{2.5.26}$$

Thus, under conditions of biexponential decay, the emission quantum yields of the fast and slow components are small. Moreover, their ratio

$$\frac{\varphi_{fast}}{\varphi_{slow}} = \frac{2N\gamma_l}{\hbar k_s} = 2\pi\gamma_l\varrho_l \tag{2.5.27}$$

is small, too, according to (2.5.18).

The number of strongly coupled levels can be defined experimentally either as the ratio of the amplitudes of the fast (C_{fast}) and slow (C_{slow}) components in (2.5.23) or by their lifetimes and quantum yields (2.5.16, 25), viz.,

$$N = \frac{C_{fast}}{C_{slow}} = \frac{\tau_{slow}\varphi_{fast}}{\tau_{fast}\varphi_{slow}} . \tag{2.5.28}$$

Now three points are worth mentioning.

i) Equation (2.5.26) is written in such a form as if it were the quantum yield of the emission from the bright state Φ_s whose total width is the sum of the radiative and non-radiative widths, the nonradiative contribution of the electronic relaxation $\Phi_s \to \Phi_l$ being proportional to the dark-state relaxation rate. In fact, however, the quantum yield of the mixed bright-dark rather than the initial bright-state emission is given in the equation.

ii) Diminishing the relaxation widths gives rise to an increasing deviation of the slow decay from the exponential behavior because of a strong variation of γ_n among the mixed states due to a breakdown of relations (2.5.11, 13) (since $\gamma_l \to 0$ while $\gamma_s \to \gamma_{\mathrm{rad}(s)} \neq 0$). The slow decay becomes strongly nonexponential when the average of the fluctuating part of γ_n in (2.5.13) grows higher than the constant part,

$$\gamma_s/N \gtrsim \gamma_l \,. \tag{2.5.29}$$

iii) In the limit of $\gamma_{\mathrm{rel}(s,\,l)} = 0$ the mean observed lifetime $\tau_{\mathrm{obs}(n)}$ coincides with the lengthened radiative lifetime (2.5.15) of state Φ_n. It is lengthened relative to the radiative lifetime of state Φ_s, $\tau_{\mathrm{rad}(s)}$, which is measured experimentally by the total oscillator strength for the $\Phi_s \leftarrow \Phi_0$ absorption in the case where the individual $\Phi_n \leftarrow \Phi_0$ transitions in the absorption spectrum are unresolved.

Now suppose that the mixed state Φ_n rather than the unperturbed bright state Φ_s is prepared by the initial excitation which demands the following conditions to be fulfilled: 1) the density of states is small and the pulse duration is large so that condition (2.4.13) is satisfied; 2) the oscillator strength of the $\Phi_n \leftarrow \Phi_0$ transition is not too low, i.e., N is not very high; 3) the dark-state relaxation is very slow, as in (2.5.18), otherwise the lifetime $\tau_{\mathrm{obs}(n)}$ is short and condition (2.1.1) for prompt initial preparation of this state is not realizable.

The third condition is obviously the clue. Recalling what was said before about the very slow relaxation limit one concludes that *if the short pulse prepares the Φ_s state decaying biexponentially, the Φ_n state can be prepared by a long excitation, but it cannot be prepared by a pulse of any duration when the Φ_s state shows a monoexponential decay*. Combining (2.1.1), (2.4.13), (2.5.3), and using (2.5.20) one obtains the following condition to prepare the initial state Φ_n:

$$v_{sl} \gg \varrho_l^{-1} \gg \hbar/\tau_{\mathrm{pulse}} \gg \hbar/\tau_{\mathrm{rel}(l)} \,.$$

When this condition is fulfilled the first stage of emission is missing since no coherent excitation of many molecular states is initiated. The intensity decay again follows the simple exponential law [second term in (2.5.23)] with the observed lifetime (2.5.19) and the quantum yield (2.5.25).

Completing the discussion of the strong coupling limit, let us formulate the main conclusions depicted in the scheme of Fig. 2.3. The luminescence decay evolves in two stages, viz., the fast decay of the luminescent (bright) state of which the rate constant depends on the intramolecular parameters only, not on the relaxation widths (that are admissible to include contributions from molecular collisions), and the slow decay of the mixed (bright and dark) molecular eigenstates, which is very sensitive to relaxation widths. Either one of these stages or both may be observed simultaneously during the experiment at $t < \tau_{\mathrm{max}}$ depending upon the excitation conditions and upon the parameters of the system. In the first (fast) stage the decay is always exponential while in the second (slow) stage the decay deviates more and more from the exponential behavior when the unperturbed-state relaxation slows down; chaotic quantum beats due to incomplete dephasing may be superimposed on the slow component at not very large numbers of strongly coupled levels. The relaxation processes in the unperturbed bright and dark states may be associated both with intramolecular energy redistribution and with intermolecular interactions. In the latter case the shape and the rate constant for the luminescence decay in vapor will depend on pressure. Increasing the pressure first suppresses the slow decay so that ratio (2.5.27) becomes large, and then turns the fast component, corresponding to the dephasing in the intermediate case with the rate constant of (2.5.14) and with a low quantum yield of (2.5.16), into the statistical limit decay with the rate constant (2.5.12) and an arbitrary quantum yield.

A final remark is that the detuning Δ is an insignificant parameter since in the strong coupling limit the levels move away from one another and the resonance, if any, disappears.

2.5.2 Experimental Criterion for the Statistical Limit

According to *Amirav* [2.149] a criterion for the transition from the intermediate case to the statistical limit for strong coupling can be given in terms of quantum yields of the two components, i.e., of experimentally measured quantities. Based on (2.5.27) and using the precise condition for the statistical limit of Sect. 2.6.2 [replacing the approximate one of (2.5.10)] and definition (2.5.5) and obtains

$$\varphi_{\text{fast}}/\varphi_{\text{slow}} \gg N + 1 , \qquad (2.5.30)$$

which differs from Amirav's condition by the large N-term.

Particular attention is worth to be given to the way in which the nature of the radiationless transition transforms in going from the statistical limit to the intermediate case. In the statistical limit the system undergoes a transition from a bright state to dark isoenergetic states followed by fast relaxation; this picture applies if the final-state relaxation is faster than the electronic transition (Fig. 2.3). On the other hand, if the relaxation is not fast enough, the transition occurs to the mixed rather than the dark states.

2.5.3 Upper Limit for Radiationless Transition Rates

The interesting question about a theoretical upper limit of the rates of intramolecular deactivation processes, and about the corresponding lower limit of the observed lifetimes of excited electronic states, arises in view of recent experimental findings [2.4, 48, 207, 208] and theoretical calculations [2.209–211] interpreted in terms of ultrafast radiationless processes occurring on a femtosecond time scale. Three different cases are to be discussed.

First, consider the statistical limit where the usual radiationless transitions exist with the rate constant k_s of (2.5.2). According to (2.5.10) this constant is confined by a maximum value,

$$k_s \ll 2\gamma_l/\hbar \equiv \tau_l^{-1} , \tag{2.5.31}$$

where τ_l^{-1} is the rate of deactivation of dark states. Now, if the transition occurs from an excited state to the ground state without predissociation, the only deactivation process in the final state is vibrational relaxation. Taking 1400 cm^{-1} as a typical figure for the frequency of the C–C stretching modes in organic molecules and supposing $2\gamma_l$ to be at least 10 times smaller one obtains $\tau_l = (5.3/140)$ ps $\simeq 40$ fs. Thus the observed lifetime τ_{obs}, which approximately coincides with the lifetime for radiationless deactivation τ_{RT}, should be much longer than the above value, i.e.,

$$\tau_{obs} \approx \tau_{RT} \gg 40 \text{ fs} \tag{2.5.32}$$

resulting in a homogeneous linewidth in the spectra below approximately 100 cm^{-1}. This is also valid for transitions between the stable excited electronic states since additional relaxation channels such as transitions to the ground state place the same limit of (2.5.31) upon the observed lifetime. This estimate is also valid in the intermediate case with a fast final-state relaxation, provided the balance equation applies (see Fig. 2.3) since the corresponding radiationless transition rate k_{sl} is still lower than k_s (Sects. 2.5.4 and 2.7).

Second, consider the intermediate case with a moderate final-state deactivation rate of (2.5.17) consistent with (2.5.21). Of the two emission components, only the fast one is observed since the slow decay has very low intensity (Fig. 2.3). The rate of the fast decay is again k_s, and it may be regarded as τ_{RT}^{-1} neglecting the irrelevant, in this case, difference between the intermediate case and the statistical limit. This quantity is determined by v_{sl} and ϱ_l alone.

To be more specific let us introduce, for intersystem crossing, the electronic (spin–orbit) matrix element $v_{el} = 10 - 0.3$ cm^{-1} and the electronic density of states $\varrho_{el} \simeq 10^{-4}$ cm^{-1}. The formal substitution into (2.5.2) gives $k_s = 10^{-2}$–10^{-5} ps^{-1}. In fact, each electronic state is "dressed" with a manifold of vibronic sublevels, and we assume the bright state to be spread uniformly among N dark states. Then ϱ_l is increased correspondingly N times, $\varrho_l = N\varrho_{el}$, and the normalization condition for the wave functions gives $v_{sl} = v_{el}/\sqrt{N}$, so k_s remains roughly unchanged. These arguments imply the following estimate for the InterSystem Crossing lifetime (ISC):

$$\tau_{ISC} \equiv 1/k_s = 100 \text{ ps–}100 \text{ ns} . \tag{2.5.33}$$

As for Internal Conversion (IC), the corresponding lifetime for closely spaced excited singlet (or triplet) levels τ_{IC} may reach a few vibration periods,

$$\tau_{IC} \gg (1400 \times 3 \times 10^{10})^{-1} c \simeq 25 \text{ fs} , \tag{2.5.34}$$

since the nonadiabatic coupling is very high, on the order of the vibrational quantum (Sect. 4.2).

Third, in the case of very slow relaxation (2.5.18) both components are seen but the fast one has a low relative quantum yield following from (2.5.27). Therefore, no information about the dephasing rate, which represents in a sense τ_{RT}^{-1}, can be gained from an experiment where only the total quantum yield is measured. As for the observed lifetime $\tau_{obs(n)}$, it contains information about relaxation processes other than the $\Phi_s \rightsquigarrow \Phi_l$ relaxation, and it is again restricted from below by 40 fs.

Ignoring the effects of state mixing may influence the evaluation of electronic-state lifetimes by quantum yields [2.49]. *Ermolaev* and *Lyubimtsev* [2.4, 48] measured the fluorescence quantum yields for the higher singlet states $S_n (2 \leq n \leq 6)$ of dye molecules in solutions and found $\varphi \simeq 10^{-6}$ for some states, while $\tau_{rad(s)} \simeq 100$ ns was obtained from integrated absorption measurements. Then (2.1.7) gives $\tau_{obs(s)} \simeq \tau_{rad(s)} \varphi = 100$ fs, which approaches the shortest times of (2.5.34). Assume, however, that the bright level is mixed to N dark levels belonging to lower electronic states, and that the emission is mainly due to the slow portion of the decay as in the third case before, i.e., $\varphi = \varphi_{slow}$. Then the observed lifetime calculated from (2.5.25) is $\tau_{obs(n)} = \tau_{slow} = N\tau_{rad(s)}\varphi$, i.e., N times higher than before.

Recently, the existence of ultrafast dephasing processes giving rise to homogeneous band widths in excess of 1000 cm^{-1} has been suggested [2.207].

2.5.4 Weak Coupling

In the case of weak coupling (2.5.4) together with condition (2.4.22), the excitation process prepares the initial state Φ_s nearly coincident with a particular molecular state Φ_n. Hence, the coherence effect due to the intramolecular dephasing does not exist (though dephasing may occur due to molecule-medium interaction [2.212]). Instead, a different king of coherence caused by the presence of the nearby state $\Phi_{l=0}$ arises for slow relaxation $\gamma_s < \gamma_l \ll \gamma_{sl}$, and for small detuning $|\Delta| \lesssim v_{sl}$ [2.171, 172] (Fig. 2.3). As a result, undamped quantum beats between Φ_s and $\Phi_{l=0}$ occur for $\gamma_l = 0$. If $\gamma_l \neq 0$, a biexponential luminescence decay modulated with the quantum beats occurs, with the modulation frequency an depth as well as the decay rate depending on detuning. The decay rates of the two components are associated with γ_s and γ_l.

At large detuning, $|\Delta| \sim \varrho_l^{-1}$, the beats essentially vanish because their amplitude is small. The decay is monoexponential enabling the radiationless transition rate k_{sl} to be introduced (now k_s of (2.5.2) is only a formal parameter). Owing to the beforementioned property of the system to undergo undamped beats in the limit of $\gamma_l \to 0$, it is obvious that the very possibility for an irreversible electronic transition is exclusively connected with the final electronic state relaxation. This is a fundamental difference between the limits of strong and weak coupling, which leads to a strong γ_l dependence of k_{sl} in contrary to k_s.

The beats also disappear at $v_{sl} \ll \gamma_l \ll \varrho_l^{-1}$ when the decay time becomes shorter than their period. The detuning is now unimportant, and the decay kinetics is described by the phenomenological expressions of Sect. 2.1 where, as before, the radiationless transition rate k_{sl} depends on γ_l.

Lastly, merging of many molecular levels in a united level with the observed lifetime (2.5.12), similar to that in the strong coupling case, occurs at very fast relaxation $\gamma_l \gg \varrho_l^{-1}$. The phenomenological equations (2.1.5–7) are again valid, and the radiationless transition rate (2.5.2) is independent of γ_l. It is worthwhile to note that, although the dynamics depend upon the contributions from a larger number of unperturbed states (rather than from two states as was the case for $\gamma_l \ll \varrho_l^{-1}$), the dephasing does not occur since there are no states Φ_n.

The following important conclusion can be drawn from these arguments. The description of time evolution of the emission intensity based on balance equation (2.1.4), which leads to (2.1.5–7) with the electronic relaxation rate constant being independent of the processes of energy redistribution and relaxation in the unperturbed states, is valid for the conditions of fast relaxation processes occurring in the final electronic state resulting in merging of the molecular levels. This conclusion is independent of the coupling strength.

The levels merge at a fast relaxation rate in the *initial* electronic state, too. However, in this case the fast relaxation not only makes it difficult to prepare the initial quasistationary state because of lifetime shortening, but also breaks down (2.1.6) since the interstate interaction is no longer present, as if turned off (Sect. 2.4.1), resulting in a quantum yield of unity.

2.6 Strong-Coupling Limit

This section and the next deal with theoretical models designed to investigate the electronic relaxation processes in excited states of polyatomic molecules, as well as with emission kinetics, whose main features were outlined, without proof, in the previous section and in Fig. 2.3.

2.6.1 The Bixon–Jortner Model

The Bixon–Jortner model [1.9, 10, 41] has played an outstanding role in developing the concepts on the nature of radiationless transitions. In the framework of this model, the survival probability (2.4.36) can be calculated explicitly, thus allowing to derive the theoretical recurrence time, the irreversibility criterion for a radiationless transition, and the transition rate constant in terms of model parameters. Another important virtue of the Bixon–Jortner theory lies in its close relation to the earlier theory of *Robinson* and *Frosch* [2.177], to be described in the next section. To elucidate this relation more explicitly seems particularly important to us because these two theories are of fundamental significance and answer basic questions regarding the mechanism of irreversibility and the role of vibrational relaxation (in general, of the total final-state relaxation including various channels). Finally, the Bixon–Jortner model, as will be shown later (Chap. 6), has much more immediate pertinence to real molecules than is commonly assumed.

To start with, consider the general problem of calculating the probability $P_{ss}(t)$ in the framework of the level model depicted in Fig. 2.2. Let us find the eigenfunctions and eigenvalues of the total Hamiltonian H. In the molecular-level model assumed, the Hamiltonian has the following nonzero matrix elements between the unperturbed excited states:

$$\langle \Phi_s | H | \Phi_s \rangle = E_s, \qquad \langle \Phi_l | H | \Phi_l \rangle = E_l, \qquad \langle \Phi_s | H | \Phi_l \rangle = v_{sl} . \tag{2.6.1}$$

The stationary-state wave function can be written in the form

$$\Phi_n = a_{sn} \Phi_s + \sum_l a_{ln} \Phi_l , \tag{2.6.2}$$

where a_{sn}, a_{ln} and v_{sl} (2.6.1) are assumed to be real. Substituting (2.6.1, 2) into the Schrödinger equation (2.3.3) gives the following coupled equations for the expansion coefficients (2.6.2):

$$(E_s - E_n) a_{sn} + \sum_l v_{sl} a_{ln} = 0 , \tag{2.6.3}$$

$$v_{ls} a_{sn} + (E_l - E_n) a_{ln} = 0 . \tag{2.6.4}$$

Expressing a_{ln} from (2.6.4) in terms of a_{sn},

$$a_{ln} = \frac{v_{ls}}{E_n - E_l} a_{sn} , \tag{2.6.5}$$

and substituting it into (2.6.3) results in the equation determining the levels E_n,

$$E_n - E_s = \sum_l \frac{v_{sl}^2}{E_n - E_l} . \tag{2.6.6}$$

Using the normalization condition for the expansion coefficients in (2.6.2),

$$a_{sn}^2 + \sum_l a_{ln}^2 = 1 , \tag{2.6.7}$$

and (2.6.5) one obtains

$$a_{sn}^2 = \left(1 + \sum_l \frac{v_{sl}^2}{(E_n - E_l)^2} \right)^{-1} . \tag{2.6.8}$$

This quantity has a simple physical meaning. Within the level model under consideration only Φ_s, among the excited levels, bears a nonzero oscillator strength for a transition from the ground state Φ_0. The interaction between Φ_s and Φ_l results in sharing the oscillator strength among the exact states Φ_n in proportion to a_{sn}^2. Owing to the conservation of the initial oscillator strength the sum of a_{sn}^2 over n must be equal to unity. Indeed, using the unitarity condition for the basis transformation (2.6.2), one obtains the more general relations

$$\sum_n a_{sn}^2 = \sum_n a_{ln}^2 = 1 , \tag{2.6.9}$$

which can also be derived from the following exact equations:

$$a_{sn}^2 = \partial E_n/\partial E_s , \qquad a_{ln}^2 = \partial E_n/\partial E_l , \tag{2.6.10}$$

$$\sum_n E_n = E_s + \sum_l E_l . \tag{2.6.11}$$

Equations (2.6.10) are obtained by taking the energy derivatives of (2.6.6) at constant values of v_{sl}; (2.6.11) (valid for any finite number of the levels) is a consequence of the trace conservation of the Hamiltonian matrix under transformation (2.6.2) (in the case of two levels this is just the Viette theorem). Differentiating (2.6.11) directly gives (2.6.9).

According to (2.4.32, 35, 36) and (2.6.2) the desired probability is equal to

$$P_{ss}(t) = |\langle \Phi_s|\exp(-\mathrm{i}Ht/\hbar)|\Phi_s\rangle|^2$$

$$= \left|\sum_n \langle \Phi_s|\exp(-\mathrm{i}Ht/\hbar)|\Phi_n\rangle \langle \Phi_n|\Phi_s\rangle\right|^2 ,$$

or

$$P_{ss}(t) = \left|\sum_n a_{sn}^2 \exp(\mathrm{i}\delta_n)\right|^2$$

$$= \sum_{n,n'} a_{sn}^2\, a_{sn'}^2 \exp[\mathrm{i}(\delta_n - \delta_{n'})] , \tag{2.6.12}$$

where $\delta_n = -E_n t/\hbar$.

The Bixon–Jortner model assumes an infinite set of equally spaced levels with a constant interaction matrix element,

$$E_l = E_s - \Delta + l\varepsilon , \qquad v_{sl} = v , \quad l = 0, \pm 1, \pm 2, \dots . \tag{2.6.13}$$

Here, ε is the spacing between the adjacent levels E_l and Δ is the detuning (2.5.1) ($|\Delta| \le \varepsilon/2$). Substituting (2.6.13) into (2.6.6, 8) and using the identities [2.213]

$$\sum_{l=-\infty}^{+\infty} \frac{1}{x-l} = \pi\cot(\pi x) , \qquad \sum_{l=-\infty}^{+\infty} \frac{1}{(x-l)^2} = \frac{\pi^2}{\sin^2(\pi x)} \tag{2.6.14}$$

one obtains

$$E_n - E_s = \pi v^2 \varrho \cot[\pi\varrho(E_n - E_s + \Delta)] , \tag{2.6.15}$$

$$a_{sn}^2 = \frac{v^2}{(E_n - E_s)^2 + (\pi v^2 \varrho)^2 + v^2} , \tag{2.6.16}$$

where $\varrho(\equiv \varrho_l) = 1/\varepsilon$ is the density of the final states Φ_l.

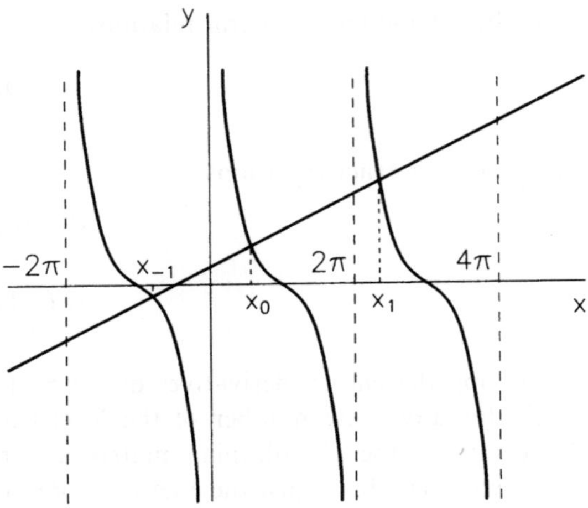

Fig. 2.4. Graphical solution to (2.6.18)

To analyze (2.6.15) qualitatively, it is appropriate to introduce the dimensionless variable

$$x = 2\pi\varrho(E - E_s + \Delta) \tag{2.6.17}$$

which takes on values x_n at $E = E_n$. Then (2.6.15) can be rewritten in the form

$$\frac{x - 2\pi\varrho\Delta}{2(\pi v\varrho)^2} = \cot(x/2)\ . \tag{2.6.18}$$

A graphical solution to this equation is given in Fig. 2.4, which demonstrates that there is a single exact level E_n between every pair of adjacent unperturbed levels E_l corresponding to integer values $x/2\pi$.

In the strong coupling limit (2.5.3), the slope of the straight line in Fig. 2.4 decreases to zero so that a large number of nearly equidistant levels E_n located approximately half-way between the adjacent unperturbed levels appear, i.e.,

$$x_n = \pi(2n + 1) + \delta x_n\ , \tag{2.6.19}$$

where δx_n is small for many levels. Bixon and Jortner adopt the additional assumption of $\delta x_n \equiv 0$, i.e.,

$$E_n = E_{n=0} + n\varepsilon, \quad n = 0,\ \pm 1, \ldots\ . \tag{2.6.20}$$

Substituting (2.6.16, 20) into (2.6.12) one obtains

$$P_{ss}(t) = \left| \sum_{n=-\infty}^{+\infty} \frac{v^2 \exp(-in\varepsilon t/\hbar)}{(\Delta_1 + n\varepsilon)^2 + (\pi v^2 \varrho)^2} \right|^2 , \tag{2.6.21}$$

where $\Delta_1 = E_{n=0} - E_s$ and a small term in the denominator is omitted due to (2.5.3).

This equation shows that $P_{ss}(t)$ is a periodic function with the period

$$T = 2\pi\hbar\varrho \,, \tag{2.6.22}$$

which is just the theoretical recurrence time in this model. Consider now the behavior of this function at short times

$$t \ll T \,. \tag{2.6.23}$$

With conditions (2.5.3) and (2.6.23), the amplitude and phase of the expression under the summation sign in (2.6.21) are slowly varying functions of n; therefore the sum may be replaced by an integral over n. The integration is performed in a simple way to give the exponential function (2.4.37) with k_s from (2.5.2).

The exponential decay (2.4.37) with the rate constant k_s (2.5.2) may evidently be observed under the condition that the maximum observation time τ_{max} is longer than the decay time $1/k_s$, while being yet very short as compared to the theoretical recurrence time T,

$$T \gg \tau_{max} > 1/k_s \,. \tag{2.6.24}$$

Substituting (2.6.22) and (2.5.2) one obtains the condition for irreversible relaxation in the Bixon–Jortner model,

$$(2\pi v\varrho)^2 \gg 1 \,, \tag{2.6.25}$$

which is a direct consequence of the strong coupling condition (2.5.3) used in calculating $P_{ss}(t)$.

Equation (2.4.37) remains valid if (2.6.20) is relaxed to a less arbitrary assumption concerning the density of the exact levels,

$$\varrho(E) = \sum_n \delta(E - E_n) \,. \tag{2.6.26}$$

With the use of (2.6.16) and (2.5.2, 3), (2.6.12) is recast as

$$P_{ss}(t) = \left| \int_{-\infty}^{+\infty} \frac{v^2 \varrho(E)\, dE}{(E - E_s)^2 + (\hbar k_s/2)^2} \exp(-\mathrm{i}Et/\hbar) \right|^2 . \tag{2.6.27}$$

Since we are interested in the behavior of the system at $t \lesssim 1/k_s$, let us divide the integration axis into intervals δE wide such that $\hbar k_s \gg \delta E \gg \varepsilon$. Within every interval, $\varrho(E)$ changes strongly while the remaining factors in the integrand of (2.6.27) do not. Therefore, $\varrho(E)$ may be replaced by an average value of $\delta v/\delta E$, where δv is the number of E_n levels between E and $E + \delta E$. Since there is one and only one level E_n between every couple of the unperturbed levels E_l, the average is clearly equal to ϱ. Calculating the integral (2.6.27) we again obtain (2.4.37). However, as distinct from the basic model, $P_{ss}(t)$ is now not a strictly periodic function.

Numerical studies of the Bixon–Jortner model [1.1, 3, 2.201, 214] have shown that already at $v\varrho = 0.5$ a distinct separation, in the time domain, between the exponential decay and the theoretical recurrence is clearly seen, and this separation survives on relaxing any of the simplifying assumptions of the model. Thus, if either the exact energies E_n are used instead of (2.6.20) or nonequidistant levels E_l instead of (2.6.13), the first recurrence appears later than in the Bixon–Jortner model, and is incomplete in the sense that $P_{ss}(t) < 1$, while the exponential decay with the rate constant (2.5.2) is still maintained at short times. Hence, these assumptions of the model are completely insignificant from a physical point of view.[6]

Relaxing the demand of (2.6.13) regarding the constant value of the matrix element does not affect the exponential decay (2.4.37) at short times (2.6.23) either, but in this case the rate constant is given by

$$k_s = \frac{2\pi}{\hbar} v_{\text{eff}}^2 \varrho_{\text{eff}} , \tag{2.6.28}$$

where v_{eff} is the effective interaction matrix element and ϱ_{eff} the effective density of final states governing the radiationless transition rate. The concept of the *effective states* was introduced by *Robinson* [1.11, 2.178, 179] in order to eliminate such final states to which the transitions are actually forbidden because of too low values of the matrix elements v_{sl}.

In the numerical calculations, a definite law for changing v_{sl} was assumed, and v_{eff}^2 was then obtained from (2.6.28) through the calculated k_s supposing that ϱ_{eff} merely coincides with the total density of final states ϱ. It was found to be equal to an average over the final states of the matrix element squared.

Applying (2.6.28) to real systems where v_{sl} is expected to fluctuate appreciably even among closely spaced levels E_l is not straightforward, as the definitions of the parameters v_{eff} and ϱ_{eff} remain unclear. In this context it is interesting to find out whether a subset of the most essential (effective) final states can be sorted out which have a certain density ϱ_{eff} and *a nearly constant value of the matrix element* v_{eff}. It will be shown in Chap. 6 that this is indeed possible in some important cases, and v_{eff} and ϱ_{eff} will be calculated explicitly in the framework of simple models for potential surfaces.

It was stated in Sect. 2.5.1 that the fast luminescence decay in the strong coupling limit at short exciting-pulse duration is caused by a loss of coherence of the initially prepared state Φ_s. Now we can substantiate this by using (2.6.12) in which the probability amplitude to find the system in the initial state is written as a sum of complex contributions from the molecular states Φ_n with phases δ_n; their amplitudes a_{sn}^2 are markedly nonvanishing for a large [due to (2.5.3)] number of levels $2N'$, with energies E_n within the width of the Lorentzian distribution (2.6.16). Taking into account that the full width at half maximum

[6] They may be significant, however, if the so-called correlation hole is observed (see below).

equals $\hbar k_s$, one obtains $N' = \hbar k_s \varrho/2$ [2.206]. It is appropriate here to introduce the quantity $N = 2\pi N'$ [2.86, 89] [see (2.5.5)], which is called *the number of strongly coupled levels* (this definition differs from that of [2.76, 168]).

Initially, at $t = 0$, the phases δ_n are strictly correlated since all of them are equal to zero, which is a consequence of coherent excitation; so $P_{ss}(0) = 1$ according to (2.6.9). At $t > 0$, dephasing begins, i.e., nonzero phase differences arise, $\delta_n - \delta_{n'} \neq 0$. The dephasing results in a partial cancellation of terms in the sum of (2.6.12), and, as a consequence, $P_{ss}(t)$ decreases exponentially (2.4.37). At $t \gg 1/k_s$ the decay curve progressively diverges from (2.4.37) since oscillations of individual terms in the sum of (2.6.12) become apparent. With increasing t, the oscillation amplitude grows larger, but in the Bixon–Jortner model, at $t = T$, the oscillations stop and the coherence restores completely, i.e., $P_{ss}(t)$ is restored to its initial value of 1 ("optical echo"). However, in a more realistic model of nonequidistant levels (with the same density ϱ) the coherence is practically not restored, even in part [2.189, 215, 216]. The coherence is then lost completely and irreversibly, and $P_{ss}(t)$ is fluctuating around a mean value of

$$
\begin{aligned}
\bar{P}_{ss} &= \sum_n a_{sn}^4 \\
&= \int_{-\infty}^{+\infty} \left(\frac{v^2}{(E - E_s)^2 + (N/2 + 1)v^2} \right)^2 \varrho(E)\, dE \\
&= \frac{N^{1/2}}{(N + 2)^{3/2}} \approx \frac{1}{N}\,,
\end{aligned}
\tag{2.6.29}
$$

which is obtained from (2.6.12) after averaging over the large phase differences nullifying the cross terms with $n \neq n'$, and after replacing $\varrho(E)$ with ϱ as in the derivation of (2.6.27).

Notice that, generally, owing to the gradual increase of the chaotic oscillation amplitude with t, the transition from (2.4.37) to (2.6.29) in the interval of

$$
1/k_s < t < T = 2N/k_s
\tag{2.6.30}
$$

develops smoothly, without a decrease of $P_{ss}(t)$ to $P_{ss}(T) = \exp(-2N) \ll \bar{P}_{ss}$ at $t = T$, which is also evident in the numerical results of [2.189, 215, 216]. However, this drop, called *correlation hole*, can be observed if the level spacings display regularity [2.105, 106].

The above-mentioned details of the $P_{ss}(t)$ behavior at $t \gg 1/k_s$ are, of course, insignificant in the limit $N \to \infty$ since then condition (2.6.24) is surely fulfilled. However, they are meaningful at finite values of N when the second stage of the decay may be observed, as will be discussed in Sect. 2.6.3.

If there exists a mechanism for a vibrational relaxation (and/or for other relaxations) in the final electronic state, it may be allowed for in a formal way by introducing halfwidths γ_l of levels E_l. The Bixon–Jortner model with a constant level broadening can also be solved [2.217–219] (see Sects. 2.6.2, 3). The major result is that, if the relaxation time is short as compared to the theoretical recurrence period (2.6.22), or equivalently,

$$\gamma_l \gg \varrho_l^{-1}, \tag{2.6.31}$$

the exponential decay law (2.4.37) with the rate constant (2.5.2) extends over all times not only in the strong coupling limit of (2.5.3) but also in the weak coupling, i.e., fully independent of the relation between v and ϱ^{-1}. This result is explained as follows (see also the end of Sect. 2.7.3).

Remember that inequality (2.6.25) ensures the fulfillment of the irreversibility criterion of (2.6.24), which can be rewritten in the form

$$2\pi\hbar/\tau_{\max} \gg \varepsilon\,. \tag{2.6.32}$$

It follows that in the Bixon–Jortner model the irreversibility arises because of the fact that the level broadening due to the limited duration of the experiment is large as compared to the level spacing, and hence the discrete spectrum of the final states "spreads" and essentially turns into a continuous one. Now, if there is a large-enough additional broadening due to the vibrational relaxation preventing the theoretical recurrences, condition (2.6.32) and its counterpart (2.6.24) naturally become unnecessary. Then the rate constant is independent of γ_l and is given by (2.5.2).

2.6.2 Inclusion of Level Broadening

In addition to what was said at the end of Sect. 2.6.1, it is interesting to investigate in detail the Bixon–Jortner model taking into account relaxation in two electronic states. This is done here in the framework of the already mentioned phenomenological approach, where real energies are replaced by complex ones, i.e., a diagonal damping matrix is introduced [2.76, 220] (off-diagonal damping is taken into account in [2.221, 222]). This approach is conventionally called *the effective Hamiltonian method*. The method will be grounded in Chap. 3 using Green's functions (see also Sect. 2.6.3).

Thus the halfwidths (2.5.7, 8) are assigned to the unperturbed levels and the corresponding halfwidths γ_n to the exact levels E_n. Performing the substitution

$$E_\nu \to \tilde{E}_\nu = E_\nu - \mathrm{i}\gamma_\nu, \quad (\nu = s, l, n) \tag{2.6.33}$$

and separating the real and imaginary parts of (2.6.15) leads to simultaneous equations

$$E_n - E_s = \frac{1}{2}\hbar k_s \frac{\sin x_n}{\cosh y_n - \cos x_n}, \tag{2.6.34}$$

$$\gamma_n - \gamma_s = \frac{1}{2}\hbar k_s \frac{\sinh y_n}{\cosh y_n - \cos x_n}, \tag{2.6.35}$$

$$y_n = 2\pi\varrho_l(\gamma_l - \gamma_n)\,. \tag{2.6.36}$$

Equation (2.6.35) shows that y_n is always of the same sign as $\gamma_n - \gamma_s$, i.e., γ_n lies between γ_s and γ_l.

Depending on the magnitude of y_n, three limiting cases are possible.

Case I: $|y_n| \ll 1$. In the linear approximation with respect to y_n the levels E_n are again determined by (2.6.15), while

$$\gamma_n = a_{sn}^2 \gamma_s + (1 - a_{sn}^2)\gamma_l \, . \tag{2.6.37}$$

This expression is not only valid in the Bixon–Jortner model but also for arbitrary E_l and v_{sl} (at constant γ_l) since it can be obtained in a straightforward way either by expanding (2.6.6) in powers of γ_v (see Sect. 3.5.1) or, regarding the increment of E_n as a function of E_s and E_l under substitution of (2.6.33), from the identity

$$\gamma_n = \frac{\partial E_n}{\partial E_s}\gamma_s + \sum_l \frac{\partial E_n}{\partial E_l}\gamma_l \tag{2.6.38}$$

with the use of (2.6.10). Substituting (2.6.37) into (2.6.36) we obtain the condition for $|y_n|$ to be small in the form

$$|\gamma_s - \gamma_l| \ll (2\pi\varrho_l)^{-1} a_{sn}^{-2} \, . \tag{2.6.39}$$

This condition is fulfilled for all n's if a_{sn}^{-2} is replaced with its lowest value at $E_n = E_s$ being equal to $(\pi v \varrho_l)^2 + 1$. Then either

$$\gamma_s < \gamma_l \ll \tfrac{1}{4}\hbar k_s \tag{2.6.40}$$

or

$$\gamma_l < \gamma_s \ll \tfrac{1}{4}\hbar k_s \tag{2.6.41}$$

is valid, where small terms are omitted on the right-hand sides of both inequalities. According to Sect. 2.5.1 these cases correspond to the limits of the fast and slow final-state relaxation as compared to the radiative decay of the initial state, while dephasing with rate constant k_s is the fastest process in both cases.

Thus, if the unperturbed-level widths are smaller than the width of the Lorentzian distribution (2.6.16), the levels E_n are not shifted but acquire halfwidths (2.6.37) equal to the sum of the unperturbed-state halfwidths weighted with the probabilities to find the system in these states, provided that it is in the state Φ_n.

Case II: $y_n \gg 1$. A large number of levels merge in a united quasistationary level $E_{n0} - i\gamma_{n0}$, where

$$E_{n0} = E_s, \qquad \gamma_{n0} = \gamma_s + \tfrac{1}{2}\hbar k_s \, . \tag{2.6.42}$$

The requirement for y_{n_0} is met for

$$\gamma_l \gg \gamma_s + \tfrac{1}{2}\hbar k_s + (2\pi\varrho_l)^{-1} . \qquad (2.6.43)$$

As will be explained in the following subsection, *this is the precise condition for the statistical limit in the present model* (it is displayed in Fig. 2.3 in a simplified form). Notice that, (i) the above derivation is independent of the coupling strength, so that level merging under condition (2.6.43) takes place both in the strong and in the weak coupling limit, which will be used in Sect. 2.7.3; (ii) the emission quantum yield may have any value lower than unity since condition (2.6.43) does not impose any limitations on the relative magnitude of radiative and nonradiative decay rates in expression (2.6.42) for the halfwidth of the united quasistationary luminescent level; (iii) in addition to the united level (2.6.42), there exists a large number of quasistationary levels close to the unperturbed levels $\tilde{E}_l$, whose contributions to the emission intensity vanish for $\gamma_l \to \infty$, whereas the contribution of the united level is independent of γ_l.

Case III: $y_n \ll -1$. Here again level merging takes place:

$$E_{n_0} = E_s , \qquad \gamma_{n_0} = \gamma_s - \tfrac{1}{2}\hbar k_s , \qquad (2.6.44)$$

under condition

$$\gamma_s \gg \gamma_l + \tfrac{1}{2}\hbar k_s + (2\pi\varrho_l)^{-1} \qquad (2.6.45)$$

independent of coupling strength. However, as distinct from (2.6.43), condition (2.6.45) implies that the second term in the halfwidth (2.6.44) is small. A formal treatment of this halfwidth results in a greater-than-unity quantum yield in the case where no relaxation occurs (i.e., where $\gamma_l = 0$ and $\gamma_s = \gamma_{\mathrm{rad}(s)}$). This seeming inconsistency is settled by taking into account contributions from the before-mentioned weakly perturbed levels close to $\tilde{E}_l$, which cancel, due to interference, the last term in (2.6.44). Thus inclusion of this term means going beyond the accuracy of the analysis. In fact, as expected in the present case, the quantum yield equals unity (or the value of (2.6.58) if other than radiative processes are active in the unperturbed initial state Φ_s).

Implications of the above analysis of the quasistationary level structure with regard to the emission decay kinetics are discussed in the following subsection.

2.6.3 Mono- and Biexponential Decays

If the state Φ_s is prepared initially, as is assumed here, the luminescence intensity is expressed by

$$I_s(t) = \tau^{-1}_{\mathrm{rad}(s)} P_{ss}(t) \qquad (2.6.46)$$

similar to (2.1.2) except that $P_{ss}(t)$ is now normalized to $P_{ss}(0) = 1$. Therefore the luminescence decay behavior is described by expression (2.6.12), which, after complex energies are introduced, becomes

$$P_{ss}(t) = \left| \sum_n g_{sn} \exp(\mathrm{i}\delta_n - \gamma_n t/\hbar) \right|^2 , \tag{2.6.47}$$

where g_{sn} stands for (2.6.8) after substitution of (2.6.33); the notation a_{sn}^2 is retained when using real energies.

The rationale behind the phenomenological approach is given in Sect. 3.5.1 where it is shown that incorporation of damping of the unperturbed levels results in a replacement of real eigenenergies E_n and coefficients a_{sn}^2 with complex poles of the Green's function, $\tilde{E}_n = E_n - \mathrm{i}\gamma_n$, and with corresponding residues g_{sn}, respectively, so that (2.6.12) takes the form of (2.6.47). After simple algebra using (2.6.14) we obtain

$$g_{sn} = \{1 + \tfrac{1}{2} N \sin^{-2}[\tfrac{1}{2}(x_n + \mathrm{i}y_n)]\}^{-1} . \tag{2.6.48}$$

In cases II and III of Sect. 2.6.2 there is only one quasistationary level $\tilde{E}_{n_0}$ contributing to the sum of (2.6.47), for which (2.6.48) gives $g_{sn_0} = 1$ since $|y_{n_0}| \gg 1$. Hence the decay is monoexponential,

$$P_{ss}(t) = \exp(-2\gamma_{n_0} t/\hbar) . \tag{2.6.49}$$

In case II of the fast *final*-state relaxation (2.6.43), the constant γ_{n_0} is given by (2.6.42) resulting in the observed decay rate (2.5.12) and in the usual phenomenological relations between the observed, radiative, and nonradiative decay rates and the emission quantum yield (2.1.6, 7). This situation is conventionally called *the statistical limit of large molecules* in the classification by *Robinson* [1.11, 2.178, 179] and *Jortner* et al. [1.8–10, 2.206, 223, 224], where the notions of the intermediate case of large molecules and the small-molecule limit (the resonance limit) were introduced.

Condition (2.6.43) for the statistical limit deserves special consideration. Originally it was supposed [1.41, 2.206] that conditions (2.6.24, 25) together with $\gamma_{\mathrm{rad}(s)} \gg \varrho_l^{-1}$ are sufficient to describe the statistical limit in the Bixon–Jortner model without relaxation, i.e., $\gamma_{\mathrm{rel}(s)} = \gamma_l = 0$ in (2.5.7, 8) (this may correspond to cases either I or III). However, as the previous analysis shows (see also cases 2b or 3 in [2.76]), it is the final-state relaxation rate that is the clue parameter so that no statistical limit can be reached if it is not large enough to ensure case II. At weak or medium coupling ($\hbar k_s \lesssim 1/\varrho_l$) the widely invoked relaxed condition $\gamma_l \gg 1/\varrho_l$ [2.149] may be sufficient provided the initial-state relaxation rate is small, which is usually the case in lower vibronic sublevels of the luminescent electronic state. However in the strong coupling limit, the full condition (2.6.43) is needed. Its physical meaning is: for the irreversible $\Phi_s \to \Phi_l$ transition to occur, the system must be able to quickly "forget" its preceding history, i.e., the characteristic time for the loss of memory [which is just $\tau_{\mathrm{rel}(l)}$ of (2.5.7)] must be shorter than both the dephasing time (2.5.6) and the recurrence period (2.6.22).

In case III [fast *initial*-state relaxation (2.6.45)] interaction between electronic states does not manifest in any way and the luminescence decay rate is equal to the decay rate of the unperturbed level E_s since $\gamma_{n0} \approx \gamma_s$ (the second term in (2.6.44) is small and can be omitted); according to (2.5.8) it is the sum of radiative and nonradiative decay rates in the initial electronic state.

In case I there are a lot of quasistationary levels $E_n - i\gamma_n$ with different halfwidths γ_n and with $g_{sn} \approx a_{sn}^2$. Therefore, the emission decay is generally nonexponential and, moreover, exhibits chaotic quantum beats. This behavior corresponds to *the strong coupling limit in the intermediate case of large molecules* [1.6, 7, 2.189, 215, 223, 225]. Despite the complicated luminescence kinetics two stages of quasiexponential decay can be separated.

The main process in the first stage at $t \lesssim 1/k_s$ is dephasing, i.e., a decrease in $P_{ss}(t)$ due to the increase of phases δ_n, whereas the individual level decay can be completely neglected, as $\gamma_n t/\hbar \lesssim \gamma_n/\hbar k_s \ll 1$ [see (2.6.37, 40, 41)]. In this approximation we again obtain (2.4.37) and (2.5.2), as in Sect. 2.6.1. If a small correction to the dephasing rate due to level decay is to be derived, γ_n may be replaced by a mean value

$$\bar{\gamma}_n = \gamma_l + \overline{a_{sn}^2}(\gamma_s - \gamma_l) , \tag{2.6.50}$$

where $\overline{a_{sn}^2} \propto 1/N$; according to the normalization condition (2.6.9), the proportionality factor is taken to be 1. Then, neglecting a small correction $\propto 1/N$, we obtain $\bar{\gamma}_n$ in the form of (2.5.13) and the decay rate in the form of (2.5.14) [after removing the factor $\exp(-\bar{\gamma}_n t/\hbar)$ from the summation of (2.6.47)].

The second stage sets in at $t \gtrsim T$, i.e., following the complete termination of dephasing at times expressed in (2.6.30). Averaging (2.6.47) over large phase differences gives

$$P_{ss}(t) = \sum_n a_{sn}^4 \exp(-2\gamma_n t/\hbar) , \tag{2.6.51}$$

or, approximately,

$$P_{ss}(t) = \bar{P}_{ss} \exp(-2\bar{\gamma}_n t/\hbar) , \tag{2.6.52}$$

where $\bar{P}_{ss}$ and $\bar{\gamma}_n$ are given by (2.6.29) and (2.5.13), respectively. This approximation is valid if γ_n only slightly fluctuates around $\bar{\gamma}_n$, which is the case for the condition opposite to (2.5.29); then the second term in (2.6.50) is small as compared to the first one.

For the slow exponential (2.6.52) to be observed, the following evident conditions have to be obeyed: the duration of the experiment must be longer than the theoretical recurrence time and the emission intensity at the beginning of the second stage (at $t \approx T$) must be higher than the minimum intensity I_{min}, detectable in the experiment. Writing down these conditions in the form $\tau_{max} \gg T = 2N/k_s$ and $P_{ss}(T)/\tau_{rad(s)} \gg I_{min}$, we obtain

$$1 \ll N \ll \tfrac{1}{2} k_s \tau_{max} , \quad N_0 , \tag{2.6.53}$$

$$\gamma_l \ll \alpha \varrho_l^{-1}, \tag{2.6.54}$$

where

$$N_0 = (I_{\min}\tau_{\mathrm{rad}(s)})^{-1}, \qquad \alpha = \frac{1}{4\pi}\ln(N_0/N), \tag{2.6.55}$$

and $\bar{\gamma}_n$ was replaced with γ_l in the left-hand side of (2.6.54). Since α is on the order of unity, condition (2.6.54) reduces to (2.5.18). These inequalities show that N is limited by the experimental conditions, and the halfwidth by the average level spacing. In the opposite case of large halfwidths [γ_s and/or $\gamma_l \gg \varrho_l^{-1}$, as in (2.6.31)] the second stage does not occur and the decay is monoexponential.

If the number of strongly coupled levels is not very high, *the slow exponential decay may be modulated with chaotic beats* because of an incomplete averaging of the cross-terms in (2.6.47). The Fourier spectrum of such beats contains lines corresponding to energy differences $E_n - E_{n'}$ with intensities $g_{sn}g_{sn'}$ [2.114, 115, 129, 166, 189].

The biexponential decay is often represented by (2.5.23) where, however, the statistical-limit expression (2.5.12) for the decay rate constant of the fast component is used instead of (2.5.14) [2.78, 86, 89, 126, 148, 150, 168]. This difference seems to be insignificant since the terms added to k_s in both expressions are small due to (2.6.40, 41). On the other hand, if conditions (2.6.40, 41) are not fulfilled (e.g., for $\gamma_l \ll \gamma_s \sim \hbar k_s/4$, as in [2.86]), the simultaneous equations (2.6.34, 35) must be solved exactly, as no small parameter permitting a Taylor expansion exists now.

If γ_n is not replaced with $\bar{\gamma}_n$, we can calculate $P_{ss}(t)$ with a higher precision using a transformation of the sum (2.6.51) to an integral like (2.6.29) and replacing $\varrho(E)$ with the constant ϱ, the latter being admissible at times such that the exponent changes little when the energy changes by ε, i.e., $(2t/\hbar)|\varepsilon\partial\gamma_n/\partial E_n| \ll 1$. Inserting (2.6.16, 37) and evaluating the energy derivative at the halfwidth of the Lorentzian distribution, $|\varepsilon\partial a_{sn}^{-2}/\partial E_n| = 2\pi$, one obtains that the said approximation is valid at times $t \ll N^2/(4\pi|\gamma_s - \gamma_l|)$. The result of the calculation is [2.76]:

$$P_{ss}(t) = \frac{1}{N}\exp(-2\gamma_l t/\hbar)f(2a\gamma_l t/\hbar), \qquad a = \frac{2(\gamma_s - \gamma_l)}{N\gamma_l}, \tag{2.6.56}$$
$$f(\tau) = \frac{2}{\pi}\int_0^1 \left(\frac{x}{1-x}\right)^{1/2} e^{-x\tau}dx, \qquad f(0) = 1.$$

This applies in a wide range of a and t. In particular for $|a| \ll 1$ (2.6.56) reduces to the slow exponential in (2.5.23) with lifetime (2.5.20), while for $a \gg 1$ [strong inequality in (2.5.29)] the exponential is multiplied by a factor of $t^{-3/2}$.

It thus proves that the slow component first decays exponentially with the rate constant $2\gamma_l/\hbar$ and then it deviates from the exponential behavior at times $t \gtrsim N\hbar/2|\gamma_s - \gamma_l| \gg T = 2N/k_s$ when the second term in (2.6.50) becomes significant.

Time integration of (2.6.46, 51) gives the quantum yield of the slow component,

$$\varphi_{\text{slow}} = \gamma_{\text{rad}(s)} \sum_n a_{sn}^4/\gamma_n \,. \tag{2.6.57}$$

If the final-state relaxation vanishes ($\gamma_l = 0$), equation (2.6.37) gives $\gamma_n = a_{sn}^2 \gamma_s$, and (2.6.57) using (2.6.9) becomes [2.206]

$$\varphi_{\text{slow}} = \gamma_{\text{rad}(s)}/\gamma_s \,. \tag{2.6.58}$$

As expected, the interstate coupling does not affect the quantum yield unless a relaxation process in the final electronic state exists.

Thus the slow decay (2.6.51, 56) is in general nonexponential and may have an arbitrary quantum yield (2.6.57, 58) lower than 1. However, if the condition opposite to (2.5.29) is obeyed, the decay is approximately exponential but the quantum yield is very low since replacing γ_n with $\bar{\gamma}_n$ and using (2.6.29) and (2.5.13) transforms (2.6.57) into (2.5.26).

Now it seems appropriate to sum up the results obtained in this and the preceding sections arranging them according to relationships between the four significant parameters γ_s, γ_l, ϱ_l^{-1}, and k_s under the limitation of the strong coupling limit, $k_s \gg \varrho_l^{-1}$ (from here constants $\hbar$ and 2π in inequalities are omitted), and instantaneous excitation. In part this has been already done in Fig. 2.3.

1) If γ_s is the largest of the significant parameters, the final electronic state is "switched off", i.e., its coupling with the initial state has not enough time to affect the decay appreciably because the lifetime of the luminescent state is too short. Therefore the observed lifetime and the quantum yield are such as if there was no coupling at all, i.e., $\tau_{\text{obs}(s)} = \hbar/2\gamma_s$ and $\varphi = \gamma_{\text{rad}(s)}/\gamma_s$.

2) If the largest parameter is γ_l, the statistical limit occurs, and the conventional equations (2.1.5–7) are valid.

3) If the largest parameter is k_s, two cases are possible.

A) For $\varrho_l^{-1} \ll \gamma_l \ll k_s$ the emission decay is monoexponential, the main decay channel being due to dephasing. The lifetime from (2.5.14) equals approximately (2.5.6), the quantum yield (2.5.16) being low. In all cases considered before the state Φ_s is prepared initially.

B) For $\gamma_l \ll \varrho_l^{-1}$, either Φ_s or Φ_n can be excited depending on the value of τ_{pulse}. *If Φ_s is excited*, the emission decays in two stages, of which the first is simply the same dephasing as in A. The second-stage emission is due to the decay of the molecular (mixed) eigenstates Φ_n. It is observed when N is not very large, otherwise the emission intensity is too low to be measured. The second-stage emission may be modulated with chaotic beats. The smooth component of the slow decay has different characteristics in the following two cases: (i) for $\gamma_s/N \ll \gamma_l$, under instantaneous excitation, the decay is purely exponential with the lifetime (2.5.19) approximately equal to $\tau_{\text{rel}(l)} = \hbar/2\gamma_l$, and with the low quantum yield of (2.5.26); (ii) for $\gamma_l \lesssim \gamma_s/N \ll \varrho_l^{-1}$, the slow decay is nonexponential and the quantum yield may take on arbitrary values. Finally, *if Φ_n is excited*, only the slow stage described before is observed.

2.7 Weak-Coupling Limit

The main feature of the weak-coupling limit (2.5.4) is an enhancement of the part the final-state relaxation plays.

As was seen in the preceding section, in the strong-coupling limit the relaxation only occurs at the second stage, provided the number of strongly coupled levels is not too large. It is clear that the role of relaxation increases with the weakening of the coupling strength and related diminishing of the number of strongly coupled levels. The only exception to this rule is the very fast relaxation (2.6.43) which affects neither kinetics nor the rate constant of the luminescence decay at any coupling strength.

For weak coupling, condition (2.6.43) takes the form of (2.6.31) (neglecting the initial-state relaxation). Since the weak coupling corresponds to small v_{sl} and small ϱ_l, this condition may be violated and level separation may become larger than level broadening. Then the luminescence kinetics will depend upon the relation between γ_l, v_{sl} and detuning (2.5.1), and the radiationless transition rate constant (if defined) will be governed not only by the intramolecular parameters but also by γ_l.

This section is devoted to the discussion of these effects, which are expected, as is clear from the previous arguments, on conditions given by inequality (2.4.22) (the condition to prepare Φ_s), and

$$v_{sl} \text{ and } \gamma_l \ll \varrho_l^{-1} . \tag{2.7.1}$$

The most important final-state relaxation channel is supposed to be the vibrational relaxation due to molecule–medium interaction. The treatment will be further extended in a straightforward manner to isolated large molecules, since the concepts of an "intrinsic quasimolecule" interacting with an "intrinsic medium" can be introduced based on the definition of the effective states (Sects. 2.9, 10).

In Sect. 2.7.1 general features of the problem of interaction between molecule and medium are considered, and the notion of an inert medium is introduced. Classical models describing the radiationless transitions under conditions of exponential decay (2.1.5) are further exposed in Sects. 2.7.2, 3, while Sect. 2.7.4 deals with the case of a nonexponential decay when the notion of a radiationless transition loses its meaning.

2.7.1 The Role of Vibrational Relaxation in the Final Electronic State

When interacting with a medium, the molecule's electronic transitions are irreversible, irrespective of the density of molecular energy levels, but in this case the phenomenon becomes much more complicated. The principal effect is that, in general, the transitions are no longer intramolecular. Since we are mainly concerned with intramolecular transitions, only such cases will be considered in this book where the transition in its nature is most closely related to intramolecular processes. This means that the breakdown of the intramolecular picture is merely due to the influence of vibrational relaxation caused by intermolecular forces on the transition-rate constant. The similarity between isolated and nonisolated molecules in these cases arises in view of the fact that intramolecular energy redistribution among different nuclear degrees of freedom, having much in common with the vibrational relaxation of a molecule in a medium, occurs also in a vibrationally excited isolated molecule. The role of the intramolecular vibrational relaxation will be considered in Sects. 2.9, 10.

Specifically the "molecule–medium" system under consideration is assumed to have the following properties: (i) the molecule has appropriate electronic levels among which the radiative and nonradiative transitions may occur; (ii) the medium does not interact with radiation in the respective spectral range; (iii) the molecule–medium interaction (possibly not weak) depends on the nuclear degrees of freedom only, and not on the electronic ones of the molecule and of the medium.

This means that the transitions under study occur among the electronic levels of which the wave functions are localized inside the molecule while the medium's electronic degrees of freedom play no role at all and may be safely ignored. Moreover, the energies and wave functions of molecular electronic states are independent of the medium's nuclear coordinates, and the electronic and vibronic interactions responsible for radiationless transitions are purely intramolecular in origin. In accordance with conventional terminology the media having these properties will be called *inert*. The role of an inert medium is evidently restricted to dissipation of the vibrational energy of the molecule by energy transfer to the medium initiated by intermolecular interactions. Involvement of electronic degrees of freedom in the molecule–medium interaction may appreciably accelerate nonradiative transitions due to the electronic energy transfer to such a noninert medium [2.7].

The effect of an inert medium on the intramolecular electronic relaxation according to the Bixon–Jortner model was discussed at the end of Sect. 2.6.1. It introduces a level broadening associated with a finite duration of experiment while the interaction with the medium causes an additional broadening due to vibrational relaxation. If the spectrum of the final states is dense enough so that separations between adjacent levels are smaller than the total level width, the transition-rate constant calculated within the Bixon–Jortner model is independent of the vibrational relaxation rate and is only defined by intramolecular parameters.

This section deals with a more general situation where the level spacings are not assumed to be small as compared to the level width. With interference effects neglected, the total transition rate constant is a sum of the partial rate constants for transitions to different final states, and each partial constant must essentially depend on the level width since the only origin of radiationless transitions in a two-level system (one initial state plus one final state) is the interaction with the medium. Such a formulation of the problem of radiationless transitions was first adopted in works by *Robinson* and *Frosch* [2.177] and *Trifonov* and *Shekhtman* [2.226].

2.7.2 The Robinson–Frosch Theory

The total Hamiltonian of the system "a molecule + an inert medium" is written as

$$\mathscr{H} = \mathscr{H}_0 + V' + V'' = H_0 + H_c + V' + V'' , \tag{2.7.2}$$

where H_0 and V' relate to an isolated molecule as defined in Sect. 2.3. H_c is the medium Hamiltonian and V'' the molecule–medium interaction with the properties listed in Sect. 2.7.1. Let χ_0 and χ_κ stand for the medium wave functions of the ground state and of the excited states at energies E_κ, respectively. According to the level diagram of Fig. 2.2 the following eigenfunctions of the zeroth-order Hamiltonian $\mathscr{H}_0$ can be constructed:

$$\Psi_s = \Phi_s \chi_0 , \tag{2.7.3}$$

$$\Psi_l = \Phi_l \chi_0 , \tag{2.7.4}$$

$$\Psi_{m\kappa} = \Phi_m \chi_\kappa ; \quad (E_{m\kappa} = E_m + E_\kappa \approx E_s) . \tag{2.7.5}$$

The function (2.7.3) describes the initially prepared state, while the functions (2.7.4, 5) pertain to the final states isoenergetic with the initial one. Other possible states are disregarded, which amounts to neglecting virtual processes via intermediate states.

The final states (2.7.4) correspond to the total electronic excitation energy transfer to intramolecular vibrations. They are directly coupled to the initial state (2.7.3) by intramolecular interaction since the respective matrix element

$$\langle \Psi_s | V' | \Psi_l \rangle = \langle \Phi_s | V' | \Phi_l \rangle = v_{sl} \tag{2.7.6}$$

does not vanish. On the other hand, the final states like (2.7.5) correspond to a partial transfer of the electronic energy to the medium, and they are not directly coupled to the initial state. Indeed, the equality

$$\langle \Psi_s | V' | \Psi_{m\kappa} \rangle = 0 \tag{2.7.7}$$

holds because of the orthogonality of χ_0 and χ_κ, and

$$\langle \Psi_s | V'' | \Psi_{m\kappa} \rangle = 0 \tag{2.7.8}$$

because of the orthogonality of the electronic wave functions at fixed nuclear coordinates (remember that V'' is independent of the electronic variables when the medium is inert). Therefore, the coupling of the states (2.7.5) to the initial state is indirect and is implemented only via the directly coupled states due to nonzero matrix elements

$$V'_{lm\kappa} = \langle \Psi_l | V' | \Psi_{m\kappa} \rangle . \tag{2.7.9}$$

A number of simplifying assumptions are introduced in the calculation of the rate constant for the radiationless transition from the initially prepared state (2.7.3) to all final states (2.7.4, 5) due to the intramolecular interaction V.

First, it is supposed that the perturbation V' has no matrix elements mixing different l-states (2.7.4), and that each l-state is mixed with its own set of a large but finite number $N-1$ of states (2.7.5) through the nonzero matrix elements (2.7.9). Then the Hamiltonian matrix $\mathscr{H}_0 + V'$ in basis (2.7.4, 5) divides into blocks that can be enumerated by the index l and diagonalized separately to give the following eigenfunctions for the l-th block corresponding to the energies E_{il}:

$$\Psi_{il} = A_{il}\Psi_l + \sum_{m\kappa} A_{im\kappa}\Psi_{m\kappa} \quad (i = 1, \ldots, N) . \tag{2.7.10}$$

The total radiationless transition-rate constant k_{st} is a sum of the partial rate constants k_{sl} for transitions to all final states (2.7.10) corresponding to a given l,

$$k_{st} = \sum_l k_{sl} . \tag{2.7.11}$$

It is further assumed that all matrix elements (2.7.9) are approximately the same, $V'_{lm\kappa} \approx \alpha$. Then the N-fold degenerate level E_l is split to form a band of states (2.7.10) with the bandwidth on the order of α. All coefficients in expansion (2.7.10) are on the same order of magnitude, so the normalization condition for functions (2.7.10) gives

$$A_{il}, A_{im\kappa} \simeq 1/\sqrt{N} . \tag{2.7.12}$$

Finally, the levels E_{il} in each band l are considered to be distributed uniformly with an average density

$$\varrho_l(E_{il}) = N/\alpha . \tag{2.7.13}$$

In order to calculate the partial transition rate k_{sl}, we first find the matrix element of the intramolecular perturbation V between the initial state (2.7.3) and the final states (2.7.10). Using (2.7.6, 7, 12) we obtain

$$\langle \Psi_s | V | \Psi_{il} \rangle = v_{sl}/\sqrt{N} . \tag{2.7.14}$$

Thus the following problem arises: we have to calculate the transition rate from the initial state Ψ_s to the final states Ψ_{il} (at fixed l) under the condition that these states are distributed uniformly at a mean density of (2.7.13), and that *the transition matrix element* (2.7.14) *is the same for all final states belonging to a given l-band*. In other words, we have faced the Bixon–Jortner problem solved in Sect. 2.6.1. Consequently we can use, in a straightforward way, expression (2.5.2) with the matrix element from (2.7.14) and the density of states from (2.7.13). The total transition rate (2.7.11) then becomes

$$k_{st} = \frac{2\pi}{\hbar} \sum_l v_{sl}^2/\alpha \, . \tag{2.7.15}$$

This is just the final result obtained by Robinson and Frosch using perturbation theory.

The limits for validity of (2.7.15) immediately follow from the fact that its derivation was based on expression (2.5.2) obtained in frames of the Bixon–Jortner model. The validity condition (2.6.25) for that expression in the present case reads

$$\left(2\pi \frac{v_{sl}}{\sqrt{N}} \frac{N}{\alpha}\right)^2 = (2\pi v_{sl}/\alpha)^2 N \gg 1 \, . \tag{2.7.16}$$

Remember that N is the total number of vibrational states of the system "molecule + medium" that are able to accept the electronic energy ΔE. This number steeply increases with ΔE so that (2.7.16) imposes essentially a lower limit on the electronic energy transferred in the transition.

The second limitation stems from the fact that the band of final states has a finite width α, whereas the Bixon–Jortner model assumes an infinite bandwidth. The difference is unimportant if the energy uncertainty associated with the finite lifetime for the electronic relaxation is less than the bandwidth. Hence, inequality

$$\hbar k_{sl} = 2\pi v_{sl}^2/\alpha \ll \alpha \tag{2.7.17}$$

should be fulfilled for each l. This means that the *intrastate vibrational relaxation within the final electronic state is fast as compared to the interstate electronic relaxation* (see Fig. 2.3), and that the matrix element of the intramolecular interaction v_{sl} is smaller than the relaxation width of level l.

The matrix element α directly determines the rate of the excess vibrational energy transfer from the molecule in the final electronic state to the medium. In the Robinson–Frosch theory the quantity $\hbar/\alpha$ is identified with the vibrational relaxation time for the molecule in the medium $\tau_{VR(l)}$, which allowed a quantitative estimate of the radiationless transition rates in real molecules, although the rather crude simplifications made in the derivation of (2.7.15) raised doubts in the reliability of such estimates. However, from a principal point of view, it is important that the rate constant obtained here depends on the vibrational relaxation parameter $\tau_{VR(l)}$.

Robinson and Frosch state that, as a matter of fact, real large molecules do not exhibit this dependence. The main argument is that the sum over l must involve not only the states strictly isoenergetic with the initial one, but also all states within the interval of

$$|E_l - E_s| \lesssim \alpha\,. \tag{2.7.18}$$

Then the number of terms in the sum is proportional to α, and the dependence of k_{st} on α disappears.

The above reasoning needs, however, a significant specification. It implicitly assumes that the interval (2.7.18) embraces a great deal of the states (2.7.4) having approximately the same matrix elements

$$v_{sl} \approx v_{\text{eff}}\,. \tag{2.7.19}$$

If the density of these states is ϱ_{eff} and their number in the interval (2.7.18) is large,

$$\varrho_{\text{eff}}\,\alpha \gg 1, \tag{2.7.20}$$

then v_{sl} in (2.7.15) can be replaced with v_{eff}, and, indeed, we arrive at an expression independent of α, viz. [1.11, 2.178],

$$k_{st} = (2\pi v_{\text{eff}}^2/\hbar\alpha)\varrho_{\text{eff}}\alpha \equiv k_s\,,$$

$$k_s = \frac{2\pi}{\hbar}\,v_{\text{eff}}^2\varrho_{\text{eff}}\,. \tag{2.7.21}$$

It is emphasized that states (2.7.4) are essentially intramolecular in nature, so ϱ_{eff} in (2.7.21) is the density of the effective intramolecular states providing the dominant contribution to the transition rate. Notice also that in condition (2.7.20) the different l-bands strongly overlap giving rise to a united and practically infinite band. As a result, restriction (2.7.17) naturally is no longer necessary.[7]

Thus we see that *it is only under conditions* (2.7.19, 20) *that the transition rate is independent of* α. In this case it is expressed by (2.7.21) which precisely coincides with (2.5.2) or (2.6.28) obtained with the Bixon–Jortner model. Just this should be expected, as condition (2.7.19) coincides with the basic assumption (2.6.13) of the Bixon–Jortner model, and condition (2.7.20) in its meaning is equivalent to (2.6.31, 32), viz., the final level broadening due to either vibrational relaxation [as in (2.7.20) and (2.6.31)] or the finite duration of the experiment [as in (2.6.32)] should be larger than the separations between the l-levels to prevent theoretical recurrences.

We have thus established that the Robinson–Frosch and Bixon–Jortner theories are absolutely identical in the case where the transition rate is independent of the vibrational relaxation. This situation obviously pertains to the strong coupling limit [cf. (2.7.20) and (2.5.3)]. As was noted in connection with

---

[7] On arriving at formula (2.7.21) *Robinson* [1.11] lost sight of inequality (2.7.20) and therefore came to the conclusion that the equation is valid at any, even low, density of the final intramolecular states (see his polemics with *Schlag* et al. [1.30]).

(2.6.28), one still has to clarify which states provide the dominant contribution to the rate constant as well as to calculate the parameters v_{eff} and ϱ_{eff}.

By themselves, the effective states are difficult to determine. But even if this problem is solved, it is not at all obvious that ϱ_{eff} will happen to be large enough as (2.7.20) demands. The high density of all final states ϱ_l in the polyatomic molecule does not yet assure a high density of the effective states since no straightforward relationship exists between these parameters. If the effective states, meeting conditions (2.7.19, 20), are impossible to be sorted out in some way (formally this corresponds to a very low density of the effective states), then the Robinson–Frosch theory predicts the transition rate constant (2.7.15) to depend significantly on the molecule–medium interaction. It will be seen later that a similar part may be played by the intramolecular interactions causing a vibrational energy redistribution in isolated molecules (i.e., the intramolecular vibrational relaxation).

In order to obtain an exhaustive solution to the problem of the electronic relaxation in polyatomic molecules, one has to derive a general expression taking into account a possible dependence of the rate of this process upon vibrational relaxation. The Robinson–Frosch theory does not provide a satisfactory solution since the derivation of the final expression (2.7.15) involved the very crude approximation that the transition rate is proportional to the average density of states (2.7.13). In fact, the rate is proportional to the density of these states that are in resonance with the initial state. Since the states are not uniformly distributed over the band, the dependence of the latter quantity on the total bandwidth α is not given by the simple expression (2.7.13).

In order to take this fact into account, let us introduce the density of states in the l-band, $r_l(E_{il})$, normalized to unity,

$$\int r_l(E_{il})dE_{il} = 1 \,. \tag{2.7.22}$$

The function $r_l(E_{il})$ depends on the energy difference $E_{il} - E_l$ and vanishes for $|E_{il} - E_l| > \alpha$. It is related to the density of states $\varrho_l(E_{il})$ by

$$r_l(E_{il}) = \varrho_l(E_{il})\, N \,. \tag{2.7.23}$$

In the Robinson–Frosch theory, $\varrho_l(E_{il})$ is equal to the constant (2.7.13) within the band, and the final expression (2.7.15) involves the average normalized density of states $r_l(E_{il}) = 1/\alpha$. For an arbitrary level distribution over the band, expression (2.7.15) should evidently be changed to

$$k_{sl} = \frac{2\pi}{\hbar} \sum_l v_{sl}^2 r_l(E_s) \,, \tag{2.7.24}$$

where $r_l(E_s)$ is the density of the final states (2.7.10), isoenergetic with the initial state (2.7.3).

It is easily seen that (2.7.24) again reduces to (2.7.21) under conditions (2.7.19, 20); otherwise, however, the transition rate depends in a complicated fashion on the vibrational relaxation.

2.7.3 The Trifonov–Shekhtman Model

The calculation of the function $r_l(E_s)$ asks for a solution of the complicated problem of vibrational relaxation in the final electronic state. In this section a phenomenological treatment is given. A general form for this function will be obtained in Chap. 3.

First consider an isolated system with two levels E_s and E_l coupled by an interaction matrix element v_{sl}. This is justified since in the weak-coupling limit under consideration the decay of the initial state Φ_s can be viewed as going via independent decay channels $\Phi_s \rightsquigarrow \Phi_l$. Let the system be in the state Φ_s at an initial time $t = 0$. The probability $P_{ss}(t)$ to find the system in that same state at $t > 0$ is given by the general expression (2.6.12), and the exact molecular energy levels E_n are obtained from (2.6.6), where the sum over l includes only one term. Solving the quadratic equation (2.6.6) gives

$$E_\pm = \tfrac{1}{2}(E_s + E_l) \pm \tfrac{1}{2}(E_s - E_l)R \,, \tag{2.7.25}$$

where

$$R = [1 + 4v_{sl}^2(E_s - E_l)^{-2}]^{1/2} \,. \tag{2.7.26}$$

The survival probability $P_{ss}(t)$ is found from (2.6.12), where $n = \pm$ and the coefficients $a_{s\pm}^2$ are taken from (2.6.8). With the use of (2.7.25, 26) one obtains

$$a_{s\pm}^2 = \frac{R \pm 1}{2R}, \qquad a_{s+}^2 + a_{s-}^2 = 1 \,. \tag{2.7.27}$$

For the isolated system the survival probability oscillates without damping since energies (2.7.25) are real.

Now we allow for level damping. If E_s is the ground vibronic sublevel of the excited electronic state and E_l is a higher vibronic sublevel of a lower (e.g., the ground) electronic state, then it is natural to suppose that the decay of the initial unperturbed state Φ_s is purely radiative while that of the final state Φ_l is purely relaxational, being caused by the interaction with the inert medium. The phenomenological way to incorporate level damping consists in that the real energies in (2.7.25, 26) are replaced with complex ones according to (2.6.33), where γ_s is the radiative halfwidth of the initial state, γ_l the relaxational halfwidth of the final state, and $n = \pm$. Virtually the detailed nature of these halfwidths is unimportant, and the above specifications are intended to imply that the strong inequality (2.5.11) is fulfilled (its significance is to be discussed at the end of this subsection).

After substituting (2.6.33) into (2.6.12) the survival probability, given now by (2.6.47), begins to decay, generally proceeding to oscillate. On the other hand, to speak of a radiationless transition in the conventional sense is only justified when the decay is purely exponential. The exponential behavior occurs for

$$v_{sl} \ll \gamma_l \tag{2.7.28}$$

coinciding with that for the validity of the Robinson–Frosch formula (see (2.7.17, where α plays the role of γ_l). This is quite natural since essentially the same problem is treated here by a different method.

The meaning of inequality (2.7.28) becomes more transparent on rewriting it in the form $2\pi\hbar/v_{sl} \gg 2\pi\hbar/\gamma_l$, where the left-hand side is the period for quantum beats and the right-hand side is proportional to the vibrational relaxation time. Thus the initial state decays exponentially provided that the beat period is short compared to the vibrational relaxation time in the final state.

Expanding (2.7.26) in powers of v_{sl} gives

$$R \approx 1 + \frac{2v_{sl}^2}{(\tilde{E}_s - \tilde{E}_l)^2}, \tag{2.7.29}$$

$$g_{s+} \approx a_{s+}^2 \approx 1, \qquad g_{s-} \approx a_{s-}^2 \approx 0, \tag{2.7.30}$$

$$\tilde{E}_+ = \tilde{E}_s + \frac{v_{sl}^2}{\tilde{E}_s - \tilde{E}_l}, \tag{2.7.31}$$

and the probability (2.6.47) becomes

$$P_{ss}(t) = \exp(-\Gamma t), \tag{2.7.32}$$

$$\Gamma = \frac{2}{\hbar}\,\mathrm{Im}\,\{\tilde{E}_+\} = \frac{2\gamma_s}{\hbar} + k_{sl}. \tag{2.7.33}$$

The first term on the right-hand side is the rate constant for the radiative decay of the s-level, and the second for the radiationless transition to the l-level. Under the neglect of interference effects, which are small in the weak coupling limit, the total transition rate k_{st} is a sum over all the final states (2.7.11). As a result, (2.7.24) is obtained, where $r_l(E_s)$ is the Lorentzian function,

$$r_l(E_s) = \frac{1}{\pi}\frac{\gamma_l}{(E_s - E_l)^2 + \gamma_l^2} \tag{2.7.34}$$

(in deriving this formula from (2.7.31, 33) we neglected γ_s as it is small compared to γ_l, see later).

Equations (2.7.24, 34) can be derived in a straightforward manner by applying stationary perturbation theory to the general equation (2.6.6) with no recourse to the two-level model. After substitution of (2.6.33) this equation is expanded in powers of v_{sl}. Let $\tilde{E}_+$ denote a quasistationary level that tends to $\tilde{E}_s$ for $v_{sl} \to 0$ (the same choice was made in writing (2.7.25), where $R > 0$). Then substituting $\tilde{E}_s$ for $\tilde{E}_+$ on the right-hand side of (2.6.6), which corresponds to the second order of the perturbation expansion, one obtains

$$\tilde{E}_+ \approx \tilde{E}_s + \sum_l \frac{v_{sl}^2}{\tilde{E}_s - \tilde{E}_l}. \tag{2.7.35}$$

This is an extension of (2.7.31) to the many-level system. After substitution into (2.6.47) using (2.7.30) one again arrives at (2.7.32) with

$$\Gamma = 2\gamma_s/\hbar + k_{st}\,, \tag{2.7.36}$$

where k_{st} is given by (2.7.24, 34).

Inequality (2.7.28) ensures the applicability of the perturbation approach even at a small detuning (2.5.1) and at the breakdown of the assumptions of weak coupling and of slow unperturbed-level damping (2.7.1) adopted here since the true small parameter used in expansions (2.7.29–31, 35) is the ratio v_{sl}/γ_l. Therefore, the formulae obtained must reduce, in some limiting cases, to the corresponding expressions of the preceding sections. Specifically, in the resonance case when the l-levels occupy a finite narrow interval such that $|E_s - E_l| \ll \gamma_l$, equations (2.7.24, 34) turn, apart from notations, to the Robinson–Frosch result (2.7.15). In the opposite case of an infinite l-band and with the conditions that v_{sl} is constant and that γ_l is larger than the E_l separations [see (2.6.31)], replacing the sum over the final states with the integral

$$\sum_l \to \int \varrho_l(E_l)\, dE_l\,, \tag{2.7.37}$$

and assuming $\varrho_l(E_l)$ is constant gives the Bixon–Jortner result (2.5.2).

Repeating once more that the exponential decay occurs independently of the coupling strength, provided (2.7.28) is fulfilled, we note that the only difference is that in the weak-coupling case (2.7.1) (a narrow band of final states) the rate constant equal to k_{st} depends on γ_l, while in the strong coupling case (a wide band) it does not, then being given by k_s of (2.5.2) or (2.6.28). In other words, enhancing the coupling strength [together with the broadening to maintain (2.7.28)] transforms the γ_l-dependent intermediate case into the γ_l-independent statistical limit. Virtually, with γ_l increasing this transformation occurs somewhat earlier, namely as soon as (2.6.31) is reached (see the end of Sect. 2.6.1 and case II in Sect. 2.6.2) while the coupling still remains weak. If v_{sl} is increasing afterwards, the weak coupling first changes to the strong one (with no apparent changes in the emission decay behavior since the statistical limit persists), and then the statistical limit converts into the γ_l-independent intermediate case (dephasing) with the same rate constant k_s since the statistical limit condition (2.6.43) is eventually violated because of the increase in k_s.

Completing the analysis of the Trifonov-Shekhtman model, it is worthwile to discuss somewhat deeper the conditions for the exponential decay in the weak-coupling limit summarized in Fig. 2.3. Equations (2.7.32, 36) evidently attest to the applicability of the balance equation. As noted in Sects. 2.1 and 2.6.3, the balance equation is valid when the back coherent effect of the l-states on the s-state is negligible, and condition (2.7.28) [see also (2.7.40)] just ensures that the effect has no time to manifest. Additionally, it would have a low amplitude in view of (2.7.30). The latter implies that the results obtained will also hold in the different limit of large detuning,

$$v_{sl} \ll |\Delta| , \tag{2.7.38}$$

since any one of inequalities (2.7.28, 38) permit applying the perturbation technique used.

Another point to elucidate is why condition (2.5.11) is necessary. Strictly speaking, expression (2.7.35) results in

$$k_{st} = \frac{2}{\hbar} \sum_l \frac{v_{sl}^2(\gamma_l - \gamma_s)}{(E_l - E_s)^2 + (\gamma_l - \gamma_s)^2} , \tag{2.7.39}$$

and equations (2.7.24, 34) are obtained by neglecting γ_s.

In order to explain the significance of (2.5.11), consider a situation where it is not true, i.e., $\gamma_l \lesssim \gamma_s$. Introducing the small parameter $\lambda \equiv v_{sl}^2/(\Delta^2 + \gamma_l^2) \ll 1$, we obtain the following estimate from (2.7.39): $\hbar|k_{st}| \sim \lambda\gamma_s \ll \gamma_s$. Here k_{st} is a small correction term in the total rate constant Γ of (2.7.36), and the quantum yield φ is close to unity. Moreover, one obtains $k_{st} < 0$ and $\varphi > 1$ for $\gamma_l < \gamma_s$. This apparent shortcoming is removed by the following reasoning (cf. case III in Sect. 2.6.2). If the small k_{st} contribution is included in Γ, then similar small terms in the sum of (2.6.47) and in the coefficients g_{sn} (e.g., $g_{s-} \sim \lambda$) must be included, too. As a result, the decay becomes slightly nonexponential, and k_{st} is no longer the $\Phi_s \to \Phi_l$ radiationless transition-rate constant; correspondingly, the balance equation is invalid as well. It is proved by a straightforward calculation that the total quantum yield allowing for all contributions on the order of λ tends to unity, as it should, for $\gamma_l \to 0$. On the other hand, in the opposite limit of $\gamma_l \gg \gamma_s$, assumed in this subsection, one has

$$\hbar k_{st} \sim \lambda\gamma_l \ll \gamma_l . \tag{2.7.40}$$

The k_{st} remains low enough to preserve the balance equation, but now it may be comparable with the radiative rate in (2.7.36). If the radiative and nonradiative contributions to Γ are on a similar order of magnitude (i.e., $\hbar k_{st} \sim \gamma_s$, which means that $\gamma_l \sim \gamma_s/\lambda$ and φ is appreciably lower than unity), then all contributions to (2.6.47) on the order of λ (or, equivalently, on the order of γ_s/γ_l) can be neglected, as was already done in deriving (2.7.34). Inequality (2.7.40) obviously reads that the interstate electronic relaxation rate is slow as compared to the vibrational intrastate relaxation, which is in line with what was said in Sect. 2.1 about the applicability of the balance equation.

2.7.4 Periodic Quantum Beats and Biexponential Decay

In the limit of (2.7.1) all dark levels $E_{l \neq 0}$ are perturbed weakly, whereas the bright level E_s and closest to it the dark level $E_{l=0}$ may undergo arbitrary perturbations depending on the detuning and on the widths. The previous subsection dealt with the case when both these levels are weakly perturbed, too. Then the balance equation turned out to apply, the decay to be exponential, and the electronic relaxation rate included (2.7.24) the contributions from all final states.

Here, the opposite case of a strong perturbation of levels E_s and $E_{l=0}$ is considered. Then the levels $E_{l\neq 0}$ may be ignored at all, and the two-level Trifonov-Shekhtman model can be used again, where the perturbed quasi-stationary levels are

$$\tilde{E}_{\pm} = \tfrac{1}{2}(\tilde{E}_s + \tilde{E}_l) \pm \tfrac{1}{4}[(\tilde{E}_s - \tilde{E}_l)^2 + v_{sl}^2]^{1/2} . \tag{2.7.41}$$

The sign is always chosen according to the rule indicated before, in the text preceding (2.7.35), and $l = 0$ is meant by l here and below. The strong perturbation arises for small detuning, $|\Delta| \ll \varrho_l^{-1}$, and for small level widths,

$$\gamma_s, \quad \gamma_l \ll v_{sl} , \tag{2.7.42}$$

while the ratio of v_{sl} and $|\Delta|$ is arbitrary. Thus the case opposite to (2.7.28, 38) is met here.

Expanding (2.7.41) in powers of $(\gamma_s - \gamma_l)/v_{sl}$ to the first order one obtains

$$E_{\pm} = E_s - \tfrac{1}{2}\Delta \pm \tfrac{1}{2}R\Delta , \tag{2.7.43}$$

where R is given by (2.7.26) and $\gamma_{\pm}$ by (2.6.37) with $n = \pm$ and $a_{s\pm}^2$ from (2.7.27).

The survival probability is calculated from (2.6.47), where $g_{s\pm}$ differs from $a_{s\pm}^2$ negligibly. Denoting $\tau_{\pm} = \hbar/2\gamma_{\pm}$ one obtains [2.224, 227]

$$\begin{aligned} P_{ss}(t) &= a_{s+}^4 \exp(-t/\tau_+) + a_{s-}^4 \exp(-t/\tau_-) \\ &\quad + 2a_{s+}^2 a_{s-}^2 \exp(-t/\tau)\cos \Omega t , \end{aligned} \tag{2.7.44}$$

$$\hbar\Omega = R|\Delta| = (\Delta^2 + 4v_{sl}^2)^{1/2} ,$$

$$\tau^{-1} = (\tau_+^{-1} + \tau_-^{-1})/2 = (\gamma_s + \gamma_l)/\hbar ,$$

$$a_{s+}^2 a_{s-}^2 = (v_{sl}/\hbar\Omega)^2 .$$

The first two terms behave monotonically while the third one describes the modulating quantum beats superimposed on the monotonic decay. The frequency and the depth of the modulation depend upon the coupling strength v_{sl} and upon the detuning Δ.

A theoretical analysis of the quantum beats was carried out by *Zewail* et al. [2.169–172] for $\gamma_s = \gamma_l$ leading to the degeneracy of the relaxation times $\tau_{\pm} = \tau$. Invoking (2.7.27) one can recast (2.7.44) to

$$P_{ss}(t) = \exp(-t/\tau)[1 - 2a_{s+}^2 a_{s-}^2 (1 - \cos \Omega t)] , \tag{2.7.45}$$

In *Zewail*'s classification this corresponds to his cases (1) and (2), where Φ_s is an optically bright state both in absorption and emission while Φ_l is a dark state. In case (1) Φ_s and Φ_l are vibronic sublevels of different electronic states (e.g., S_1 and T_1, or S_1 and S_0), while in case (2) they belong to one and the same electronic state (say, S_1). The state Φ_l is dark because of either electronic or vibrational selection rules. In case (2) this is assured by a special choice of such parts of the absorption and emission spectra where the oscillator strengths for the $\Phi_0 \leftrightarrow \Phi_l$ transitions are negligible as compared to $\Phi_0 \leftrightarrow \Phi_s$.

At exact resonance $\Delta = 0$ the modulation depth amounts to 100%,

$$P_{ss}(t) = \exp(-t/\tau)\cos^2(v_{sl}t/\hbar) . \tag{2.7.46}$$

With increasing $|\Delta|$ the modulation frequency is growing while the depth decreases, the latter becoming negligible for $\varrho_l^{-1} \gtrsim |\Delta| \gg v_{sl}$; in parallel, one of the exponentials in (2.7.44) disappears, and again the monoexponential decay of (2.7.32) is setting in with the rate constant (2.7.33) (or rather that of (2.7.36) after contributions from other dark states are included).

The emission quantum yield is found by substituting (2.7.44) into (2.6.46) and integrating over time [see definition (2.1.3)] neglecting the contribution of the oscillating term, which is on the same order of magnitude as the terms omitted in the derivation of (2.7.43) and (2.6.37). Then

$$\varphi = \varphi_+ + \varphi_- ,$$

$$\varphi_\pm = \frac{\gamma_{\mathrm{rad}(s)}}{\gamma_s + (a_{s\pm}^{-2} - 1)\gamma_l} a_{s\pm}^2 . \tag{2.7.47}$$

Considering that $a_{s\pm}^{-2} > 1$ according to (2.7.27), and that $\gamma_{\mathrm{rad}(s)} < \gamma_s$ from definition (2.5.8) one obtains $\varphi \leq 1$, the equality occurring only when no vibrational relaxation is present, i.e., when $\gamma_{\mathrm{rel}(s)} = \gamma_{\mathrm{rel}(l)} = 0$ so that $\gamma_s = \gamma_{\mathrm{rad}(s)}$ and $\gamma_l = 0$. Despite the fact that the quantum yield is generally lower than unity, the concept of a radiationless transition with a calculable rate constant cannot be introduced in a reasonable way since the phenomenological equations (2.1.5–7) are violated.

Of interest is also the situation when the ratio

$$\kappa = v_{sl}^2/\Delta^2 \ll 1 \tag{2.7.48}$$

is small, as before, but the emission intensity on the order of κ^2 of the total one is measurable. Then (2.7.44) predicts a biexponential decay, provided that γ_s and γ_l are very different in magnitude, with a modulation of the *fast* component by *periodic* quantum beats (compare this to the strong coupling limit in Sect. 2.6.3). Now we consider the biexponential decay in more detail aiming at a comparison with the theory of the biexponential decay in the strong coupling limit of Sect. 2.6.3.

Suppose that the quantum beats are impossible to be observed because of a low experimental time resolution δt; specifically, inequalities $\tau_\pm \gg \delta t \gg \Omega^{-1}$ are assumed. Then, averaging (2.7.44) over the time intervals δt at every value of t nullifies the beats, and the luminescence intensity (2.6.46) takes the form

$$I_s(t) = \tau_{\mathrm{rad}(s)}^{-1}[a_{s+}^4 \exp(-t/\tau_+) + a_{s-}^4 \exp(-t/\tau_-)] . \tag{2.7.49}$$

Now (2.7.26) is expanded to the first order in κ to give $R = 1 + 2\kappa$, which is further substituted into (2.7.27) and (2.6.37). The result is

$$\begin{aligned} a_{s+}^2 &= 1-\kappa, \qquad a_{s-}^2 = \kappa, \\ \gamma_+ &= \gamma_s + \kappa\gamma_l, \qquad \gamma_- = \kappa\gamma_s + \gamma_l . \end{aligned} \tag{2.7.50}$$

Which one of the two components in (2.7.49) is fast and which is slow is governed by the relative value of γ_s and γ_l. The case when $\gamma_l \gg \gamma_s$ is not important here, since, as is easily established, the fast component, besides existing only at short times, has a low amplitude and a low quantum yield, which makes the decay essentially monoexponential in agreement with the preceding subsection. Therefore, only the case of $\gamma_s \gg \gamma_l$ has to be considered.

Actually we focus on the limiting case of no vibrational relaxation. According to definitions (2.5.7, 8) this amounts to $\gamma_s = \gamma_{\mathrm{rad}(s)}$ and $\gamma_l = 0$. Substituting (2.7.50) into (2.7.49) brings the biexponential decay into the form of (2.5.23), where the parameters N, τ_{fast} and τ_{slow} can be introduced in a formal manner as

$$N \equiv C_{\mathrm{fast}}/C_{\mathrm{slow}} = a_{s+}^4/a_{s-}^4 = \kappa^{-2} \gg 1 \tag{2.7.51}$$

(rather than $N = 1/\kappa$, as given in [1.3, 2.110]) and

$$\begin{aligned} \tau_{\mathrm{fast}} &= \tau_+ = \tau_{\mathrm{rad}(s)} , \\ \tau_{\mathrm{slow}} &= \tau_- = \tau_{\mathrm{rad}(s)}\sqrt{N} . \end{aligned} \tag{2.7.52}$$

The corresponding quantum yields (2.7.47) become

$$\begin{aligned} \varphi_{\mathrm{fast}} &= \varphi_+ = 1-\kappa , \\ \varphi_{\mathrm{slow}} &\doteq \varphi_- = \kappa . \end{aligned} \tag{2.7.53}$$

The experimentally measurable quantities (2.7.51–53) meet (2.5.28). Note that $\varphi = 1$, as expected in the case of no relaxation.

There are two points to illustrate the difference between the biexponential decays in the weak and strong coupling limits. First, the slow-to-fast lifetime ratio from (2.7.52) is $\sqrt{N}$, while in the strong coupling limit (case I of Sect. 2.6.2) equations (2.5.13, 14, 19) result in

$$\tau_{\mathrm{slow}}/\tau_{\mathrm{fast}} = Nk_s\tau_{\mathrm{rad}(s)} + 1 \gg N . \tag{2.7.54}$$

Hence, if N is defined experimentally by (2.5.28), either as the fast-to-slow amplitude ratio or through the lifetimes and quantum yields, and if $N \gg 1$ is found, then the relaxation lifetime ratios in the two cases will be quite different.

Second, the quantum yields of the two components also behave differently. Equations (2.7.53) and (2.5.16, 26) imply that the dominant portion of the radiation is emitted via the fast component for weak coupling and via the slow component for strong coupling.

Formula (2.7.44) holds valid when the state Φ_l also has a nonzero oscillator strength for a radiative transition to Φ_0 (for instance, it may be a vibronic sublevel belonging to the same electronic state as Φ_s); it is only necessary that the $\Phi_l \to \Phi_0$ emission be spectrally distinguishable from the $\Phi_s \to \Phi_0$ emission.

Then it makes sense to consider, along with (2.6.46), the emission from the l-states

$$I_l(t) = \tau_{\mathrm{rad}(l)}^{-1} P_{ls}(t) , \qquad (2.7.55)$$

again assuming that Φ_s is prepared initially, i.e., $P_{ls}(0) = 0$. Using the same algebra as in the derivation of (2.6.12) we obtain for real energies

$$\begin{aligned} P_{ls}(t) &= |\langle \Phi_l | \exp(-\mathrm{i}Ht/\hbar) | \Phi_s \rangle|^2 \\ &= \left| \sum_{n=\pm} \frac{v_{sl} a_{sn}^2}{E_n - E_l} \exp(-\mathrm{i}E_n t/\hbar) \right|^2 , \end{aligned}$$

where (2.6.4) was employed. Substituting $\tilde{E}_n$ for E_n in the exponent and invoking (2.7.27, 43), results in

$$P_{ls}(t) = a_{s+}^2 a_{s-}^2 [\exp(-t/\tau_+) + \exp(-t/\tau_-) - 2\exp(-t/\tau)\cos\Omega t] . \qquad (2.7.56)$$

The observed emission from the optically dark (in absorption) state Φ_l populated via the coupling with the bright state Φ_s rather than through a direct absorption may show quantum beats with 100% amplitude modulation, provided that either the halfwidths of the states Φ_s and Φ_l coincide, $\gamma_s = \gamma_l$ [*Zewail*'s case (3)], or a resonance occurs, $\Delta = 0$; in both cases the population of Φ_l after exciting Φ_s is given by

$$P_{ls}(t) = 2a_{s+}^2 a_{s-}^2 \exp(-t/\tau)(1 - \cos\Omega t) . \qquad (2.7.57)$$

A comparison of (2.7.56, 44) shows that both kinds of the emission oscillations are at equal frequencies and amplitudes, but they are phase-shifted by π. The total population of the Φ_s and Φ_l states from (2.7.44, 56),

$$P_{ss}(t) + P_{ls}(t) = a_{s+}^2 \exp(-t/\tau_+) + a_{s-}^2 \exp(-t/\tau_-) , \qquad (2.7.58)$$

decays monotonically. On the other hand, the total intensity $I_s(t) + I_l(t)$ may still oscillate because of a difference in the radiative lifetimes $\tau_{\mathrm{rad}(s)}$ and $\tau_{\mathrm{rad}(l)}$. If there is no such a difference, the oscillations may still be present in the case where the less restrictive condition $2v_{sl} > |\gamma_s - \gamma_l|$ is fulfilled, instead of (2.7.42) [2.228]. Also, the population beats arise if off-diagonal matrix elements of the damping matrix for the unperturbed states do not vanish [2.221] (only the diagonal elements γ_s and γ_l are considered here).

The situation described in (2.7.55–58) relates to the physical process of the Intramolecular Vibrational energy Redistribution (IVR) within a given electronic state. In the weak-coupling limit under condition of strong perturbation, when only two vibronic states interact efficiently, such a redistribution generates the periodic beats in the emission intensity. In view of the apparent partial reversibility of the process it is called *restricted* IVR [2.170–172, 229]. In order to experimentally observe the beats, the bands associated with the $\Phi_s \rightarrow \Phi_0$ and $\Phi_l \rightarrow \Phi_0$ transitions have to be separated in the emission spectrum; then the time variations in the intensities of these bands are given by (2.7.44, 56). As was indicated before, the two types of beats are shifted in their phases by π and have different modulation depths. For example, at equal widths of the unperturbed levels the emission of the dark state shows the 100% modulation while that of the bright state is modulated at a less than 100% depth and depends on the coupling strength [cf. (2.7.45, 57)].

When the vibrational energy (E_{vib}) increases, the separation between vibronic levels decreases and the weak coupling condition (2.5.4) is eventually violated: more and more levels are involved in the interaction and the beat pattern becomes complicated as the beat spectrum contains difference frequencies between each pair of the levels. At large E_{vib} a transition occurs to the intermediate case behavior and then to the statistical limit where the intramolecular vibrational energy redistribution acquires the irreversibility features and where the emission intensity decay follows the biexponential or monoexponential law in full analogy with the electronic radiationless transitions discussed in Sects. 2.5–7 [2.170–172, 228–235]. Theoretical and experimental studies of the anthracene fluorescence beats due to interaction of several vibronic sublevels of the S_l state were performed in the above-cited papers by Zewail et al.

It is once more emphasized in concluding this section that the results obtained will not suffer if other possible relaxation channels (e.g., internal conversion or intersystem crossing in a third electronic state, dissociation, etc.) are considered, in which case only the meaning of the phenomenological constants γ_s and γ_l is changed.

2.8 Time-Dependent Perturbation Theory

Up to this point the time perturbation method in its standard form has not yet been resorted to. In the Bixon–Jortner model the exact expression (2.5.2) was derived with the only simplification given by (2.6.20) which does not affect $P_{ss}(t)$ in the regime of exponential decay, and its exact result was then used to obtain the Robinson–Frosch formula (2.7.15).

The method of Sect. 2.6 is restricted by the Bixon–Jortner model and does not apply in the general case of variable matrix elements v_{sl}. Usually the transition rate is calculated by the time perturbation method, which needs however a refinement when applied to systems with discrete energy spectra of the final states such as isolated molecules.

The first-order time-perturbation method results in the well-known expression for the transition rate,[8]

$$k_{st} = \frac{2\pi}{\hbar} \sum_l v_{sl}^2 \delta(E_s - E_l) , \tag{2.8.1}$$

[8] As previously, the subscript "t" of the rate constant implies an influence of factors of non-intramolecular nature like time in this section (see later) or level broadening in the preceding one, and it is omitted when their influence disappears as in (2.7.21).

where the sum is treated as the integral of (2.7.37) over the final states in which $\varrho_l(E_l)$ is the density of states with energy E_l. Usually the replacement (2.7.37) is assumed to be always legitimate at a high enough density of states $\varrho_l(E_s)$ (this quantity is everywhere denoted merely as ϱ_l). Then (2.8.1) reduces to the Fermi "golden rule"

$$k_{sl} = k_s = \frac{2\pi}{\hbar} v_{sl}^2 \varrho_l , \tag{2.8.2}$$

where v_{sl} is taken at $E_l = E_s$. However, a more refined criterion has to be derived to assure the validity of (2.8.2).

To this end we shall make a recourse to the starting expression obtained from perturbation theory for the transition probability per unit of time,

$$k_{sl} = \frac{2\pi}{\hbar} \sum_l v_{sl}^2 \Delta_t(E_s - E_l) , \tag{2.8.3}$$

$$\Delta_t(E) = \frac{2\hbar \sin^2(Et/2\hbar)}{\pi E^2 t} . \tag{2.8.4}$$

Now we have to ascertain the conditions for which the quantity (2.8.3) is independent of t and reduces to (2.8.2).

To begin with, consider, as an example, a system with a continuous spectrum of the l-states for which the sum in (2.8.3) is understood in the sense of (2.7.37).[9] As is known from textbooks on quantum mechanics, in this case (2.8.1) is deduced from (2.8.3) at "long enough times", $t \to \infty$. Proceeding to this limit has the following meaning.

Let v_{sl} be a smooth function of E_l and δE a characteristic energy scale for its variation at $E_l \approx E_s$. The dominant contribution to the integral (2.8.3) comes from the states

$$|E_s - E_l| \lesssim 2\pi\hbar/t \tag{2.8.5}$$

since the function $\Delta_t(E)$ has a sharp peak at $E = 0$ with the width of $2\pi\hbar/t$. Obviously, the "long enough" times are

$$t \gg 2\pi\hbar/\delta E \tag{2.8.6}$$

since then the matrix element v_{sl} for states (2.8.5) is nearly constant so that v_{sl}^2 can be safely replaced with its value at $E_l = E_s$ and taken off the integral sign. The latter step is equivalent to the replacement of $\Delta_t(E)$ with $\delta(E)$ resulting in (2.8.1, 2) (for simplicity $\varrho_l(E_l)$ is assumed to be constant in a wide enough energy range around E_s).

[9] Here $\varrho_l \propto N \to \infty$ (N is the total number of states) and $v_{sl} \propto 1/\sqrt{N} \to 0$ so that (2.8.2) is finite (cf. Sect. 2.7.2).

The situation just described is typical of, for instance, scattering and predissociation problems where the matrix element v_{sl} actually depends only little on E_l so that condition (2.8.6) is fulfilled at the times of physical interest. In contrast, this is generally not true for the problem of a radiationless intramolecular transition not involving a dissociation or an ionization of the molecule, because even closely spaced levels E_l may have very different sets of vibrational quantum numbers and, hence, very different Franck–Condon factors defined in Sect. 4.5. However, aiming at a comparison of the present approach to those of previous sections, here we shall confine ourselves to the case where the matrix element v_{sl} varies insignificantly between adjacent levels, i.e.,

$$\varrho_l \delta E \gg 1 \,. \tag{2.8.7}$$

Consider a system with a discrete spectrum E_l satisfying this condition. At times long enough in the sense of (2.8.6), the quantity v_{sl}^2 can again be replaced with its value at $E_l = E_s$ and taken off the summation sign in (2.8.3). The remaining expression turns into ϱ_l, provided that E_l can be regarded as a continuous variable and substitution (2.7.37) is permitted. This is so under the condition that the total width of the strongest peak of the function $\Delta_t(E)$, $2\pi\hbar/t$, embraces a large number of discrete levels E_l, i.e., if

$$t \ll 2\pi\hbar\varrho_l \,. \tag{2.8.8}$$

Thus, in the case of the discrete spectrum, the transformation of (2.8.3) into (2.8.1, 2) via (2.7.37) is only possible under conditions (2.8.6–8).

An additional restriction on t comes from the condition of validity of the time-perturbation approach in which depopulation of the initial state is disregarded. The depopulation is unimportant if

$$t \ll \tau_{\mathrm{ER}} = 1/k_s \,, \tag{2.8.9}$$

where τ_{ER} is the electronic relaxation time. However, if it is proved independently, i.e., without the use of the perturbation method, that the probability $P_{ss}(t)$ to find the system in the initial state at times (2.8.8) falls off by the exponential law (2.4.37), τ_{ER} may be evaluated by the perturbation theory, and the times beyond the limit of (2.8.9), particularly $t \gtrsim \tau_{\mathrm{ER}}$, may be considered. Then $\tau_{\mathrm{ER}} \ll 2\pi\hbar\varrho_l$ has to be fulfilled to satisfy (2.8.8). Summarizing the above arguments, we arrive at the following validity conditions for (2.8.2) [1.4, 5, 2.236, 237]:

$$2\pi\hbar/\delta E \ll \tau_{\mathrm{ER}} \ll 2\pi\hbar\varrho_l \,. \tag{2.8.10}$$

It is easy to see that (2.8.8) is precisely equivalent to the Freed–Jortner irreversibility criterion (2.4.41) since we are interested in times $t \lesssim \tau_{\max}$. It is also evident that the perturbation result (2.8.2) coincides with the exact one (2.5.2) of the Bixon–Jortner model. This is explained by the fact that the validity conditions for the two formulas are identical. Indeed, the right-hand inequality of (2.8.10) is the same as condition (2.6.24) for the exponential decay of $P_{ss}(t)$ in the Bixon–Jortner model (it is just the Freed–Jortner criterion for that model with τ_{ER} appearing as $\tau_{\max}$). The left-hand inequality is satisfied identically since $\delta E = \infty$ for the Bixon–Jortner model.

The principal result obtained here is condition (2.8.10) for the applicability of the simple formula (2.8.2) of the time-perturbation theory. In problems dealing with continuous spectra this condition is quite unimportant since $\varrho_l \to \infty$ and δE is large due to the fact that the matrix element v_{sl} is a smoothly varying function of energy. A different situation may arise in the problem of an intramolecular radiationless transition. The final states are vibrationally excited molecular states and v_{sl}^2 is proportional to the overlap integral squared of the vibrational wave functions for the initial and final electronic states (the Franck–Condon factor). Every vibronic level with the energy E_l not very close to the dissociation limit can be characterized by a definite set of vibrational quantum numbers assigned to either normal or local modes (see Sect. 2.10). Since the frequencies of various modes are not connected by any multiplicity relations, any two adjacent levels E_l generally have quite different sets of quantum numbers and consequently quite different electronic–vibrational functions Φ_l. In polyatomic molecules the potential surfaces for the lowest electronic states corresponding to one-electron excitation are displaced little with respect to each other. So, the overlap of the functions Φ_s and Φ_l mainly occurs in the classically forbidden domain, hence being exponentially small. Therefore, two different sets of quantum numbers will give rise to very different Franck–Condon factors, thus to different matrix elements v_{sl}. Such a situation is characterized by the relation $\varrho_l \delta E \lesssim 1$, which contradicts the validity conditions for both the Bixon–Jortner model and (2.8.2) of perturbation theory. Obviously, the definition of the average matrix element entering (2.8.2) loses its significance at such a strong variation of the partial v_{sl}. Thus the problem arises to calculate the transition rate in the general case not restricted by condition (2.8.7).

A conceivable way to solve this problem based on a separation of a set of effective states having nearly constant matrix elements was mentioned at the end of Sect. 2.6.1. Generally one may proceed as follows. The manifold of the stationary final states isoenergetic with the initial one is split into groups in such a way that the matrix element would change little enough among the states belonging to a given group. Let $\varrho_l^{(\lambda)}$ designate the density of states in the group λ, $v_{sl}^{(\lambda)}$ be the average matrix element, and δE_λ a characteristic energy scale for the variation of v_{sl} with E_l among the states of this group. Then the rate constant can be written as the sum over the groups,

$$k_{sl} = \frac{2\pi}{\hbar} \sum_{\lambda} v_{sl}^{(\lambda)2} \varrho_l^{(\lambda)} \tag{2.8.11}$$

if condition

$$2\pi\hbar/\delta E_\lambda \ll \tau_{\mathrm{ER}} \ll 2\pi\hbar \varrho_l^{(\lambda)} , \tag{2.8.12}$$

similar to (2.8.10), is fulfilled for all λ's. In going from (2.8.1) to (2.8.11) we have, in fact, performed a partial summation over l within every group λ, the summation being replaced by the integration due to the slow variation of the matrix element among the states belonging to a given group. In the particular case where the matrix element is approximately the same for all final states (i.e., $\delta E_\lambda \to \infty$) only one term with $\varrho_l^{(\lambda)} = \varrho_l$ remains in the sum over λ, and (2.8.11) reduces to (2.8.2).

The following remarks are made with regard to (2.8.11) and inequalities (2.8.12). Expression (2.8.3), derived by first-order perturbation theory, has the meaning of the radiationless transition rate constant under the condition that it is independent of t on the time scale of physical interest; otherwise the phenomena originating from the intramolecular coupling have no bearing on radiationless transitions. In as much as the radiationless transition is a manifestation of irreversibility in the system under consideration, the above condition is also an irreversibility criterion. In the present problem the physically significant times are $t \simeq \tau_{\mathrm{ER}}$ where expression (2.8.3) is indeed independent of t and takes the form of (2.8.11) provided that inequalities (2.8.12) are met. Therefore, (2.8.12) is simultaneously the irreversibility criterion, which is somewhat more restrictive as compared to the Freed–Jortner one and reduces to the latter for $\delta E_\lambda \to \infty$. This is quite natural because (2.4.40) for the theoretical recurrence time obtained in the Bixon–Jortner model with the constant matrix element (2.6.13) was used to derive the Freed–Jortner criterion (2.4.41).

Notice that the general expression (2.8.11) can be recast in the form similar to (2.6.28),

$$k_{sl} = \frac{2\pi}{\hbar} \overline{v_{sl}^2} \varrho_l \,, \tag{2.8.13}$$

where ϱ_l is, as usual, the total density of final states and

$$\overline{v_{sl}^2} = \sum_\lambda v_{sl}^{(\lambda)2} (\varrho_l^{(\lambda)}/\varrho_l) \tag{2.8.14}$$

is an average of the transition matrix element squared. As was mentioned in Sect. 2.6.1, (2.8.13, 14) are confirmed by numerical calculations [2.201].

The derivation of the general formula (2.8.11) given before is not constructive for it does not provide a means to divide the final states manifold into groups, so the physical meaning of the parameters $v_{sl}^{(\lambda)}$ and $\varrho_l^{(\lambda)}$ remains unclear. The issue of classification of the final states is discussed in Sect. 2.10.

2.9 Comparison of Various Expressions for the Transition Rate

Now it is appropriate to compare various expressions derived by different methods for the transition rate in isolated molecules as well as in molecules interacting with inert media.

The formulas obtained are divided into two groups. The first one includes formulae (2.5.2), (2.7.21) and (2.8.2) that coincide with each other apart from notations. The conditions for applicability of these expressions are naturally the same as well, and read as follows.

First, the transition matrix element v_{sl} should be a slow function of the final state energy E_l. The precise condition is given by (2.8.7) and by the left-hand inequality (2.8.10). Second, a broadening of the final states should exist exceeding the average level spacing and thus preventing the possibility of theoretical recurrences. The nature of the broadening may be quite diverse. In the Bixon–Jortner model for an isolated molecule this broadening is just a quantum uncertainty in energy caused by a finite electronic relaxation time, τ_{ER} (or, which is the same, by a finite experiment duration τ_{max}), and the corresponding condition has the form of (2.6.24). This same meaning is embedded in the right-hand inequality of (2.8.10) in the perturbation theory for the isolated molecule. On the other hand, the broadening due to vibrational relaxation becomes more significant for a molecule interacting with an inert medium. The condition of the large enough broadening has the form of (2.6.31) in the Bixon–Jortner model with constant broadening of final states, and the form of (2.7.20) when the Robinson–Frosch method is employed. It is worthwhile to note that *with the said conditions being fulfilled, the transition rate is not affected by coupling with the medium and coincides with the transition rate in the isolated molecule*.

The second group involves (2.7.15, 24) and (2.8.11), which pertain to a more general case not assuming the slow variation of the matrix element. The Robinson–Frosch expression (2.7.15) for a molecule in an inert medium predicts that in this case the transition rate depends on the molecule–medium interaction but the specific dependence given by this expression is incorrect in as much as too rough approximations were made in its derivation. It was shown at the end of Sect. 2.7.2 in which way this expression can be refined, and a more accurate formula (2.7.24) was obtained which is governed not only by the total bandwidth (i.e., by the vibrational relaxation rate) but also by the distribution of the density of states over the band.

Equation (2.8.11) for isolated molecules derived at the end of the previous section bears some resemblance to (2.7.24) for a molecule in a medium. The following physical arguments can be given to show this not to be an artifact. The Robinson–Frosch theory was based on a classification of all final states of the system "molecule + medium" into two types. The states assigned to the first type are directly coupled with the initial state through a nonzero interaction matrix element, and are purely intramolecular states corresponding to a transfer of the total electronic transition energy to intramolecular vibrations. In turn, every such state interacts vibrationally with a large number of states of the second type which are responsible for a partial transfer of the electronic energy to the medium, and therefore a whole band of final states corresponds to a given directly coupled state. In the sum of (2.7.24) each term is the transition rate to a separate band, and it is proportional to the density of states in this band.

The method to classify the final states according to their coupling strengths with the initial state can be immediately applied to the case of isolated molecules. The final states take different parts in the radiationless transition because of the assumed strong variation of the matrix element with energy. By analogy with Sect. 2.7.2 the concepts of strongly coupled states pertaining to a certain "quasimolecule" and of an "inert medium" embracing the weakly coupled states can be introduced. Then each term in the sum of (2.8.11) proves to be associated with a certain strongly coupled "physically correct" *nonstationary* vibrational state of the molecule (it is stationary state of the isolated quasimolecule) interacting with a great number of weakly coupled states (the "medium states"), and the partial transition-rate constant is governed by the density of states in the corresponding band. In this terminology, the intramolecular redistribution of the vibrational energy among the correct states of the molecule as a whole is nothing else but the vibrational relaxation of the quasimolecule in the medium. Thus, the intramolecular energy redistribution in an isolated molecule plays the same role as the vibrational relaxation of the molecule in a medium. These concepts are further elaborated in the next section.

2.10 Characterization of Final States of an Isolated Molecule

The final state wave functions Φ_l we employed before to derive the transition rate expressions are eigenfunctions of the vibronic Hamiltonian of the final electronic state

$$H_{r'}(q)\,\Phi_l = E_l\Phi_l . \tag{2.10.1}$$

The knowledge of the wave functions Φ_l enables one, of course, to calculate the matrix elements v_{sl} and hence the radiationless transition-rate constant. However, in view of a large number of the final vibrational states of the isolated molecule (all the more of the molecule in a medium), the same large (often huge) number of the matrix elements v_{sl} would be needed to calculate only one quantity of physical interest.

This difficulty is typical of the problems pertaining to large systems if exact microscopic equations [in the present case equation (2.10.1)] containing exhaustive information on the system are chosen as a starting point. The usual way to circumvent this difficulty is to give up to utilize detailed information, but instead, to introduce a few averaged parameters. Exactly in this manner we acted in Sect. 2.8 when deriving (2.8.11), where groups of the states Φ_l were sorted out having approximately the same matrix elements

$$v_{sl} \approx v_{sl}^{(\lambda)} , \tag{2.10.2}$$

λ being the number of the group. Each group of states is characterized by two parameters, viz., the average matrix element $v_{sl}^{(\lambda)}$ and an average density of states in that group $\varrho_l^{(\lambda)}$. The total number of the groups N depends on how rapidly the matrix element v_{sl} varies among the E_l levels in close vicinity of the initial level E_s. For instance, if v_{sl} is nearly a constant, one has $N = 1$, and the transition-rate constant is given by (2.8.2) involving two parameters only.

Let us now elucidate the physical significance of separating the final states into groups, which will simultaneously provide a recipe to perform this procedure. As we shall see, such a separation is essentially equivalent to a change to a new basis strongly truncated as compared to the initial one, Φ_l. In other words, an incomplete set of wave functions pertaining to some nonstationary states is introduced, in place of the complete set of the wave functions of the stationary states, to describe the nuclear motion. The transformation to the new basis is appropriately performed in two stages.

First we divide, in the meanwhile arbitrarily, the original vibronic Hamiltonian into the sum of

$$H_{r'}(q) = H_{r'}^{(0)}(q) + W_{r'}(q) , \tag{2.10.3}$$

and choose Φ_λ as the eigenfunctions of the "zeroth-order" Hamiltonian

$$H_{r'}^{(0)}(q)\Phi_\lambda = E_\lambda \Phi_\lambda . \tag{2.10.4}$$

From a mathematical point of view it does not matter at all which one of the two full sets of wave functions, Φ_λ or Φ_l, is used to calculate the rate constant for the radiationless transition to all final states. However, one can choose the zeroth-order Hamiltonian in such a way that physically these two sets will be nonequivalent. A similar situation was already faced in Sect. 2.4.1 in the discussion about the state initially prepared by an elementary act of photoexcitation, so the results obtained there can be immediately applied in the present case. Namely, we choose such a zeroth-order Hamiltonian that Φ_λ describe the vibronic states "initially prepared" by the radiationless transition.

The primary event in the course of the radiationless transition is a "jump" of the system from one potential surface to another, this one here representing the elementary act of excitation of the vibrational degrees of freedom of the molecule during the electronic transition. Precisely, as in Sect. 2.4.1, the duration of the elementary excitation event τ is needed to assess the nature of the initially prepared vibronic state [as a matter of fact, to make a correct choice of the representation (2.10.3)]. The quantity $1/\tau$ should not be confused with the total rate constant for the radiationless transition k_{st}. Thus, if the jump goes one way only, k_{st} is the number of jumps per unit of time, whereas τ is the duration of one jump. The permissibility to invoke these classical concepts to describe such a quantum process as the tunneling radiationless transition stems from the quasiclassical nature of the nuclear wave functions in the underbarrier region (see Chap. 5). The latter also lays the foundation for the method of complex classical trajectories for calculating the probabilities of tunneling transitions [1.13, 2.238].

In polyatomic molecules the potential surfaces of the ground electronic state and of the lowest excited ones are usually little displaced with respect to each other, and the initial state Φ_s is often placed near the minimum of the corresponding surface. Therefore, even at room temperature, the radiationless transition is a tunneling one, and τ is the underbarrier tunneling time between the two potential surfaces. In this situation it is reasonable to suppose that τ is much less than k_{st}^{-1} but greater than the molecular oscillation period. Then the figure of

$$\tau \approx 10^{-13} - 10^{-12}\,\mathrm{s} \tag{2.10.5}$$

can serve as a rough guide.

Following the arguments of Sect. 2.4.1 we choose the Hamiltonian decomposition of (2.10.3) in such a way that the symbolic inequality

$$W_{r'}(q) \ll 2\pi\hbar/\tau = 30\text{–}300\,\mathrm{cm}^{-1} \tag{2.10.6}$$

is fulfilled in full analogy with (2.4.22). This inequality means that the perturbation $W_{r'}(q)$ is considered as being "turned off" during the jump, so the zeroth-order Hamiltonian $H_{r'}^{(0)}(q)$ should be used to determine the physically correct states, i.e., the states that are prepared as a consequence of surface hopping. The correct states Φ_λ are obviously nonstationary, and therefore, at times

$$t \sim 2\pi\hbar/W_{r'}(q) \gg \tau \tag{2.10.7}$$

when the perturbation $W_{r'}(q)$ is being "turned on", the second stage of the radiationless transition involving transitions between different states Φ_λ sets in.

The new basis of the functions Φ_λ is complete, too, so in this sense it is in no way better than the original basis Φ_l. Therefore let us make a step further in assuming that the vibronic Hamiltonian turns out to be separable after removing $W_{r'}(q)$; in other words, such a full set of the vibrational coordinates q_μ can be chosen that the zeroth-order Hamiltonian is decomposed into the sum of the Hamiltonians for the independent vibrations q_μ,

$$H_{r'}^{(0)}(q) = \sum_\mu H_{\mu r'}^{(0)}(q_\mu)\,. \tag{2.10.8}$$

The Hamiltonian $H_{r'}(q)$ satisfying conditions (2.10.3, 6, 8) can be called approximately separable. This property of the vibronic Hamiltonian playing a fundamental role in the theory of radiationless transitions is discussed in more detail at the end of this section.

In as much as the vibronic Hamiltonian is not the same in different electronic states, generally different sets of separable coordinates q_μ correspond to the initial and final electronic states. However, up to now this effect of the so-called mode mixing (the *Duchinsky* effect [2.239]) is practically not enough investigated. Therefore, as is usually done, we shall neglect it assuming the coordinate system q_μ to be defined in a unique way for both electronic states.

Upon the electronic transition, an energy ΔE is transferred to vibrations with such a distribution among the various degrees of freedom q_μ that the energy conservation law is fulfilled.,

$$\Delta E = \sum_\mu \Delta E_\mu \,, \tag{2.10.9}$$

where ΔE_μ is the energy transferred to the μ-th mode. The ratio

$$p_\mu = \Delta E_\mu / \Delta E \tag{2.10.10}$$

characterizes the accepting ability of the μ-th mode, or its activity in the radiationless transition, as we shall term it. The μ-th mode activity, p_μ, is a measure of the strength of its coupling with the electronic degrees of freedom.

Generally not all the modes are good acceptors of electronic energy. One can usually sort out a relatively small number of active modes (good acceptors with $p_\mu \lesssim 1$), whereas all the remaining modes are passive (poor acceptors with $p_\mu \ll 1$). The dominating contributions to the transition rate clearly come from such final states Φ_λ in which the quantum numbers of the poor acceptors are nearly the same as in the initial state Φ_s; these final states are strongly coupled with the initial state. The number of the strongly coupled states is significantly lower than the total number of final states, and they form the desirable basis Φ_λ shortened as compared with the original basis of the stationary states Φ_l. In order to find the strongly coupled states, it is sufficient to solve (2.10.4) with the Hamiltonian (2.10.8) retaining only the active modes in the sum over μ.

Now we are in a position to present the classification of the states Φ_l explaining the rationale behind dividing them into the λ-groups. We define, in the usual manner, the matrix elements of the intramolecular coupling as

$$v_{s\lambda} = \langle \Phi_s | V | \Phi_\lambda \rangle \,. \tag{2.10.11}$$

As said before, they are appreciably nonvanishing for strongly coupled states only. On the other hand, each strongly coupled state interacts with a large group of weakly coupled states in resonance with it. Taking the vibrational interaction $W_{r'}(q)$ into account [i.e., solving (2.10.1)] one obtains a whole band of stationary states Φ_l, the matrix element $v_{s\lambda}$ being "distributed" uniformly over all the band states; just for this reason the matrix elements v_{sl} are nearly constant among all the states of a given band. Thus the superscript λ in (2.8.11) enumerates the strongly coupled states, while the subscript l does the same for the stationary states belonging to the λ-band.

These arguments are easily seen to duplicate, to a large extent, the discussion of Sect. 2.7.2 regarding molecules in inert media. At this point a very profound analogy, emphasized at the end of the previous section, emerges between the electronic relaxation processes in isolated and nonisolated polyatomic molecules. Being weakly coupled with the electronic degrees of freedom, the passive modes play essentially the same part as the nuclear degrees of freedom of the inert media. Therefore, it is natural to invoke the notion of a "quasimolecule" to which the electronic degrees of freedom and those of the active modes coupled to them are assigned, as well as of an "inert medium" involving the passive modes. The vibrational interactions between the quasimolecule and the intramolecular medium give rise to the vibrational relaxation of the quasimolecule and the corresponding broadening of its levels (the "intramolecular environmental broadening" by *Robinson* [1.11, 2.178]).

Similar ideas for optical transitions were elaborated by *Neporent* [2.71] to account for diffuse vibronic spectra of polyatomic molecules. He proved the diffusiveness of the spectra to persist in going from solutions to rarefied vapors in which the period of time between collisions is three orders of magnitude as high as the excited state lifetime, the diffusiveness hence being principally on account of intramolecular processes. Neporent introduced the concept of "active" vibrations changing their states in the optical transition, and of an "internal solvent" whose interaction with the active vibrations produces a broadening of the corresponding vibronic levels, in consequence of which the spectra of the complex molecules are diffuse, provided that a large enough vibrational energy is stored in either the initial or the final electronic state. Surely, this broadening mechanism is not unique [1.22] but it is the most universal one, being based on a property essentially common to all polyatomic molecules, viz., the rapid increase in the density of vibrational levels with vibrational energy.

Usually the number of active modes is relatively not very high, so that the energy levels of the quasimolecule in the transition region are spaced wide enough for ΔE being not too large. Then the argument similar to that given in Sect. 2.4.1 leads us to the conclusion that a *nonstationary vibrationally excited molecular state*, which can be qualified as a stationary state of the isolated quasimolecule (or generally as a superposition of several stationary states) is prepared after the jump between potential surfaces.

The above discussion completely holds valid for a molecule in an inert medium, too, as the vibrational relaxation time is usually longer than that of (2.10.5) and consequently the interaction of the molecule with the medium is also turned off during the jump. The only difference concerns the definition of the medium which now includes both the real physical environment and the passive intramolecular vibrations.

In summary one can state that if the jump duration is short as compared to the characteristic times for both the intra- and intermolecular vibrational relaxations, then both the intra- and intermolecular couplings are turned off at times $t \lesssim \tau$, and the initially prepared vibrational state is a state of the isolated quasimolecule. At $t \gg \tau$ these interactions are gradually turned on resulting in a relaxation of the initial strongly nonequilibrium distribution over the states of the quasimolecule.

Now it is clear that (2.7.24) and (2.8.11) for a molecule in a medium and for an isolated molecule, respectively, coincide identically since, in fact, they pertain to one and the same physical situation, provided the difference in the relaxation mechanisms for the strongly coupled states is disregarded. The superscript λ in (2.8.11) obviously identifies the states of the quasimolecule with energies E_λ and wave functions Φ_λ, while $\varrho_l^{(\lambda)}$ is the density of levels E_l within the band formed due to the vibrational interaction of the level E_λ with the resonant "medium" levels, and must be calculated at the energy $E_l = E_s$. This quantity is on the order of the total density of states ϱ_l, so that the right-hand inequality (2.8.12) virtually coincides with the Freed–Jortner irreversibility criterion. The parameters $v_{sl}^{(\lambda)}$ in (2.8.11) are connected with $v_{s\lambda}$ of (2.10.11) by relation

$$v_{sl}^{(\lambda)} = v_{s\lambda} \sqrt{N_\lambda} , \qquad (2.10.12)$$

where N_λ is the number of states in the λ-band [cf. (2.7.14)]. Defining also the band density of states normalized to unity,

$$r_\lambda(E_l) = \varrho_l^{(\lambda)} / N_\lambda , \qquad (2.10.13)$$

recasts (2.8.11) to

$$k_{sl} = \frac{2\pi}{\hbar} \sum_\lambda v_{s\lambda}^2 r_\lambda(E_s) , \qquad (2.10.14)$$

in full analogy with (2.7.24).

The argument of the function $r_\lambda(E_l)$ may be regarded as a continuous variable since the discrete nature of the levels is unimportant owing to condition (2.8.12), stating that the energy uncertainty originating from the limited electronic relaxation time overlaps the mean separation between the levels E_l. Thus $r_\lambda(E)$ can be treated as a "shape function" for the level E_λ broadened by vibrational relaxation. The accurate *ab initio* calculation of this function needs solving the vibrational relaxation problem, which is beyond the scope of this book, while the phenomenological approach of Sect. 2.7.3 gives the Lorentzian function (2.7.34) (see also Sect. 3.4).

In conclusion we discuss the issue of the separation of variables in the vibrational Hamiltonian. First of all we note that if the Hamiltonian is not nearly separable, the case of $N = 1$ mentioned in the text following (2.10.2) occurs, in which the transition rate is given by (2.8.2). In other words, the matrix elements v_{sl} of a nonseparable Hamiltonian for the levels E_l close to the initial level E_s are on a similar order of magnitude, and all the final states provide comparable contributions to the transition rate. While we are not able to prove this statement in a mathematically rigorous manner, it is quite obvious from the physical point of view. In fact, the nonseparability of the Hamiltonian implies that all degrees of freedom are strongly coupled with each other, and consequently there is not any reason for the matrix elements to be essentially different among various final states.

Despite the fact that only two parameters, viz., the mean interaction matrix element and the total density of final states govern the transition rate for the nonseparable Hamiltonian, the calculation of the transition rate is an excessively difficult task. First, the multidimensional potential surfaces with nonseparable variables are hard to be described analytically; second, there are no methods to calculate the matrix elements (the multidimensional Franck–Condon factors) and the density of final states. Instead of calculating these parameters one can directly seek for the transition rate itself by the method of complex classical trajectories [1.13, 2.238], but this method is also quite complicated and, additionally, is not elaborated enough.

Fortunately the vibrational Hamiltonian is nearly separable in many practically important cases. For example, data obtained from high-resolution fluorescence excitation spectra of pyrazine [2.114–116] evidence strong fluctuations of v_{sl} among the mixed singlet–triplet levels contained within very narrow energy intervals. The chaotic changes of v_{sl} in going from level to level testify that anharmonic interactions do not mix the triplet levels very much, and the levels preserve their specificity [2.117].

The separability is not, of course, an absolute property but depends on the amount of the vibrational energy in the final electronic state [2.240] or, which is the same, on the electronic energy gap ΔE between the initial and final states; in particular the very choice of the separating variables q_μ is also controlled by ΔE.

At low ΔE the normal coordinates must be taken as the q_μ's, and then $H^{(0)}_{\mu r'}(q_\mu)$ is the Hamiltonian for a harmonic oscillator, while $W_{r'}(q)$ is the anharmonic interaction. Thus, in aromatic hydrocarbons the S_1–T_1 energy gap does not usually exceed the value of $(10–15) \times 10^3 \text{cm}^{-1}$, which amounts to 3–5 quanta of the C–H or 10–15 quanta of the C–C vibrations. At such quantum numbers $W_{r'}(q)$ approximately obeys inequality (2.10.6), thus justifying the vibrational Hamiltonian representation in the form of (2.10.3, 8).

On the other hand, the so-called *local mode* representation where bond lengths are chosen as the q_μ's is appropriate at high enough energies ΔE. The local mode approximation in the theory of molecules was proposed by *Henry* and *Siebrand* [2.241] who applied it to describe high overtones of the C–H vibrations in the IR spectrum of benzene [2.242]. According to the Henry–Siebrand theory the benzene molecule absorbs the IR and visible radiation in the range of $(15–20) \times 10^3 \text{cm}^{-1}$ in such a manner as if it contained six independent local anharmonic oscillators that mimic the C-H bond vibrations. This point of view was confirmed by further experimental and theoretical studies [2.243–252]. At present the most popular is a combined model where some modes are treated as local and some as normal [2.253–257]. The upper limit on the coupling between the local vibrations can be estimated by the C–H overtone bandwidths which amounts to a few tens (up to a hundred) of cm^{-1} (the actual homogeneous bandwidth is still lower [2.258]). The interaction term $W_{r'}(q)$ has to be on the same order of magnitude or lower, hence condition (2.10.6) is again fulfilled. Further discussion of the noninteracting vibrations approximation will be given in Sect. 4.6.

2.11 Small, Large and Intermediate Molecules

The qualitative theory of radiationless transitions outlined in this chapter enables us to classify the molecules according to the properties of their energy spectra and of the intramolecular interactions. Since these properties are closely related to the number of atoms in the molecule, it is generally accepted, following *Robinson* [2.178, 179], to make a distinction between small, large and intermediate molecules.

The small molecules are characterized by large mean separations between vibrational–rotational levels E_l (see Fig. 2.2). If the interaction matrix element v_{sl} is large enough to destroy a possible $E_s - E_l$ resonance, the exact molecular levels E_n are also widely spaced, and, according to Sect. 2.4.1, the coherent excitation of the molecule produces a stationary state except for excessively short exciting pulses on a subpicosecond time scale [2.17, 259–265]. The absorption and emission spectra of such molecules consist of lines whose widths originate from the finite radiative lifetimes of the stationary states E_n. If, alternatively, v_{sl} is small and there is a level E_l in close resonance with the initial level E_s, then the coherent excitation of the neighboring states E_n is possible giving rise to a nonexponential decay of the emission intensity (see Sect. 2.7.4) and related phenomena [2.266]. In any case, the emission quantum yield for isolated small molecules is unity since the unperturbed levels E_s and E_l have no intrinsic nonradiative widths due to the lack of intrinsic vibration–rotation continua associated with other electronic states; hence the radiationless transitions in isolated small molecules do not exist. They can only occur due to interactions with the environment, their rate constants being significantly affected by the medium-induced vibrational relaxation (of course, other relaxation channels may also be important[10]). If γ stands for the corresponding level width, then the basic condition for the small molecule limit is

$$\gamma \ll 1/\varrho_l . \tag{2.11.1}$$

Usually, the coherent bandwidth of the exciting radiation also does not exceed the level separation. A further comment to this condition in connection with Fig. 2.3 will be given later in this section.

The mean separation between the levels E_l rapidly decreases with increasing number of atoms. The limiting case of *large molecules* is subdivided into the statistical limit and the intermediate case. *The statistical limit of large molecules* is characterized by the following properties.

1. The excitation of a separate stationary molecular state is in principle impossible due to the large density of final states ϱ_l, isoenergetic to the initial one; the nonstationary state Φ_s is prepared by the coherent excitation (see Sect. 2.4.1 and later).

2. The Freed–Jortner irreversibility criterion (2.4.41) is fulfilled. According to this criterion, the E_l-level broadening due to the finite experiment duration exceeds the mean level spacing. Then the irreversible evolution of the initially prepared state of the isolated molecule is assured since the spectrum of the final states is essentially continuous (its discreteness and the associated possibility for the system to return to the initial state do not manifest at the times of physical

---

[10] For brevity, we often refer to the vibrational relaxation, meaning by this any process depopulating the Φ_l states except for the $\Phi_s \to \Phi_l$ transition under study.

interest). Thus the large molecules behave irreversibly like statistically large systems.

3. The balance equation (2.1.4) is valid and the radiationless transition rate is independent of the vibrational and other relaxational processes in the two electronic states. In particular the rate does not depend on whether the molecule is isolated or interacts with an inert medium (e.g., it does not depend on pressure in vapors).

The third property significantly narrows the class of molecules to be assigned to the statistical limit. Even when the state Φ_s is prepared initially and its subsequent decay is exponential, still either the balance equation may be invalid or, if it applies, the radiationless transition rate may depend upon the vibrational relaxation. In fact, as was shown in the previous section, the rate constant is generally affected by the intra- and intermolecular relaxation. This is formally expressed in that (2.10.14) for the transition rate involves the relaxational quasimolecule level broadening γ via the level shape function similar to (2.7.34) [see (3.4.17)]. However, the conditions have previously been analysed for the said dependence to disappear. In Sect. 2.7 this was shown to be the case if there exists a sufficiently dense manifold of the effective final states providing the dominant contributions to the transition rate and having approximately equal transition matrix elements [see (2.7.19–21)]. The similar conclusion is reached by analyzing (2.10.14) for a quasimolecule in an inert medium. The above-mentioned effective states clearly pertain to the quasimolecule. If the broadening of the corresponding levels exceeds their separations

$$\gamma \gg 1/\varrho_{\mathrm{eff}}\ , \tag{2.11.2}$$

and if the matrix element is nearly constant for all effective states, then (2.10.14) is again converted to (2.7.21). Therefore, both for a molecule in a medium and for an isolated molecule *inequality* (2.11.2) *is the main condition for the statistical limit* (see later for further comments). In both case *the transition rate in the statistical limit is given by* (2.7.21) *and is independent of the vibrational relaxation* as well as of other possible relaxation processes deactivating the two communicating electronic states.

Condition (2.11.2) needs a refinement in connection with the results of Sects. 2.6.2, 3 obtained in the framework of the Bixon–Jortner model with equal broadening of all levels. Preliminarily we note that the effective states playing the major role in the statistical limit obey the basic assumption of this model regarding the constancy of the matrix element; in other words, all states in this model are effective, i.e., $v_{sl} \equiv v_{\mathrm{eff}}$ and $\varrho_l = \varrho_{\mathrm{eff}}$. Therefore, the Bixon–Jortner model can safely be applied to the statistical limit.

According to the results mentioned, *the statistical limit ensues from a level merging*, with properties 1 to 3 being a direct consequence of the collapse in the energy spectrum. The merging sets in with condition (2.6.43) requiring a large density of states and a large level broadening. There, replacing γ_l and ϱ_l with the respective parameters γ and ϱ_{eff} for the effective states and omitting the initial state width, one obtains the statistical limit condition in the form corresponding to the representation of Fig. 2.3, viz.,

$$\gamma \gg \tfrac{1}{2}\hbar k_s + (2\pi\varrho_{\text{eff}})^{-1} , \qquad (2.11.3)$$

where k_s is given by (2.7.21).

In the weak coupling limit, $v_{\text{eff}} \ll 1/\varrho_{\text{eff}}$, the first term on the right-hand side is small with respect to the second, and (2.11.3) reduces to (2.11.2). Notice that the second nonstatistical limit situation of the two mentioned immediately after property 3 above is possible at an intermediate value of γ when (2.11.2) is violated but condition (2.7.28) for the balance equation is still valid. It is thereby stressed once more that the balance equation may describe the behavior of the nonstatistical limit molecules when no level merging occurs.

On the other hand, in the opposite limit of strong coupling, the first nonstatistical situation is possible, again at an intermediate value of γ, when (2.11.2) is fulfilled but (2.11.3) is not. This situation only mimics, while veri-similarly, the statistical limit in that the luminescence decays by the exponential law (2.1.5) with the electronic relaxation rate constant given by the same statistical limit expression (2.7.21). However, as was discussed in Sects 2.5.1 and 2.6.3, the balance equation does not apply, and the total luminescence decay rate is not equal to the sum of the radiative and nonradiative rates [(2.1.6) is invalid]. In fact, the radiative contribution proves to be relatively small and the rate's nonadditivity may be disregarded provided that the radiative contribution is neglected. For this reason, this "false" statistical limit may safely be regarded as the "true" one.

Thus, with these remarks in mind, we regard (2.11.2) as the basic condition for the statistical limit.

Strictly speaking, two kinds of broadening (apart from the medium-induced ones) ought to be distinguished: the broadening of the stationary states of the isolated molecule due to the finite experiment duration (Sect. 2.4.2) and that of the nonstationary effective states produced by intramolecular interactions. The former enters (2.1.1) while the latter enters (2.11.2). Nevertheless, the common notation γ is used in both cases as this cannot give rise to confusion.

If condition (2.11.2) is not satisfied but the molecule exhibits properties 1 and 2, *the intermediate case of large molecules* occurs, which is characterized by

$$1/\varrho_{\text{eff}} \gg \gamma \gg 1/\varrho_l . \qquad (2.11.4)$$

The molecules obeying this condition behave irreversibly with regard to the $\Phi_s \rightsquigarrow \Phi_l$ transition but, as was already noted, either they display the nonexponential luminescence decay or their electronic relaxation rate is given by the general expression (2.10.14) and is essentially controlled by the vibrational relaxation due to both the intra- and intermolecular vibrational energy redistribution. Notice that in the former case the $\Phi_s \rightsquigarrow \Phi_l$ transitions are irreversible only in the asymptotic limit of $t \to \infty$ when they eventually end up in the ineffective final states, but the transitions at finite times when the effective states have not yet ceased to participate are *partly reversible*, which just gives rise to the nonexponential behavior. In contrast, in the latter case of the purely exponential decay the full irreversibility reigns at all times since the effective states rapidly decay to the ineffective ones averting the recurrence to the initial state.

It should be emphasized that no distinct border can be drawn unambiguously between the intermediate case of large molecules and the case of small molecules because it depends exactly on what is meant by small and large molecules. Their differentiation according to (2.11.1, 4) can be treated as follows.

The irreversibility of the $\Phi_s \rightsquigarrow \Phi_l$ transition in the intermediate case ensuing from (2.11.4) implies that the effective final states are intrinsically broadened due to their interactions with a dense manifold of the ineffective states. As for the small molecules, their final levels have no such widths since the full manifold of the final states is scarce.

While being physically transparent, these definitions may seem to be insufficient. For example, no intermediate case in the sense of (2.11.4) exists in the Bixon–Jortner model where all the final states are effective and hence $\varrho_{\text{eff}} = \varrho_l$. On the other hand, it makes sometimes sense to expand the framework of the intermediate case to include the molecules which have nonzero contributions to the halfwidths γ_s and γ_l due to interactions of the initial and final electronic states, respectively, with vibronic quasicontinua belonging to other electronic states; historically, just this definition was introduced in early works and widely used later on. Then, as indicated in Fig. 2.3, the small molecule limit occurs when no such intramolecular continua are present at all so that the halfwidths of the unperturbed levels arise only as a result of collisions. With this new definition, the Bixon–Jortner model of Sect. 2.6.2 exhibits both the small molecule limit and the intermediate case, but they have identical properties as demonstrated by Fig. 2.3. Actual differences in the properties arise when only a small fraction of the manifold of the final states belong to the effective states.

Another important feature of the intermediate case to tell it from the small molecule limit is that the level spacing has a value, not very large and not very small, enabling one to smoothly pass from a stationary Φ_n-state excitation (characteristic of the small molecule case) to a nonstationary Φ_s-state excitation (characteristic of the statistical limit) by simply varying the longevity of the excitation pulse within the experimentally accessible limits. The corresponding molecules are often said to have an *intermediate level structure*. In fact, however, this definition covers only a small fraction of the intermediate-case molecules since the mentioned intermediate value of the level spacing must belong to a very narrow interval of about 10 to 100 states/cm^{-1}.

Lastly, the definitions of small and large molecules may be chosen such as to pay a tribute to the original intuitional (and very physical, indeed) thinking of them as being very close in their properties to atoms and condensed matter, respectively. From this point of view the possibility to prepare an exact molecular state Φ_n, a mixture of Φ_s and Φ_l, is considered as characteristic of the small molecule limit independently of whether that state has an intrinsic width or not.

Clearly, this again returns us to the first definition (2.11.1). On the other hand, the exponential decay of the initially prepared Φ_s-state with a greater-than-radiative decay rate due to the coupling with the dense manifold of the Φ_l-states is attributed to large molecules. However, no discrimination between the statistical limit and the intermediate case is proposed by this approach in the important situation where the decay rate is sensible to the zero-state relaxations.

It is hardly appropriate to further proceed with this discussion. The arguments given seem to be quite sufficient to easily manipulate with these notions. We only emphasize once more the following basic features of our choice embodied in (2.11.1–4): 1) the small molecules have large E_l-level separations; 2) the statistical limit exhibits the exponential emission decay with the rate constant of (2.7.21) independent of the relaxational parameters; 3) the intermediate case implies the irreversible $\Phi_s \rightarrow \Phi_l$ transition with either partial recurrences at finite times or the relaxation-dependent decay rate.

In the statistical limit the luminescence decay follows the simple exponential law (2.1.5) (see also Chap. 3), and there are quite general methods to calculate the radiationless transition rate (see Chap. 6). On the other hand, in the cases of small and intermediate molecules the issue of the luminescence decay behavior is not resolved in such a unique way because it is governed by a number of factors, viz., the density of final states, the intramolecular coupling strength, and the spectral features of the exciting radiation. These factors determine the initially prepared state as well as the route of its subsequent evolution. As evidenced by experiment and by numerical analysis of mathematical models [1.1, 3] as well as by the results of Sects. 2.5–7, there may show up both the purely exponential decay with a lifetime longer than the purely radiative one found from integrated absorption spectra of triatomic [2.267–269] and polyatomic [2.270, 271] molecules (see Sect. 2.5), the biexponential decay, the intensity beats [2.266] and other effects [1.1]. Therefore, the problem of the radiationless transitions in small and intermediate molecules has to be formulated differently depending on the specific physical situation.

The advent of the statistical limit might be expected to be unavoidable with an unlimited increase of the number of atoms in the molecule. Indeed, the density of the final states ϱ_l steadily increases with the number of atoms (at a given energy ΔE) so that a large enough molecule always exhibits properties 1 and 2 and behaves like a statistical system. As for property 3, the matter is much more complicated since the parameters γ and ϱ_{eff} entering condition (2.11.4) pertain to the quasimolecule and hence they will not be too sensitive to the molecular size at a sufficiently large number of atoms. Therefore, some situations are possible where even macroscopically large systems do not belong to the statistical limit, understood in the sense outlined before.

On the other hand, it is obvious that the parameters γ and ϱ_{eff} rapidly increase with ΔE, provided that the number of atoms is large enough; additionally, as will be seen in Chap. 6, ϱ_{eff} increases with the strength of the electronic–vibrational coupling causing displacements and distortions of the potential energy surfaces of the initial and final electronic states. Consequently, condition (2.11.2) for the statistical limit is in essence the requirement of a large electronic energy gap ΔE and of a strong electronic–vibrational coupling.

A variety of methods were developed to calculate the radiationless transition rates in the statistical limit. Importantly, in the case where the potential surfaces for the initial and final electronic states are known, one can in principle indicate explicitly the effective states, i.e., these final states that provide dominant contributions to the statistical limit radiationless transition–rate constant, as well as to evaluate the parameters v_{eff} and ϱ_{eff} and to establish the identity of the results calculated by these methods with expression (2.7.21). The knowledge of ϱ_{eff} together with a reasonable estimate for γ enables a straightforward verification of the validity of condition (2.11.2). This program will be really effectuated in Sect. 6.1 for the simplest potential surfaces.

The existence of the effective states having a constant matrix element and determining the transition rate in the statistical limit sheds new light on the Bixon–Jortner model. Up to now this model is conventionally regarded in the literature as a classical example of a problem permitting an exact solution to elucidate the specificity of the irreversibility in the systems with discrete spectra but having no bearing on real molecules. *In fact, however, the Bixon–Jortner model furnishes an accurate expression for the radiationless transition–rate constant in the statistical limit* provided that only the effective final states rather than all of them are taken into account. In this case the main condition (2.6.25) for the applicability of the Bixon–Jortner model falls away since, according to (2.11.2), a sufficiently fast vibrational relaxation occurs (see the end of Sect. 2.6.1).

An important feature of the radiationless transition rate in the statistical limit is its exponential dependence on the electronic energy,

$$k = A \exp(-\Delta E/\hbar\omega_*) \,, \tag{2.11.5}$$

where A and ω_* are slow functions of ΔE. Such a behavior is due to the fact that the matrix element squared in (2.7.21) is proportional to the Franck–Condon factor. In large molecules the electronic–vibrational coupling is usually not very strong giving rise to only slight displacements and distortions of the initial and final electronic potential energy surfaces. Therefore, if the initial state lies near the minimum of the corresponding potential surface, the wave functions overlap mainly in the classically forbidden region, and the Franck–Condon factor exponentially depends on the width of the barrier between the surfaces, the width in turn being proportional to ΔE. Equation (2.11.5) is conventionally called *the energy gap law* [2.177, 272, 273].

Figure 2.5 shows the experimental dependence of the rate constant for the $T_1 \rightsquigarrow S_0$ intersystem crossing in aromatic hydrocarbons upon ΔE [1.1, 3, 5, 2.184, 274, 275] demonstrating the energy gap law, and Table 2.1 displays the same dependence for the $S_1 \rightsquigarrow T_1$ transition [1.28, 29, 2.276] where no regular variation with ΔE is observed. The energy gap law is also observed for the $S_2 \rightsquigarrow S_1$ internal conversion in azulene and its derivatives [1.1, 277] as well as in other molecules [2.278, 279]. These data show that the $T_1 \rightsquigarrow S_0$

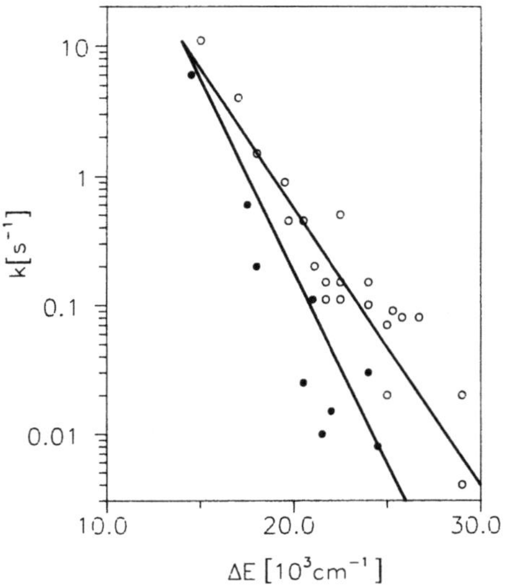

Fig. 2.5. Plot of the $T_1 \rightarrow S_0$ intersystem crossing rate constant vs the energy gap ΔE in aromatic hydrocarbons (condensed phase) [2.184, 274]: ○ C_xH_y; ● C_xD_y

Table 2.1. The $S_1 \rightarrow T_1$ intersystem crossing rate dependence upon the energy gap ΔE in aromatic hydrocarbons (low-temperature condensed phases) [2.276]

Compound	ΔE [10^3 cm^{-1}]	k_{ISC} [10^7 s^{-1}]
Benzene	8.3	0.4
Naphthalene	10.5	0.15
Anthracene	11.7	13.8
Phenanthrene	6.7	1.44
Pyrene	9.0	≤ 0.022
Perylene	10.3	1.0
1,2-Benzanthracene	9.5	1.34
Crysene	7.7	2.32
Coronene	4.4	0.47
Tetracene	10.8	12.1

tranisition in aromatic molecules can be assigned to the statistical limit, whereas the $S_1 \rightarrow T_1$ transition *in the same molecules* cannot. The latter is also partly confirmed by some evidence attesting a manifestation of a pressure dependence of the $S_1 \rightarrow T_1$ transition rate in benzene [2.280, 281], anthracene, naphthalene, and other compounds [1.30, 2.282–286] in the vapor phase, while in some cases this dependence may be due to the vibrational relaxation in the initial electronic state [2.287]. A violation of the energy gap law may also manifest as a significant reduction of the $S_2 \rightarrow S_1$ nonradiative transition rate at small $\Delta E_{S_2S_1}$ giving rise to a competition from the side of the spin-forbidden $S_1 \rightarrow T_1$ transition with a larger energy gap [2.288].

Figure 2.5 clearly demonstrates the normal deuterium isotope effect on the transition rate: the process slows down when hydrogen atoms in the molecule are replaced with deuterium. The deuteration lowers the characteristic frequency ω_* by a factor of

$$\sigma = \omega_*^H/\omega_*^D = (\mu_*^D/\mu_*^H)^{1/2}, \tag{2.11.6}$$

where μ_* is the reduced mass of the oscillators with frequency ω_*. If these oscillators are assumed to be associated with the oscillations of the C–H bond lengths and the reduced C–H mass is taken as μ_*^H, then the theoretical value of σ is 1.33 while the experimental one measured by the slopes of the curves in Fig. 2.5 is 1.36 [1.5, 28, 2.183, 184, 276]. Being that the change of the pre-exponential factor in (2.11.5) is of minor importance, (2.11.5) explains Siebrand's rule of Sect. 2.2 stating that the deuteration reduces the radiationless transition rate in the statistical limit. Experiments on the effect of the full and partial deuteration upon the phosphorescence lifetimes of aromatic hydrocarbons show that just high frequency C–H vibrations are the best acceptors of electronic energy at large ΔE [2.35, 74, 243, 289–295]. In other molecules the N–H and O–H vibrations may also participate [2.296, 297]. A medium-sized [2.298] or small [2.299–301] deuteration effect implies that other than C–H modes are responsible for the $T_1 \rightsquigarrow S_0$ intersystem crossing in these cases [2.302, 257], and the same is true of the $S_1 \rightsquigarrow S_0$ internal conversion [2.6]. The deuterium isotope effect in phosphorescence and fluorescence was observed in many molecules [2.50, 51, 303, 304]. The effect often depends on the position of the substituent in the molecule [2.305].

The $S_1 \rightsquigarrow T_1$ transition in aromatic molecules is attributed to the intermediate case [1.1, 2.237, 285]. Here, the transition rate is given by (2.10.14) and appreciably depends upon the intra- and intermolecular vibrational relaxation. Another important distinction of the intermediate case from the statistical limit is that the dependence of the transition rate upon ΔE is not governed by the simple equation (2.11.5) because of an irregular dependence of separate contributions to (2.10.14) on ΔE; specifically, the deuteration effect is no longer represented by the simple Siebrand's rule and the *inverse deuterium isotope effect*, i.e., an increase of the transition rate by deuteration can be observed [1.42, 53, 2.306–312].

In the majority of the theoretical papers devoted to calculations of the $S_1 \rightsquigarrow T_1$ intersystem crossing rate constant in aromatic hydrocarbons, these molecules were supposed to belong to the statistical limit (see, e.g., [2.54, 313–315]). The rationale behind this was the repeatedly expressed opinion that property 3 of the statistical limit is a direct consequence of property 2 [1.1, 2.204, 316]. As was seen, the latter is not true and the large number of atoms in the molecule is an insufficient condition for the statistical limit.

Notice that the classification given here relates, as a matter of fact, to certain pairs of molecular electronic states rather than to the molecule as a whole since the total density of the final states and the density of the effective ones are functions of both the number of atoms and the energy gap. Hence, different pairs of the electronic states in one and the same large polyatomic molecule may belong to different cases (including the case of small molecules).

3. Luminescence Intensity as a Function of Time and the Radiationless Transition Rate

This chapter describes a general method for calculating the amplitudes of the probabilities to find the system in the excited state versus time that control the experimentally measured quantity of interest, viz., the intensity of the luminescence excited by a short enough light pulse or by some other external perturbation.

The main goal of the theory is to provide a solid basis to the phenomenological equations (2.1.4, 5) and the ensuing relations (2.1.6, 7). The importance of this problem stems from the fact that the radiationless transition rate is usually calculated by time-dependent perturbation theory which only applies at short times as compared to the electronic relaxation period when the decrease in the amplitude of the initial state may still be neglected. Therefore, for the quantity calculated in this way to actually have the meaning of the radiationless transition-rate constant, it is necessary to prove that under definite conditions, typically met in physical systems, the luminescence kinetics is described by (2.1.4, 5) at all times.

The second task of the theory is to derive a general expression for the radiationless transition-rate constant incorporating the relaxation phenomena. The electronic relaxation of polyatomic molecules was shown in the previous chapter always to be conditioned by a mechanism for broadening of discrete energy levels, but the broadening was included, in framework of qualitative theory, only phenomenologically. The general treatment of the present chapter immediately produces the Lorentzian-type broadening (2.7.34) and enables one to specify the parameter γ_l in terms of microscopic parameters.

3.1 Formulation of the Problem

The theory develops in line with the quantum theory of atomic radiation [3.1]. A closed system composed of the molecule, the medium, and the electromagnetic field without external sources is considered. It is supposed that at time $t = 0$ the system is prepared in a certain initial state corresponding to an electronically excited state of the molecule and to the ground states of both the medium and the field, and that the excitation process, being fast enough, does not affect the subsequent evolution of the system. The problem is to solve the time-dependent

Schrödinger equation with a time-independent Hamiltonian,

$$i\hbar \frac{\partial \Psi(t)}{\partial t} = H\Psi(t)\,, \tag{3.1.1}$$

under the initial conditions

$$\Psi(t) = \begin{cases} 0 & \text{at } t < 0\,, \\ \Psi(+0) & \text{at } t = +0\,. \end{cases} \tag{3.1.2}$$

Figure 3.1 shows the level diagram and the level couplings essential for the present problem. This diagram completely specifies an approximate Hamiltonian. The level E_0 is the system's ground state. The matrix elements of the total Hamiltonian between the ground state and excited states are assumed to vanish unless an interaction with an external source is present, so we can introduce a Hamiltonian H and a wave function $\Psi(t)$ related to the excited states only. In correspondence to Fig. 3.1 we define

$$H = H_0 + V + W_1 + W_2\,, \tag{3.1.3}$$

where the zeroth-order Hamiltonian has the eigenfunctions Φ_s, Φ_λ, Φ_A, and Φ_B with the eigenvalues E_s, E_λ, E_A, and E_B, respectively; V is the intramolecular interaction, W_1 and W_2 are the interactions of the molecule with the field and the medium, respectively. The interactions are specified by the following non-vanishing matrix elements:

$$\begin{aligned} v_{s\lambda} &= \langle \Phi_s | V | \Phi_\lambda \rangle\,, \\ w_{As} &= \langle \Phi_A | W_1 | \Phi_s \rangle\,, \\ w_{\lambda B} &= \langle \Phi_\lambda | W_2 | \Phi_B \rangle\,. \end{aligned} \tag{3.1.4}$$

Each of the levels E_s and E_λ are assigned to the excited states of the molecule and

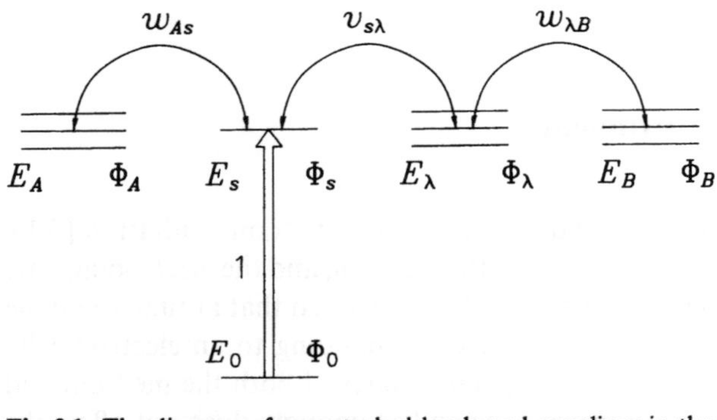

Fig. 3.1. The diagram of unperturbed levels and couplings in the system "molecule + medium + field". (1) excitation with an external source

to the ground states of the medium and the field. The level E_s relates to the initial electronic state. It has a non-vanishing oscillator strength for the transition from the ground state and is populated by excitation from an external source[1] so that the initial condition (3.1.2) is of the form

$$\Psi(+0) = \Phi_s\,. \tag{3.1.5}$$

The levels E_λ (with λ running over a discrete set of values) belong to the vibrationally excited final electronic state.

As compared with Fig. 2.2 we changed the notation of the molecular levels (E_λ for E_l) in view of the following circumstance. As discussed in Sect. 2.10, if the vibronic Hamiltonian permits a separation of variables, the notions of a "quasimolecule" may be introduced as a system of active intramolecular vibrations interacting with the optical electrons of the molecule, and of a "medium" involving both the passive vibrations and the actual physical medium (whose nuclear degrees of freedom are passive, too, according to the definition of the inert medium given in Sect. 2.7.1). Then, by E_λ one should mean the energy levels of the quasimolecule, whereas E_l, as before, stands for the energy levels of the molecule as a whole. But, if the active vibrations are impossible to be sorted out (all the states E_l are uniform with respect to their coupling to the initial state E_s), the definition of the quasimolecule loses its significance, and E_λ merely coincides with E_l.

Within this approach, all the results of the formal treatment to be developed in this chapter equally apply to both isolated molecules and the molecules in media, and the difference between the two systems is only embodied in the physical meaning and in the orders of magnitude of the parameters entering the theory.

The levels E_A are related to the ground states of both the quasimolecule and the medium and to one-photon states of the field, while the levels E_B are related to vibrationally relaxed states of the quasimolecule and vibrationally excited states of the medium; A and B are the corresponding sets of quantum numbers running over a (quasi)continuous manifold of values ($B = m\kappa$, within the notations of Sect. 2.7.2).

The excited states with nearly the same energy are only included in Fig. 3.1, as the couplings are independent of time. The omission of all other excited states amount to a neglect of contributions from the transitions via intermediate states and corresponds to the first-order time-independent perturbation theory and degenerate levels.

Just as in the quantum theory of atomic radiation, the solution to the problem (3.1.1, 2) is sought by the method of Laplace transformation.

---

[1] To be specific, we speak of the photoexcitation of the system from its ground state. Generally E_0 is an arbitrary state of the system coupled with E_s during a short enough lapse of time.

3.2 Laplace Transformation, Green's Functions and Resonant States

The formal solution to the problem (3.1.1, 2) is written in the form

$$\Psi(t) = U(t)\,\Psi(+0)\,, \tag{3.2.1}$$

where

$$U(t) = \theta(t)\exp(-\mathrm{i}Ht/\hbar) \tag{3.2.2}$$

is the evolution operator and $\theta(t)$ is the Heaviside step function,

$$\theta(t) = \begin{cases} 0 & \text{at } t < 0\,, \\ 1 & t \geq 0\,. \end{cases} \tag{3.2.3}$$

Notice that the evolution operator obeys the inhomogeneous equation

$$\left(\mathrm{i}\hbar\frac{\partial}{\partial t} - H\right)U(t) = \delta(t)$$

and therefore it is the retarded Green's function operator for the Schrödinger equation.

Let Φ_j be an arbitrary complete set of orthonormalized wave functions. The matrix elements of the evolution operator,

$$c_{jk}(t) = \langle\Phi_j|\,U(t)|\Phi_k\rangle\,, \tag{3.2.4}$$

fulfil the conditions

$$c_{jk}(t) = \begin{cases} 0 & \text{at } t < 0 \qquad (3.2.5) \\ \delta_{jk} & \text{at } t = +0\,, \qquad (3.2.6) \end{cases}$$

while at $t > 0$ they obey the simultaneous equations

$$\mathrm{i}\hbar\dot{c}_{jk}(t) = \sum_n H_{jn}c_{nk}(t)\,, \tag{3.2.7}$$

where $H_{jn} = \langle\Phi_j|\,H\,|\Phi_n\rangle$ are the Hamiltonian matrix elements over the basis chosen.

We now introduce the imaginary Laplace transformation of the matrix elements (3.2.4) by

$$G_{jk}(E) = \frac{1}{\mathrm{i}\hbar}\int_0^\infty c_{jk}(t)\exp(\mathrm{i}Et/\hbar)\,dt\,. \tag{3.2.8}$$

These functions are analytical all over the upper half-plane of the complex variable E because the $c_{jk}(t)$ are restricted in magnitude. Equations (3.2.8) specify the Green's function operator $G(E)$, which is the Laplace transform of the evolution operator (3.2.2).

The transformation (3.2.8) enables us to reduce the simultaneous differential equations (3.2.7) for $c_{jk}(t)$ to a set of algebraic equations for their Laplace transforms. We multiply (3.2.7) with $\exp(\mathrm{i}Et/\hbar)$ and integrate over t from zero to infinity. Using the initial condition (3.2.6) and assuming $\mathrm{Im}\{E\} > 0$, one obtains the system of algebraic equations

$$\sum_n (E\delta_{jn} - H_{jn})G_{nk}(E) = \delta_{jk} \tag{3.2.9}$$

that are equivalent to the operator equation for the Green's function,

$$(E - H)G(E) = 1 \ . \tag{3.2.10}$$

The solution to (3.2.7) with the initial condition (3.2.6) is given by the inverse transform

$$c_{jk}(t) = -\frac{1}{2\pi\mathrm{i}} \int_{\mathrm{i}\eta-\infty}^{\mathrm{i}\eta+\infty} G_{jk}(E) \exp(-\mathrm{i}Et/\hbar)\, dE \ , \tag{3.2.11}$$

where $\eta > 0$. Condition (3.2.5) is met automatically as at $t < 0$ the integration path in (3.2.11) may be closed in the upper half-plane where the Green's function is analytical.

If the wave function at the initial time is expanded in the series

$$\Psi(+0) = \sum_k a_k \Phi_k \ ,$$

then the wave function (3.2.1) at any time is represented by

$$\Psi(t) = \sum_{jk} c_{jk}(t) a_k \Phi_j \ , \tag{3.2.12}$$

and the probability to find the system in the state Φ_j at time t is

$$P_j(t) = \left| \sum_k c_{jk}(t) a_k \right|^2 \ . \tag{3.2.13}$$

Thus the problem is reduced to solving the operator equation (3.2.10) and to a subsequent integration in (3.2.11) using an analytical continuation of the Green's function into the lower half-plane of the complex variable E.

If the Φ_j's are chosen to be the eigenfunctions of the Hamiltonian H, the solution to (3.2.9) has the form

$$\mathrm{G}_{jk}(E) = \frac{\delta_{jk}}{E - E_j} \ ,$$

and (3.2.11, 12) give, for $t \geq 0$,

$$c_{jk}(t) = \delta_{jk} \exp(-\mathrm{i}E_j t/\hbar) , \tag{3.2.14}$$

$$\Psi(t) = \sum_j a_j \exp(-\mathrm{i}E_j t/\hbar)\Phi_j , \tag{3.2.15}$$

which, of course, could have been immediately obtained from (3.2.2, 4, 12) as well.

As was outlined in Sect. 2.4.1, even when no field is present, the use of the basis of eigenfunctions of the Hamiltonian H is only justified in the case of a small number of discrete excited states. Then, if there is no occasional E_j-level degeneracy, and a separate stationary state Φ_j can be prepared by the excitation process, $|a_j|^2$ represents the experimentally measured oscillator strength for the transition to the state Φ_j. And if there is a degeneracy and the excitation prepares a superposition of the degenerate states, (3.2.15) describes quantum beats of the luminescence intensity [1.7, 2.18, 221, 223, 225, 3.2, 3] which may also be observed in experiment under proper conditions. In other words, in the case considered, (3.2.15) describes real physical phenomena, and the Green's function method is exactly equivalent to solving the Schrödinger equation since the poles of the Green's function necessary to evaluate the integral in (3.2.11) coincide with the energies of the stationary states.

On the contrary, in the presence of a large number of discrete states or of a continuous spectrum, (3.2.15) has no physical meaning and only gives a formal solution to the problem because the stationary states of such a system are unobservable. In this case the excitation process produces some nonstationary state of which the characteristics are just the quantities measured experimentally. In the absorption spectrum such a state is observed as a resonance, therefore it is conventionally called *the resonant state*; the parameters of that state observed in experiment are the peak position and the width of the corresponding spectral feature. The aim of the theory is to investigate the properties of the resonant states, so an appropriate basis set of functions should be used in solving (3.2.9).

Those functions were shown in Sect. 2.4.1 to be the eigenfunctions of a certain zeroth-order Hamiltonian. We shall assume the Hamiltonian splitting in (3.1.3) to satisfy the necessary restrictions. Namely, it is supposed that the excitation process is fast enough and that only the $E_s \leftarrow E_0$ transition has a nonzero oscillator strength. Then the nonstationary state Φ_s is prepared initially, and the problem is to study the evolvement of that state in time. For this reason, the eigenfunctions of the zeroth-order Hamiltonian depicted in Fig. 3.1 will be employed as a basis set. With this choice of the basis, the pole of the Green's function $G_{ss}(E)$ gives both the position and the width of the resonant state originating from the level E_s.

3.3 Computation of the Green's Functions

In the basis of the zeroth-order Hamiltonian eigenfunctions with the use of the initial condition (3.1.5), (3.2.12, 13) take the form

$$\Psi_s(t) = \sum_j c_{js}(t)\Phi_j ,$$
$$P_{js}(t) = |c_{js}(t)|^2 , \tag{3.3.1}$$

where j takes on the values s, λ, A, B. In particular, we shall be interested in the probabilities $P_{ss}(t)$ and $\sum_A P_{As}(t)$. The former is the probability to observe the system in the initial state while the latter is the probability of a transition with photon emission, or the number of photons per one excited molecule emitted up to time t. The luminescence intensity per one excited molecule, which is proportional to the photon counting rate, is given by

$$I_s(t) = \frac{d}{dt}\sum_A P_{As}(t) . \tag{3.3.2}$$

In order to calculate these probabilities, one has to know the Green's functions G_{ss} and G_{As} (the argument E is dropped). After substituting expansion (3.1.3) into (3.2.10) one obtains

$$(E - H_0 - V - W_1 - W_2)G = 1 . \tag{3.3.3}$$

The simplest way to solve (3.3.3) is to write out a few equations containing the G-matrix elements of interest, taking the matrix elements of both sides of (3.3.3) and using the nonzero interaction matrix elements of (3.1.4). Taking the matrix element As gives:

$$(E - E_A)G_{As} - w_{As}G_{ss} = 0 , \tag{3.3.4}$$

whence

$$G_{As} = \frac{w_{As}}{E - E_A}G_{ss} . \tag{3.3.5}$$

In order to calculate G_{ss}, we take the matrix elements ss, λs and Bs:

$$(E - E_s)G_{ss} - \sum_\lambda v_{s\lambda}G_{\lambda s} - \sum_A w_{sA}G_{As} = 1 , \tag{3.3.6}$$

$$(E - E_\lambda)G_{\lambda s} - v_{\lambda s}G_{ss} - \sum_B w_{\lambda B}G_{Bs} = 0 , \tag{3.3.7}$$

$$(E - E_B)G_{Bs} - \sum_\lambda w_{B\lambda}G_{\lambda s} = 0 . \tag{3.3.8}$$

Let us put

$$G_{\lambda s} = x_{\lambda s} G_{ss} \,. \tag{3.3.9}$$

Substituting (3.3.5, 9) into (3.3.6) we obtain

$$G_{ss} = \left(E - E_s - \sum_{\lambda} v_{s\lambda} x_{\lambda s} - \sum_{A} \frac{|w_{sA}|^2}{E - E_A} \right)^{-1} . \tag{3.3.10}$$

Elimination of G_{Bs} from (3.3.7, 8) yields for the unknown function $x_{\lambda s}$ the integral equation

$$(E - E_{\lambda}) x_{\lambda s} = v_{\lambda s} + \sum_{\lambda'} R_{\lambda\lambda'} x_{\lambda' s} \tag{3.3.11}$$

with the kernel

$$R_{\lambda\lambda'} = \sum_{B} \frac{w_{\lambda B} w_{B\lambda'}}{E - E_B} \,. \tag{3.3.12}$$

An approximate solution of (3.3.11) is obtained by neglecting the off-diagonal elements of $R_{\lambda\lambda'}$,

$$x_{\lambda s} = \frac{v_{\lambda s}}{E - E_{\lambda} - R_{\lambda\lambda}} \,. \tag{3.3.13}$$

In order to estimate the accuracy of this approximation, we compare the sum

$$\sum_{\lambda' \neq \lambda} R_{\lambda\lambda'} x_{\lambda' s} = \sum_{\lambda' \neq \lambda} \frac{R_{\lambda\lambda'} v_{\lambda' s}}{E - E_{\lambda'} - R_{\lambda'\lambda'}} \tag{3.3.14}$$

neglected in the right-hand side of (3.3.11) with the leading term $v_{\lambda s}$. Obviously, (3.3.14) will only provide an insignificant renormalization of the interaction $v_{\lambda s}$ if the off-diagonal matrix elements $R_{\lambda\lambda'}$ are sufficiently small, i.e., when either

$$|R_{\lambda\lambda'}| \ll |R_{\lambda'\lambda'}| \tag{3.3.15}$$

or

$$|R_{\lambda\lambda'}| \ll |E - E_{\lambda'}| \,. \tag{3.3.16}$$

holds.

The smallness of the off-diagonal elements (3.3.15) may arise from the fact that $R_{\lambda\lambda'}$ depends on the phases of the molecule–medium interaction [these are the phases of the complex-valued matrix elements $w_{\lambda B}$ in (3.3.12)], and if the number of the B-states is very large, averaging over the rapidly varying phases occurs in the sum of (3.3.12). In this case the solution (3.3.13) corresponds to the random-phase approximation.

Condition (3.3.15) can be expected to be well-satisfied in the presence of a real medium with a continuous spectrum of the E_B-levels. Now consider an isolated molecule, and let λ enumerate the levels of the quasimolecule. Here, "the medium" is formed by the passive intramolecular vibrations so that the subscript B in (3.3.12) runs over a series of discrete values, and it is generally not clear whether condition (3.3.15) will be fulfilled. Therefore, turn to inequality (3.3.16).

In the case of an isolated molecule, $R_{\lambda\lambda'}$ represent the broadenings and the shifts of the quasimolecule levels induced by anharmonic interactions with the passive vibrations, and E in inequality (3.3.16) is a pole of the Green's function (3.3.10). The $x_{\lambda s}$ appreciably contribute to (3.3.10), provided that the pole E is close to E_λ where $x_{\lambda s}$ is large. Thus (3.3.16) reduces to the demand that the anharmonic broadenings and shifts of the quasimolecule levels be small as compared to the level spacings. The quasimolecule may have a very low number of degrees of freedom, and those spacings will be of the order of the vibrational quanta. The arguments given support the assumption that approximation (3.3.13) is reasonable.

On inserting (3.3.13) into (3.3.10) one obtains

$$G_{ss} = [E - E_s - R_s(E) - R(E)]^{-1}, \qquad (3.3.17)$$

$$R_s(E) = \sum_A \frac{|w_{sA}|^2}{E - E_A}, \qquad (3.3.18)$$

$$R(E) = \sum_\lambda \frac{|v_{s\lambda}|^2}{E - E_\lambda - R_{\lambda\lambda}(E)}, \qquad (3.3.19)$$

$$R_{\lambda\lambda}(E) = \sum_B \frac{|w_{\lambda B}|^2}{E - E_B}. \qquad (3.3.20)$$

Remind that the levels E_A are assigned to the ground state of the quasimolecule and the one-photon states of the field, i.e., $E_A = E_0 + \hbar\omega_A$, where $\hbar\omega_A$ is the photon energy and A is the set of its quantum numbers (the energy, the propagation direction n_A, and the polarization e_A). Introducing the density of the photon states, $\varrho_1(E_A)$, rewriting (3.3.18) in the explicit form, and separating out the energy integration gives

$$R_s(E) = \sum_{n_A, e_A} \int \frac{|w_{sA}|^2}{E - E_A} \varrho_1(E_A) dE_A. \qquad (3.3.21)$$

The energy E_B is written as $E_B = E_{\lambda'} + E_\kappa$, where $E_{\lambda'}$ are the vibrationally relaxed levels of the quasimolecule ($E_{\lambda'} < E_\lambda$) and E_κ are the vibrationally excited levels of the medium. Consequently, the sum over B in (3.3.20) is the sum over λ' and the integral over E_κ,

$$R_{\lambda\lambda}(E) = \sum_{\lambda'} \int \frac{|w_{\lambda\lambda'\kappa}|^2}{E - E_{\lambda'} - E_\kappa} \varrho_2(E_\kappa) dE_\kappa, \qquad (3.3.22)$$

where $\varrho_2(E_\kappa)$ is the density of the medium states. If the medium states are not specified by the energy E_κ alone, a summation over all other quantum numbers is implied here as well.

3.4 Evolution of the Initial State and the Luminescence Intensity

Without couplings the pole of the Green's function (3.3.17) is at $E = E_s$. If the couplings are weak enough, and if $R_s(E)$ and $R(E)$ are slowly varying functions near E_s (see Sect. 3.5), the pole's displacement can be approximately estimated by replacing those functions with their values at $E = E_s$. To put it precisely, one has to force $E = E_s + i\eta$ with $\eta \to +0$, for those functions are defined in the upper half-plane. Then the Green's function's pole is

$$E = E_s + R_s(E_s + i\eta) + R(E_s + i\eta) \,. \tag{3.4.1}$$

At real E the integral in (3.3.21) over the real E_A-axis (circumventing the pole at $E_A = E$ from below) can be represented in the form

$$R_s(E + i\eta) = D_s(E) - \frac{i\hbar}{2}\Gamma_s(E) \,, \tag{3.4.2}$$

where $D_s(E)$ is the principal value of the integral (3.3.21) and

$$\begin{aligned}\Gamma_s(E) &= \frac{2\pi}{\hbar}\sum_{n_A,e_A}\int |w_{sA}|^2\,\delta(E - E_A)\varrho_1(E_A)\,dE_A \\ &= \frac{2\pi}{\hbar}\sum_{n_A,e_A}|w_{sA}|^2\,\varrho_1(E)\end{aligned} \tag{3.4.3}$$

(in the latter expression w_{sA} is taken at $E_A = E$). The $R_{\lambda\lambda}(E)$, too, can be represented in a similar form for real E,

$$R_{\lambda\lambda}(E + i\eta) = D_\lambda(E) - \frac{i\hbar}{2}\Gamma_\lambda(E) \,, \tag{3.4.4}$$

where $D_\lambda(E)$ is the principal value of the integral in (3.3.22) and

$$\begin{aligned}\Gamma_\lambda(E) &= \frac{2\pi}{\hbar}\sum_{\lambda'}|w_{\lambda\lambda'\kappa}|^2\,\delta(E - E_{\lambda'} - E_\kappa)\varrho_2(E_\kappa)\,dE_\kappa \\ &= \frac{2\pi}{\hbar}\sum_{\lambda'}|w_{\lambda\lambda'\kappa}|^2\,\varrho_2(E - E_{\lambda'})\end{aligned} \tag{3.4.5}$$

($w_{\lambda\lambda'\kappa}$ is calculated at $E_\kappa = E - E_{\lambda'}$). Substituting (3.4.2–5) and (3.3.19) into (3.4.1) one obtains

$$E = E_s + D_s(E_s) - \frac{i\hbar}{2}\Gamma_s(E_s) + \sum_\lambda \frac{|v_{s\lambda}|^2}{E_s - E_\lambda - D_\lambda(E_s) + (i\hbar/2)\Gamma_\lambda(E_s)}. \qquad (3.4.6)$$

The quantities $D_s(E_s)$ and $D_\lambda(E_s)$ yield insignificant level shifts, which can safely be absorbed into E_s and E_λ, respectively. The same is true of the real part of the sum in (3.4.6). The resulting Green's function (3.3.17) takes the form

$$G_{ss} = \left(E - E_s + \frac{i\hbar}{2}\Gamma\right)^{-1}, \qquad (3.4.7)$$

$$\Gamma = \Gamma_s(E_s) + k_{st}, \qquad (3.4.8)$$

$$k_{st} = \sum_\lambda \frac{|v_{s\lambda}|^2 \Gamma(E_s)}{(E_s - E_\lambda)^2 + [(\hbar/2)\Gamma_\lambda(E_s)]^2}. \qquad (3.4.9)$$

Equations (3.4.3, 8) show that $\Gamma_s(E_s)$ is the radiative decay rate constant for the level E_s,

$$\tau_{\mathrm{rad}(s)}^{-1} = \Gamma_s(E_s), \qquad (3.4.10)$$

whereas k_{st} is the rate constant of the radiationless transition from the state E_s to all the final states E_λ broadened by the coupling with the medium; the subscript t implies, as in Chap. 2, that the rate constant is affected by zero-level relaxations entering via $\Gamma_\lambda(E_s)$.

A trivial calculation of the contour integral in (3.2.11) gives the amplitude

$$c_{ss}(t) = \exp\left(-\frac{i}{\hbar}E_s t - \frac{1}{2}\Gamma t\right), \qquad (3.4.11)$$

hence, the probability to find the system in the initial state decreases by the exponential law (2.7.32).

Now we calculate the luminescence intensity. We recast (3.3.2) with the use of (3.3.1) in the form

$$I_s(t) = 2\,\mathrm{Re}\left\{\sum_A \dot{c}_{As}^*(t)\, c_{As}(t)\right\}. \qquad (3.4.12)$$

Elimination of the time derivative using the Schrödinger equation (3.2.7) and the level diagram in Fig. 3.1, where all nonvanishing Hamiltonian matrix elements are included, gives

$$I_s(t) = -\frac{2}{\hbar}\,\mathrm{Im}\left\{\sum_A w^*_{As} c^*_{ss}(t) c_{As}(t)\right\}. \tag{3.4.13}$$

The amplitude $c_{As}(t)$ is given by integral (3.2.11) with the Green's function (3.3.5), so that (3.4.13) takes the form

$$I_s(t) = \frac{2}{\hbar}\,\mathrm{Im}\left\{c^*_{ss}(t)\frac{1}{2\pi \mathrm{i}}\int_{\mathrm{i}\eta-\infty}^{\mathrm{i}\eta+\infty} R_s(E) G_{ss}(E)\,\mathrm{e}^{-\mathrm{i}Et/\hbar}\,dE\right\}, \tag{3.4.14}$$

where $R_s(E)$ is given by (3.3.18). Within the same accuracy as in the calculation of the Green's function G_{ss} one can replace $R_s(E)$ with its value at $E = E_s + \mathrm{i}\eta$ and remove it from the integral sign. The remaining integral equals to $-c_{ss}(t)$, and taking (3.4.2) and (2.7.32) into account one obtains the final result,

$$I_s(t) = \Gamma_s(E_s) P_{ss}(t) = \tau^{-1}_{\mathrm{rad}\,(s)}\,\mathrm{e}^{-\Gamma t}. \tag{3.4.15}$$

Thus, the luminescence intensity decays exponentially with the observed lifetime $\tau_{\mathrm{obs}\,(s)} = 1/\Gamma$, which leads to the phenomenological equation (2.1.6) after using (3.4.8, 10). Since $I_s(t)$ is the luminescence intensity per one excited molecule, the emission quantum yield (2.1.3) is

$$\varphi_s = \int_0^\infty I_s(t)dt = \tau_{\mathrm{obs}\,(s)}/\tau_{\mathrm{rad}\,(s)}, \tag{3.4.16}$$

which coincides with the phenomenological equation (2.1.7)

Equations (3.4.9) gives a general representation for the radiationless transition rate constant, which can also be written in the form of (2.10.14) with

$$r_\lambda(E_s) = \frac{1}{\pi}\,\frac{(\hbar/2)\,\Gamma_\lambda(E_s)}{(E_s - E_\lambda)^2 + [(\hbar/2)\,\Gamma_\lambda(E_s)]^2}. \tag{3.4.17}$$

Thus, the level distribution within the band which originated from the level E_λ is described by the Lorentzian with the half-width $\gamma_\lambda = \hbar\Gamma_\lambda(E_s)/2$. Notice that $\Gamma_\lambda(E_s)$ is generally not the vibrational relaxation rate of the level E_λ since, according to (3.4.5), the relaxation rate is $\Gamma_\lambda(E_\lambda)$. The study of γ_λ as a function of E_λ with the use of some special models shows [3.4] that γ_λ rapidly tends to zero with increasing $|E_s - E_\lambda|$. Therefore, virtually, the summation in (3.4.9) is only over the states with the energies E_λ close to E_s.

The above derivation of the phenomenological relations is based on the obvious statement that the Green's function has the isolated pole (3.4.1) at a not very strong interaction. On the other hand, the same relations follow from the balance equation which thus turns out to be uniquely related to the existence of that pole. Hence, the said condition of the not very strong interaction is nothing else as the validity condition for the balance equation discussed in Chap. 2 and depicted in Fig. 2.3. In terms of the quasimolecule parameters it reads $v_{s\lambda} \ll \gamma_\lambda$ or $v_{s\lambda} \ll (\gamma_\lambda/\varrho_\lambda)^{1/2}$ in the weak or strong coupling limit, respectively.

3.5 Resonant States

In Sect. 3.2 the concept of the resonant states defined by Green's function poles was introduced, and the resonant state associated with the zero level E_s was found in Sect. 3.4. Let us now investigate which transformations the resonant states are subject to in going from small molecules to intermediate and large ones. This issue was already addressed in Sect. 2.6.2 in frames of the effective Hamiltonian method. According to Sect. 2.11, the above cases are characterized by different relations (2.11.1, 2, 4) between the level width and the density of states in the molecule, and the problem is to investigate the poles of the Green's function taking account of these relations.

The choice of the Green's function representation (i.e., of the basis functions or, which is the same, of the zeroth-order Hamiltonian) is dictated by the physical initial conditions. In Sect. 3.4 the Green's function was calculated in the basis of the functions shown in Fig. 3.1 since the state Φ_s was assumed there to be prepared initially. In the small molecule limit it is the stationary state of the isolated molecule Φ_n that is produced by the initial excitation (see Sect. 2.4.1), and therefore, (3.2.10) for the Green's function ought to be solved in the basis of the functions Φ_n. However, for the sake of uniformity, we shall always investigate the Green's function G_{ss}, which is quite sufficient for our purposes. Indeed, G_{ss} is a linear superposition of the functions G_{nn}, so that the poles of G_{nn} are simultaneously the poles of G_{ss} (see Sect. 3.5.1).

The general expression for the function G_{ss} is given by (3.3.17). As in Sect. 3.4 it is simplified by replacing $R_s(E)$ and $R_{\lambda\lambda}(E)$ with their values at $E = E_s + i\eta$ and by omitting the insignificant level shifts $D_s(E_s)$ and $D_\lambda(E_s)$. However now, as distinct from (3.4.1), the same procedure with regard to the function $R(E)$ cannot be carried out because in the case of small molecules the singularities of this function in a vicinity of the real axis have to be allowed for, i.e., $R(E)$ is not a slowly varying function. Thus, the starting expression for G_{ss} has the form

$$G_{ss} = \left(E - E_s + i\gamma_s - \sum_\lambda \frac{|v_{s\lambda}|^2}{E - E_\lambda + i\gamma_\lambda} \right)^{-1}, \tag{3.5.1}$$

where $\gamma_{s,\lambda} = \hbar\Gamma_{s,\lambda}(E_s)/2$.

The rationale behind the assumption of the slow variation of $R_s(E)$ and $R_{\lambda\lambda}(E)$ is as follows. In studies of the intramolecular processes, it is often not necessary to allow for whatever selectivity of the interactions of the molecule with the electromagnetic field and with the inert medium. Then, it is natural to neglect the selectivity completely, and instead to describe these interactions using a minimum set of parameters which might be regarded as phenomenological constants. The nonselectivity of the interactions means that the matrix elements and densities of states in (3.3.21, 22) are slowly varying in a wide enough energy region around E_s. Then, $R_s(E)$ and $R_{\lambda\lambda}(E)$ are also slowly varying functions of E inside the same region and can be replaced by constant values; as a result, the interactions with the field and medium give rise to the appearance of the unperturbed-level widths which are further not evaluated, but are rather treated as fixed phenomenological constants. In contrast to this behavior, the intramolecular interactions are allowed to be quite selective with regard to the parameter values and symmetry species of the participating vibrational modes excited at a given vibronic level, which gives rise to large-amplitude fluctuations of the interaction matrix elements among neighbouring levels [2.117].

The level broadening due to the limited duration of the experiment τ_{max}, $\gamma_{max} = \hbar/2\tau_{max}$, introduced in Sects. 2.4–6 in discussing the irreversibility in the case of strictly isolated molecules, has not yet been considered here. Now let us allow for this additional broadening invoking the following formal procedure. We introduce the additional factor $\exp(-t/2\,\tau_{max})$ on the right-hand side of (3.2.11) for the amplitudes $c_{jk}(t)$ and require it to be close to unity during the times of physical interest, i.e., at t on the order of the radiative lifetime $\tau_{rad\,(s)}$, and of the lifetimes of the resonant states. This restriction can be written in the form

$$-\mathrm{Im}\{E\}, \gamma_s \gg \gamma_{max} , \tag{3.5.2}$$

where E is the complex energy of the resonant state. Equation (3.2.11) shows that introducing the above factor leads to the replacement of $E \rightarrow E + \mathrm{i}\gamma_{max}$ in all subsequent formulae. Performing such a replacement in (3.5.1) we obtain the following equation for the Green's function poles:

$$E = E_s - \mathrm{i}(\gamma_s + \gamma_{max}) + \sum_{\lambda} \frac{|v_{s\lambda}|^2}{E - E_\lambda + \mathrm{i}(\gamma_\lambda + \gamma_{max})} . \tag{3.5.3}$$

Thus the total widths of the initial and final states are $\gamma_s + \gamma_{max}$ and $\gamma_\lambda + \gamma_{max}$, respectively. Equation (3.5.3) derived from the first principles lays the foundation of the phenomenological method of the effective Hamiltonian used in Chap. 2.

3.5.1 Small Molecules

In the case of small molecules the E_λ's are to be understood as the levels E_l of the isolated molecule so that (3.5.3) for the poles of the Green's function takes the form

$$E = E_s - \mathrm{i}(\gamma_s + \gamma_{max}) + \sum_{l} \frac{|v_{sl}|^2}{E - E_l + \mathrm{i}(\gamma_l + \gamma_{max})} . \tag{3.5.4}$$

The small molecules are characterized by (2.11.1) requiring the total widths $\gamma_s + \gamma_{max}$ and $\gamma_l + \gamma_{max}$ to be small as compared to the spacing between the levels E_l,

$$\gamma_s + \gamma_{max}, \gamma_l + \gamma_{max} \ll 1/\varrho_l . \tag{3.5.5}$$

In the vapor phase the width γ_l is proportional to pressure since collisions are the only physical mechanism to broaden the final states. At low pressures it may be comparable with the radiative width of the initial state. We shall say that *the molecule is nearly isolated* if condition $\gamma_l \ll \gamma_s$ holds, and that *the molecule is embedded in a medium* in the opposite case of $\gamma_l \gg \gamma_s$. The latter is generally met in high-pressure vapors and in condensed phases.

We consider first the case of weak coupling, $|v_{sl}| \ll |E_s^{(0)} - E_l^{(0)}|$, where

$$E_s^{(0)} = E_s - \mathrm{i}(\gamma_s + \gamma_{\max}), \qquad E_l^{(0)} = E_l - \mathrm{i}(\gamma_l + \gamma_{\max}) . \tag{3.5.6}$$

It is easy to see that here the resonant states are associated with the unperturbed levels E_s and E_l since (3.5.4) may be solved by the method of successive approximations with $E_s^{(0)}$ and $E_l^{(0)}$ as the zeroth-order approximation. The approximate complex energy of the main resonant state associated with the level E_s is found by means of the substitution of $E_s^{(0)}$ for E on the right-hand side of (3.5.4) [cf. (3.4.1)] which gives

$$E = E_s^{(0)} + \sum_l \frac{|v_{sl}|^2}{E_s^{(0)} - E_l^{(0)}}, \tag{3.5.7}$$

while the resonances originating from the levels E_l are disregarded since the residues of the Green's function (3.5.1) in the corresponding poles are small; it is a simple matter to verify that the oscillator strengths to excite these resonances are negligible, too. Equations (3.5.6, 7) give the following width of the principal resonance:

$$-\mathrm{Im}\{E\} = \gamma_s + \gamma_{\max} - \sum_l \frac{|v_{sl}|^2 (\gamma_s - \gamma_l)}{(E_s - E_l)^2 + (\gamma_s - \gamma_l)^2} . \tag{3.5.8}$$

Here, $\gamma_{\max}$ may be omitted in view of condition (3.5.2).

The luminescence intensity $I_s(t)$ after the excitation of the principal resonance is calculated in the same way as in Sect. 3.4. We define the radiative lifetime according to (3.4.10) and the observed decay time of the luminescence,

$$\tau_{\mathrm{obs}(s)} = -(\hbar/2)\,\mathrm{Im}\{E\} . \tag{3.5.9}$$

For a nearly isolated molecule one obtains from (3.5.8)

$$-\mathrm{Im}\{E\} = \left(1 - \sum_l \frac{|v_{sl}|^2}{(E_s - E_l)^2 + \gamma_s^2}\right)\gamma_s \approx \gamma_s . \tag{3.5.10}$$

As expected, the observed lifetime coincides with the radiative one and the quantum yield (3.4.16) is unity (it looks as being slightly greater than unity but, as explained in Sects. 2.6.2, 3, it becomes unity after the contributions from the weak E_l-resonances are included). For a molecule in a medium (3.5.8) gives

$$-\mathrm{Im}\{E\} = \gamma_s + \tfrac{1}{2}\hbar k_{sl} , \tag{3.5.11}$$

where

$$k_{sl} = \frac{2}{\hbar} \sum_l \frac{|v_{sl}|^2 \gamma_l}{(E_s - E_l)^2 + \gamma_l^2} \tag{3.5.12}$$

is the rate constant for the radiationless transition from the initial state E_s to all final states E_l. This result was derived earlier in Sect. 2.7.3.

The matrix elements, the level widths, and the density of the final vibronic states evaluated from the experimental measurements on the $S_1 \to S_0$ fluorescence emission of formaldehyde [3.5, 6] present evidence for the occurrence of the weak coupling limit in this molecule. An effective Hamiltonian with nonzero off-diagonal elements of the damping matrix for the manifold of the final states was used to interpret the data, resulting in a more complicated expression than (3.5.11, 12) since the latter is obtained with a neglect of the off-diagonal damping.

We proceed now to the case of strong coupling $|v_{sl}| \gtrsim 1/\varrho_l$. Calculations of the matrix elements and the densities of states performed in [3.7] for the molecules C_3, BO_2, ClO_2, and CS_2 demonstrated the validity of the strong-coupling condition in these triatomic systems. The occurence of the strong-coupling limit in larger polyatomic molecules such as pyrazine was directly proved in measurements of ultrahigh-resolution spectra [2.116–126] (see also the review articles [2.123, 166, 167, 205]).

In this case the small parameters are the level widths, and solutions to (3.5.4) can be obtained by its formal expansion in powers of γ_s and $\gamma_l + \gamma_{\max}$. The equation of the zeroth-order approximation,

$$E = E_s + \sum_l \frac{|v_{sl}|^2}{E - E_l}, \tag{3.5.13}$$

is easily seen to be identical to (2.6.6) for the eigenvalues E_n of the full molecular Hamiltonian. Hence the resonant states are close to the stationary states (eigenstates) of the isolated molecule. The levels E_n can be found graphically as was done in Fig. 2.4 for the Bixon–Jortner model. It is easily verified that there always lies a single exact level E_n between every two adjacent unperturbed levels E_l, and that under the strong-coupling condition the differences $|E_n - E_l|$ are on the order of the level spacings between the unperturbed levels.

In the next approximation the solution to (3.5.4) is sought in the form $E = E_n - i\gamma_n$, where n is the number of the resonance. Substituting this into (3.5.4) and expanding the sum in powers of γ_n and $\gamma_l + \gamma_{\max}$ to the first order one obtains after a simple algebra [2.76] [cf. (2.6.37)]

$$\gamma_n = |a_{sn}|^2 \gamma_s + \sum_l |a_{ln}|^2 \gamma_l + \gamma_{\max}, \tag{3.5.14}$$

where a_{sn} and a_{ln} are given by (2.6.5, 8) and obey the normalization conditions (2.6.7, 9). Here again, $\gamma_{\max}$ turns out to be insignificant in view of condition

(3.5.2). From (3.5.5) and (2.6.7) it follows that

$$\gamma_n \ll 1/\varrho_n \,. \tag{3.5.15}$$

Notice that in the weak coupling limit (3.5.14) turns into (3.5.11), provided the E_s-level does not happen to fall in a sporadic resonance with some E_l-level. For strong coupling, the condition for applicability of (3.5.14) proves to be less restrictive than (3.5.5), as was demonstrated in Sect. 2.6.2 for the similar equation (2.6.37) in the Bixon–Jortner model.

Let us calculate $I_n(t)$, the luminescence intensity on excitation of the n-th resonance. This needs the knowledge of the Green's function G_{nn} because the initially prepared state is the stationary state Φ_n. This function can be obtained by straightforward solving of (3.2.10) in the basis of the Φ_n's similar to how it was done in Sect. 3.3; in frames of that method, account has to be taken of the fact that in the case of small broadening under consideration the off-diagonal elements $G_{nn'}$ will be small as well as the off-diagonal terms with subscript l arising when use is made of expansion (2.6.2). One may, however, act simpler by utilizing (2.6.2) and writing down G_{ss} in the form

$$\begin{aligned} G_{ss} &\equiv \langle \Phi_s | G(E) | \Phi_s \rangle \\ &= \sum_{nn} \langle \Phi_s | \Phi_n \rangle G_{nn} \langle \Phi_n | \Phi_s \rangle \simeq \sum_n |a_{sn}|^2 G_{nn} \,. \end{aligned} \tag{3.5.16}$$

In this approximation we neglect the quantum beats and the fast dephasing considered in Sects. 2.5, 6. By definition, $E_n - \mathrm{i}\gamma_n$ is a pole of Green's function; hence, to determine G_{nn} it is sufficient to find the residue

$$g_{nn} \equiv \mathrm{Res}\{G_{nn}\}_{E = E_n - \mathrm{i}\gamma_n} \equiv (dG_{nn}^{-1}/dE)^{-1}_{E = E_n - \mathrm{i}\gamma_n} \,.$$

As follows from (3.5.16), it is expressed in terms of the residue of G_{ss} by

$$g_{sn} \equiv \mathrm{Res}\{G_{ss}\}_{E = E_n - \mathrm{i}\gamma_n} = |a_{sn}|^2 g_{nn} \,. \tag{3.5.17}$$

A trivial calculation, using the explicit expression for G_{ss}, gives

$$g_{sn} = \left(1 + \sum_l \frac{|v_{sl}|^2}{[E_n - E_l - \mathrm{i}(\gamma_n - \gamma_l)]^2}\right)^{-1} \simeq |a_{sn}|^2 \,, \tag{3.5.18}$$

where in view of (3.5.5, 15), the widths γ_n and γ_l are neglected as compared to the level separations. Inserting (3.5.18) into (3.5.17) gives $g_{nn} = 1$, and the desired Green's function is

$$G_{nn} = (E - E_n + \mathrm{i}\gamma_n)^{-1} \,. \tag{3.5.19}$$

The luminescence intensity on excitation of the n-th resonant state is calculated in the same way as in Sect. 3.4, and equals to

$$I_n(t) = \tau^{-1}_{\mathrm{rad}(n)} \exp(-t/\tau_{\mathrm{obs}(n)}) \,, \tag{3.5.20}$$

$$\tau_{\mathrm{rad}(n)} = \tau_{\mathrm{rad}(s)}/|a_{sn}|^2, \qquad \tau_{\mathrm{obs}(n)} = \hbar/2\gamma_n \,. \tag{3.5.21}$$

The emission quantum yield (2.1.3) is

$$\varphi_n = \frac{\tau_{\mathrm{obs}\,(n)}}{\tau_{\mathrm{rad}\,(n)}} = \frac{1}{1 + k_{nt}\tau_{\mathrm{rad}\,(n)}}, \tag{3.5.22}$$

where

$$k_{nt} = \frac{2}{\hbar} \sum_l |a_{ln}|^2 \gamma_l \tag{3.5.23}$$

is the rate consant for the radiationless transition from the state Φ_n to the states Φ_l. It may be recast in the form like (3.5.12) by introducing the matrix elements for this transition, $v_{nl} = \langle \Phi_n | V | \Phi_l \rangle$. Inserting expansion (2.6.2) here and remembering that the intramolecular interaction V has nonzero matrix elements between different electronic states only, one obtains $v_{nl} = a_{ns} v_{sl}$. Then, after account is taken of (2.6.5), (3.5.23) becomes

$$k_{nt} = \frac{2}{\hbar} \sum_l \frac{|v_{nl}|^2}{(E_n - E_l)^2} \gamma_l \,. \tag{3.5.24}$$

As is distinct from (3.5.12), no γ_l^2 term is present in the denominator [this term is insignificant now because in the strong-coupling limit the differences $|E_n - E_l|$ are always on the order of the E_l-level separations, the latter exceeding the level widths according to (3.5.5)], and that Φ_n rather than Φ_s is the initial state. As long as γ_l is proportional to pressure in the vapor phase we again arrive at the conclusion that at low pressures the radiationless transition rate in small molecules tends to zero, and the quantum yield to unity.

To summarize, for strong coupling the resonant states are the stationary states of the isolated molecule broadened by their couplings with the field and the medium. The radiative lifetimes of these states, $\tau_{\mathrm{rad}\,(n)}$, are longer than $\tau_{\mathrm{rad}\,(s)}$, i.e., they are longer than the radiative lifetime of the main resonance for weak coupling. The quantity $|a_{sn}|^2$ in (3.5.21) has a simple meaning. Let d be the dipole moment operator, and $f_s \propto |\langle \Phi_0 | d | \Phi_s \rangle|^2 \propto \tau_{\mathrm{rad}\,(s)}^{-1}$ be the oscillator strength for the $E_s \leftarrow E_0$ transition which governs $\tau_{\mathrm{rad}\,(s)}$ through Einstein's relations; the $E_l \leftarrow E_0$ oscillator strengths are assumed to vanish. The $E_n \leftarrow E_0$ oscillator strengths are $f_n \propto |\langle \Phi_0 | d | \Phi_n \rangle|^2 \propto \tau_{\mathrm{rad}\,(n)}^{-1}$. They are related with f_s by equations $f_n = |a_{sn}|^2 f_s$ and $\sum_n f_n = f_s$ which follow from (2.6.2) and from the conditon of completeness of the basis set Φ_n. Now it is seen that the relation between the radiative lifetimes given in (3.5.21) is a consequence of the above relation between the corresponding oscillator strengths. Thus, the lengthening of the resonant states' lifetimes relative to $\tau_{\mathrm{rad}\,(s)}$ is caused by the fact that in the presence of the intramolecular coupling the $E_s \leftarrow E_0$ oscillator strength is distributed over the numerous $E_n \leftarrow E_0$ transitions [1.5, 8, 2.206, 223, 224, 3.8].

Exciting of a particular resonance is possible under conditions $\Delta\omega \ll 1/\varrho_n$ $\tau \ll \tau_{\mathrm{obs}\,(n)}$, where $\Delta\omega$ and τ stand for the total spectral width and the exciting pulse duration, respectively, and $1/\varrho_n$ is a mean separation between the levels E_n

(see Sect. 2.4.1). In this case the decay of the luminescence intensity follows the exponential law (3.5.20).

If the above conditions are not fulfilled, the luminescence displays a more sophisticated behavior. Consider, for example, the situation specified by conditions $\Delta\omega_\tau \ll 1/\varrho_n \ll \Delta\omega_{\rm incoh}$, where $\Delta\omega_\tau$ and $\Delta\omega_{\rm incoh}$ are the coherent and incoherent contributions to $\Delta\omega$. In this case, as discussed at the end of Sect. 2.4.1, an incoherent superposition of the resonant states Φ_n is produced with the initial populations proportional to the product $i_n f_n$, where i_n is the exciting radiation intensity at the position of the n-th resonance. The total emission intensity at time t is then given by the expression

$$I(t) \propto \sum_n i_n f_n I_n(t) , \tag{3.5.25}$$

where $I_n(t)$ is defined in (3.5.20). The situation $1/\varrho_n \ll \Delta\omega_\tau$ leading to quantum beats and/or biexponential decay is unlikely to occur in small molecules because ϱ_n is low (see, however, the next section).

3.5.2 Intermediate Case

Spacings between the molecular resonances decrease with increasing density of final states, ϱ_l, and when they become lower than the total level halfwidths,

$$1/\varrho_l \ll \gamma_l + \gamma_{\rm max} , \tag{3.5.26}$$

the separate resonances merge, i.e., conversion of the resonant states occurs resulting in a pronounced qualitative transformation of the molecular spectrum.

Two qualitatively different situations are possible now. The first one is such that the apparent merging of the resonances can be still removed by changing the external conditions to restore the small molecule behavior. For instance, if the density of states is not so large and the relation $1/\varrho_l \gg \gamma_{\rm max}$ is valid, whereas (3.5.26) is fulfilled only due to a large value of γ_l (condensed matter or high-pressure vapors), then in low-pressure vapors a back transformation must occur to the small molecule case discussed in Sect. 3.5.1 [1.1, 11, 2.178, 179].

Another possibility for such a kind of transformation may arise in connection with variations of the observation time scale. Namely, let the constraint for small molecules (3.5.5) be met and the experimental apparatus enable observations at times t much shorter than the maximum experiment longevity, $\tau_{\rm max}$. The corresponding apparent level halfwidth, $\gamma_t = \hbar/2t$, may then overlap the level gaps thus producing a seemingly join-up of the levels which, however, disappears at longer times. This manifests as the biexponential luminescence decay of Sect. 2.6.3.

The second situation occurs if inequality

$$1/\varrho_l \ll \gamma_{\rm max} \tag{3.5.27}$$

is fulfilled. Then the above conversion of the resonant states is an intrinsic property of the given molecule independent of external factors. Obviously, (3.5.27) is identical with the Freed–Jortner irreversibility criterion (2.4.41) for isolated molecules. Considering this fact we may interpret in simple physical terms the phenomenon of the conversion of resonant states.

The resonant states of a small molecule are the stationary states (the eigenstates) of the isolated molecule broadened due to its interactions with the field and the environment. Therefore, the resonances' widths vanish when no external forces are present. This means that the molecule excited by an external radiation source in a narrow spectral range covering one resonance is prepared in a stationary state and resides in that state infinitely long. Upon allowing for its interaction with the electromagnetic field, the probability to survive in the initial state is degrading in time, the decay rate being equal to the radiative transition rate, and the quantum yield to unity. Such a behavior signals the lack of any relaxation processes in small molecules noninteracting with their environment.

In the limit of (3.5.27) the molecular energy spectrum becomes essentially continuous. Taking into account the large number of the molecular vibrational degrees of freedom (of which a consequence is just the high value of ϱ_l) as well as the strongly varying mode activities with respect to the electronic transition, one can introduce the notions of a quasimolecule involving a comparatively low number of active modes, and of a "medium" embracing the passive modes (see Sect. 2.10). The quasimolecule has the discrete electronic–vibrational levels E_s and E_λ, and the density of final states isoenergetic with the initial state, ϱ_λ, is small as compared to ϱ_l: $\varrho_\lambda \ll \varrho_l$. The medium is characterized by a high density of the vibrational states of the same order of magnitude as ϱ_l, so its energy spectrum is continuous due to condition (3.5.27).

The interactions of the quasimolecule with the passive intramolecular vibrations as well as with the real external medium give rise to a broadening of the levels E_λ, the respective halfwidths being denoted by γ_λ. Obviously,

$$\gamma_\lambda \gg \gamma_l \tag{3.5.28}$$

owing to the presence of an additional (intramolecular) vibrational relaxation mechanism arising as a consequence of the continuity of the spectrum of the passive intramolecular vibrations. As distinct from γ_l, the widths γ_λ do not vanish at low pressures and, as a rule, largely exceed the radiative width,

$$\gamma_\lambda \gg \gamma_s \,. \tag{3.5.29}$$

This reasoning shows that in the case of (3.5.27) the resonant states are associated with the quasimolecule levels broadened because of the interactions with the field, the passive intramolecular vibrations, and the external medium. When all the external interactions are switched off, the widths of these resonances do no longer vanish. This means that the excitation of an individual resonance promotes the isolated molecule into a nonstationary state, and the probability to find the molecule in that state falls off exponentially with a lifetime defined by the width of the resonance. Such a behavior can be classified as a radiationless transition; in other words, on condition (3.5.27) the isolated molecule excited at a given resonance behaves irreversibily, like a statistically large sysem. Thus we are led to the conclusion that the conversion *of the resonant states in the limit of* (3.5.27) *is accompanied by the advent of the irreversibility in the isolated molecule.*

It is also possible to introduce the quasimolecule when the less restrictive condition (3.5.26) is fulfilled since that condition, too, assures the continuity of the spectrum of the passive vibrations and provides a possibility to introduce the level widths γ_λ. Here, the conversion of the resonant states is connected with the fact that the intramolecular vibrational relaxation comes into play when going from the small-molecule limit to the case of (3.5.26).

In the general case (3.5.26) the resonant states are deduced from solutions to (3.5.3). The types of these solutions strongly depend on the relation between γ_λ and $1/\varrho_\lambda$. Here it is supposed that the inequality

$$\gamma_\lambda + \gamma_{\max} \ll 1/\varrho_\lambda \tag{3.5.30}$$

holds, where the left-hand side is the total halfwidth of the level E_λ including the broadening due to the finite duration of the experiment. Then we meet the intermediate case whose definition is given by (2.11.4), where γ is to be understood as $\gamma_\lambda + \gamma_{\max}$ and ϱ_{eff} is the density of the molecular effective states providing the major contributions to the radiationless transition rate and having nearly equal matrix elements for the transitions from the initial state (see Sects. 2.6, 7). The effective states are obviously pertinent to the quasimolecule as it involves, by definition, all the active modes. Hence one has $\varrho_{\text{eff}} \lesssim \varrho_\lambda$. Combining this with (3.5.26, 28, 30) one obtains $1/\varrho_l \ll \gamma_\lambda + \gamma_{\max} \ll 1/\varrho_{\text{eff}}$ which coincides with condition (2.11.4).

Formally, (3.5.3) for the poles of the Green's function is of the same form as (3.5.4) for the small molecules differing from the latter by the level notations only. Comparing (3.5.29, 30) with (3.5.5) and with the definitions in the text following it shows that the intermediate case can be treated as the case of a small quasimolecule in a medium. Therefore the poles of the Green's function are obtained in the same way as in Sect. 3.5.1. Let us summarize the main results for the intermediate case without repeating the corresponding derivations.

For weak coupling, $|v_{s\lambda}| \ll |E_s - E_\lambda + \mathrm{i}\gamma_\lambda|$, there exists one dominant resonance associated with the bright level E_s,

$$E = E_s - \mathrm{i}(\gamma_s + \tfrac{1}{2}\hbar k_{sl}), \tag{3.5.31}$$

where the E_s-level shift and the small term $\gamma_{\max}$ are neglected. This resonance was found in Sect. 3.4 [see (3.4.7–9)]. The intensity of the luminescence excited via the resonant state (3.5.31) decays by the exponential law (3.4.15) with the emission quantum yield (3.4.16) below unity. The radiationless transition rate constant k_{sl} is given by (3.4.9); *it is affected by the intra- and intermolecular vibrational relaxations in a complicated manner.* Other resonances connected with the levels E_λ are weak.

For strong coupling, $|v_{s\lambda}| \gtrsim 1/\varrho_\lambda$, there are a great deal of resonances related to the stationary states of the quasimolecule, E_ν. The complex energies of these resonances are $E_\nu - i\gamma_\nu$, where γ_ν is given by (3.5.14) with changing the notations of $n, l \to \nu, \lambda$. Depending upon the excitation conditions, the emission intensity either decays by the exponential law (3.5.20) with the lengthened lifetime of (3.5.21), or is described by a sum of exponentials (3.5.25); the dephasing, the biexponential behavior, and the quantum beats considered in Sects. 2.6, 7 are possible as well, provided that $\Delta\omega_\tau$ is comparable with the separations between the neighboring resonances.

Thus, by the kind of effects predicted, the intermediate case of large polyatomic molecules is in many respects similar to the small molecule limit. The effect of the lifetime lengthening was experimentally discovered in triatomic molecules NO_2, SO_2, and CS_2 [3.9–13] , and later in naphthalene, 3,4-benzopyrene [3.14] and benzophenone [3.15]. All the experiments were carried out in low-pressure vapors where molecules may be considered as isolated, and more recent supersonic molecular beam results are also available [2.267–271].

In the naphthalene molecule the second excited singlet state S_2 lies higher than the first (S_1) by $\Delta E \approx 3000\ \mathrm{cm}^{-1}$. For such a narrow energy gap the density of the S_1 vibrational sublevels at the origin of the S_2 vibrational spectrum is extremely low, whereas the matrix element of the intramolecular interaction is large since the Franck–Condon factor is large at low ΔE. Additionally, the S_2 oscillator strength for the absorption from the ground state, S_0, is high as compared to the S_1 state, which also favors the observation of the strong-coupling effects in the intermediate case. Using a narrow-band excitation of the lowest vibrational sublevels of the naphthalene S_2 state (in a solid matrix at low temperatures) *Hochstrasser* [3.16] revealed a whole spectrum of sharp absorption lines corresponding to the stationary states E_n originating from mixing of the S_2 sublevels with the vibrationally excited S_1 states. *Lawrance* and *Knight* [2.9] observed a similar $S_2 - S_4$ level mixing in azulene.

The S_2 state of 3,4-benzopyrene is also separated from S_1 by about $3000\ \mathrm{cm}^{-1}$, while benzophenone has a triplet level T_1 lying below S_1 by the same value of $\Delta E \approx 3000\ \mathrm{cm}^{-1}$. Upon excitation of these molecules at S_2 and S_1, respectively, the luminescence intensity decays nonexponentially; in particular, there is evidence for the biexponential behavior and quantum beats in these and some other molecules [2.76, 80, 85, 86, 89, 90, 3.15] while not all such results were confirmed by subsequent studies [2.95, 3.17, 18]. The magnetic field effect on the lifetime and the intensity of the glyoxyal fluorescence in low pressure vapors [3.19] is presumably explained by the weak coupling effect in the intermediate case.

3.5.3 Statistical Limit

Previous treatments [1.6, 14, 3.20] assumed that the molecule level merging in the limit of (3.5.26) immediately produces the statistical limit. In contrast, we regard this merging only as a sign of the intrinsic irreversibility giving rise to a broadening of the quasimolecule levels. Within this approach the statistical limit is characterized by a large density of the final states of the quasimolecule and by the large broadening of its levels,

$$\gamma_\lambda \gg 1/\varrho_\lambda \,. \tag{3.5.32}$$

Whereas the transition from small to intermediate molecules is promoted by a mere increase in the number of atoms, which immediately leads to an increase in the total density of the final states of the molecules, ϱ_l, the transition from the intermediate case to the statistical limit is of a more complicated nature because in the limit of infinitely increasing ϱ_l the density of states of the quasimolecule, ϱ_λ, remains restricted and eventually becomes fully independent of the number of atoms in the molecule being defined by the magnitude of the intramolecular coupling V and the energy gap ΔE between the initial and final electronic states. More specifically, ϱ_λ increases with V and ΔE (see Chap. 6). Then it follows that the intermediate-case–statistical-limit transition is a transition to the case of strong coupling and a large energy gap in molecules with large enough number of atoms.

Increasing ϱ_λ eventually results in merging of the resonances related to the levels of the quasimolecule, so, once more, we face conversion of resonant states. Similarly to the intermediate case, the conversion is caused by the appearance of irreversibility, this time of the irreversibility of the electronic transition *in the isolated quasimolecule*. As a result, a single united resonance is formed, which can be found in the framework of the Bixon–Jortner model with broadened final states discussed in Sect. 2.6.2. Invoking this model is justified by that it adequately describes the statistical limit, where the effective states obeying the main assumptions of the model are of greatest importance.

Within the model mentioned, the broadening is uniform among the levels, $\gamma_\lambda \equiv \gamma$, and all the states of the quasimolecule are effective, i.e., $\varrho_\lambda \equiv \varrho_{\text{eff}}$ and $v_{s\lambda} \equiv v_{\text{eff}}$. Then (3.5.3) for the poles of the Green's function takes the form [cf. (2.6.15)]

$$E = E_s - \mathrm{i}\gamma_s + (\hbar k_s/2)\cot[\pi\varrho_{\text{eff}}(E - E_s + \Delta + \mathrm{i}\gamma)] \,, \tag{3.5.33}$$

where k_s is given by (2.7.21) and the small term $\gamma_{\max}$ is omitted.

Apart from notations, (3.5.33) is identical with (2.6.34, 35), and condition (3.5.32) with (2.11.2). Using the results of Sect. 2.6.2 we obtain the complex energy of the united resonant state [1.1],

$$E = E_s - \mathrm{i}(\gamma_s + \hbar k_s/2) \,, \tag{3.5.34}$$

and the condition for the statistical limit,

$$\gamma \gg \gamma_s + \hbar k_s/2 + (2\pi\varrho_{\text{eff}})^{-1} \,. \tag{3.5.35}$$

This condition means that the broadening of the final states of the quasimolecule must exceed not only the level separations, as was initially assumed in (3.5.32), but also the total (radiative + nonradiative) width of the resonant state. Despite this fact, as discussed in Sect. 2.11, we consider (3.5.32) written in the form of (2.11.2) as the major condition for the statistical limit.

Condition (3.5.35) is easily satisfied for highly excited vibrational sublevels of the final electronic state since they undergo fast intramolecular relaxations. At the same time the limit of $\gamma \to 0$ is also of interest, since it corresponds to the Bixon–Jortner model for an isolated molecule. It was studied in [2.217]. The main result is that the condition for the exponential decay of the probability $P_{ss}(t)$ with the rate constant (2.7.21) can be written in the form

$$\gamma + \hbar k_s/2 \gg 1/\varrho_{\mathrm{eff}} , \tag{3.5.36}$$

which is an extension of (2.11.2) to arbitrary γ's, and in the light of the discussion of Sects. 2.5, 6 can be considered as a generalization of the strong-coupling condition (2.5.5). In particular, at $\gamma = 0$ (3.5.36) reduces to the net strong-coupling condition (2.6.25) in the Bixon–Jortner model for isolated molecules. The obvious distinction between (3.5.35) and (3.5.36) reflects the fact previously mentioned that the conditions for the statistical limit and for the exponential decay are not the same (for details see Sect. 2.11).

In summary, in the statistical limit there exists one dominating resonant state realted to the bright level E_s. When excited at this resonance, the luminescence decreases by the exponential law (3.4.15) with the quantum yield (3.4.16) below unity and with the radiationless transition rate constant k_s of (2.7.21). The rate constant *is independent of the intra- and intermolecular vibrational relaxations.*

On the basis of the results of Sect. 2.8 the radiationless transition rate in the statistical limit can be formally represented in the form of Fermi's golden rule,

$$k_s = \frac{2\pi}{\hbar} \sum_{\lambda} |v_{s\lambda}|^2 \delta(E_s - E_\lambda) . \tag{3.5.37}$$

Indeed, if the summation in (3.5.37) is restricted to the effective states only, then, due to slow variations of the matrix elements assumed in the model, the sum can be replaced by an integral, as in (2.7.37), which again leads to (2.7.21).

3.6 Method of Projection Operators

When solving (3.2.10) for the Green's function we used the zeroth-order Hamiltonian represented by the level diagram of Fig. 3.1, where only isoenergetic excited states of the system are included. Such an approximation corresponds to the first-order time-independent perturbation theory for degenerate levels where the Hamiltonian matrix is diagonalized over the restricted basis set of the wave functions of the degenerate states, thereby ignoring indirect couplings among these states via intermediate states arising in the higher orders of the perturbation expansion.

We consider now a Hamiltonian of the most general form, $H = H_0 + V$, where H_0 is a zeroth-order Hamiltonian that has a good deal of states, which are off-resonant with the initial state Φ_s, in addition to the resonant states indicated in Fig. 3.1; by V in this section we shall mean the sum of all interactions in the system "molecule + medium + field". The problem is to establish the general structure of the Green's function G_{ss} and to obtain a compact equation for its poles which could be solved by an iterative procedure [3.21].

We begin with the following preliminary remark. If the Hamiltonian H has a purely discrete spectrum, it is completely characterized by the energies and wave functions of the stationary states (eigenstates) that are obtained by solving the time-independent Schrödinger equation using the standard perturbation method. On the other hand, if a continuous spectrum is present, of highest interest are resonant (rather than stationary) states that are defined as poles of the Green's function. Therefore the problem formulated above consists in constructing a more general perturbation procedure enabling calculations of complex energies of the resonant states to any order in the interaction V.

The most appropriate method to solve this problem is the method of projection operators. Let us define the operator P for projecting into the subspace L built on the function Φ_s, and the operator Q for projecting into the orthogonal subspace $\tilde{L}$,

$$P = |\Phi_s\rangle\langle\Phi_s|, \qquad Q = 1 - P\,. \tag{3.6.1}$$

They have the obvious properties,

$$P^2 = P, \qquad PQ = QP = 0\,. \tag{3.6.2}$$

According to definition (3.6.1), the following identity holds for any operator T:

$$PTP = T_{ss}P\,, \tag{3.6.3}$$

where T_{ss} is the average value of T over the state Φ_s. Notice as well that P and Q commute with H_0, as Φ_s is an eigenfunction of the zeroth-order Hamiltonian. Therefore, owing to the orthogonality between P and Q, one has

$$QH_0P = PH_0Q = 0\,. \tag{3.6.4}$$

We start with (3.2.10) for the Green's function. It is right-hand-side multiplied by P and recast in the form $(E - H)(P + Q)\,GP = P$. Further, left-hand-side multiplying this equation first by P and then by Q and utilizing (3.6.2–4) one obtains the simultaneous equations for G_{ss} and QGP [1.1, 2.210, 3.22–24],

$$(E - H_{ss})G_{ss}\,P - PVQ(QGP) = P\,, \tag{3.6.5}$$

$$-QVPG_{ss} + Q(E - H)Q(QGP) = 0\,. \tag{3.6.6}$$

We introduce an auxiliary operator $\tilde{G}$ only defined over the subspace $\tilde{L}$ and obeying the equation

$$Q(E - H)Q\tilde{G} = 1 . \tag{3.6.7}$$

Multiplying (3.6.6) by $\tilde{G}$ from the left one obtains

$$QGP = \tilde{G}QVPG_{ss} , \tag{3.6.8}$$

and then, after substituting (3.6.8) into (3.6.5), one finds the Green's function sought for:

$$G_{ss} = [E - E_s - R_{ss}(E)]^{-1} , \tag{3.6.9}$$

where $R_{ss}(E)$ is the matrix element of the operator

$$R = V + VQ\tilde{G}QV . \tag{3.6.10}$$

The poles of the Green's function (3.6.9) are roots of the equation

$$E = E_s + R_{ss}(E) . \tag{3.6.11}$$

Let g stand for the zeroth-order Green's operator,

$$(E - H_0)g = 1 , \tag{3.6.12}$$

and $\tilde{g}$ for the zeroth-order operator related to $\tilde{G}$,

$$Q(E - H_0)Q\tilde{g} = 1 . \tag{3.6.13}$$

The operator g, like H_0, commutes with Q, and therefore, $\tilde{g}$ is identical with the projection of g into the subspace $\tilde{L}$:

$$\tilde{g} = QgQ = gQ , \tag{3.6.14}$$

which is a straightforward consequence of (3.6.12, 13)[2]. Combining (3.6.7, 13) it is easy to obtain, in a standard fashion, the Dyson equation for $\tilde{G}$,

$$\tilde{G} = \tilde{g} + \tilde{g}QVQ\tilde{G} . \tag{3.6.15}$$

Substituting $\tilde{G}$ in the form of (3.6.15) into (3.6.10) and taking (3.6.14) into account one obtains the equation for R [3.21]

$$R = V + VgQR . \tag{3.6.16}$$

Equations (3.6.9, 11, 16) provide the solution to the problem under consideration.

[2] However, $\tilde{G} \neq QGQ$ since generally G does not commute with Q.

In the case of the discrete spectrum the poles of the Green's function coincide with the eigenvalues of the total Hamiltonian H, and therefore (3.6.11) defines the exact energies of the stationary states. If V is a small perturbation, then $R_{ss}(E)$ can be found to any accuracy by solving (3.6.16) with an iterative procedure. Equation (3.6.11) is also solved by successive approximations, with $E = E_s$ as the zeroth-order approximation. The diagonal matrix element of the operator R determines a "shift" of the exact energy E with respect to the unperturbed energy E_s. For this reason R is called *the level-shift operator.*

Let us show in which way the results of Sect. 3.3 are obtained from the above equations. The zeroth-order Hamiltonian is determined by the level scheme of Fig. 3.1, and V in (3.6.16) should be replaced with $V + W_1 + W_2$ [see (3.1.3)]; nonvanishing matrix elements of each term are given in (3.1.4). For the marix elements of R one derives from (3.6.16)

$$R_{ss} = \sum_{\lambda} \frac{v_{s\lambda}}{E - E_\lambda} R_{\lambda s} + \sum_{A} \frac{w_{sA}}{E - E_A} R_{As} , \tag{3.6.17}$$

$$R_{\lambda s} = v_{\lambda s} + \sum_{B} \frac{w_{\lambda B}}{E - E_B} R_{Bs} , \tag{3.6.18}$$

$$R_{Bs} = \sum_{\lambda} \frac{w_{B\lambda}}{E - E_\lambda} R_{\lambda s}, \qquad R_{As} = w_{As} . \tag{3.6.19}$$

It follows from (3.6.18, 19) that the function $x_{\lambda s} \equiv R_{\lambda s}/(E - E_\lambda)$ satisfies (3.3.11) to which an approximate solution is given by (3.3.13). As a result, after simple algebra one comes to the Green's function of (3.3.17).

Now we suppose that some states Φ_m, off-resonant with Φ_s, exist in addition to the resonant states depicted in Fig. 3.1, and that the intramolecular interaction V has nonvanishing matrix elements

$$v_{sm} = \langle \Phi_s | V | \Phi_m \rangle, \qquad v_{m\lambda} = \langle \Phi_m | V | \Phi_\lambda \rangle . \tag{3.6.20}$$

Then the solution to (3.6.16) can be easily obtained by treating (3.6.20) as a perturbation. The contribution of the matrix elements (3.6.20) can be significant when the direct interaction $v_{s\lambda}$ between the initial and final states is weak because of some selection rules.

Formula (3.5.37) for the radiationless transition rate in the statistical limit corresponds to the first-order approximation in v_{sm} and $v_{m\lambda}$. The method developed here enables a derivation of a general expression for the transition rate constant accurate to any order in the intramolecular coupling.

In view of the fact that k_s is independent of the vibrational relaxation in the statistical limit we may ignore the levels E_B in Fig. 3.1, i.e., we may force $W_2 = 0$. Inclusion of the molecule–field interaction W_1 is not difficult, as it only gives rise to the radiative width of the initial state E_s. Indeed, if W_1 is small and only the states E_A being resonant with E_s may be taken into account, inclusion of W_1 always adds to R_{ss} a term representing the radiative shift of the level E_s. This is easily verified by writing out the matrix elements ss and As of the operator equation (3.6.16), as was done above.

So, we consider a fully isolated quasimolecule having the states Φ_s, Φ_λ and Φ_m. In the statistical limit the spectrum of the final states can be regarded as being continuous (see Sect. 3.5.3). Then it can be shown [3.21] that the operator $R(E)$ as a function of E is analytical over the complex plane with a cut along the real axis where the energy levels are situated. To calculate the amplitude $c_{ss}(t)$ by means of the inverse Laplace transformation (3.2.11), one needs the values of this function at the upper side of the cut, i.e., the values of $R(E + \mathrm{i}\eta)$ for real E and $\eta \to +0$.

We shall assume V to be not very large and the poles of the Green's function to be near E_s. This is true on condition that the shift of the level E_s is not accompanied with any significant change in the density of final states isoenergetic to it. Then $R_{ss}(E)$ in the Green's function (3.6.9) can be substituted by $R_{ss}(E_s + \mathrm{i}\eta)$ when calculating the integral of (3.2.11) for $c_{ss}(t)$. The real part of $R_{ss}(E_s + \mathrm{i}\eta)$ determines the E_s-level shift while the imaginary part gives the radiationless transition rate constant

$$k_s = (-2/\hbar)\langle\Phi_s|\mathrm{Im}\{R(E_s + \mathrm{i}\eta)\}|\Phi_s\rangle \,. \tag{3.6.21}$$

For the operator R a relation can be deduced similar to the optical theorem of scattering theory. For real E (3.6.10, 7) give

$$\begin{aligned}\mathrm{Im}\{R(E + \mathrm{i}\eta)\} &= VQ[\tilde{G}(E + \mathrm{i}\eta) - \tilde{G}(E - \mathrm{i}\eta)]QV/2\mathrm{i} \\ &= VQ\tilde{G}(E - \mathrm{i}\eta)[\tilde{G}^{-1}(E - \mathrm{i}\eta) \\ &\quad - \tilde{G}^{-1}(E + \mathrm{i}\eta)]\tilde{G}(E + \mathrm{i}\eta)QV/2\mathrm{i} \\ &= -\eta VQ\tilde{G}(E - \mathrm{i}\eta)\tilde{G}(E + \mathrm{i}\eta)QV \,.\end{aligned}$$

Substituting here $\tilde{G}$ in the form of (3.6.15) and using (3.6.10) again one obtains

$$\mathrm{Im}\{R(E + \mathrm{i}\eta)\} = -\eta R(E - \mathrm{i}\eta)Q\tilde{g}(E - \mathrm{i}\eta)\tilde{g}(E + \mathrm{i}\eta)QR(E + \mathrm{i}\eta) \,.$$

Taking the limit of $\eta \to +0$, one obtains the desired relation

$$\mathrm{Im}\{R(E + \mathrm{i}\eta)\} = -\pi R^+(E + \mathrm{i}\eta)Q\delta(E - H_0)QR(E + \mathrm{i}\eta) \,.$$

Substituting it into (3.6.21) gives

$$k_s = (2\pi/\hbar)\langle\Phi_s|R^+(E_s + \mathrm{i}\eta)Q\delta(E_s - H_0)QR(E_s + \mathrm{i}\eta)|\Phi_s\rangle \,. \tag{3.6.22}$$

According to definition (3.6.1) and the level notations adopted, the operator Q can be expanded in the sum of two terms,

$$Q = \sum_\lambda |\Phi_\lambda\rangle\langle\Phi_\lambda| + \sum_m |\Phi_m\rangle\langle\Phi_m| \,,$$

of which only the first one provides a nonzero contribution to (3.6.22) because the states E_m in the second term are off resonance with E_s and the delta function vanishes. In this way the final result is obtained,

$$k_s = \frac{2\pi}{\hbar} \sum_{\lambda} |R_{s\lambda}(E_s + i\eta)|^2 \delta(E_s - E_\lambda) . \tag{3.6.23}$$

In the lowest order in V (3.6.16) merely gives $R = V$, and expression (3.6.23) reduces to (3.5.37).

4. Matrix Elements of Intramolecular Interactions

In order to calculate the radiationless transition rate, the matrix elements of the intramolecular interactions between the initially prepared state and the final states have to be known. The purpose of this chapter is to give account of the physical approximations and the methods involved in calculations of the matrix elements.

We begin with a definition of the zeroth-order Hamiltonian based on the adiabatic, or Born–Oppenheimer, approximation. A meaningful way to introduce this approximation used in Sect. 4.1 is to deduce, by means of application of the variational principle, a set of coupled equations, where zeroth-order terms and perturbations are separated in an obvious fashion. Section 4.2 provides order-or-magnitude estimates for nonadiabatic coupling terms. A different, while equivalent, way to perform separation of the electronic and nuclear motion, which relies on the diabatic, rather than adiabatic, picture of motion is presented in Sect. 4.3, where a singularity of the nonadiabatic coupling in the vicinity of any real or complex term crossing is demonstrated explicitly within a model involving only two electronic states. Section 4.4 treats in detail various interstate perturbations responsible for nonradiative transitions. Some additional approximations commonly used in treatments of radiationless transitions are discussed in Sects. 4.5 and 4.6. The analysis of the perturbations given in this chapter is completed in Sect. 4.7, where importance of various coupling terms for internal conversion and intersystem crossing is assessed, and the corresponding selection rules associated with these couplings are formulated. The final Sect. 4.8 contains reference formulas for vibronic overlap integrals which enter the matrix element of any interstate perturbation explicitly.

4.1 Adiabatic Approximation

Remind that available initial and final states are fixed by a zeroth-order Hamiltonian H_0, and the total Hamiltonian H is represented as a sum of H_0 and an intramolecular interaction V, that is of interest here. In Sects. 2.3 and 2.4.1 physical arguments were given showing that in polyatomic molecules the adiabatic Hamiltonian should be taken as H_0, while the nonadiabatic and spin–orbit as well as (possibly) Coriolis and other small terms, mixing different adiabatic states, should be included in V. The main goal of this section is an explicit determination of H_0 and V' on the basis of the adiabatic (Born–Oppenheimer) approximation.

There are a lot of monographs and review articles treating in detail both the physical principles underlying the adiabatic approximation and the corresponding mathematical framework (see, e.g., [1.15, 16, 20, 22, 24, 25, 4.1–3]). Our treatment in its main features follows the traditional one. While the adiabatic approximation results in a separation of the electronic and nuclear motions, the ensuing wave functions are of a rather complicated form, and further simplifications are usually introduced depending on the particular problems. Various approximations commonly invoked in the theory of radiationless transitions are to be discussed in the subsequent sections.

The total molecular Hamiltonian has the form

$$H = T_{\mathrm{el}}(x) + T_{\mathrm{nucl}}(q) + U(x, q) + U_{\mathrm{so}}(s, x, q)\,, \tag{4.1.1}$$

where x and $q = \{q_\mu\}$ denote the sets of the electronic and nuclear coordinates, respectively, $T_{\mathrm{el}}(x)$ is the operator of the electrons' kinetic energy,

$$T_{\mathrm{nucl}}(q) = -\sum_{\mu} \frac{\hbar^2}{2M_\mu} \frac{\partial^2}{\partial q_\mu^2} \tag{4.1.2}$$

is the operator of the nuclei's kinetic energy,

$$U(x, q) = U_{\mathrm{el}}(x) + U_{\mathrm{nucl}}(q) + U_{\mathrm{el\text{-}nucl}}(x, q) \tag{4.1.3}$$

is the total potential energy of the molecule, $U_{\mathrm{so}}(s, x, q)$ is the spin–orbit interaction (s are spin variables).

In the Born–Oppenheimer theory [4.1] the energy and the wave function of the molecule are consistently expanded in powers of the small parameter

$$\kappa = (m/M)^{1/4} \tag{4.1.4}$$

involving the ratio of the electronic (m) and nuclear (M) masses, and the adiabatic approximation results from the lowest order of this peculiar perturbation theory, where the operator of the nuclear kinetic energy serves as the perturbation. However, the expansion is quite cumbersome, and usually a more refined method is employed where an exact system of equations identically equivalent to the Schrödinger equation for the molecule is derived, which is very appropriate for developing the adiabatic approximation. Below we follow this approach.

First, consider the physical ideas laid in the foundation of the adiabatic approximation. On the ground of a large difference between the electronic and nuclear masses, the mean velocity of the electrons' motion is expected to greatly exceed that of the nuclei's motion. In view of this fact the notions of the "fast" (electronic) and "slow" (nuclear) subsystems can be introduced. Concerning the fast subsystem, the adiabatic approach assumes that the distribution of the electronic density is governed only by instantaneous values of the nuclear coordinates, being independent of the nuclei's momenta (to be more specific, of their masses); in other words, it is assumed that this distribution can be described by an electronic wave function $\varphi(x, q)$ evaluated at fixed values of the nuclear coordinates, i.e., under complete neglect of the nuclei's motion.

In the framework of the classical description of the motion of the nuclei along a multidimensional trajectory $q(t)$, the fast subsystem moves in a slowly varying potential $U[x, q(t)]$ created by the "slow" subsystem. In this case, the above assumption signifies that the period of the nuclear motion is long as compared to the characteristic times of the motion of the fast subsystem and that the electronic density distribution adiabatically follows the drift of the nuclear coordinates, i.e., at any time t the distribution is controlled only by instantaneous values of the coordinates of the nuclei and is independent of their velocities. Hence, the average distribution of the electron density is bound to be given by

$$n(x) = \overline{|\varphi(x, q)|^2} = \lim_{T \to \infty} \frac{1}{T} \int_0^T |\varphi(x, q(t))|^2 dt \, , \tag{4.1.5}$$

where time averaging is along the classical trajectory of the nuclear motion.

In the quantum description of the slow subsystem the above-said assumption implies that the average distribution of the electron density is expected to have the form

$$\begin{aligned} n(x) &= \overline{|\varphi(x, q)|^2} \\ &= \int |\varphi(x, q)|^2 |\chi(q)|^2 dq, \quad \left(dq = \prod_\mu dq_\mu \right), \end{aligned} \tag{4.1.6}$$

where the first factor in the integrand represents, as above, the electron density distribution at fixed q, while the second one is the density of the quantum probability distribution of the q values.

In order to obtain the mean electronic density in the form of (4.1.5) or (4.1.6), one needs first to determine the wave functions of the fast subsystem $\varphi(x, q)$ at fixed q values. These *adiabatic electronic wave functions* obey the equation

$$[T_{\mathrm{el}(x)} + U(x, q)] \, \varphi_r(x, q) = \mathscr{E}_r(q) \, \varphi_r(x, q) \, , \tag{4.1.7}$$

where the subscript r enumerates eigenstates of the electronic subsystem in the potential $U(x, q)$ created by the fixed nuclei, and $\mathscr{E}_r(q)$ stand for the energies of these states as functions of the nuclear configuration q (*the adiabatic electronic terms* of the molecule).

The totality of the functions $\varphi_r(x, q)$ form a full orthogonal set in the space of the electronic variables x since these functions are eigenfunctions of an Hermitian operator, the Hamiltonian of the electrons in the potential of the immovable nuclei. They can be normalized by the condition

$$\int \varphi_r^*(x, q) \, \varphi_{r'}(x, q) \, dx = \delta_{rr'} \, . \tag{4.1.8}$$

Further, according to (4.1.5, 6), either a trajectory for the classical motion or the wave function of the slow subsystem in the average potential acting from the side of the fast subsystem need to be found. Under such a treatment, of particular interest is a description of *nonadiabatic transitions*, i.e., the transitions between the adiabatic electronic states due to a breakdown of the adiabatic approximation. The classical description of the nuclear motion is commonly used in the theory of nonadiabatic transitions in atomic and molecular collisions [2.238, 4.4, 5]. However, in the case of intramolecular transitions the quantization of the nuclear motion needs be allowed for.

The total molecular wave function $\Psi(x, q)$ thought of as a function of x at fixed q can be represented in the form of the expansion

$$\Psi(x, q) = \sum_r \varphi_r(x, q)\chi_r(q) \tag{4.1.9}$$

with the coefficients $\chi_r(q)$ being functions of q. Following the variational principle [1.40] one has to choose these coefficients such that the first variation of the functional

$$\int \Psi^*(x, q) H \Psi(x, q)\, dx\, dq \tag{4.1.10}$$

vanishes under the additional restriction of

$$\int \Psi^*(x, q) \Psi(x, q)\, dx\, dq = 1 \tag{4.1.11}$$

to fix the normalization of the wave function.

If no special limitations are imposed on the form of $\Psi(x, q)$, the variational principle immediately entails the Schrödinger equation. We shall take the wave function in the form of expansion (4.1.9) and shall vary $\chi_r(q)$. Being that the expansion is quite general and does not narrow the class of the functions to be variates, the ensuing system of equations for $\chi_r(q)$ combined with (4.1.7) for $\varphi_r(x, q)$ is fully identical to the Schrödinger equation for $\Psi(x, q)$.

Applying the Lagrange formalism we write

$$\delta \int \Psi^*(x, q)(H - E)\Psi(x, q) dx\, dq = 0\,, \tag{4.1.12}$$

where E is the Lagrangian multiplier. We substitute here (4.1.1, 9) and perform the integration over dx. Using (4.1.7, 8) we obtain

$$\delta \sum_{rr'} \int \chi_r^*(q) \{[\mathscr{E}_r(q) - E]\delta_{rr'} + T_{rr'}(q) + U_{rr'}(s, q)\} \chi_{r'}(q) dq = 0\,, \tag{4.1.13}$$

where

$$T_{rr'}(q) = \int \varphi_r^*(x, q) T_{\text{nucl}}(q) \varphi_{r'}(x, q) dx \tag{4.1.14}$$

is a differential operator acting on $\chi_{r'}(q)$, and

$$U_{rr'}(s, q) = \int \varphi_r^*(x, q) U_{\text{so}}(s, x, q) \varphi_{r'}(x, q) dx\,. \tag{4.1.15}$$

With the use of (4.1.2) the operator (4.1.14) is represented as

$$T_{rr'}(q) = T_{\text{nucl}}(q)\delta_{rr'} + L_{rr'}(q) + K_{rr'}(q)\,, \tag{4.1.16}$$

$$L_{rr'}(q) = -\sum_{\mu} \frac{\hbar^2}{M_\mu} C^{\mu}_{rr'}(q) \frac{\partial}{\partial q_\mu}\,, \tag{4.1.17}$$

$$C^{\mu}_{rr'}(q) = \int \varphi_r^*(x, q) \frac{\partial \varphi_{r'}(x, q)}{\partial q_\mu} dx\,, \tag{4.1.18}$$

$$K_{rr'}(q) = -\sum_{\mu} \frac{\hbar^2}{2M_\mu} \int \varphi_r^*(x, q) \frac{\partial^2 \varphi_{r'}(x, q)}{\partial q_\mu^2} dx\,. \tag{4.1.19}$$

Operators (4.1.17, 19) acting in the space of the nuclear coordinates can be regarded as matrix elements of certain operators L and K (in the space of x, q) between the electronic wave functions

$$\begin{aligned} L_{rr'}(q) &= \int \varphi_r^*(x, q) L\, \varphi_{r'}(x, q) dx\,, \\ K_{rr'}(q) &= \int \varphi_r^*(x, q) K\, \varphi_{r'}(x, q) dx\,. \end{aligned} \tag{4.1.20}$$

Doing the variation in (4.1.13) gives the following system of equations for the unknown functions $\chi_r(q)$:

$$\begin{aligned} &[T_{\text{nucl}}(q) + \mathscr{E}_r(q) - E]\chi_r(q) = \\ &\quad -\sum_{r'}[L_{rr'}(q) + K_{rr'}(q) + U_{rr'}(s, q)]\chi_{r'}(q)\,. \end{aligned} \tag{4.1.21}$$

This system of equations is exact, and in conjunction with (4.1.7) is equivalent to the Schrödinger equation for the total wave function $\Psi(x, q)$. Just this system will serve us as a starting point to develop the adiabatic approximation.

Within the adiabatic approximation the mean distribution of the electronic density

$$n(x) = \int |\Psi(x, q)|^2 dq \tag{4.1.22}$$

must be given by an integral of the form of (4.1.6). Turning to (4.1.9) one sees that this will be the case provided separate terms in expansion (4.1.9) are already good approximations to some exact wave functions

$$\Psi_r(x, q) \approx \varphi_r(x, q)\chi_r(q)\,, \tag{4.1.23}$$

i.e., the index of the electronic state r is a good quantum number. From (4.1.21) it follows that mixing of different adiabatic electronic states predominantly comes from the nuclear motion (the mixing through $U_{rr'}(s, q)$ might be removed by absorbing, from the outset, the spin–orbit coupling in the electronic Hamiltonian, thereby redefining both the set of functions $\varphi_r(x, q)$ and the operators L and K). Since such a mixing destroys the adiabatic approximation

(4.1.23), the operator $L + K$ nondiagonal in the electronic index is called *the operator of non-adiabaticity*. The adiabatic approximation is valid only under the condition that the operator of non-adiabaticity is a small perturbation.

Assuming this condition is fulfilled, and omitting the right-hand side of (4.1.21) (the spin–orbit coupling is always supposed to be small) one obtains the following equation for $\chi_r(q)$:

$$H_r(q)\chi_r(q) = E\chi_r(q) , \tag{4.1.24}$$

$$H_r(q) = T_{\text{nucl}}(q) + \mathscr{E}_r(q) . \tag{4.1.25}$$

It is seen that $\chi_r(q)$ can be treated as a wave function for the nuclear motion in the electronic state r because (4.1.24) is the Schrödinger equation with Hamiltonian (4.1.25). Such an interpretation is in line with the starting assumption of (4.1.6) since $|\chi_r(q)|^2$ is indeed the probability distribution of the q values.

As follows from (4.1.25), *the electronic term* $\mathscr{E}_r(q)$ plays the role of the nuclear potential energy, which is the reason also to call $\mathscr{E}_r(q)$ *the adiabatic potential*, or *the potential surface*, of the r-th electronic state. The adiabatic potential (and the nuclear wave function, too) depends upon the electronic state. If one substitutes the potential energy of the molecule in the form of (4.1.3) into (4.1.7), premultiplies the equation by $\varphi_r^*(x, q)$, and integrates over dx, then the following expression is obtained for the adiabatic potential:

$$\begin{aligned}\mathscr{E}_r(q) = {} & \int \varphi_r^*(x, q)[T_{\text{el}}(x) + U_{\text{el}}(x)]\, \varphi_r(x, q)\, dx + U_{\text{nucl}}(q) \\ & + \int U_{\text{el nucl}}(x, q)|\varphi_r(x, q)|^2 dx .\end{aligned} \tag{4.1.26}$$

The first term is a sum of the mean kinetic energy of the electrons and the mean energy of interelectronic interaction, the second is the internuclear potential interaction energy, and the third is the mean potential at the point q produced by the distribution of the electronic density at the given q.

If the electronic state r is stable, the adiabatic potential $\mathscr{E}_r(q)$ has a minimum in a particular configuration $q = q_r$ which is the equilibrium configuration of the molecule in the state r. In standard treatments of the nuclear motion the value of the potential energy at equilibrium is taken to be zero. Thus the role of the potential energy in the electronic state r is played by

$$U_r(q) = \mathscr{E}_r(q) - \mathscr{E}_r(q_r) , \tag{4.1.27}$$

and the value of the adiabatic energy at the minimum point

$$\varepsilon_r = \mathscr{E}_r(q_r) \tag{4.1.28}$$

is the purely electronic energy of the given state.

The vibrational[1] energies ε_{rm} and the wave functions $\chi_{rm}(q)$ are found from the Schrödinger equation[2]

$$[T_{\text{nucl}}(q) + U_r(q)]\chi_{rm}(q) = \varepsilon_{rm}\chi_{rm}(q)\,. \tag{4.1.29}$$

Since $U_r(q)$ differs from $\mathscr{E}_r(q)$ by the constant (4.1.28) only, the functions $\chi_{rm}(q)$ also obey (4.1.24), with the corresponding values of the Lagrangian multiplier E being equal to

$$E_{rm} = \varepsilon_r + \varepsilon_{rm}\,. \tag{4.1.30}$$

Hence the Lagrangian multiplier is the total energy of the molecule equal to the sum of the purely electronic and vibronic energies. The total molecular wave function corresponding to the energy E_{rm} is

$$\Psi_{rm}(x, q) = \varphi_r(x, q)\chi_{rm}(q)\,. \tag{4.1.31}$$

The functions $\chi_{rm}(q)$ with a given r and different m's are mutually orthogonal and permit a normalization by the condition

$$\int \chi^*_{rm}(q)\chi_{rm'}(q)\,dq = \delta_{mm'}\,. \tag{4.1.32}$$

Notice that the functions $\chi_{rm}(q)$ and $\chi_{r'm'}(q)$ with $r \neq r'$ are not orthogonal to each other since they are eigenfunctions of different operators.

The main result of the adiabatic approximation is a separation of the electronic and nuclear motions, which manifests in the representation of the total wave function as a product (4.1.31) of the electronic and nuclear components, and the total molecular energy as a sum (4.1.30) of the electronic and vibronic energies. In consequence, the electronic-state label r and the number of the vibrational level m form a full set of quantum numbers to specify the wave function of the molecule.

The essential feature of the adiabatic separation of variables is the dependence of the electronic wave function upon the nuclear coordinates and the vibrational wave function upon the label of the electronic state caused by the strong coupling between the fast and slow subsystems. Therefore, the terms

---

[1] In order to emphasize the fact that the vibrational energy depends upon the electronic state (which is indicated by the subscript r) it is frequently called the *vibronic* energy. Similar meaning is attached to the term *vibronic states*, and *vibronic* (electronic–vibrational) *coupling* is a coupling between the vibronic states caused by the nuclear motion.

[2] The translational motion of the molecule as a whole is supposed to be eliminated by the choice of the coordinate origin at the center of mass. If low-amplitude molecular vibrations around the equilibrium configuration q_r are considered, then one can approximately separate out the rotations, too [1.40], and consider q as the set of the vibrational coordinates only (see Sect. 4.4.3). In the general case of arbitrary displacements from the equilibrium position, q is a full set of the nuclear coordinates of the molecule in the center-of-mass system, while ε_{rm} and $\chi_{rm}(q)$ are energies and wave functions of the mixed vibration–rotation states.

"purely electronic energy" and "purely vibrational energy" should not be interpreted too literally; the energy of each of the subsystems is essentially dependent upon the interaction with the other one, and this dependence vanishes only in the limit of weak interaction.

No matter how accurate the adiabatic approximation is, the functions (4.1.31) comprise a full orthonormal set. Let us write the Schrödinger equation in the matrix form using the basis set (4.1.31). The total wave function is represented as expansion

$$\Psi(x, q) = \sum_{rm} c_{rm}\, \varphi_r(x, q)\, \chi_{rm}(q)\,, \tag{4.1.33}$$

which coincides with that of (4.1.9) if one puts

$$\chi_r(q) = \sum_{m} c_{rm}\, \chi_{rm}(q)\,. \tag{4.1.34}$$

The $\chi_r(q)$'s obey the simultaneous equations (4.1.21). Substituting (4.1.34) into (4.1.21), premultiplying (4.1.21) by $\chi^*_{rm}(q)$, and performing integration over dq one arrives at the following set of equations for the coefficients c_{rm}:

$$(E_{rm} - E)c_{rm} = -\sum_{r'm'} (L_{rmr'm'} + K_{rmr'm'} + U_{rmr'm'})c_{r'm'}\,, \tag{4.1.35}$$

where

$$\begin{aligned} L_{rmr'm'} &= \int \chi^*_{rm}(q)\, L_{rr'}(q)\, \chi_{r'm'}(q)\, dq \\ &= \int \Psi^*_{rm}(x, q)\, L\, \Psi_{r'm'}(x, q)\, dx\, dq\,, \end{aligned} \tag{4.1.36}$$

etc.

At this point each term in the decomposition (2.3.2) of the total molecular Hamiltonian can be given a more precise meaning. Namely, in the basis of (4.1.31) the matrix elements of the zeroth-order Hamiltonian H_0 and the intramolecular interaction V are calculated by the following formulae:

$$\langle rm|H_0|r'm'\rangle = \delta_{rr'} \int \chi^*_{rm}(q)\, H_r(q)\chi_{r'm'}(q)\, dq = E_{rm}\delta_{rr'}\delta_{mm'}\,, \tag{4.1.37}$$

$$\begin{aligned} v_{rmr'm'} &\equiv \langle rm|V|r'm'\rangle \\ &= L_{rmr'm} + K_{rmr'm'} + U_{rmr'm'}\,. \end{aligned} \tag{4.1.38}$$

Thus, H_0 is the vibronic Hamiltonian, whereas the nonadiabatic, spin–orbit, and other possible couplings comprise the perturbation V.

While one can always go over, in a formal way, to the representation used here, actually this only makes sense when the adiabatic approximation applies, i.e., when the system's states of interest are adequately described by the wave functions (4.1.31). Evidently this requires the matrix elements of the nonadiabaticity operator to be small enough. The limits of applicability of the adiabatic approximation are to be discussed in Sect. 4.2.

Several remarks are worth to be made concerning various aspects of the adiabatic approximation. In the absence of the external magnetic field, the electronic wave functions can be chosen as real [1.40]. Then, from (4.1.18, 8) at $r = r'$, one obtains

$$C^{\mu}_{rr}(q) = \frac{1}{2}\frac{\partial}{\partial q_\mu}\int \varphi_r^2(x, q)\,dx \equiv 0\,, \tag{4.1.39}$$

i.e., L has no diagonal matrix elements in the electronic index. As for the operator K, its electronic matrix elements $K_{rr'}(q)$ are merely functions of q, and the diagonal element $K_{rr}(q)$ in (4.1.21) can be absorbed into the adiabatic potential $\mathscr{E}_r(q)$. The nuclear–mass-dependent correction term to the adiabatic potential

$$\delta\mathscr{E}_r(q) = K_{rr}(q) \tag{4.1.40}$$

is often omitted (see, however, [4.6]). Corresponding estimates will be carried out in the next section. As for the off-diagonal elements of the operators L and K, they are just responsible for the violation of the adiabatic approximation.

The off-diagonal element $L_{rr'}(q)$ defined in (4.1.17, 18) can be recast in a different form. We take the derivative of (4.1.7) with respect to q_μ, premultiply it by $\varphi_{r'}(x, q)$ with $r' \neq r$, and integrate over dx. Interchanging r and r' we obtain

$$C^{\mu}_{rr'}(q) = -\frac{1}{\Delta\mathscr{E}}\int \varphi_r(x, q)\frac{\partial U(x, q)}{\partial q_\mu}\,\varphi_{r'}(x, q)\,dx\,, \tag{4.1.41}$$

where

$$\Delta\mathscr{E} = \mathscr{E}_r(q) - \mathscr{E}_{r'}(q) \tag{4.1.42}$$

is the *adiabatic splitting* of the electronic terms. Thus for $r' \neq r$

$$L_{rr'}(q) = \frac{1}{\Delta\mathscr{E}}\sum_\mu \frac{\hbar^2}{M_\mu}\left[\int \varphi_r(x, q)\frac{\partial U(x, q)}{\partial q_\mu}\,\varphi_{r'}(x, q)\,dx\right]\frac{\partial}{\partial q_\mu}\,. \tag{4.1.43}$$

The adiabatic method treated here is based on the decomposition of the molecular system into the fast (electronic) and slow (nuclear) subsystems. For the sake of argument all the electrons of the molecule were assigned to the fast subsystem, and the "bare" nuclei of the comprising atoms to the slow one. However, from the physical point of view it is appropriate to slightly modify the definitions of the fast and slow subsystems considering differences in the binding energies of individual groups of electrons [1.16, 4.7].

Usually, there exists a relatively small group of electrons whose excitation energies are in the optical spectral range; these are the so-called *optical electrons*. The lowest excited electronic states of the molecules, to which in fact the theory of radiationless transitions is directed, correlate with transitions of just the optical electrons into higher molecular orbitals. The binding energies of the rest of electrons are significantly larger so that the state of the strongly-bound electrons is practically not affected by the optical excitation of the molecule; it is all the more not affected by the nuclear motion. For these reasons the molecule can be treated as a system composed of the optical electrons and of the atomic cores having no internal degrees of freedom [4.8].

With such definitions of the fast and slow subsystems, by x and q one should mean the sets of coordinates of the optical electrons and the atomic cores, respectively. The behavior of the internuclear interaction potential $U_{\mathrm{nucl}}(q)$ is also significantly changed. Whereas previously it was purely repulsive, now the appearance of one or several configurations $q_0 = \{q_{0\mu}\}$ is possible where the potential minimizes. The expansion of the potential near a stable configuration up to the second order in displacements along the *normal coordinates* has the form

$$U_{\mathrm{nucl}}(q) = \frac{1}{2}\sum_{\mu} M_\mu \omega_\mu^2 (q_\mu - q_{0\mu})^2 , \tag{4.1.44}$$

where M_μ and ω_μ are effective masses and frequencies of low-amplitude vibrations near the equilibrium positions; the potential origin is placed upon the configuration q_0.

Let us see now which is the shape of the adiabatic potential (4.1.26) in the vicinity of the stable configuration. The spin–orbit contribution is again neglected, and the electron–nucleus interaction potential is expanded to the first order in $q - q_0$,

$$U_{\mathrm{el\text{-}nucl}}(x, q) = U_{\mathrm{el\text{-}nucl}}(x, q_0) + \sum_{\mu} A_\mu(x)(q_\mu - q_{0\mu}) , \tag{4.1.45}$$

where the zeroth-order term $U_{\mathrm{el\text{-}nucl}}(x, q_0)$ and

$$A_\mu(x) = \partial U_{\mathrm{el\text{-}nucl}}(x, q)/\partial q_\mu|_{q=q_0} \tag{4.1.46}$$

are purely electronic operators.

At small displacements the second term in (4.1.45) can be regarded as a small perturbative correction to the Hamiltonian of the optical electrons in the field of the motionless atomic cores fixed at the equilibrium configuration

$$H_{\mathrm{el}}(x) = T_{\mathrm{el}}(x) + U_{\mathrm{el}}(x) + U_{\mathrm{el\text{-}nucl}}(x, q_0) . \tag{4.1.47}$$

Now we suppose that the unperturbed energies and the electronic wave functions are known, i.e., the Schrödinger equation

$$H_{\mathrm{el}}(x)\varphi_r(x, q_0) = \mathscr{E}_r(q_0)\varphi_r(x, q_0) \tag{4.1.48}$$

is solved. Then, in the first order of the perturbation theory (4.1.26) takes the form

$$\mathscr{E}_r(q) = \mathscr{E}_r(q_0) + U_{\mathrm{nucl}}(q) + \sum_{\mu} A_{\mu r}(q_\mu - q_{0\mu}) , \tag{4.1.49}$$

where

$$A_{\mu r} = \int |\varphi_r(x, q_0)|^2 A_\mu(x) dx \tag{4.1.50}$$

is the average value of the derivative (4.1.46) in the r-th electronic state. Inserting (4.1.44) transforms (4.1.49) into

$$\mathscr{E}_r(q) = \varepsilon_r + \frac{1}{2}\sum_\mu M_\mu \omega_\mu^2 (q_\mu - q_{\mu r})^2 , \tag{4.1.51}$$

where $q_{\mu r} = q_{0\mu} - A_{\mu r}/M_\mu \omega_\mu^2$ and $\varepsilon_r = \mathscr{E}_r(q_0) - \sum_\mu (A_{\mu r}^2/2M_\mu \omega_\mu^2)$ are new equilibrium positions of the normal vibrations and the value of the adiabatic potential at the point of its minimum, respectively. Thus, within the approximations adopted, the interaction of the optical electrons with the atomic cores results in displacing the equilibrium configuration and in lowering the electronic energy; both these effects are dependent upon the electronic state.

In the next order of the perturbation theory as well as upon incorporating the quadratic terms in expansion (4.1.45), there appear effects of vibrational frequency changes and of normal coordinates' "mixing" [1.16] (*the Duchinsky effect* [2.239], see Sect. 4.6).

If the configuration q_0 possesses a certain symmetry, the vibrations q_μ are classified as *totally symmetric or non-totally symmetric* according to whether the symmetry of the molecule is lowered or not at $q_\mu \neq q_{\mu 0}$. A general theorem of group theory [1.40] states that for non-totally symmetric vibrations the integral in (4.1.50) disappears identically. Therefore, the equilibrium positions of the non-totally symmetric vibrations do not shift between the electronic states of the molecule. As for the frequency changes and the Duchinsky effect, these always occur for both the totally and non-totally symmetric vibrations.

A simple analysis of the second-order perturbative corrections to the adiabatic potential (4.1.49) shows that the electronic–vibrational interaction "mixes" the vibrations of the same symmetry species only. This is equally true of any other intramolecular interaction. Thus the Duchinsky effect only occurs among the vibrations of the same symmetry [2.239].

4.2 Accuracy of the Adiabatic Approximation

In order to assess the accuracy of the adiabatic approximation one has to estimate the matrix elements of the operators L and K off-diagonal in the electronic index, and to calculate the nonadiabatic corrections to the energies and wave functions (4.1.30, 31). In this analysis an important part is played by the Born–Oppenheimer parameter (4.1.4). Therefore the relation of this parameter to certain physical characteristics of molecules is first to be established.

Let a denote a typical size of the electronic cloud; by the order of magnitude it is the average separation between the nuclei. By the uncertainty principle, the electronic energy $\mathscr{E}$ is

$$\mathscr{E} = \hbar^2/ma^2 , \tag{4.2.1}$$

where m is the electron mass; just such an order of magnitude is typical of the energy gap between the ground and lowest excited electronic states of molecules.

In stable electronic states the nuclei are involved in vibrational motion that can be characterized by a frequency ω. The fundamental assumption for the adiabatic approach about the slowness of the nuclear motion is equivalent to the requirement that the period of the nuclei's vibration $2\pi/\omega$ be long as compared to the typical time scale for the electronic motion $2\pi\hbar/\mathscr{E}$. Hence the adiabaticity condition for the nuclear motion can be represented as inequality

$$\hbar\omega/\mathscr{E} \ll 1 , \tag{4.2.2}$$

which is always well fulfilled for the lowest electronic states.

The small parameter (4.2.2) is expressed in a straightforward manner via the Born–Oppenheimer parameter because the slowness of the nuclear motion is exclusively a consequence of the small electron–nucleus mass ratio. Indeed, the vibrational frequency squared is the ratio of the stiffness of the molecule (the second derivative of the nuclear potential energy) to the nucleus' mass M. The condition for the molecule to be in equilibrium implies the nuclear potential energy to be on the order of the electronic energy $\mathscr{E}$. Considering that the length scale for electronic energy variations is a and using (4.2.1) one obtains

$$\omega^2 = \frac{\mathscr{E}}{Ma^2} = \frac{m}{M}\left(\frac{\mathscr{E}}{\hbar}\right)^2 . \tag{4.2.3}$$

Hence the parameter (4.2.2) is related to κ by

$$\hbar\omega/\mathscr{E} = \kappa^2 . \tag{4.2.4}$$

The parameter κ also defines the ratio of the amplitude of the zero-point nuclear vibrations

$$b = (\hbar/M\omega)^{1/2} \tag{4.2.5}$$

to the interatomic separation a:

$$b/a = (\hbar/M\omega a^2)^{1/2} = (m\mathscr{E}/M\hbar\omega)^{1/2} = \kappa . \tag{4.2.6}$$

The first-order correction to the wave function (4.1.31) obtained by perturbation theory in the non-adiabaticity operator is

$$\delta\Psi_{rm} = \sum_{\substack{r' \neq r \\ m'}} \frac{L_{r'm'rm} + K_{r'm'rm}}{E_{rm} - E_{r'm'}} \Psi_{r'm'} , \tag{4.2.7}$$

while the correction term to the energy (4.1.30) is the diagonal electronic matrix element $K_{rr}(q)$, which produces no coupling in (4.1.21) and consequently does not break down the adiabatic approximation. Being that $K_{rr}(q)$ is simply a function of coordinates, this quantity represents an additional nuclear–mass-dependend term (4.1.40) to the adiabatic potential. The second-order correction to the energy is

$$\delta E_{rm} = \sum_{\substack{r' \neq r \\ m'}} \frac{|L_{r'm'rm} + K_{r'm'rm}|^2}{E_{rm} - E_{r'm'}} . \tag{4.2.8}$$

Let us estimate the matrix elements of L and K. Since only their absolute magnitudes are of interest, the signs in the approximate relations given below are chosen arbitrarily.

Let d stand for the length scale for variations of $\varphi_r(x, q)$ as a function of q (for simplicity, only one coordinate is considered and subscript μ is omitted). Then, in the integrals defining $C_{r'r}(q)$ and $K_{r'r}(q)$, one can use the estimate $\partial^n \varphi_r(x, q)/\partial q^n \approx \varphi(x, q)/d^n$, where $\varphi(x, q)$ is a function localized (at a given q) inside the same domain of the variables x as $\varphi_r(x, q)$, but not orthogonal to the wave functions of other electronic states. The above relation should be understood in an integral sense, viz.,

$$\int \varphi_{r'}(x, q) \frac{\partial^n \varphi_r(x, q)}{\partial q^n} dx \approx \frac{1}{d^n} \int \varphi_{r'}(x, q)\, \varphi(x, q) dx .$$

The overlap integral of the electronic functions on the right-hand side is assumed to be unity because the small overlap renders the nonadiabatic effects negligible. Then (4.1.17–19) result in

$$C_{r'r} \approx \frac{1}{d} \quad (r' \neq r) , \tag{4.2.9}$$

$$L_{r'r} \approx \frac{\hbar^2}{Md} \frac{\partial}{\partial q} \quad (r' \neq r) , \tag{4.2.10}$$

$$K_{r'r'} \approx \frac{\hbar^2}{2Md^2} . \tag{4.2.11}$$

In order to evaluate d we turn to (4.1.41). Considering that the molecular potential energy $U(x, q)$ is on the same order of magnitude as the electronic energy $\mathscr{E}$, and that the length scale for its variations is a, we can perform the following substitution in the integrand: $\partial U(x, q)/\partial q \approx \mathscr{E}/a$, and to force the electronic overlap integral to be unity, as was done above. The result is

$$C_{r'r} \approx \frac{1}{a} \frac{\mathscr{E}}{\Delta\mathscr{E}} \quad (r' \neq r) , \tag{4.2.12}$$

where $\Delta\mathscr{E}$ is the adiabatic splitting of the electronic terms (4.1.42). From

(4.2.9, 12) it follows that

$$d \approx a\Delta\mathscr{E}/\mathscr{E} \ . \tag{4.2.13}$$

As expected, the length scale for variations of the electronic function of the ground electronic state is merely the interatomic separation since the energy separation to the first excited state is $\Delta\mathscr{E} \approx \mathscr{E}$. In contrast, the rate of change of $\varphi_r(x, q)$ versus q is enhanced for the excited states which have $\Delta\mathscr{E} < \mathscr{E}$.

Equations (4.2.11, 13) provide the following estimate for the correction (4.1.40) to the adiabatic potential:

$$\delta\mathscr{E}_r \approx \tfrac{1}{2}\kappa^4\mathscr{E}(\mathscr{E}/\Delta\mathscr{E})^2 \ , \tag{4.2.14}$$

where $\Delta\mathscr{E}$ should be interpreted as the energy separation between the term r and the nearest term r'.

Combining (4.2.10, 11, 13) leads to the following expressions for the off-diagonal matrix elements of interest:

$$L_{r'm'rm} \approx \frac{\hbar^2}{Ma}\left(\frac{\mathscr{E}}{\Delta\mathscr{E}}\right)\int \chi_{r'm'}(q)\,\frac{\partial\chi_{rm}(q)}{\partial q}\,dq \ , \tag{4.2.15}$$

$$K_{r'm'rm} \approx \frac{\hbar^2}{2Ma^2}\left(\frac{\mathscr{E}}{\Delta\mathscr{E}}\right)^2 \int \chi_{r'm'}(q)\,\chi_{rm}(q)\,dq \ . \tag{4.2.16}$$

The nuclear wave functions involved here strongly depend on q, so there always exists a configuration q_* providing the main contribution to these integrals. For instance, if the electronic terms $\mathscr{E}_r(q)$ and $\mathscr{E}_{r'}(q)$ are little displaced relative to one another and the quantum numbers m and m' are low, then the main contribution comes from the domain of the potential minima; but if the number m' is high and $E_{rm} = E_{r'm'}$, the dominant contribution is provided by a point of term intersection located deep in the classically forbidden region on the real axis or in the complex plane (see Chap. 5). In (4.2.15, 16) we took $\Delta\mathscr{E}$ beyond the integral sign at $q = q_*$ assuming $E_{rm} \neq E_{r'm'}$; the $E_{rm} = E_{r'm'}$ case is discussed in Sect. 4.5.

Let b_* be a length scale for the change of $\chi_{rm}(q)$ at q near q_*. Then, in the integrand of (4.2.15) one can put approximately

$$\partial\chi_{rm}(q)/\partial q \approx \chi_{rm}(q)/b_* \ . \tag{4.2.17}$$

Introducing the notation

$$S_{r'm'rm} = \int \chi_{r'm'}(q)\,\chi_{rm}(q)\,dq \tag{4.2.18}$$

for the overlap integral of the vibrational functions and utilizing (4.2.1–6), one obtains the following estimates for the matrix elements (4.2.15, 16):

$$L_{r'm'rm} \approx \kappa^3\mathscr{E}(b/b_*)(\mathscr{E}/\Delta\mathscr{E})S_{r'm'rm} \ , \tag{4.2.19}$$

$$K_{r'm'rm} \approx \tfrac{1}{2}\kappa^4\mathscr{E}(\mathscr{E}/\Delta\mathscr{E})^2 S_{r'm'rm} \ . \tag{4.2.20}$$

The ratio of the off-diagonal matrix elements of the operators K and L is

$$K_{r'm'rm}/L_{r'm'rm} \approx \kappa(b_*/b)(\mathscr{E}/\Delta\mathscr{E})/2 \,. \tag{4.2.21}$$

One can show that always

$$b_* \leq b \,. \tag{4.2.22}$$

For example, if q_* belongs to the classically allowed region, the b_* is merely the de Broglie wavelength

$$b_* = \hbar/(2M\varepsilon_{rm})^{1/2} = b(\hbar\omega/2\varepsilon_{rm})^{1/2} , \tag{4.2.23}$$

and inequality (4.2.22) is satisfied since the vibrational energy ε_{rm} is larger than or equal to the energy of the zero-point vibrations. And, if q_* is located deep in the classically forbidden region (see Chap. 5), b_* is still lower. Indeed, far away from the turning point, i.e., at $|q| \approx |q_*| \gg (2\varepsilon_{rm}/M\omega^2)^{1/2}$, the wave function decays exponentially, $\chi_{rm}(q) \propto \exp(-q^2/2b^2)$, and from definition (4.2.17) one obtains

$$b_* = b^2/|q_*| \ll b(\hbar\omega/2\varepsilon_{rm})^{1/2} \,. \tag{4.2.24}$$

From (4.2.21, 22) it follows that the matrix elements $K_{r'm'rm}$ can be neglected, provided that the electronic term separation is not too small. Notice, however, that the remaining operator L is non-Hermitian (only the full non-adiabaticity operator $L + K$ is Hermitian, see Sect. 4.4.1). Therefore, in calculations of physical quantities under a neglect of the operator K, the skew-Hermitian part of L must be dropped, too, since it is evidently on the same order of magnitude as the operator K.

Now we can estimate the correction terms (4.2.7, 8) to the wave function and the energy of the rm state. We restrict ourselves to the case where the quantum number m is small and only one electronic state contributes to the sums over r'. The properties of the overlap integral (4.2.18) are to be discussed in Sect. 4.8 and in Chap. 5. Here, we only use the fact that at small displacements and shape changes of the electronic terms $\mathscr{E}_r(q)$ and $\mathscr{E}_{r'} q)$, inherent to polyatomic molecules, the overlap integral is near unity for $m' \approx m$ and severely falls off with increasing $|m' - m|$. Therefore, we can take

$$\delta\Psi_{rm} \approx (L_{r'm'rm}/\Delta\mathscr{E})\Psi_{r'm'} , \tag{4.2.25}$$

$$\delta E_{rm} \approx |L_{r'm'rm}|^2/\Delta\mathscr{E} , \tag{4.2.26}$$

where the vibrational state m' is in close vicinity of the minimum of $\mathscr{E}_{r'}(q)$ and $E_{rm} - E_{r'm'} \approx \Delta\mathscr{E}$. Since both vibrational functions $\chi_{rm}(q)$ and $\chi_{r'm'}(q)$ are localized within domains with the size of about b, and the domains are not largely shifted with respect to one another, one has $S_{r'm'rm} \approx 1$, $b_* \approx b$, and

$$L_{r'm'rm} \approx \kappa^3\mathscr{E}(\mathscr{E}/\Delta\mathscr{E}) \,. \tag{4.2.27}$$

The substitution of these equations into (4.2.25, 26) gives the final result

$$\delta\Psi_{rm} \approx \kappa^3(\mathcal{E}/\Delta\mathcal{E})^2\Psi_{r'm'}\,, \tag{4.2.28}$$

$$\delta E_{rm} \approx \kappa^6\mathcal{E}(\mathcal{E}/\Delta\mathcal{E})^3\,. \tag{4.2.29}$$

Thus, if $\Delta E \approx \mathcal{E}$, the corrections to the wave function, the adiabatic potential [see (4.2.14)] and the energy are on the order of κ^3, κ^4, and κ^6, respectively [1.22].

Let us perform some numerical estimates. Assume $a = 1$ Å, $\kappa = 0.1$ and $\Delta\mathcal{E} = \mathcal{E}/2$. From (4.2.1, 14, 29) we obtain $\mathcal{E} \approx 7\,\text{eV} \approx 6 \times 10^4\text{cm}^{-1}$, $\delta\mathcal{E}_r \approx 10\,\text{cm}^{-1}$ and $\delta E_{rm} \approx 0.5\,\text{cm}^{-1}$. In (4.2.19) the factor in front of the overlap integral,

$$L_{r'r} \approx \kappa^3\mathcal{E}(b/b_*)(\mathcal{E}/\Delta\mathcal{E})\,, \tag{4.2.30}$$

is conventionally called *the electronic matrix element of the non-adiabaticity operator*. It is estimated as

$$L_{r'r} \approx 100(b/b_*)\,\text{cm}^{-1}\,. \tag{4.2.31}$$

This factor can be on the order of the vibrational quantum. Numerical calculations of this quantity are carried out in [1.23, 4.9–11]. Equations (4.2.23, 24, 31) show that the electronic factor is enhanced when moving q_* away from the classical region, which occurs for increasing values of m' (see Chap. 5).

Obviously, the accuracy of the adiabatic approximation is higher, the larger the spacing between the electronic terms. As the term separations steeply decrease with increasing electronic excitation energy, it is evident that the nonadiabatic contributions to the energy and to the wave function of the nondegenerate ground state are smaller than for the excited states that always exhibit the effects of degeneration or quasidegeneration.

In the presence of the electronic (quasi)degeneration in certain nuclear configurations q, the gaps between some electronic terms $\mathcal{E}_r(q)$ and $\mathcal{E}_{r'}(q)$ may be negligibly small, and the nuclear motion in the vicinity of such configurations is no longer adiabatic. In fact, at small $\Delta\mathcal{E}$ the electronic subsystem has the long characteristic period $2\pi\hbar/\Delta\mathcal{E}$, and the adiabaticity condition takes the form

$$\hbar\omega/\Delta\mathcal{E} \ll 1 \tag{4.2.32}$$

instead of (4.2.2). If this condition is not fulfilled, the electronic states r and r' are strongly mixed by the nuclear motion. The exception is the case of crossing of terms belonging to different symmetry species, in which case $L_{rr'}(q) \equiv K_{rr'}(q) \equiv 0$ since the integrals over the electronic coordinates in (4.1.18, 19) vanish identically. To avoid confusion, we emphasize that the term symmetry is defined with regard to those nonunit symmetry elements that survive at *arbitrary* q's. Only di- and triatomic molecules have such symmetry elements. The symmetry group of a molecule with a larger number of atoms contains the unit element alone under arbitrary displacements of its atoms so that all terms have the same (full) symmetry.

In the presence of the crossing of terms having the same symmetry species in molecules with the number of atoms larger than two[3] the electronic matrix elements of the operators L and K do not disappear; moreover they are singular functions of q in the vicinity of the term crossing [1.40]. Consider, for instance, $L_{rr'}(q)$. According to (4.1.43) this matrix element is singular, provided that the matrix elements of the potential energy derivatives over the nuclear coordinates do not disappear all simultaneously,

$$\lim_{q \to q_*} \int \varphi_r(x, q) \frac{\partial U(x, q)}{\partial q_\mu} \varphi_{r'}(x, q) dx \neq 0 , \tag{4.2.33}$$

where q_* is any term-crossing point (by a general theorem [1.40], the points q_* fill up a hypersurface of $s - 2$ dimensions, where s is the number of the vibrational degrees of freedom). If this degeneration is accidental, i.e., not caused by an appearance of additional molecular symmetry elements at the configuration q_*, then there is no reason for the matrix elements (4.2.33) to vanish. But, if q_* is a symmetric configuration, some of them disappear, namely, the matrix elements of the derivatives with respect to the totally symmetric coordinates, i.e., the coordinates whose variations do not lower the molecular symmetry of the configuration q_*. Nevertheless, the Jahn–Teller theorem [1.40, 4.3] states that nonzero elements are bound to exist[4] (these are the derivatives over non-totally symmetric coordinates lifting the degeneracy) so that in this case, too, $L_{rr'}(q)$ will be singular at $q = q_*$. Turning to the definition (4.1.17, 18) we see that singularities of $L_{rr'}(q)$ are associated with those of the electronic functions in the crossing region (see Sect. 4.3). It is therefore clear from (4.1.19) that in this region $K_{rr'}(q)$ will be singular as well.

Therefore, in the region near a term crossing the non-adiabaticity operator cannot be regarded as a small perturbation and it is impossible to adiabatically separate the electronic and nuclear motions as in (4.1.30, 31). Approximate molecular wave functions have to be searched in the form of (4.1.9), where the sum is over all the electronic states close in energy, and the coefficients $\chi_r(q)$ obey the set of simultaneous equations (4.1.21), which just define the molecular energy levels in the domain of the term crossing. It is seen that in the presence of electronic degeneration an essential change of the molecular energy spectrum occurs at energies near the term crossing. This change is caused by the Jahn–Teller effect, i.e., by the instability of symmetric molecular configurations

---

3 In diatomic molecules the terms of the same symmetry species do not cross [1.40].

4 An exception is a linear molecule where the Renner effect can exist [4.12]. The argument given in [4.13] that the nonadiabatic coupling is nonsingular at the term crossing is based on an application of L'Hôpital's rule to (4.1.41), thus incorrectly implying the matrix elements in (4.2.33) to vanish all simultaneously.

with electronic degeneration. The influence of the Jahn–Teller effect on radiationless transitions was studied in [1.2, 15, 4.14, 15]. In this monograph only nondegenerate electronic states are discussed.

4.3 Crude Adiabatic Approximation

We consider again the set of exact equations (4.1.21) neglecting the spin–orbit interaction. Then the non-adiabaticity operator is the only operator mixing different *adiabatic states* (4.1.31). This *nonadiabatic coupling* is of dynamic nature, as it arises due to the effect of nuclear motion on the electronic wave functions.

In the crude adiabatic approximation, by means of the basis set, one takes the so–called *diabatic states*, which are mixed by *the adiabatic coupling*, whereas the nonadiabatic coupling between them vanishes. The adiabatic coupling is purely static having no relation to the nuclear motion.

Let q_0 be a fixed nuclear configuration. The adiabatic electronic functions $\varphi_r(x, q)$ at $q = q_0$ given by solutions to (4.1.48) form a full orthonormal set in the space of the electronic variables x so that they can be employed as a basis to develop the total wave function. The electronic states described by the wave functions $\varphi_r(x, q_0)$, *independent of the nuclear coordinates*, are called diabatic states. Thus, instead of (4.1.9), we let

$$\Psi(x, q) = \sum_r \varphi_r(x, q_0)\,\tilde{\chi}_r(q)\,, \tag{4.3.1}$$

where $\tilde{\chi}_r(q)$ are coefficients to be determined. With this choice of the basis the nuclear kinetic energy operator does not act on the electronic wave functions any more, and the nonadiabatic coupling disappears.

Calling on the variational method, as in Sect. 4.1, we arrive at the following set of simultaneous equations for the coefficients $\tilde{\chi}_r(q)$:

$$[T_{\text{nucl}} + \tilde{\mathscr{E}}_r(q) - E]\,\tilde{\chi}_r(q) = -\sum_{r' \neq r} \tilde{L}_{rr'}(q)\,\tilde{\chi}_{r'}(q)\,, \tag{4.3.2}$$

where $\tilde{\mathscr{E}}_r(q) = \mathscr{E}_r(q_0) + \tilde{L}_{rr}(q)$ is *the diabatic potential*, and the electronic matrix elements of the operator $\tilde{L}$ are given by

$$\tilde{L}_{rr'}(q) = \int \varphi_r(x, q_0)\,[U(x, q) - U(x, q_0)]\,\varphi_{r'}(x, q_0)\,dx\,. \tag{4.3.3}$$

Thus, the coupling between the diabatic states is generated by the electrostatic electron–nucleus interaction, and hence, is purely adiabatic, not connected with the nuclear motion. Equation (4.3.3) shows that this coupling will be weak at a small nuclear displacement $q - q_0$ only. Emphasizing that the set of equations (4.3.2) in the diabatic basis is fully equivalent to the system (4.1.21) in the adiabatic basis, a non-equivalence of these two basis sets only emerges when using the perturbation approach, and manifests in different splittings of the total Hamiltonian into a zeroth-order one and a perturbation.

The crude adiabatic approximation is widely used in the theory of atomic and molecular collisions in which, by means of q_0, a configuration is taken where the colliding species are remote from each other, and the difference $U(x, q) - U(x, q_0)$ is small nearly everywhere along the trajectory of their motion [4.4]. The use of this approximation in the theory of molecular radiationless transitions [4.16] has no solid foundation because the above difference is not small when q_0, as is the common practice, is chosen in the classically accessible region; this is due to the fact that the tunneling radiationless transitions are dominated by the configuration q_* deep in the classically forbidden region (or even in the complex region, see Chap. 5). The crude adiabatic approximation is discussed from different points of view in reviews [1.1–4, 10, 30].

In as much as the diabatic functions comprise a full set in the space of the electronic variables, the adiabatic functions can be developed in a series over this set, and since the diabatic functions are independent of the nuclear coordinates, the whole nuclear–coordinate dependence will be absorbed by the coefficients of the aforementioned development.

According to the results of Sect. 4.2, the adiabatic functions, and hence the coefficients of their development over the diabatic ones, are singular in the region where the electronic terms of the same symmetry species cross. Another proof to this fact will be given here in frames of a two-state model, with account being taken of not only real but also complex crossings. It is essentially similar to the argument given in the book of *Landau* and *Lifshitz* [1.40, Chap. 7], where the nuclei were supposed to move along a classical trajectory $q(t)$; thereby the electronic terms were functions of time t, and the question was about singularities of the adiabatic functions and the electronic terms at the "moment" of the term crossing, $t = t_*$.

Let $\psi_1(x)$ and $\psi_2(x)$ stand for real diabatic wave functions of two electronic states (they are assumed to be orthogonal and normalized to unity), and $h_{ik}(q)$ for the matrix elements, in the diabatic basis, of the Hamiltonian (4.1.1) minus the nuclear kinetic energy (and the spin–orbit interaction)

$$h_{ik}(q) = \int \psi_i(x)[T_{el}(x) + U(x, q)]\psi_k(x)dx \, . \qquad (4.3.4)$$

According to Sect. 4.1, one has to diagonalize the matrix $h_{ik}(q)$ in order to obtain the adiabatic functions $\varphi_1(x, q)$ and $\varphi_2(x, q)$ (the matrix is nondiagonal since terms of the same symmetry are considered). The transformation from the diabatic basis $\{\psi_1, \psi_2\}$ to the adiabatic one $\{\varphi_1, \varphi_2\}$ maintaining the orthonormality of the functions is a transformation of rotation by an angle α; with the arguments omitted, this transformation is written as

$$\varphi_1 = \psi_1 \cos\alpha + \psi_2 \sin\alpha, \qquad \varphi_2 = -\psi_1 \sin\alpha + \psi_2 \cos\alpha \, , \qquad (4.3.5)$$

where α is a function of the nuclear coordinates. The diagonalization of the matrix h_{ik} produces the adiabatic electronic terms

$$\mathscr{E}_{1,2} = \tfrac{1}{2}(h_{11} + h_{22}) \pm \tfrac{1}{2}[(h_{11} - h_{22})^2 + 4h_{12}^2]^{1/2} \equiv \bar{\mathscr{E}} \pm \Delta\mathscr{E}/2 \tag{4.3.6}$$

and the rotation angle

$$\alpha = \tfrac{1}{2}\arctan[2h_{12}/(h_{11} - h_{22})] \,. \tag{4.3.7}$$

We are interested in the adiabatic-term crossing points q_* that are zeros of the adiabatic splitting, i.e., roots of equation $\Delta\mathscr{E} = 0$. Assuming that q_* is a first-order zero of the function $\Delta\mathscr{E}^2$ we see that the term-crossing point is a branch point for these terms.

In the case of two or more nuclear coordinates there exist real crossing points [1.40] defined by the pair of equations $h_{12} = 0$ and $h_{11} = h_{22}$. The limiting value of α at $q \to q_*$ is also real, and is in the interval from 0 to $\pi/4$, but it depends upon the way in which q approaches q_*. Such a kind of singularity is well known in the theory of the Jahn–Teller effect [1.15, 4.3, 12].

In the case of a single nuclear coordinate there are no real intersection points because the above two equations cannot be satisfied simultaneously by varying only one real parameter q. There exist, however, complex intersection points, and just these points govern the transition probabilities between the adiabatic states in the case under consideration. They are found from equation

$$h_{11} - h_{22} = \pm 2\mathrm{i}h_{12} \tag{4.3.8}$$

From (4.3.7, 8) it is seen that $\alpha \to \pm\mathrm{i}\infty$ at $q \to q_*$ so that the adiabatic functions (4.3.5) are again singular at the crossing point. We now define the new functions

$$\psi_\pm = (\psi_1 \pm \mathrm{i}\psi_2)/\sqrt{2} \,. \tag{4.3.9}$$

They possess the following properties

$$\int \psi_\pm^2 \, dx = 0, \qquad \int \psi_+ \psi_- \, dx = 1 \tag{4.3.10}$$

(cf. [1.40, Chap. 7]). For the adiabatic functions we have from (4.3.5, 9)

$$\begin{aligned} \varphi_1 &= \frac{1}{\sqrt{2}} \mathrm{e}^{-\mathrm{i}\alpha}(\psi_+ + \psi_- \mathrm{e}^{2\mathrm{i}\alpha}) \,, \\ \varphi_2 &= \frac{1}{\mathrm{i}\sqrt{2}} \mathrm{e}^{-\mathrm{i}\alpha}(\psi_+ - \psi_- \mathrm{e}^{2\mathrm{i}\alpha}) \,. \end{aligned} \tag{4.3.11}$$

Choosing the minus sign on the right-hand side of (4.3.8) we derive from (4.3.9)

$$\mathrm{e}^{2\mathrm{i}\alpha} \sim (\Delta\mathscr{E}^2)^{1/2} \sim (q - q_*)^{1/2} \,. \tag{4.3.12}$$

i.e., the coefficients of ψ_- in (4.3.11) are small as compared to those of ψ_+ at $q \to q_*$. However, this does not signify that the ψ_- contributions can be neglected in the vicinity of q_*. Indeed, consider, using the basis of (4.3.11), the electronic matrix elements of an arbitrary operator A acting on the nuclear variable only. In view of (4.3.10), the contribution of the largest term ψ_+^2 in the matrix element is identically zero and only the cross terms $\psi_+\psi_-$ survive. Thus, for the off-diagonal element one has

$$\int \varphi_1 A \varphi_2 dx = \frac{1}{2\mathrm{i}}(\mathrm{e}^{\mathrm{i}\alpha} A e^{-\mathrm{i}\alpha} - \mathrm{e}^{-\mathrm{i}\alpha} A e^{\mathrm{i}\alpha}) . \tag{4.3.13}$$

In particular, for the operators of nonadiabatic coupling $A = \partial/\partial q$ and $\partial^2/\partial q^2$, one obtains from (4.3.12, 13)

$$\begin{aligned} \int \varphi_1 \frac{\partial \varphi_2}{\partial q} dx &= -\frac{\partial \alpha}{\partial q} = \frac{\mathrm{i}}{4(q - q_*)} , \\ \int \varphi_1 \frac{\partial^2 \varphi_2}{\partial q^2} dx &= -\frac{\partial^2 \alpha}{\partial q^2} = \frac{-\mathrm{i}}{4(q - q_*)^2} . \end{aligned} \tag{4.3.14}$$

Thus the electronic matrix element of the nonadiabatic coupling has a pole at the complex intersection point of the adiabatic terms. It is interesting to note that its behavior near the pole is independent of any molecular parameters.

4.4 Coupling Operators

4.4.1 Nonadiabatic Coupling

The definitions (4.1.17, 18, 36) provide the following expression for the matrix elements of the operator L between the adiabatic states (4.1.31):

$$L_{rmr'm'} = -\sum_\mu \frac{\hbar^2}{M_\mu} \int C^\mu_{rr'}(q) \chi_{rm}(q) \frac{\partial \chi_{r'm'}(q)}{\partial q_\mu} dq , \tag{4.4.1}$$

where the wave functions are all selected real. The full non-adiabaticity operator is the sum $L + K$, but the estimates made in Sect. 4.2 show that the matrix elements of K are by a factor of κ as small as those of L. Therefore, we shall neglect the operator K and call L the non-adiabaticity operator.

Strictly speaking, L is a non-Hermitian operator because the real matrix element (4.4.1) is not equal to the corresponding element of the transposed operator,

$$L_{r'm'rm} = -\sum_\mu \frac{\hbar^2}{M_\mu} \int C^\mu_{r'r}(q) \chi_{r'm'}(q) \frac{\partial \chi_{rm}(q)}{\partial q_\mu} dq . \tag{4.4.2}$$

Indeed, this can be recast with the use of the identity

$$C^{\mu}_{r'r}(q) \equiv - C^{\mu}_{rr'}(q) , \tag{4.4.3}$$

which is derived by differentiating (4.1.8) and taking into account that the electronic functions are real. Insertion of (4.4.3) into (4.4.2) and integration by parts gives the relation

$$\begin{aligned} L_{r'm'rm} &= -\sum_{\mu} \frac{\hbar^2}{M_\mu} \int \chi_{rm}(q) \frac{\partial}{\partial q_\mu} [C^{\mu}_{rr'}(q) \chi_{r'm'}(q)] dq \\ &= L_{rmr'm'} - \sum_{\mu} \frac{\hbar^2}{M_\mu} \int \chi_{rm}(q) \chi_{r'm'}(q) \frac{\partial}{\partial q_\mu} C^{\mu}_{rr'}(q) dq , \end{aligned} \tag{4.4.4}$$

demonstrating that the operator L is non-Hermitian.

Remind, however, that the additional smallness of the neglected operator K, as compared to L, arises because of the double differentiation of the electronic wave function that is a relatively slowly varying function of q, as compared to $\chi_{rm}(q)$. The second term on the right-hand side of (4.4.4) also involves the two-fold differentiation of the electronic functions. Hence, the difference $L_{r'm'rm} - L_{rmr'm'}$ is on the same order of magnitude as $K_{rmr'm'}$, and must also be omitted. Thus, in the approximation adopted (up to the order of κ^3, inclusive) the operator L may be regarded as Hermitian.

The non-adiabaticity operator has nonzero matrix elements for transitions between states of equal multiplicity only since it does not act on spin variables. It plays the major part in the processes of internal conversion. Hence, *the internal conversion is a nonadiabatic transition between states with the same total spin* (see Sect. 4.7). Its matrix elements were estimated numerically in Sect. 4.2.

4.4.2 Spin–Orbit Coupling

Intersystem crossing, i.e., a transition between states of different multiplicity is promoted by the spin–orbit coupling. The operator of the spin–orbit interaction is

$$U_{\text{so}}(s, x, q) = \sum_a \boldsymbol{B}(\boldsymbol{x}_a, x', q) \cdot \boldsymbol{s}_a , \tag{4.4.5}$$

$$\boldsymbol{B}(\boldsymbol{x}_a, x', q) = -\frac{e\hbar}{2m^2c^2} \boldsymbol{E}(\boldsymbol{x}_a, x', q) \times \boldsymbol{p}_a , \tag{4.4.6}$$

where $\boldsymbol{x}_a$, $\boldsymbol{p}_a$ and $\boldsymbol{s}_a$ are the radius-vector, the operators of momentum and spin of the a-th electron (spin is in units of $\hbar$), respectively, $\boldsymbol{E}(\boldsymbol{x}_a, x', q)$ is the electric field acting on the a-th electron from the side of the nuclei and other electrons, x' the set of coordinates of other electrons, q the set of nuclear coordinates to be regarded as fixed parameters in discussions of purely electronic matrix elements; the summation is over all electrons.

The general expression for the spin–orbit operator is greatly simplified in frames of the Hartree–Fock self-consistent field approximation [1.40, 4.17] if in the electronic Hamiltonian the true potential $U(\boldsymbol{x}_a, x', q)$ felt by the a-th electron is approximated by a mean potential

$$U(\boldsymbol{x}_a, q) = \langle U(\boldsymbol{x}_a, x', q) \rangle_{x'} , \tag{4.4.7}$$

where averaging is over the distribution of other electrons in a given electronic state assuming this distribution to be insensitive to the location of the given electron $\boldsymbol{x}_a$. The true field of (4.4.6) is correspondingly replaced with an average field $\boldsymbol{E}(\boldsymbol{x}_a, q)$ and the many-electron operator (4.4.5) converts into the sum of one–electron operators,

$$U_{\text{so}} \equiv U_{\text{so}}(s, x, q) = \sum_a \boldsymbol{b}_a \cdot \boldsymbol{s}_a , \tag{4.4.8}$$

$$\boldsymbol{b}_a = \langle \boldsymbol{B}(\boldsymbol{x}_a, x', q) \rangle_{x'} , \tag{4.4.9}$$

each acting on the coordinates of its own electron only.

In consequence the Hamiltonian of electrons at fixed nuclei is represented as a sum of one-electron Hamiltonians, and the wave function as a sum of the Slater determinants of one-electron functions $\psi_p(\boldsymbol{x}_a)|\sigma_a\rangle$, where p labels one-electron states (orbitals), $\psi_p(\boldsymbol{x}_a)$ is a coordinate factor (for brevity its dependence on q is not indicated explicitly), and $|\sigma_a\rangle$ a spin factor ($\sigma_a = \pm 1/2$).

In the ground state S_0 the bottom orbitals are all filled (they bear two electrons each) while the top ones are empty. Let $p = 1$ correspond to the highest filled orbital and $p = 2$ to the lowest empty one; we shall suppose that no orbital degeneration occurs. The lowest excited S_1 and T_1 states on the Jabłoński diagram (see Fig. 2.1) correspond to the lowest-energy one-electron excitation, i.e., to the transition of one electron from orbital 1 to orbital 2. For calculations of the matrix elements of the perturbation (4.4.8) between the states S_0, S_1 and T_1 only, it is sufficient to treat a system of two electrons $a = 1$ and 2, in two orbitals $p = 1$ and 2, since the remaining orbitals contribute nothing to these matrix elements. In this way the spin–orbit operator condenses to a sum of two terms

$$U_{\text{so}}(s, x, q) = \boldsymbol{b}_1 \cdot \boldsymbol{s}_1 + \boldsymbol{b}_2 \cdot \boldsymbol{s}_2 . \tag{4.4.10}$$

To proceed further we use symmetry properties of the wave functions of the states under consideration. According to general theorems of quantum mechanics [1.40, 4.17] the wave function of a two–electron system with definite values of the total spin S and its projection σ onto a quantization axis factors into the product of a spin part $\varphi_{S\sigma} = |S\sigma\rangle$ and a coordinate part $\varphi_r(x, q) = |r\rangle (r = S_0, S_1, T_1)$, the latter being symmetric in the singlet states with respect to permutation of the two electrons and antisymmetric in the triplet state (the former vice versa). In consequence, the operator (4.4.10) is fully specified by three (rather than six) real vector constants (their dependence on q is temporarily disregarded),

$$c_{rr'} \equiv \langle r|\boldsymbol{b}_1|r'\rangle = (-1)^{S+1}\langle r|\boldsymbol{b}_2|r'\rangle, \qquad r' = T_1 , \tag{4.4.11}$$

one for each of three orbitally nondegenerate electronic states $r = S_0, S_1$ and T_1 with the total spin $S = 0$ or 1 (in the presence of electronic degeneration the number of these constants naturally grows higher). The electronic matrix elements of this operator defined in (4.1.15) are expressed at $r' = T_1$ via the above constants as

$$U_{r'r'}(s, q) \equiv U_{T_1T_1}(s, q) = \boldsymbol{c}_{T_1T_1} \cdot \boldsymbol{S} , \tag{4.4.12}$$

$$U_{rr'}(s, q) \equiv U_{rT_1}(s, q) = \boldsymbol{c}_{rT_1} \cdot \boldsymbol{G}, \qquad r = S_0, S_1 , \tag{4.4.13}$$

$$\boldsymbol{S} = \boldsymbol{s}_1 + \boldsymbol{s}_2, \qquad \boldsymbol{G} = \boldsymbol{s}_1 - \boldsymbol{s}_2 . \tag{4.4.14}$$

The constant $\boldsymbol{c}_{T_1T_1}$ contributes (usually insignificantly as compared to the spin–spin interaction) to the fine splitting of the triplet state while $\boldsymbol{c}_{S_0T_1}$ and $\boldsymbol{c}_{S_1T_1}$ define the probabilities of the $T_1 \to S_0$ and $S_1 \to T_1$ intersystem crossings.

The constants $\boldsymbol{c}_{rr'}$ can be expressed in terms of the matrix elements of the one-electron operators $\boldsymbol{b}_a$,

$$\boldsymbol{B}_{pp'} \equiv \int \psi_p(\boldsymbol{x}_a)\boldsymbol{b}_a\psi_{p'}(\boldsymbol{x}_a)d\boldsymbol{x}_a ,$$

providing the explicit expressions for the coordinate wave functions through the one-electron orbitals,

$$\begin{aligned} |S_0\rangle &= \psi_1(\boldsymbol{x}_1)\psi_1(\boldsymbol{x}_2) , \\ |r\rangle &= [\psi_1(\boldsymbol{x}_1)\psi_2(\boldsymbol{x}_2) + (-1)^S\psi_1(\boldsymbol{x}_2)\psi_2(\boldsymbol{x}_1)]/\sqrt{2}, \qquad r = S_1, T_1 , \end{aligned} \tag{4.4.15}$$

are inserted in (4.4.11). An elementary calculation yields

$$\begin{aligned} &\boldsymbol{c}_{S_0T_1} = -\boldsymbol{B}_{12}/\sqrt{2}; \qquad \boldsymbol{c}_{S_1T_1} = (\boldsymbol{B}_{11} - \boldsymbol{B}_{22})/2; \\ &\boldsymbol{c}_{T_1T_1} = (\boldsymbol{B}_{11} + \boldsymbol{B}_{22})/\sqrt{2} . \end{aligned}$$

These formulae enable calculations of the orbital constants (4.4.11) for the spin–orbit operator as soon as the Hartree–Fock problem is solved, i.e., the self-consistent one-electron orbitals and the average electric field felt by the electrons at every fixed nuclear configuration are found.

In the zeroth-order approximation (no spin–orbit coupling) the triplet electronic state T_1 is triply degenerate by the orientations of spin $\boldsymbol{S}$, i.e., by the values of the quantum number σ of the spin projection S_z onto an arbitrary quantization axis z. The spin factor of the wave function has already been said to be a common eigenfunction of the operators $\boldsymbol{S}^2$ and S_z. Let α_a and β_a denote the spin factors of the one-electron wave functions with the spin projections onto the z-axis equal to $\sigma_a = +1/2$ and $-1/2$, respectively. Then the spin functions

$|S\sigma\rangle$ for a two-electron system are

$$|S0\rangle = [\alpha_1 \beta_2 + (-1)^{S+1} \alpha_2 \beta_1]/\sqrt{2}, \qquad S = 0, 1 ,$$
$$|11\rangle = \alpha_1 \alpha_2, \qquad |1-1\rangle = \beta_1 \beta_2 ; \tag{4.4.16}$$

the signs of these functions as well as those in (4.4.15) are arbitrary.

In the first-order approximation to the energy the unperturbed triplet level is split. As prescribed by perturbation theory for degenerate levels the unperturbed wave functions are taken in the form of linear combinations of functions (4.4.16) with $S = 1$ and various σ's to diagonalize the perturbation operator (4.4.12). The equivalent way to construct the correct unperturbed functions for the spin–orbit perturbation consists in a due choice of the quantization axis since a rotation of the quantization axis precisely corresponds to a linear transformation of functions (4.4.16).

The operator (4.4.12) represents the energy of the spin $\boldsymbol{S}$ in an effective intramolecular magnetic field $\boldsymbol{H} \propto \boldsymbol{c}_{T_1T_1}$ whose direction is now taken as the quantization axis, the z-axis of the molecular coordinate system. Then

$$U_{T_1T_1}(s, q) = |\boldsymbol{c}_{T_1T_1}| S_z , \tag{4.4.17}$$

and the correct unperturbed functions are those of (4.4.16), where α_a and β_a now stand for the one-electron spin functions defined with this new choice of the quantization axis. Thus in the first-order approximation a splitting is obtained of the unperturbed triplet level into three components with the energies $|\boldsymbol{c}_{T_1T_1}|\sigma$ corresponding to the values of $\sigma = 0, \pm 1$. We emphasize that molecular rotations have not yet been included here (see Sect. 4.4.3) so that the molecular body-fixed coordinate system is at the same time a space-fixed system and as the laboratory one.

The matrix elements of the operator (4.4.13) for the intersytem crossing between the triplet sublevels and the singlet states will emerge in the first-order approximation to the wave function. Let $\boldsymbol{e}_i$ stand for the directional unit vectors of the molecular coordinate system, and $\boldsymbol{c}_{rr'i}$ for the Cartesian components of the vector $\boldsymbol{c}_{rr'}$ in this coordinate system, $\boldsymbol{c}_{rr'} = \boldsymbol{c}_{rr'i}\boldsymbol{e}_i$ (the summation over the repeated index $i = x, y, z$ is implied). We define the first-rank spherical tensor components $c_{rr'}^{(\sigma)}(\sigma = 0, \pm 1)$ in the molecular coordinate system via the Cartesian components by the formulae [4.18, 19]

$$c_{rr'}^{(\pm 1)} = \mp 2^{-1/2}(c_{rr'x} \pm \mathrm{i} c_{rr'y}), \qquad c_{rr'}^{(0)} = c_{rr'z} , \tag{4.4.18}$$

and similarly the tensor components $G^{(\sigma)}$. The following relations are easily established by a straightforward calculation:

$$\boldsymbol{c}_{rr'} \cdot \boldsymbol{G} = \sum_{\sigma = 0, \pm 1} c_{rr'}^{(\sigma)} (-1)^\sigma G^{(-\sigma)} , \tag{4.4.19}$$

$$\langle 00|(-1)^\sigma G^{(-\sigma)}|1\sigma'\rangle = \delta_{\sigma\sigma'} . \tag{4.4.20}$$

Using these relations to calculate the matrix elements of (4.4.13) with functions (4.4.16) one obtains the final result

$$\langle 00|U_{rr'}(s,q)|1\sigma'\rangle = c_{rr'}^{(\sigma')} ,$$
$$r = S_0, S_1, \qquad r' = T_1 , \tag{4.4.21}$$

or, in other words, the purely electronic matrix elements of the spin–orbit interaction between a singlet state and a triplet-state sublevel with a definite total spin projection σ onto the intramolecular magnetic field direction is given by the corresponding component of the spherical tensor (4.4.18).

Equation (4.4.21) allows definite conclusions regarding the intersystem crossing probabilities, provided that the equilibrium configuration of the molecule has a high enough symmetry (see Sect. 4.7). These conclusions are valid on condition that molecular rotation is negligible, which corresponds, in the language of the theory of molecular spectra, to the coupling in Hund's case A [1.40] (see Sect. 4.4.3). Additionally, the states S_0, S_1 and T_1 were supposed to be orbitally nondegenerate, which holds for a majority of polyatomic molecules. The orbital degeneration, often occurring in higher electronic states, alters both the splitting diagram for the unperturbed triplet levels, which is described by a spin Hamiltonian more complicated in nature than (4.4.17), and the scheme of the singlet–triplet couplings that is constructed by conventional rules of the group theory applied to the matrix elements of the axial vector (4.4.9) [1.40, 45].

The set of exact simultaneous equations (4.1.21) show that the operator (4.4.12) furnishes a spin–orbit contribution to the adiabatic potential $\mathscr{E}_{T_1}(q)$ giving rise to a fine triplet term splitting. However, in polyatomic molecules the spin–orbit coupling is generally quite small, about 1–50 cm^{-1} [1.23, 31, 32, 45–51, 4.9–12, 20, 21] (it is still smaller in aromatic molecules, about tenths of cm^{-1} [1.45–49, 4.22]) so that inclusion of the spin effects into the electronic energy may be of significance at helium temperatures only [1.30]. By the estimate (4.2.31) it is also small compared to the nonadiabatic coupling inducing transitions between states of the same multiplicity. For these reasons the spin–orbit coupling is generally taken into account only in calculations of intramolecular interaction matrix elements for transitions between states of different multiplicities induced by the operator (4.4.13). Its matrix elements are written as

$$U_{rm\sigma r'm'\sigma'} = \int \chi_{rm}(q)\langle S\sigma|U_{rr'}(s,q)|S'\sigma'\rangle\chi_{r'm'}(q)dq . \tag{4.4.22}$$

They are expressed in terms of the parameters defined later in (4.4.31). Power experimental techniques are developed for directly measuring these matrix elements [1.31, 2.99, 115–123, 4.23].

As follows from (4.4.21) the dependence of the spin–orbit matrix elements upon spin variables leads to unequal populations among the triplet sublevels created by the $S_1 \rightarrow T_1$ transition and to different depopulation rates of these sublevels in the course of the $T_1 \rightarrow S_0$ transition [1.30, 4.24–31]. These effects, too, are insignificant at room temperature. Therefore, the said dependence is often disregarded and the above matrix elements are replaced with the spin-averaged matrix elements defined as

$$|U_{rmr'm'}|^2 = \sum_{\sigma\sigma'} |U_{rm\sigma r'm'\sigma'}|^2 \, . \tag{4.4.23}$$

The radiationless transition-rate constant involves just this quantity that arises due to summation over the final spin states and averaging over the initial ones in the case where both the density of final states and the population probability for the initial states are spin-insensitive.

Rotational effects on radiationless transitions mostly become observable in gas-phase supersonic-jet cooling of molecules and with the use of a narrow-band excitation [2.135] (see Sect. 2.2 for detailed references). In the next subsection the rotational matrix elements are calculated without averaging over the spin variables.

4.4.3 Rotational Matrix Elements

With the nuclear motion taken into consideration, the total wave function in the adiabatic approximation has the form of the product (4.1.31) of the electronic wave function at fixed nuclear coordinates (its coordinate part $\varphi_r(x, q)$ has to be multiplied by a spin part $\varphi_{S\sigma}$, see Sect. 4.4.2) and the nuclear function describing the vibrational–rotational motion of the molecule. The nuclear function obeys (4.1.29), where q stands for a set of vibrational and rotational variables.

When the atoms undergo low-amplitude vibrations around their equilibrium positions one can approximately separate the vibrations from rotation. To this end a molecular (rotating, body-fixed) coordinate system (x, y, z) rigidly bound to the equilibrium configuration of the molecule is introduced; the principal tensor-of-inertia axes are conventionally selected as the coordinate axes to diagonalize the rotational part of the kinetic energy T_{rot}. Its orientation with respect to the laboratory system is specified by the set of three Euler angles ϑ, which are just rotational variables, and by means of vibrational coordinates one takes either the Cartesian displacements of atoms from their equilibrium positions or linear combinations thereof, the normal coordinates, diagonalizing the quadratic form of the potential energy.

Obviously, an infinitesimal rotation of the molecular frame at any given moment of time leads to a redefinition (an infinitesimal increment) of the Euler angles and the normal coordinates. This uncertainty in orientation of the molecular frame relative to the laboratory one is removed by imposing the additional Eckart conditions [1.40, 4.19, 32–36] that nullify most of the total angular momentum related to the rotation of the molecule as a whole.

In the nuclear Hamiltonian (4.1.25), the potential energy is independent of the rotational variables. In what follows we change the notation $q \to (q, \vartheta)$, meaning that $q = \{q_\mu\}$ will now stand for the set of the normal vibrational coordinates only. Then the kinetic energy in (4.1.25, 29) $T_{\mathrm{nucl}}(q)$ is replaced with the sum of vibrational, rotational and vibrational–rotational (Coriolis) energies $T_{vib}(q) + T_{rot}(\vartheta) + T_{Cor}(q, \vartheta)$, where T_{vib} is given by the right-hand part of (4.1.2). The electronic wave functions are to be written now as $\varphi_r(x, q, \vartheta)$, and the derivatives over the new q's and ϑ's in (4.1.17–19) yield the operators of vibrational and rotational non-adiabaticity, respectively. With the new meaning of q_μ, (4.1.17–19) are matrix elements of the purely vibrational non-adiabaticity operators L and K, while the purely rotational non-adiabaticity is usually disregarded (see below). Since the vibration and rotation are not completely separated, there is also a rotation-assisted vibrational nonadiabatic coupling which is included in T_{Cor}.

The Coriolis energy is normally quite small and is disregarded here (see Sect. 4.4.4). It is worthwhile to note that the smallness of the Coriolis coupling is a direct consequence of the choice of the molecular frame as dictated by the Eckart conditions that minimize that coupling [4.19]. Under any different choice of the molecular frame the vibration–rotation coupling would generally not be small, and the separation of vibrations and rotation would be impossible.

Thus the unperturbed nuclear wave function, being an eigenfunction of $T_{vib} + U_r + T_{rot}$, falls into the product of a vibrational function $\chi_{rm}(q)$ (m is a set of vibrational quantum numbers) obeying (4.1.29), where T_{nucl} is replaced with T_{vib} and a rotational function $\chi_\tau(\vartheta)$, which is an eigenfunction of T_{rot} with an eigenvalue ε_τ, where τ is a set of rotational quantum numbers.

In the case of the internal conversion the perturbation is produced by the non-adiabaticity operator that, as follows from Sect. 4.1, is virtually independent of the rotational coordinates (apart from a small above-mentioned rotation-induced vibronic coupling included in T_{Cor}). Indeed, if the electronic subsystem is capable of being instantly adjusted to changes in the nuclear configuration, then the electronic wave function only depends on the relative positions of the atoms rather than on the molecule orientation with respect to the laboratory coordinate system; a dependence on ϑ arises only because of inertia of the electronic subsystem, but this effect (the effect of *the orbit–rotation coupling* [4.37–39]) is proportional to $m/M = \kappa^4$, i.e., is of a higher order of smallness than the vibrational non-adiabaticity [see (4.2.27)]. Physically, the rotation proves to be still more "adiabatic" than the vibration. For these reasons the rotation does not influence the internal conversion rate essentially, and the most prominent rotational effect should manifest in the intersystem crossing through the spin–orbit coupling. The corresponding matrix elements are discussed later in this subsection.

For the sake of argument a symmetric top will be considered. Then

$$T_{rot} = \frac{1}{2}\hbar^2 \sum_i I_{ii}^{-1} N_i^2 = AN^2 + BN_z^2 , \tag{4.4.24}$$

where $A = \hbar^2 I_{xx}^{-1}/2$ and $B = \hbar^2(I_{zz}^{-1} - I_{xx}^{-1})/2$ are rotational constants, N^2 and N_z the operators of the rigid-body angular momentum squared and the momentum projection onto the top axis chosen as the z-axis of the molecular frame

(N_i in units of $\hbar$) and I_{ii}^{-1} the inverse moment of inertia eigenvalues; the internal magnetic field that appears in the spin Hamiltonian of (4.4.12) is directed along the z-axis since it is the only direction of preference characteristic of a symmetric top without electronic degeneration.

Let us make a convention to place a tilde over the projection operator and over its eigenvalue when the projection is taken on the z-axis of the laboratory frame. Three quantities, $\boldsymbol{N}^2$, N_z and $\tilde{N}_z$ form a full set of commutative operators in the space of functions depending on the Euler angles (see (A.11, 16) of the Appendix A). Their common eigenfunctions, corresponding to the quantum numbers N, K and $\tilde{K}$, are the Wigner functions $D^{(N)}_{K\tilde{K}}(\vartheta)$ [1.40, 4.18, 19,32], which are chosen as rotational wave functions with the quantum numbers $\tau = NK\tilde{K}$ and the eigenenergies

$$\varepsilon_{NK} = AN(N+1) + BK^2 \,. \tag{4.4.25}$$

The normalized rotational wavefunctions are

$$\chi_{NK\tilde{K}}(\vartheta) = \left(\frac{2N+1}{8\pi^2}\right)^{1/2} D^{(N)}_{K\tilde{K}}(\vartheta) \,. \tag{4.4.26}$$

Proceeding to the discussion of the total electronic–nuclear wave function we remind that simultaneous inclusion of rotation and spin–orbit coupling engenders a necessity of distinguishing between different Hund's coupling cases depending on the relation between the rotation and spin–orbit energies. The rotational constants of polyatomic molecules are small, on the order of tenths of cm^{-1}, but the spacings between the levels (4.4.25) increase with N and K. Therefore Hund's case B occurs at high enough values of N and K, while the case A may be expected at low quantum numbers.

In case A the spin–orbit energy is large, and it is included in the unperturbed Hamiltonian, whereas T_{rot} is treated perturbatively. The operators $\boldsymbol{N}^2$ and N_z^2 entering T_{rot} commute with $\boldsymbol{S}^2$, so T_{rot} does not induce intersystem crossings (this is also true of the asymmetric top). We come to the conclusion that in Hund's case A the rotation does not affect the intersystem crossing[5], and the spin–orbit matrix elements are given by (4.4.21, 22).

In Hund's case B it is the rotational energy operator (4.4.24) that is assigned to the unperturbed Hamiltonian while the spin–orbit interaction (4.4.8) serves as a perturbation promoting transitions between the electronic–vibrational–rotational (rovibronic) levels. The problem consists in constructing the correct unperturbed functions and evaluating the matrix elements of U_{so} between them.

The total wave function can be obtained in the form of a product of nuclear, electronic coordinate and spin functions, or in the form of a linear combination of such products belonging to states of the same energy. In Sect. 4.4.2 the spin functions were chosen to be common eigenfunctions of the total spin squared $\boldsymbol{S}^2$,

[5] More precisely, the rotation effect appears only via higher-order cross terms involving both U_{so} and T_{rot}.

and the spin projection onto the molecular quantization axis S_z. With the rotation included, $\boldsymbol{S}^2$ is still conserved because it commutes with the unperturbed Hamiltonian $T_{\text{vib}} + U_r + T_{\text{rot}}$ independent of the spin variables[6], but S_z is not, since S_z does not commute with $\boldsymbol{N}^2$, hence with T_{rot}. Physically, this stems from the fact that the vectors $\boldsymbol{S}$ and $\boldsymbol{N}$ are fixed with respect to the laboratory frame, whereas the molecular quantization axis (the top axis) precesses around $\boldsymbol{N}$ so that S_z changes. The reader interested in formal proofs to this and subsequent statements regarding the commutative properties of the operators is referred to Appendix A (A.16).

As the full set of commutative operators in the space of functions of the rotational and spin variables we choose the operators $\boldsymbol{N}^2$, N_z, $\tilde{N}_z$, $\boldsymbol{S}^2$, and $\tilde{S}_z$. This set is convenient in that the total eigenfunction of the unperturbed Hamiltonian is represented in the form of the product

$$|r; rm; NK\tilde{K}; S\tilde{\sigma}\rangle = \varphi_r(x, q)\, \chi_{rm}(q)\, \chi_{NK\tilde{K}}(\vartheta)\, \varphi_{S\tilde{\sigma}}\,, \tag{4.4.27}$$

where $\tilde{\sigma}$ stands for an eigenvalue of $\tilde{S}_z$ and $\varphi_{S\tilde{\sigma}}$ is given by (4.4.16), where α_a and β_a are replaced with $\tilde{\alpha}_a$ and $\tilde{\beta}_a$, the eigenfunctions of the operator $\tilde{s}_{az}$.

The quantum numbers r, m, N, K, and S define the energy of an unperturbed rovibronic level. Energy gaps between vibronic states are large (on the order of hundreds and thousands cm^{-1}) as compared to the spin–orbit interaction energy. Therefore the quantum numbers r, m and S are "good" in the sense that perturbed levels are also assigned with definite values of these numbers equal to those for the corresponding unperturbed level. In Hund's case B the quantum numbers N and K are good as well by a similar reason.

In contrast, the quantum numbers $\tilde{K}$ and $\tilde{\sigma}$ are "poor". States with given r, m, N, K, and S are degenerated in $\tilde{K}$ and $\tilde{\sigma}$, i.e., in orientations of the rotation and spin momenta in the laboratory frame, and the operator U_{so} has nonzero matrix elements off-diagonal in $\tilde{K}$ and $\tilde{\sigma}$ since U_{so} does not commute with $\tilde{N}_z$ and $\tilde{S}_z$ (A.18). As prescribed by perturbation theory for degenerate levels, by means of the unperturbed functions one must take such linear combinations of functions (4.4.27) with various $\tilde{K}$ and $\tilde{\sigma}$ (the rest of quantum numbers being fixed) that U_{so} becomes diagonal in two novel quantum numbers specifying the correct functions of the zeroth-order approximation. Obviously, the correct functions correspond to mixed spin–rotation states.

We define the operator of the total angular momentum $\boldsymbol{J} = \boldsymbol{N} + \boldsymbol{S}$. The operators $\boldsymbol{J}^2$ and $\tilde{J}_z$ commute with U_{so}; they also commute with each other and with the operators $\boldsymbol{N}^2$, $\boldsymbol{S}^2$ and N_z [see (A.16, 17)]. Therefore, common eigenfunctions of $\boldsymbol{S}^2$, $\boldsymbol{J}^2$, $\boldsymbol{N}^2$, N_z, and $\tilde{J}_z$ with the quantum numbers S, J, N, K, and $\tilde{P}$ are the correct unperturbed functions wanted. On the ground of the above-mentioned commutation rules the operator U_{so} has no off-diagonal matrix

[6] Additionally, the unperturbed Hamiltonian has an electronic part which involves the spin–dependent exchange interaction lifting the total-spin degeneracy. The exchange operator also commutes with $\boldsymbol{S}^2$, and possible spin–rotational interactions are too weak as compared to the exchange energy to break down the conservation of the total spin.

elements between states differing in J or $\tilde{P}$. The matrix element of U_{so} diagonal in J and in the rest of quantum numbers yields a splitting of the unperturbed degenerate level. Evidently, J is a good quantum number. As for $\tilde{P}$, the matrix elements are independent of it since the level degeneration is still preserved with respect to the total momentum orientation in the laboratory frame. For this reason $\tilde{P}$ is not regarded as a good quantum number.

It should be emphasized that U_{so} does not commute with $\boldsymbol{S}^2$, $\boldsymbol{N}^2$ and N_z, and therefore, has off-diagonal matrix elements between states with different S, N and K. We are interested in transitions with a change in S, whereas those changing N and K at a given S, and hence responsible for an intrastate rotational relaxation, will not be dealt with.

Employing the momentum addition rule by the coupling scheme of $\boldsymbol{J} = \boldsymbol{N} + \boldsymbol{S}$ with projections in the laboratory frame, we develop the correct spin–rotation wave functions in the basis (4.4.27); omitting the sign factor and the common electronic–vibrational factor $\varphi_r(x, q)\,\chi_{rm}(q)$, we obtain

$$|SJNK\tilde{P}\rangle = (2J+1)^{1/2} \sum_{\tilde{K}\tilde{\sigma}} \begin{pmatrix} N & S & J \\ \tilde{K} & \tilde{\sigma} & -\tilde{P} \end{pmatrix} \chi_{NK\tilde{K}}(\vartheta)\,\varphi_{S\tilde{\sigma}} \,. \tag{4.4.28}$$

After integrating over the electronic space variables and using the definition (4.1.15) the desired matrix element of the spin–orbit operator (4.4.10) between the unperturbed functions (4.4.28) assumes the form

$$\langle rm;\, SJNK\tilde{P}|\, U_{rr'}(s, q, \vartheta)\,|r'm';\, S'J'N'K'\tilde{P}'\rangle \,,$$

where $U_{rr'}(s, q, \vartheta)$ is given by (4.4.12, 13) (here q is replaced with q, ϑ). Inserting these relations and integrating over the nuclear vibrational coordinates q (now we recall that the $\boldsymbol{c}_{rr'}$ are functions of q, and indicate this explicitly) yield expressions

$$\langle 1JNK\tilde{P}|\,\boldsymbol{c}_{T_1mT_1m'}\cdot \boldsymbol{S}\,|\,1J'N'K'\tilde{P}'\rangle \tag{4.4.29}$$

and

$$\langle 0JNK\tilde{P}|\,\boldsymbol{c}_{rmT_1m'}\cdot \boldsymbol{G}\,|\,1J'N'K'\tilde{P}'\rangle \tag{4.4.30}$$

for $r = S_0, S_1$, where

$$\boldsymbol{c}_{rmr'm'} = \int \chi^*_{rm}(q)\,\boldsymbol{c}_{rr'}(q)\chi_{r'm'}(q)\,dq \tag{4.4.31}$$

for $r = S_0, S_1, T_1$ and $r' = T_1$. The quantities (4.4.31) are vectors rigidly bound to the molecule. Their scalar products with $\boldsymbol{S}$ and $\boldsymbol{G}$ commute with $\boldsymbol{J}^2$, which is verified in the same way as the commutation relation (A.17) for U_{so}. Therefore the matrix elements in (4.4.29) and (4.4.30) are diagonal in J. The former define the unperturbed-level splittings and the probabilities of vibration–rotation transitions within the electronic state T_1, while the latter control the probabilities of singlet–triplet transitions with a change in the rotational state.

Being interested in the latter only, we consider (4.4.30). We substitute the development (4.4.28) for the triplet function; as for the singlet function, the development contains only one term, $\chi_{NK\tilde{P}}(\vartheta)\varphi_{0\tilde{0}}\delta_{JN}$. The scalar product is written by analogy with (4.4.19) via the spherical tensor components in the laboratory frame (we place tildes on them). In this way (4.4.30) converts into

$$\delta_{JN}(2J'+1)^{1/2}\sum_{\tilde{K}'\tilde{\sigma}'\tilde{\sigma}}\begin{pmatrix} N' & 1 & J' \\ \tilde{K}' & \tilde{\sigma}' & -\tilde{P}' \end{pmatrix}$$

$$\times\langle NK\tilde{P}|\tilde{c}^{(\tilde{\sigma})}_{rmr'm'}|N'K'\tilde{K}'\rangle\langle 0\tilde{0}|(-1)^{\tilde{\sigma}}\tilde{G}^{(-\tilde{\sigma})}|1\tilde{\sigma}'\rangle ,$$

where $\tilde{c}^{(\tilde{\sigma})}_{rmr'm'}$ is defined similar to (4.4.31). The spin factor can be calculated by (4.4.20) valid for arbitrary choice of the quantization axis. In order to calculate the rotation factor, the spherical tensor components $\tilde{c}^{(\tilde{\sigma})}_{rmr'm'}$ are written as linear combinations of its components in the molecular frame $c^{(\sigma)}_{rmr'm'}$; as is well known, the corresponding coefficients are Wigner's functions [1.40]

$$\tilde{c}^{(\tilde{\sigma})}_{rmr'm'}=\sum_{\sigma}D^{(1)}_{\sigma\tilde{\sigma}}(\vartheta)\,c^{(\sigma)}_{rmr'm'} .$$

The above spherical tensor components in the molecular frame are independent of ϑ since they are composed of the Cartesian components of the body-fixed vector $c_{rmr'm'}$. Therefore, after the substitution of the wave functions (4.4.26) and the replacement of $D^{(N)*}_{K\tilde{P}}(\vartheta)$ with $(-1)^{K-\tilde{P}}D^{(N)}_{-K-\tilde{P}}(\vartheta)$, the rotation factor transforms to a known integral of the product of three Wigner functions [1.40, 4.18], and reads

$$\langle NK\tilde{P}|\tilde{c}^{(\tilde{\sigma})}_{rmr'm'}|N'K'\tilde{K}'\rangle=(-1)^{K-\tilde{P}}[(2N+1)(2N'+1)]^{1/2}$$

$$\times\sum_{\sigma}c^{(\sigma)}_{rmr'm'}\begin{pmatrix} N & 1 & N' \\ -K & \sigma & K' \end{pmatrix}\begin{pmatrix} N & 1 & N' \\ -\tilde{P} & \tilde{\sigma} & \tilde{K}' \end{pmatrix} .$$

The final step consists in the calculation of the sum over $\tilde{K}'$ and $\tilde{\sigma}'$ based on the orthogonality condition for the Wigner 3-J symbols. The eventual expression for the spin–orbit matrix element between the singlet ($r=S_0$ or S_1) and triplet ($r'=T_1$) states, to within an insignificant sign factor, is thus:

$$\langle r;\, rm;\, 0JNK\tilde{P}|U_{so}(s,x,q)|r';\, r'm';\, 1J'N'K'\tilde{P}'\rangle$$

$$=\delta_{JN}\delta_{JJ'}\delta_{\tilde{P}\tilde{P}'}(2N'+1)^{1/2}\sum_{\sigma}c^{(\sigma)}_{rmr'm'}\begin{pmatrix} N & 1 & N' \\ -K & \sigma & K' \end{pmatrix} . \qquad (4.4.32)$$

A more elegant, but less transparent, derivation was proposed by *Howard* and *Schlag* [2.198][7] who used, instead of (4.4.28), an expansion in the basis of the

[7] The rotation effect on the intersystem crossing in naphthalene [2.197] was not confirmed in a later study [4.40].

eigenfunctions of the operators $\boldsymbol{S}^2, \boldsymbol{J}^2, S_z, J_z$, and $\tilde{J}_z$ whose commutativity follows from (A.14, 16). In this expansion, one again invokes the momenta addition rule, but this time by the coupling scheme of $\boldsymbol{N} = \boldsymbol{J} + (-\boldsymbol{S})$ and with projections in the molecular rather than the laboratory frame [4.38, 41–44]:

$$|SJNK\tilde{P}\rangle = (2N+1)^{1/2} \sum_{P\sigma} \begin{pmatrix} J & S & N \\ P & \sigma & -K \end{pmatrix} \chi_{JP\tilde{P}}(\vartheta)\varphi^*_{S\sigma} ,$$

where

$$\varphi^*_{S\sigma} = (-1)^{S-\sigma}\varphi_{S,-\sigma}$$

are eigenfunctions of the operator $(-\boldsymbol{S})_z$, as obtained by applying the time-reversal operation [1.40]. This expansion can also be deduced immediately from (4.4.28) by substituting

$$\varphi_{S\tilde{\sigma}} = \sum_{\sigma} D^{(S)}_{\sigma\tilde{\sigma}} \varphi_{S\sigma} ,$$

and then performing the summation over $\tilde{K}\tilde{\sigma}$ using the well-known properties of the Wigner functions and 3-J symbols [1.40, 4.18]. The advantage of this approach is in that once the operator (4.4.13) is transformed to the form of (4.4.19), the rotation factor is simply

$$\langle JP\tilde{P}|J'P'\tilde{P}'\rangle = \delta_{JJ'}\delta_{PP'}\delta_{\tilde{P}\tilde{P}'} ,$$

for $c^{(\sigma)}_{rmr'm'}$ is independent of ϑ, and we again recover (4.4.32). This equation will be used in Sect. 4.7 in the discussion of the selection rules for the intersystem crossing.

Similar considerations may be used to derive the rotational matrix elements for internal conversion between triplet states [2.197, 198].

4.4.4 Coriolis Coupling

The vibrational motion of the molecule creates a vibrational angular momentum [4.45]

$$N_i^{(\mathrm{vib})} = -\mathrm{i} \sum_{\mu\mu'} \zeta^{i}_{\mu\mu'} \left(q_\mu \frac{\partial}{\partial q_{\mu'}} - q_{\mu'} \frac{\partial}{\partial q_\mu} \right), \tag{4.4.33}$$

where $\zeta^{i}_{\mu\mu'}$ are dimensionless Coriolis constants with typical values of 0.3–0.03, and the summation runs over all different pairs of normal coordinates. Now it is the total rotational angular momentum $\boldsymbol{N}^{(\mathrm{tot})} = \boldsymbol{N}^{(\mathrm{rb})} + \boldsymbol{N}^{(\mathrm{vib})}$ that is conserved for the molecule as a closed system rather than the rigid-body one, $\boldsymbol{N}^{(\mathrm{rb})}$ (denoted as $\boldsymbol{N}$ in Sect. 4.4.3), and $\boldsymbol{N}^{(\mathrm{vib})}$ separately. The kinetic energy of the rigid-body rotation is again given by (4.4.24), where $\boldsymbol{N}^{(\mathrm{tot})} - \boldsymbol{N}^{(\mathrm{vib})}$ is now substituted for $\boldsymbol{N}^{(\mathrm{rb})}$. Writing $\boldsymbol{N}^{(\mathrm{tot})}$ simply as $\boldsymbol{N}$ and opening the braces we obtain the first term in the form of (4.4.24), while the second is the Coriolis coupling [1.40, 4.32]

$$T_{\mathrm{Cor}} = -\hbar^2 \sum_i I_{ii}^{-1} N_i N_i^{(\mathrm{vib})} . \tag{4.4.34}$$

Thus T_{Cor} arises due to an interaction between the total and vibrational momenta. Inspection of (4.4.33, 34) shows that the Coriolis tensor $\zeta^i_{\mu\mu'}$ couples the μ-th and μ'-th vibrations through the rotation around the i-th molecular axis provided that the product of the vibrations' symmetry species contains the rotation one (Jahn's rule [1.40, 4.32]).

The Coriolis coupling operates not only in a given electronic state but also between different states of the same multiplicity. This is just due to the non-adiabaticity effect of the form (4.1.17, 18). Taking the interstate ($r \neq r'$) electronic–vibrational matrix element between the adiabatic functions (4.1.31) and invoking the orthogonality condition (4.1.8) we obtain [4.38]

$$\langle r; rm | T_{\mathrm{Cor}} | r'; r'm' \rangle = \mathrm{i}\hbar^2 \sum_{i\mu\mu'} I_{ii}^{-1} \zeta^i_{\mu\mu'} N_i$$

$$\times \langle r; rm | q_\mu \frac{\partial \varphi_{r'}(x, q)}{\partial q_{\mu'}} - q_{\mu'} \frac{\partial \varphi_{r'}(x, q)}{\partial q_\mu} | r'; r'm' \rangle . \tag{4.4.35}$$

It is supposed that a change in the moments of inertia between the two electronic states can be neglected.

The matrix element in (4.4.35) is still an operator (V_{Cor}) acting on the rotational variables. As such, it can be written in a simplified form,

$$V_{\mathrm{Cor}} = \sum_i V_i N_i \equiv V^{\parallel}_{\mathrm{Cor}} + V^{\perp}_{\mathrm{Cor}} ,$$

$$V^{\parallel}_{\mathrm{Cor}} = V_z N_z , \qquad V^{\perp}_{\mathrm{Cor}} = V_+ N_+ + V_- N_- , \tag{4.4.36}$$

where $N_\pm = N_x \pm \mathrm{i} N_y$ and N_z are angular momentum operators, while V_i are constants defined through (4.4.35). The Coriolis coupling is conventionally divided into the parallel and perpendicular parts involving the angular momentum projections parallel and perpendicular to the top axis. The corresponding nonzero matrix elements are [4.45, 46]

$$\langle NK | V^{\parallel}_{\mathrm{Cor}} | NK \rangle = V_z K , \tag{4.4.37}$$

$$\langle NK | V^{\perp}_{\mathrm{Cor}} | NK \pm 1 \rangle = V_\pm [N(N+1) - K(K \pm 1)]^{1/2} . \tag{4.4.38}$$

The formulae (4.4.36–38), but with a different meaning of the parameters V_i, hold true for the remaining part of the Coriolis coupling that rotationally mixes vibronic states within a given electronic state [4.47, 48], and hence provides a rotation-dependent contribution to the rate constant for *Intramolecular* intra-state *Vibrational* energy *Redistribution* (often abbreviated as IVR) [2.140b, c, 149, 234, 4.49–53].

It is unknown whether the interstate Coriolis coupling may appreciably affect the radiationless transitions in the statistical limit since *Novak* and *Rice* [4.37, 38] estimated the effect to be small. On the contrary, the rotationally dependent fluorescence quantum yield of the intermediate-case molecule pyrazine was explained by the interstate S_1–S_0 Coriolis coupling [2.122, 143], and numerous singlet–triplet perturbations in the ultrahigh-resolution fluorescence excitation spectra of pyrimidine by the T_1–S_0 Coriolis coupling [4.54].

On the other hand, *Riedle* et al. [4.55–61] found evidence that the intrastate-Coriolis-induced IVR in S_1 is responsible for the lack of some K-lines in high-resolution fluorescence excitation spectra of benzene, i.e., it has an indirect effect on the radiationless transition rate [4.62–64]. The intrastate coupling in S_1 is believed to be responsible for temperature effects on the intensity redistribution between sharp lines and diffuse bands in the emission spectrum of anthracene [4.65, 66], for IVR in *p*-difluorobenzene [4.67] and for fluorescence depolarization of those molecules [4.68], while the intrastate coupling in T_1 induces the above-mentioned rotational effect in pyrazine [2.149, 151, 165].

It is evident that, although the Coriolis effects are reliably observed, it is not a simple matter to decide definitely in each case whether the inter- or intrastate Coriolis coupling, or both are present.

4.5 Condon Approximation

A widely used approximation in the theory of radiationless transitions is a factorization of the matrix elements of any operator $A(x, q)$ in the adiabatic basis (4.1.31) into a product of the form

$$A_{rmr'm'} = A_{rr'} S_{rmr'm'} , \tag{4.5.1}$$

where $A_{rr'}$ is a purely electronic factor independent of (or weakly dependent on) the vibrational quantum numbers and $S_{rmr'm'}$ is the overlap integral (4.2.18) of the vibrational wave functions of the initial and final electronic states. In the approximation (4.5.1) the effect of nuclear motion on the nonradiative transition rate is described by *the Franck–Condon factor*

$$F_{rmr'm'} = S^2_{rmr'm'} \tag{4.5.2}$$

since the transition rate is proportional to the intramolecular matrix element squared.

The factorization of the intramolecular interaction matrix elements in the form of (4.5.1) is implemented on the basis of the Condon approximation and, in the case of the non-adiabaticity operator, some additional approximations.

The Condon approximation was adopted from the theory of optical spectra of molecules and solids [1.15, 16, 22, 24, 25, 34, 4.2, 32]. The probability of an optical electronic–vibrational transition is governed by the matrix element of the dipole moment operator $d(x, q)$ between vibronic sublevels of two electronic states r and r',

$$d_{rmr'm'} = \int \chi_{rm}(q)\, d_{rr'}(q)\, \chi_{r'm'}(q)\, dq \,, \tag{4.5.3}$$

where

$$d_{rr'}(q) = \int \varphi_r(x, q)\, d(x, q)\, \varphi_{r'}(x, q)\, dx \tag{4.5.4}$$

is the electronic dipole moment matrix element as a function of nuclear coordinates.

Figure 4.1 shows the behavior of the vibrational functions in the case of a single nuclear coordinate q. A vibronic sublevel m of the excited electronic state r next to the ground sublevel is taken as the initial state. The function $\chi_{rm}(q)$ is appreciably nonvanishing near the equilibrium configuration q_r, approximately within the amplitude of zero-point vibrations b, and rapidly falls off at $|q - q_r| > b$. The final state wave functions $\chi_{r'm'}(q)$ are peaked at the turning points $q^{\pm}_{r'm'}$ and oscillate between them. Therefore, the matrix element (4.5.3) is the largest for the vertical transition indicated by the arrow in Fig. 4.1, i.e., the electronic transition occurs at a fixed nuclear position (in our example this position corresponds to the minimum of the upper term and to a turning point for the classical motion with the energy $E_{r'm'}$ in the lower term). This is a specific example of the Franck–Condon principle for optical transitions [4.69–71].

From Fig. 4.1 it is obvious that the dominant contribution to the integral (4.5.3) comes from q near q_r,

$$|q - q_r| \lesssim b \,, \tag{4.5.5}$$

whereas the characteristic length scale for changing $d_{rr'}(q)$ versus q is the interatomic separation $a = b/\kappa \gg b$ [see (4.2.6)]. For this reason the electronic

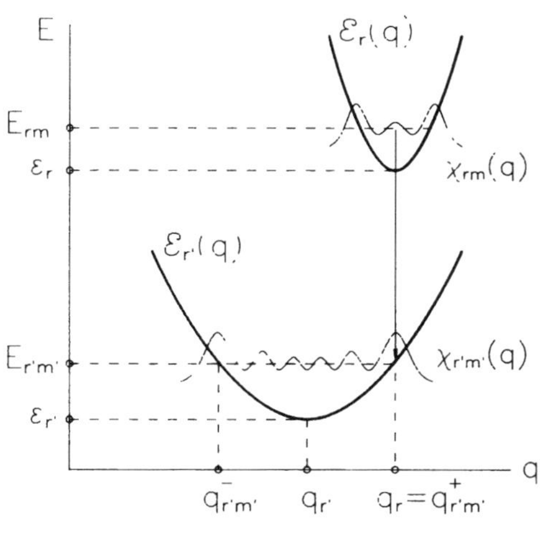

Fig. 4.1. Scheme of a vertical (optical) transition

matrix element (4.5.4) may be approximated by the constant

$$d_{rr'}(q) \approx d_{rr'}(q_r) = \int \varphi_r(x, q_r) d(x, q_r) \varphi_{r'}(x, q_r) dx \ , \tag{4.5.6}$$

independent of the nuclear coordinate and equal to its value at $q = q_r$. This is just the Condon approximation that immediately entails an expression like (4.5.1) for the dipole-moment matrix elements.

If the equilibrium molecular configuration in the initial electronic state, $q = q_r$, is symmetric (i.e., the symmetry group contains nonunit elements), (4.5.6) establishes selection rules for the optical transitions. In particular, the $r \to r'$ transition may happen to be forbidden by symmetry in the Condon approximation (the matrix element (4.5.6) is zero). Then higher-order terms of the expansion of the electronic matrix element (4.5.4) in $q \to q_r$ need to be incorporated. In this case the breakdown of the Condon approximation relaxes the forbiddeness to bring about vibronically-induced electronic transitions (*the Herzberg–Teller effect*) [4.32, 72]. *Kovarskiĭ* [1.20] showed that deviations from the Condon approximation essentially affect the probabilities of allowed optical transitions, too, provided that the transition occurs between the levels of which one is close to a term intersection point.

Obviously, the Condon approximation in the theory of optical transitions corresponds to an expansion of the electronic matrix element (4.5.4) in powers of $q - q_r$ to the zeroth-order term. Hence, the accuracy of this approximation is estimated by the relative magnitude of the next term of the expansion

$$d_{rr'}(q) \approx d_{rr'}(q_r) + (q - q_r) \left[\frac{d}{dq} d_{rr'}(q) \right]_{q = q_r} . \tag{4.5.7}$$

The length scale for variations of the dipole moment $d(x, q)$ against q is a, and that for the electronic wave functions is d, as given in (4.2.13). Employing (4.5.5) and the above estimate for a, one obtains that by the order of magnitude the ratio of the second term to the first one is $b/d = \kappa \mathscr{E} / \Delta \mathscr{E}$; in other words, (4.5.7) is nothing else but an expansion in the Born–Oppenheimer parameter. It is thus concluded that in the theory of optical transitions the Condon approximation is always applicable as soon as the adiabatic approximation is, since it is a straightforward consequence of the adiabatic approximation.

In the theory of radiationless transitions the Condon approximation for the matrix elements of the non-adiabaticity and spin–orbit operators corresponds to replacing the functions $C^{\mu}_{rr'}(q)$ and $c_{rr'}(q)$ in (4.4.1, 31) with constants independent of nuclear coordinates,

$$C^{\mu}_{rr'}(q) \approx C^{\mu}_{rr'} \ , \tag{4.5.8}$$

$$c_{rr'}(q) \approx c_{rr'} \ . \tag{4.5.9}$$

The Condon approximation (4.5.8, 9), as well as that in the theory of optical transitions, is justified by the slow change of the electronic matrix elements as compared to the nuclear wave functions. By analogy with (4.5.6) the approximation (4.5.8, 9) is often supposed to correspond to an expansion of the electronic matrix elements in powers of $q - q_r$, so that the constants $C^{\mu}_{rr'}$ and $c_{rr'}$ are merely the values of the matrix elements at $q = q_r$. From this it is concluded that the selection rules for both the radiative and nonradiative transitions are governed by one and the same molecular symmetry group, viz., the symmetry group of the equilibrium configuration q_r in the initial electronic state [1.4, 10]. However, as we shall see later, the constants $C^{\mu}_{rr'}$ and $c_{rr'}$ have quite a different significance since the dominant contribution to the integrals in (4.4.1, 31) is provided by such values of the nuclear coordinates that are in the classically forbidden or even complex domain rather than near the equilibrium configuration. This feature must be borne in mind when dealing with the problem of the selection rules for nonradiative transitions (see Sects. 4.7, 5.4.4).

We now consider the same situation as above: there is only one coordinate q, and a low-lying vibrational sublevel m of the upper electronic term is taken as the initial state. Of importance for a radiationless transition are such final states m' that satisfy the energy conservation law

$$E_{rm} = E_{r'm'} . \tag{4.5.10}$$

The behavior of the functions $\chi_{rm}(q)$ and $\chi_{r'm'}(q)$ is schematically depicted in Fig. 4.2. The function $\chi_{r'm'}(q)$ is quasiclassical as it corresponds to a large vibrational excitation. It is peaked at the turning points $q^{\pm}_{r'm'}$, and randomly oscillates in the interval in-between, in particular, in that same region where the function $\chi_{rm}(q)$ is appreciably nonzero (its derivative appearing in the integrand of (4.4.1) exhibits the same behavior). In other words, the ranges of the most probable q values in the initial and final electronic states (connected by the

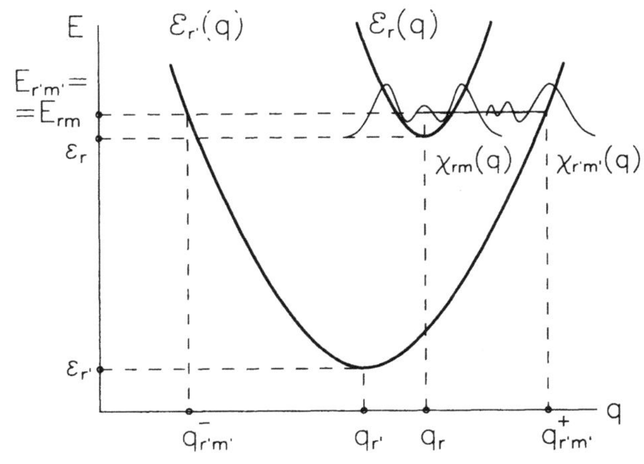

Fig. 4.2. Scheme of a horizontal (nonradiative) transition

horizontal arrow in Fig. 4.2) do not overlap[8], which essentially differs from the situation in the case of optical transitions shown in Fig. 4.1. Hence, it is clear that all physically accessible q values yield nearly the same exponentially small contributions to the integrals (4.4.1, 31).

A general method to calculate integrals of quasiclassical functions is outlined in the book by *Landau* and *Lifshitz* [1.40], and will be used in Chap. 5 to calculate the overlap integrals. The integration path is moved to the complex q-plane to circumvent the turning points for the classical motion in both terms. Far away from the turning points the functions $\chi_{rm}(q)$ and $\chi_{r'm'}(q)$ are approximated as

$$\chi(q) \sim \exp\left[\frac{\mathrm{i}}{\hbar}\int p(q)dq\right], \tag{4.5.11}$$

where

$$p(q) = \{2M[E - \mathscr{E}(q)]\}^{1/2} \tag{4.5.12}$$

is the classical momentum, and then the integrals are solved by the saddle-point method. The saddle point q_* whose vicinity provides the major contribution to the integral along the distorted contour is found from

$$E_{rm} - \mathscr{E}_r(q) = E_{r'm'} - \mathscr{E}_{r'}(q), \tag{4.5.13}$$

which is the condition for the classical momenta (4.5.12) to have the same value in the initial and final electronic states. For the strictly horizontal transition (4.5.10), the q_* is a crossing point of the electronic terms. At a small term displacement, the point q_* lies far in the classically inaccessible region or in the complex plane provided that the terms do not cross at real q's; in both cases q_* lies in a nonphysical region where momenta (4.5.12) are complex.

Notice that the same technique can be also applied to evaluate the matrix elements (4.5.3) for the optical transitions, provided that the final state corresponds to a high vibrational quantum number m'. Then the saddle point is defined by the same condition (4.5.13), which, in this case, conveys the classical Franck–Condon principle: the optical transition occurs at fixed nuclear momenta and positions (at q_*). Thus, in the example of Fig. 4.1, the optical transition is accomplished at $q_* = q_r$, and the classical momenta at the transition point are close to zero. By this analogy (4.5.13) may be interpreted as a generalization of the Franck–Condon principle to the radiationless transitions (see Sect. 5.4); however, in the case of the radiationless transitions, the nuclear "coordinates" and "momenta" take on nonphysical values [1.39].

After the saddle point is found, the integral over the nuclear coordinates may be approximated, within an exponential accuracy, by its integrand's value at

---

[8] The case of large displacements of the electronic terms, where q_r is close to a turning point, is not considered.

that point. In this way one obtains for the overlap integral [1.40]

$$S_{rmr'm'} \sim \exp\left(-\frac{1}{\hbar}\,\mathrm{Im}\left\{\int^{q_*}\{2M[E_{rm}-\mathscr{E}_r(q)]\}^{1/2}dq\right.\right.$$

$$\left.\left. -\int^{q_*}\{2M[E_{r'm'}-\mathscr{E}_{r'}(q)]\}^{1/2}dq\right\}\right), \qquad (4.5.14)$$

where the root branch is setup in such a way that the expression under the Im sign are positive, and the lower integration limits may be chosen at arbitrary points within the classically allowed region.

Extension of these results to multidimensional cases is a complicated problem. We only mention that the quasiclassical wave function is of the form [1.40, 4.73, 74]

$$\chi(q) \sim \exp\left[\frac{i}{\hbar}\sum_\mu \int p_\mu(q)dq\right], \qquad (4.5.15)$$

where the action integral is over a trajectory of classical motion with a given energy and the saddle point q_* is defined by the condition that all the momentum components $p_\mu(q)$ are the same in two electronic states; in particular, q_* obeys (4.5.13). The overlap integral is given by an equation patterned after (4.5.14), where the imaginary part is taken from the difference of the actions calculated along complex classical trajectories in the two electronic terms starting in the classically allowed region and ending at q_* [1.13, 40, 2.238] (see Sect. 5.7.3).

Let us now bring out the meaning of the parameters $C^\mu_{rr'}$ and $c_{rr'}$ as well as establish a general form of the matrix elements (4.4.1, 23). The matter is the most trivial for the spin–orbit interaction. Replacing the slowly varying function $c_{rr'}(q)$ in (4.4.31) with its value at $q=q_*$ and removing it from the integral sign yields $c_{rmr'm'}=c_{rr'}(q_*)S_{rmr'm'}$, and by inserting (4.4.21) into (4.4.22) with the use of definition (4.4.18), the singlet–triplet matrix element becomes $U_{rm\sigma r'm'\sigma'}=c^{(\sigma')}_{rr'}(q_*)S_{rmr'm'}$. In this way one obtains the Condon form (4.5.1) of the spin-averaged matrix elements (4.4.23)

$$U_{rmr'm'} = U_{rr'}S_{rmr'm'}\,, \qquad (4.5.16)$$

$$U_{rr'} = |c_{rr'}(q_*)|\,. \qquad (4.5.17)$$

Thus the constant $c_{rr'}$ in (4.5.9) is $c_{rr'}(q_*)$ rather than $c_{rr'}(q_r)$. Obviously, the Condon approximation (4.5.16) may turn out to be invalid if more than one point q_* provides comparable contributions to the integrals of (4.4.22, 31).

In the case of the non-adiabaticity operator, account must be taken of the singularity of the function $C^\mu_{rr'}(q)$ in the term crossing region (see Sect. 4.3). Let us single out the singularity by writing $C^\mu_{rr'}(q)=B^\mu_{rr'}(q)f_{rr'}(\Delta\mathscr{E})$, where $B^\mu_{rr'}(q)$ is a regular function in the vicinity of q_* and $f_{rr'}(\Delta\mathscr{E})$ is a dimensionless function that assumes the values on the order of unity when the adiabatic term splitting $\Delta\mathscr{E}$ is on the order of the electronic energy gap

$$\Delta E = \varepsilon_r - \varepsilon_{r'} , \tag{4.5.18}$$

and is singular at $\Delta\mathscr{E} = 0$. In evaluating the integral of (4.4.1) between the resonant states (4.5.10) the integration path is drawn in such a way as to circumvent the singular point q_*, which, according to (4.5.13), is simultaneously a saddle point for the integrand. The singularity of $f_{rr'}(\Delta\mathscr{E})$ has no effect on the exponential factor (4.5.14) and only results in an enhancement of the preexponential. For this reason, when calculating the integral, the function $f_{rr'}(\Delta\mathscr{E})$ can be replaced by the constant $f_{rr'}(\Delta\mathscr{E}) \approx f_{rr'}$ that allows for the change in the preexponential factor. To calculate the latter accurately is a complicated problem (see Sect. 5.7), though $f_{rr'}$ can always be evaluated by an order of magnitude provided that the terms' behavior near the crossing point is known. Generally one may only state that $f_{rr'} > 1$. The resultant expression for the constant $C^{\mu}_{rr'}$ in (4.5.8) is

$$C^{\mu}_{rr'} = B^{\mu}_{rr'}(q_*) f_{rr'} . \tag{4.5.19}$$

Now it is seen that *the constant* $C^{\mu}_{rr'}$ *is not the value of the function* $C^{\mu}_{rr'}(q)$ *at* $q = q_r$, *but is rather determined by the behavior of this function in the vicinity of* q_*. Note that this constant depends (through q_*) upon the vibrational quantum numbers. Equation (4.5.19) will be used in Sect. 4.7 in a qualitative discussion of the selection rules for nonradiative transitions.

The above-stressed difference between (4.5.19) and $C^{\mu}_{rr'}(q_r)$ is just responsible for the acceleration of nonradiative transitions due to the non-Condon effect revealed by *Kovarskiĭ* [4.75, 76]. Numerical calculations as well as analytical studies to check the importance of the non-Condon corrections were carried out in [4.77–83]. The importance of the term crossing in the nonradiative processes [2.209–212] as well as in calculations of the Franck–Condon factors for other processes [4.84] was also emphasized. Note that a similar role is played by conic intersections in nonadiabatic collisions [4.85].

In the Condon approximation the matrix element (4.4.1) becomes

$$L_{rmr'm'} = -\sum_{\mu} \frac{\hbar^2}{M_\mu} C^{\mu}_{rr'} \int \chi_{rm}(q) \frac{\partial \chi_{r'm'}(q)}{\partial q_\mu} dq . \tag{4.5.20}$$

Sometimes it is further simplified to the form of (4.5.1)

$$L_{rmr'm'} = L_{rr'} S_{rmr'm'} . \tag{4.5.21}$$

For instance, if the nuclear motion in the final state is quasiclassical along all nuclear coordinates, then differentiating (4.5.15) over q_μ yields

$$\partial \chi_{r'm'}(q)/\partial q_\mu \sim \frac{1}{\hbar} p_\mu(q) \chi_{r'm'}(q) .$$

Inserting this into (4.5.20) and replacing the momentum with its value at $q = q_*$ we find the expression of the type of (4.5.21). Notice that the quantity $\hbar/|p_\mu(q_*)|$ is a characteristic length scale for variations of $\chi_{r'm'}(q)$ around q_*, and since the classical momenta in both terms are the same at $q = q_*$ (4.5.13), it represents the parameter b_* defined in (4.2.17) to characterize the behavior of $\chi_{rm}(q)$ around q_*.

4.6 Model of Noninteracting Oscillators

In the general case of nonseparable variables the determination of the vibrational wave functions and the overlap integrals is impossible. Therefore, the assumption is introduced that with a specific choice of the set of nuclear coordinates $q = \{q_\mu\}$ the vibrational Hamiltonians (4.1.25) of two electronic states are simultaneously represented as sums of the Hamiltonians of noninteracting one-dimensional vibrations (see Sect. 2.10)

$$H_r(q) = \sum_\mu H_{\mu r}(q_\mu), \qquad H_{r'}(q) = \sum_\mu H_{\mu r'}(q_\mu) . \tag{4.6.1}$$

The approximation (4.6.1) as well as the Condon one come from the theory of optical electronic–vibrational transitions where it was given a solid-enough foundation. We consider a transition from an initial state rm close to the ground vibrational state $r0$. As we saw in Sect. 4.5, the matrix elements of the dipole moment are dominated in this case by the magnitudes of the vibrational wave functions near the minimum $q_r = \{q_{\mu r}\}$ of the adiabatic potential $\mathscr{E}_r(q)$ (4.5.5). Therefore, these functions can be found by expanding the adiabatic potentials about q_r. Keeping the lower terms of the expansions we arrive at *the model of noninteracting harmonic oscillators*

$$\mathscr{E}_r(q) = \varepsilon_r + \sum_\mu \frac{1}{2} M_\mu \omega_{\mu r}^2 (q_\mu - q_{\mu r})^2 , \tag{4.6.2}$$

$$\mathscr{E}_r(q) - \mathscr{E}_{r'}(q) = \Delta\mathscr{E} + \sum_\mu g_\mu (q_\mu - q_{\mu r}) , \tag{4.6.3}$$

where ε_r is a purely electronic energy of the r-state, $\Delta\mathscr{E}$ the difference of the adiabatic terms at $q = q_r$, and g_μ are constants. The coordinate system q_μ is selected such as to simultaneously diagonalize two quadratic forms, $\mathscr{E}_r(q)$ and T_{vib} [the vibration kinetic energy operator is given by (4.1.2)]. In other words, q_μ are normal coordinates for the r-th electronic state, M_μ and $\omega_{\mu r}$ the effective masses and frequencies of the normal vibrations.

Equations (4.6.2, 3) define the adiabatic potential for the second electronic state,

$$\mathscr{E}_{r'}(q) = \varepsilon_{r'} + \sum_{\mu} \frac{1}{2} M_\mu \omega_{\mu r'}^2 (q_\mu - q_{\mu r'})^2 , \tag{4.6.4}$$

$$\varepsilon_{r'} = \varepsilon_r - \Delta\mathscr{E} - \sum_{\mu} (g_\mu^2 / 2 M_\mu \omega_{\mu r}^2) , \tag{4.6.5}$$

$$q_{\mu r'} = q_{\mu r} + g_\mu / M_\mu \omega_{\mu r}^2, \qquad \omega_{\mu r'} = \omega_{\mu r} . \tag{4.6.6}$$

In the harmonic approximation (4.6.2–4) the vibrational Hamiltonians of both electronic states assume the form of (4.6.1), and the variables q_μ separate. The normal coordinates q_μ correspond to noninteracting harmonic oscillators having the same masses and frequencies in the electronic states r and r' while the equilibrium positions are shifted with respect to each other as is seen from (4.6.6). The vibrational wave functions and the total energies of the states rm and $r'm'$ are

$$\chi_{rm}(q) = \prod_{\mu} \chi_{rm_\mu}(q_\mu) , \tag{4.6.7}$$

$$E_{rm} = \varepsilon_r + \sum_{\mu} \hbar \omega_{\mu r}(m_\mu + 1/2) , \tag{4.6.8}$$

$$\chi_{r'm'}(q) = \prod_{\mu} \chi_{r'm'_\mu}(q_\mu) , \tag{4.6.9}$$

$$E_{r'm'} = \varepsilon_{r'} + \sum_{\mu} \hbar \omega_{\mu r'}(m'_\mu + 1/2) , \tag{4.6.10}$$

where m and m' are written explicitly as the sets of the quantum numbers of individual vibrations $\{m_\mu\}$ and $\{m'_\mu\}$. The overlap integral of the functions (4.6.7, 9) factorizes as

$$S_{rmr'm'} = \prod_{\mu} S_{rm_\mu r'm'_\mu} , \tag{4.6.11}$$

where

$$S_{rm_\mu r'm'_\mu} = \int \chi_{rm_\mu}(q_\mu) \chi_{r'm'_\mu}(q_\mu) dq_\mu \tag{4.6.12}$$

is the overlap integral for the μ-th oscillator, which is easy to derive in an analytical form with the help of the available harmonic oscillator wave functions (see Sect. 4.8).

The quadratic members of Taylor's expansion (4.6.3) are of the form

$$\sum_{\mu\nu} g_{\mu\nu}(q_\mu - q_{\mu r})(q_\nu - q_{\nu r}) . \tag{4.6.13}$$

They cause frequency changes of the normal vibrations in the state r' as well as the Duchinsky effect, i.e., "entangling" of the normal coordinates. Of course, the adiabatic potential $\mathscr{E}_{r'}(q)$ incorporating the quadratic terms (4.6.13) may also be diagonalized (simultaneously with T_{vib}), but in a new coordinate system $\{q_{\mu'}\}$, the potential $\mathscr{E}_r(q)$ will no longer be diagonal, as it is generally impossible to diagonalize the three quadratic forms T_{vib}, $\mathscr{E}_r(q)$ and $\mathscr{E}_{r'}(q)$ simultaneously [4.86]. Therefore, when the quadratic terms are allowed for, two electronic states have different sets of normal coordinates interrelated via a linear transformation, the Duchinsky transformation,

$$q_{\mu'} = \sum_{\mu} a_{\mu'\mu} q_{\mu} + b_{\mu'} , \tag{4.6.14}$$

where $b_{\mu'} = 0$ for nontotally symmetric vibrations, and the transformation entangles only the coordinates belonging to the same symmetry species. In the case of Duchinsky's mixing, the overlap integral does not split into the product (4.6.11).

The Duchinsky effect of the normal mode mixing was discovered in optical spectra of quite a few molecules [2.42, 4.58, 87–93]. The theory was developed and applied mostly to the optical transitions, too [4.94–104]. The rotation-assisted Duchinsky effect was treated in [4.105].

The situation becomes yet more complicated with the inclusion of the higher-order (anharmonic) terms of the expansion (4.6.2) since the variables are nonseparable even in a single electronic state.

Thus, the separation of variables in (4.6.1) is obtained, provided that the normal coordinates are chosen as q_μ and the Duchinsky effect is neglected. Such an approximation in the theory of optical transitions is justified by the fact that the main contribution to the integrals of the vibrational functions comes from q near q_r, and the small values of $q - q_r$ allow for utilizing the approximate relations (4.6.2–4).

The approximation (4.6.2–4) is applied in the theory of nonradiative transitions as well. However, as was outlined in Sect. 4.5, in this case, the q values near q_*, lying far from q_r, are of major importance. This essentially restricts the range of validity of the model of noninteracting harmonic oscillators. Evidently, the anharmonicity and the Duchinsky effect are the stronger, the greater the electronic energy (4.5.18) transferred to vibrations in the radiationless transition.

To assess at which ΔE these effects are yet not prominent, the cubic term of the expansion of the one-dimensional adiabatic potential $\beta(q - q_r)^3$ is to be considered. Here, β is estimated as $\beta = \mathscr{E}/a^3$ since at $|q - q_r| \approx a$ every member of the expansion becomes as high as the electronic energy (4.2.1). We impose the limitation that the anharmonic energy at $q = q_*$ is small as compared to the vibrational quantum since otherwise the approximation of the normal vibrations loses its significance. The corresponding inequality, with the use of (4.2.4, 6), can be written as

$$|q_* - q_r| \ll b\kappa^{-1/3} . \tag{4.6.15}$$

In the harmonic model, q_* is expressed in terms of ΔE, of a displacement of the potential minima $\Delta q = q_r - q_{r'}$ and of the frequency change $\Delta\omega = \omega_r - \omega_{r'}$. Let first $\Delta\omega = 0$. Then (4.6.2, 4) yield

$$q_* - q_r = \frac{\Delta E}{M\omega^2 \Delta q} - \frac{1}{2}\Delta q . \tag{4.6.16}$$

In polyatomic molecules the displacements of the adiabatic potentials minima in low-lying electronic states typically do not exceed the amplitude of zero-point vibrations, $|\Delta q| \lesssim b$. On the other hand, the electronic energy gap must be larger than the vibrational quantum

$$\Delta E \gg \hbar\omega = M\omega^2 b^2 , \tag{4.6.17}$$

since we are using the adiabatic approximation. Then it follows that the second term in (4.6.16) can be dropped. Substituting (4.6.16) into (4.6.15), and using (4.2.4, 5) we obtain

$$\Delta E/\mathscr{E} \ll \kappa^{5/3} |\Delta q|/b , \tag{4.6.18}$$

i.e., ΔE must be very small as compared to the typical electronic energy (4.2.1).

Conditions (4.6.17, 18) restrict Δq from below, $|\Delta q| \gg \kappa^{1/3} b$. And if Δq is small, the frequency change has to be allowed for. At $\Delta q = 0$ and at a small relative frequency change we have $|q_* - q_r| = (\Delta E/M\omega|\Delta\omega|)^{1/2}$, and condition (4.6.15) leads to the inequality

$$\Delta E/\mathscr{E} \ll \kappa^{4/3} |\Delta\omega|/\omega .$$

In particular, this means that, as follows from (4.6.17) and (4.2.4), the relative frequency change should not be too small, i.e., $|\Delta\omega|/\omega \gg \kappa^{2/3}$. Thus the anharmonic energy will be small, provided that the electronic energy gap (4.5.18) is small as compared to (4.2.1), whereas the displacement of the equilibrium positions and/or the frequency change are sufficiently large.

For transitions from the excited electronic states to the ground state, the first condition usually does not apply, but it may apply to the transitions between the excited states, as the energy gaps are significantly narrower. On the contrary, the second condition is met the better, the larger ΔE, since usually $\Delta\omega$ and Δq increase with ΔE. In organic molecules the anharmonic corrections are expected to be insignificant for both the internal conversion processes from the highest to the lowest excited states and the $S_1 \rightsquigarrow T_1$ intersystem crossing, whereas for the $S_1 \rightsquigarrow S_0$ and $T_1 \rightsquigarrow S_0$ transitions the harmonic model appears to be inappropriate [4.106–108].

The Duchinsky effect as well as the anharmonicity destroy the approximation of noninteracting vibrations. The former is on the same order of magnitude as the frequency change because both these effects originate from the same quadratic terms (4.6.13) [1.16]. If the displacement of the equilibrium position is nonzero and the frequency change is small, then all the quadratic-term effects can be neglected; otherwise, the frequency change and the Duchinsky effect need be both taken into account [4.94, 95]. Despite this, because of practical difficulties only, the frequency change is often treated with no regard to the Duchinsky effect.

At large $|q_* - q_r|$, the model of normal vibrations does not hold. In this case it is appropriate to turn to *a local mode approximation* [1.3, 5, 2.241], where bond lengths are taken as the coordinates q_μ (see Sect. 2.10). As an example, consider the benzene molecule C_6H_6. Suppose that at the initial moment of time one of the hydrogen atoms is displaced far away from the equilibrium position of a given C–H bond (this may be a result of a radiationless transition or some external perturbation). In view of the strong anharmonicity the vibration frequency of this excited bond will differ significantly from the frequencies of other (unexcited) C–H bonds. The lack of a resonance slows down the energy transfer from the excited to the unexcited bond. If the transfer rate is low enough, the initially displaced hydrogen atom will oscillate long with a large amplitude, whereas the vibration amplitudes of all the rest of the atoms of the molecule remain small. Such a vibration can be termed local since the excitation is localized at one bond. This is in complete analogy with the local vibrations in solids, whose amplitudes rapidly decrease when moving away from the localization domain.

Within the model of local vibrations, the vibrational Hamiltonians of two electronic states are again written as (4.6.1), but now the individual vibrations are anharmonic. Note that the kinetic energy of local vibrations is, strictly speaking, no longer a diagonal quadratic form of momenta; this property is lost in going from atomic to relative (bond lengths) coordinates. Hence the approximation (4.6.1) corresponds now to a neglect of cross terms not only in the potential but also in the kinetic energy.

The potential energy of a local vibration in each of two electronic states is frequently approximated with the Morse functions

$$\begin{aligned} \mathscr{E}_r(q) &= \varepsilon_r + D_r\{1 - \exp[-\alpha_r(q - q_r)]\}^2 , \\ \mathscr{E}_{r'}(q) &= \varepsilon_{r'} + D_{r'}\{1 - \exp[-\alpha_{r'}(q - q_{r'})]\}^2 , \end{aligned} \tag{4.6.19}$$

where $D_{r,r'}$ are dissociation limits, $q_{r,r'}$ the equilibrium bond lengths and $\alpha_{r,r'}$ the Morse range parameters determining the steepness of the potentials in the states r and r'.

The model of noninteracting (normal or local) vibrations entails a factorization of the wave functions (4.6.7, 9). Then, the matrix elements of the nonadiabaticity operator (4.5.20) assumes the form (4.5.21) with

$$L_{rr'} = -\sum_\mu \frac{\hbar^2 C^\mu_{rr'}}{M_\mu S_{rm_\mu r'm'_\mu}} \int \chi_{rm_\mu}(q_\mu) \frac{\partial \chi_{r'm'_\mu}(q_\mu)}{\partial q_\mu} dq_\mu .$$

While the electronic matrix element depends upon the vibrational quantum numbers, this dependence is much weaker than that of the overlap integral (4.6.12). Therefore, $L_{rr'}$ is usually treated as a parameter.

4.7 Mechanisms and Selection Rules for Radiationless Transitions

The mechanisms of radiationless transitions are classified by the type of coupling between the initial and final states, and each mechanism has its own selection rules.

As was shown in Sect. 3.6, the coupling between the initial ($s = rm$) and final ($\lambda = r'm'$) states is effectuated by the matrix elements of the level-shift operator R. If only the states of the same energy are allowed for, R is merely the operator of the intramolecular coupling

$$V = L + U_{so} + T_{Cor} , \tag{4.7.1}$$

which comprises the nonadiabatic, spin–orbit and Coriolis couplings. This corresponds to the first-order perturbation treatment of the operator V. In the second order, other states $r''m''$, nonresonant with the initial state, need be incorporated. The matrix elements for the operator R follow from (3.6.16):

$$R_{rmr'm'} = v_{rmr'm'} + \sum_{\substack{r''m'' \\ r'' \neq r, r'}} \frac{v_{rmr''m''} v_{r''m''r'm'}}{E_{rm} - E_{r''m''}} , \tag{4.7.2}$$

where $E = E_{rm}$ is assumed in the denominator [see (3.6.23)] and the imaginary term $i\eta$ is omitted, since the denominator does not vanish.

The first term in (4.7.2) is the matrix element of the intramolecular interaction V between the resonant states, while the second involves the matrix elements of V between nonresonant states; accordingly, we speak of the *resonant* and *nonresonant couplings*. The nonresonant coupling becomes essential when the resonant coupling is weak, or even is at all absent because of selection rules.

In the case of the internal conversion, i.e., a transition between states of equal multiplicities, the matrix element of the non-adiabaticity operator is nonzero. Its electronic factor, estimated in Sect. 4.2, is on the order of the energy of the vibrational quantum, which largely exceeds the magnitude of the spin–orbit and Coriolis couplings in polyatomic molecules. Therefore, when calculating the matrix elements of V between states of equal multiplicities, one may neglect the spin–orbit and Coriolis couplings. The nonresonant nonadiabatic coupling is also weak as compared to the resonant coupling. Therefore, for the internal conversion one has simply

$$R_{rmr'm'} = L_{rmr'm'} . \tag{4.7.3}$$

Thus, *the mechanism of internal conversion is characterized by the resonant nonadiabatic coupling*. In other words, *the internal conversion is a nonadiabatic transition*, i.e., a transition induced by the non-adiabaticity of the nuclear motion. In the Condon approximation (4.5.21) the selection rules for the internal conversion are governed by the selection rules for the electronic matrix element $L_{rr'}$ and the Franck–Condon factor (4.5.2).

Additionally to the nonadiabatic mechanism (4.7.3), the *mechanisms of the spin–orbit-induced triplet–triplet internal conversion* and *the Coriolis-induced singlet–singlet conversion* are provided by the resonant coupling in the first-order perturbation theory. These mechanisms are responsible for small but rotation-dependent contributions to the internal conversion rate constants.

The mechanisms of intersystem crossing were discussed by *Henry* and *Siebrand* [1.4, 5, 2.236, 276, 4.109–111]. The transition matrix element of the non-adiabaticity operator vanishes between states of different multiplicity since L does not operate on spin variables. Therefore, in the first order of the perturbation theory, the intersystem crossing is caused by *the adiabatic spin–orbit coupling*

$$R^{(1)}_{rmr'm'} = U_{rmr'm'} \,. \tag{4.7.4}$$

If the Condon approximation (4.5.16) is invoked, one has *the mechanism of the direct spin–orbit coupling*.[9] The selection rules for this mechanism are determined by those for the electronic matrix element (4.5.17) and for the Franck-Condon factor. If $U_{rr'} = 0$ by symmetry, one has to allow for a non-Condon contribution, i.e., the next term in expansion (4.5.9) of the function $c_{rr'}(q)$ near $q = q_*$. This is *the mechanism of the vibration-induced, or vibronic spin–orbit coupling* [1.31, 2.99, 4.24, 25, 112–116] similar to the Herzberg-Teller mechanism for optical transitions.

The direct spin–orbit coupling mechanism is highly sensitive to the presence of a heavy atom in the molecule since the spin–orbit interaction strength increases with atomic number. An essential enhancement of the intersystem-crossing rate after the implantation of a heavy atom in the molecule, *the internal heavy-atom effect*, was observed for a few molecules [4.112, 117–120]. Another way to accelerate the intersystem crossing is based on the so-called *external heavy-atom effect*, where the acceleration is caused by collisions with a heavy-atom partner. In this case one has *the collision-induced mechanism for intersystem crossing* [1.12, 2.50, 109, 4.121–130]. The external heavy-atom effect is also seen in van der Waals complexes of organic molecules with rare gas atoms [2.122, 164], in solid matrices where the heavy atom is inserted into the host lattice [2.290, 4.131–135], and in solutions [2.44c]. The lack of the heavy-atom

---

[9] A different meaning of this term is also found in literature [1.31], implying a one step $S_1 \rightsquigarrow T_1$ process, as opposed to a two-step $S_1 \rightsquigarrow T_x \rightsquigarrow T_1$ process, involving an intermediate (unknown) triplet state T_x.

effect [4.136] is treated as an indication that the intersystem crossing is of little importance.

In the second order of the perturbation theory for the intersystem crossing process, the nonresonant spin–orbit and nonadiabatic interactions have to be included simultaneously since the states of the same multiplicity as either r or r' always exist among the intermediate states r''. Again, neglecting the matrix elements of U_{so} between states of equal multiplicity we obtain

$$R^{(2)}_{rmr'm'} = \sum_{\substack{r''m'' \\ r'' \neq r, r'}} \frac{U_{rmr''m''} L_{r''m''r'm'} + L_{rmr''m''} U_{r''m''r'm'}}{E_{rm} - E_{r''m''}} . \tag{4.7.5}$$

This is the *combined* (spin–orbit plus nonadiabatic) *nonresonant coupling mechanism.* In the same way, other combined mechanisms emerge, for instance, the vibronic spin–orbit plus nonadiabatic coupling [4.41]. When rotation is taken into account, there appear the spin–rotation and electronic–orbit-rotation mechanisms [4.41, 42], as well as the spin–orbit-Coriolis mechanism [4.37, 38].

The combined second-order mechanism (4.7.5) may compete with the direct first-order one (4.7.4) because the spin–orbit coupling is small as compared to the nonadiabatic one. If the matrix elements in (4.7.4, 5) are on the same order of magnitude, cross terms arise in the factor $|R_{rmr'm'}|^2$ entering the radiationless transition-rate constant, and the above relatively simple classification of the intersystem crossing mechanisms loses its significance. Further discussion of the internal conversion and intersystem crossing mechanisms is given in [1.4, 5, 30, 42, 52, 2.202, 236, 276, 308, 4.109–111, 137].

Selection rules for a particular mechanism establish for which $r \rightsquigarrow r'$ transitions the corresponding matrix elements do not vanish. As discussed earlier in this chapter, the total wave function of a two-electron molecular system in the adiabatic approximation is represented as a product of the electron–spin, electron–orbit, vibration, and rotation functions, or as a linear combination of such products. Correspondingly, one may speak of the spin, orbit, vibration, and rotation (or angular momentum) selection rules.

The spin selection rules were discussed before in this subsection. The nonadiabatic and Coriolis couplings mix only the states of similar multiplicity $\Delta S = 0$, while the spin–orbit coupling mixes states of equal (except for singlets) or different multiplicity, $\Delta S = 0$ and ± 1, where ΔS is a change of the total spin S in the radiationless transition.

The orbital selection rules for the electronic matrix elements $C^{\mu}_{rr'}(q)$ and $c_{rr'}(q)$ are governed by the symmetry group of the configuration q. For example, $C^{\mu}_{rr'}(q)$ is nonzero provided that the direct product of the irreducible representations (the symmetry species), by which the electronic wave functions $\varphi_r(x, q)$ and $\varphi_{r'}(x, q)$ and the nuclear operator $\partial/\partial q_\mu$ transform, contains the unit representation [1.15, 24, 40].

In the Condon approximation (4.5.8, 9) the orbital selection rules are determined by the symmetry group of the configuration q_* where the electronic terms cross[10] since, according to (4.5.19, 17), the electronic matrix elements $C^{\mu}_{rr'}$ and $U_{rr'}$ are expressed in terms of the corresponding functions at $q = q_*$. *This is essentially different from the widely used assumption that the selection rules are determined by the symmetry group of the equilibrium configuration* q_r, that is, the same symmetry group as for the optical transitions [1.4, 5, 10]. The latter assumption is valid provided that only totally symmetric vibrations are active in the radiationless transition since in this case the symmetry of the configuration q_* is precisely the same as that of the equilibrium one, q_r. For nontotally symmetric active vibrations this assumption may hold only approximately provided q_* is not too far from q_r, $|q_* - q_r| \ll d$, where d, the length scale for variations of the electronic wave functions, is given by (4.2.13). This results in some limitations on the electronic energy transferred to vibrations in the course of the radiationless transition, and on the change of the parameters of the adiabatic potentials, which are easy to deduce from the above inequality similarly to how it was done in Sect. 4.6 (see (4.6.15) and what follows).

For transitions between states of equal multiplicity the electronic matrix element of the non-adiabaticity operator $L_{rr'}$ is always nonvanishing. Indeed, the set of coordinates q_μ is complete in the sense that it embraces all irreducible representations of the symmetry groups of any molecular configuration, that is, among those coordinates one can always find such one that transforms by a given arbitrarily chosen representation. For this reason the constants $C^{\mu}_{rr'}$, which are the electronic matrix elements of the operator $\partial/\partial q_\mu$, cannot disappear all at a time. In other words, *no orbital selection rules exist for the non-adiabaticity operator*, hence for the internal conversion processes. The vibrations μ for which $C^{\mu}_{rr'} \neq 0$ are called *promoting modes* [1.5, 2.316, 4.138].

The discussion of the electronic spin–orbit matrix element is appropriate to begin with El-Sayed's rule of Sect. 2.2 [1.31, 51] stating that the intersystem crossing rate is by 2 to 4 orders of magnitude higher for the $n\pi^* \leftrightarrow \pi\pi^*$ than for the $n\pi^* \leftrightarrow n\pi^*$ and $\pi\pi^* \leftrightarrow \pi\pi^*$ transitions. This pertains to the aromatic hydrocarbons whose planarity leads to a cancellation of one-center integrals in the spin–orbit electronic matrix element [1.45] so that it becomes less than 1 cm^{-1} for transitions between similar configurations ($\pi\pi^* \leftrightarrow \pi\pi^*$), while it is about 50 cm^{-1} if different configurations are involved ($\pi\sigma^* \leftrightarrow \pi\pi^*$ or $n\pi^* \leftrightarrow \pi\pi^*$) [1.3, 23, 29–34, 4.22, 139]. El-Sayed's rule leads to a large difference in the populations of the triplet spin sublevels of impurities in solids at low temperatures, i.e., to a large spin polarization and alignment [1.31, 2.291, 4.140]. The spin–orbit matrix element is enhanced appreciably if the planarity is broken down, e.g., as a result of the formation of an eximer molecule [4.141] or of participation of out-of-plane vibrations [4.113].

---

[10] Although q_* takes on nonphysical values, one may speak of a symmetry group of the "configuration" q_* in the same sense as of any real configuration of the molecule. Namely, this is the group of electron coordinate transformations keeping the potential energy of the molecule unchanged at the fixed value of $q = q_*$.

The orbital selection rules for the spin–orbit operator (4.4.8) coincide with those for the axial vector (4.4.9). Consider them in more detail invoking the results of Sect. 4.4.

In the discussion of the intersystem crossing due to the spin–orbit coupling, the rotation should in general be taken into account. It is essential in Hund's case B ($T_{rot} \gg U_{so}$), of course if the experimental conditions are such that the initial excitation is selective enough the discriminate between the individual rotational states.

We start, however, with Hund's case A, when the rotation is insignificant and the spin–orbit matrix element is given by (4.4.21), where σ' is the T_1 triplet-state total-spin projection on the body-fixed quantization axis directed along the intramolecular self-consistent magnetic field acting on the electron spins, and $c_{rr'}^{(\sigma')}$ are the spherical components of the purely electronic matrix element of the axial vector between the orbitally nondegenerate singlet (r) and triplet (r') states [see definition (4.4.11)]. If the molecular configuration at which the constant $\boldsymbol{c}_{rr'}$ is calculated (see Sect. 4.5) does not show any symmetry, the vectors $\boldsymbol{c}_{rr'} = (c_{rr'x}, c_{rr'y}, c_{rr'z})$ and $\boldsymbol{c}_{T_1T_1} = (0, 0, |\boldsymbol{c}_{T_1T_1}|)$ are not parallel and the spherical components (4.4.18), hence the matrix elements (4.4.21), too, are all nonvanishing. If there is a center of inversion, they are nonzero only for transitions between states of the same parity. The presence of other symmetry properties imposes further restrictions on the intersystem crossing transitions [1.45].

For example, a molecule with spherical-top symmetry in nondegenerate electronic states has no axis of preference along which a vector like $\boldsymbol{c}_{rr'}$ or $\boldsymbol{c}_{T_1T_1}$ could be directed. Therefore, $\boldsymbol{c}_{T_1T_1} = \boldsymbol{c}_{rr'} = 0$, i.e., there are neither triplet state splittings nor intersystem crossings induced by the direct spin–orbit mechanism. In a formal way this orginates from the spherical symmetry of the coordinate electronic wave functions of the nondegenerate states which lead to vanishing of the matrix elements (4.4.11) of a vector. The intersystem crossings by the direct mechanism in a spherical-top molecule are only allowed with an involvement of orbitally degenerate states.

A molecule with a symmetric-top configuration in a nondegenerate electronic state has the internal magnetic field and the vector $\boldsymbol{c}_{rr'}$ both directed along the only preferential direction, the top axis. Hence, $c_{rr'x} = c_{rr'y} = 0$, and from definition (4.4.18) we have $c_{rr'}^{(\sigma')} = 0$ at $\sigma' \neq 0$. On the basis of (4.4.21) we come to the conclusion that only one of three triplet sublevels, viz., that with $\sigma' = 0$, efficiently takes part in the singlet–triplet transitions. Thus this sublevel alone is preferentially populated during the $S_1 \rightsquigarrow T_1$ transition, and it is just the same level to be rapidly depopulated via the $T_1 \rightsquigarrow S_0$ transition. Note that these arguments do not pertain to the aromatic molecules that have anomalously weak spin–orbit coupling [1.45–49, 4.22], so spin–spin interactions become significant [4.140].

In Hund's case B the selection rules come from (4.4.32), and they establish the limitations on changes in the rotational quantum numbers J, N and K. The quantum number J is conserved as the total angular momentum commutes with the spin–orbit operator, while the restrictions imposed on changes of the quantum numbers of the rigid-body momentum and its projection on the molecular quantization axis stem from the known properties of Wigner's 3-J symbols [4.18], or, which is the same, from the momentum addition rules.

In (4.4.32) the so-called *parallel* spin–orbit coupling (the term with $\sigma' = 0$) and the *perpendicular* one (the terms with $\sigma' = \pm 1$) are distinguished. Using the properties of the 3-J symbols we obtain the selection rules [2.197, 198, 4.38] for the parallel component,

$$\Delta J = \Delta N = \Delta K = 0 \, , \tag{4.7.6}$$

and for the perpendicular one,

$$\Delta J = 0, \qquad \Delta N = \pm 1, \qquad \Delta K = \pm 1 \, . \tag{4.7.7}$$

Generally both components are present and the united selection rule (4.7.6, 7) applies.

Additional limitations are imposed on the rotation quantum number changes by the molecular symmetry. For instance, in the case of a symmetric top, if the transition is between nondegenerate electronic states with an involvement of totally symmetric vibrations only, one has, as before, $c^{(\sigma')}_{rmr'm'} = 0$ at $\sigma' \neq 0$. It follows that here the perpendicular component is absent and the selection rules are given by (4.7.6).

In the case of internal conversion, the rotational selection rules result from the fact that the nonadiabatic coupling arising from the nonadiabaticity of the vibrational motion is independent of the Euler angles, hence it is diagonal in the rotational quantum numbers J, N and K, and is independent of them. The internal conversion may also be induced by the spin–orbit and Coriolis couplings. In the case of the singlet–singlet conversion, we have (4.7.6) for the parallel component of the Coriolis perturbation and

$$\Delta N = 0, \qquad \Delta K = \pm 1 \tag{4.7.8}$$

for the perpendicular one (4.4.36–38). The Coriolis contribution to the rate constant is usually neglected but one should bear in mind that it is rotation-dependent and increases rapidly with N. *Riedle* et al. [4.55–61] found that the Coriolis-induced vibrational coupling *within the* S_1 *state* is responsible for the rotation effects in the benzene spectra in the "channel 3" region; the very interesting problem of the "channel 3" in benzene is not considered here, and the reader is referred to literature [2.1, 64, 165, 4.142–159], where the "channel-three-like" behavior of other molecules is also discussed [4.160–162].

Finally, the vibrational selection rules are considered. The vibrations providing major contributions to the overlap integral are called *accepting vibrations* since they are the best acceptors of the electronic energy (see Sect. 2.10). The accepting properties of a given mode are the better, the stronger the shape distortion and the equilibrium-position displacement of the adiabatic potential $\mathscr{E}_{\mu r}(q_\mu)$ between the electronic states. In particular, if the shape and the equilibrium position undergo no changes between two states, i.e., the difference

$\mathscr{E}_{\mu r}(q_\mu) - \mathscr{E}_{\mu r'}(q_\mu)$ is a constant, then the functions $\chi_{rm_\mu}(q_\mu)$ and $\chi_{r'm'_\mu}(q_\mu)$ are mutually orthogonal, and the strict selection rule $m_\mu = m'_\mu$ holds for the Franck–Condon factor. However, in general, no rigorous selection rules exist for the Franck–Condon factors.

At small equilibrium-position displacements and potential-shape distortions the different vibrational modes can be classified by their accepting properties. Then, if a labels poor accepting modes, this approximate selection rule is valid: $m_a \approx m'_a$.

A peculiar selection rule can be formulated for an anharmonic oscillator (see Sect. 5.3.3). When applied to the molecules containing C–H bonds (particularly to the aromatic hydrocarbons) it reads: at large values of the electronic energy ($\Delta E \gtrsim 2.5$ eV) the radiationless transition is allowed, provided that the C–H bond length in the higher electronic state is shorter than in the lower, otherwise it is forbidden with a quenching factor on the order of 10^2–10^4.

4.8 Overlap Integrals for Harmonic and Morse Oscillators

The dependence of the matrix element of the intramolecular interaction on the vibrational quantum numbers and on the adiabatic potential parameters of the initial and final electronic states mainly arises through the overlap integral of the vibrational wave functions. Within the approximation of non-interacting oscillators it is represented as the product (4.6.11) of the one-dimensional overlap integrals (4.6.12) that in some important cases can be evaluated accurately.

First, consider the harmonic oscillator. The nuclear potential of the upper electronic state and the energy of the m-th vibronic state are written in the form (4.6.2, 8), where the subscript μ will be omitted. The normalized wave function is

$$\chi_{rm}(q) = \left(\frac{1}{2^m m!\, b_{r0}\sqrt{\pi}}\right)^{1/2} H_m(x_r)\exp\left(-\frac{1}{2}x_r^2\right), \tag{4.8.1}$$

where $H_m(x)$ is the Hermite polynomial, $b_{r0} = (\hbar/M\omega_r)^{1/2}$ is the amplitude of zero-point vibrations and

$$x_r = (q - q_r)/b_{r0} \tag{4.8.2}$$

is the dimensionless coordinate. The corresponding expressions for the lower electronic state r' are obtained by replacing rm with $r'm'$ in the above formulas.

It is appropriate to introduce the frequency ratio $f = \omega_r/\omega_{r'}$, the displacement of the equilibrium position $\Delta q = q_r - q_{r'}$ and the Huang–Rhys dimensionless parameter $\xi = (\Delta q/b_{r0})^2/2$, that defines the Stokes' shift in the optical spectra [1.15, 16, 22, 24, 25, 4.2, 163].

We calculate the overlap integral (4.2.18) at $m' \geqslant m$. After the integration variable q is replaced by $x_{r'}$, it is recast in the form

$$S_{rmr'm'} = \left(\frac{\sqrt{f}}{2^{m+m'} m! \, m'! \, \pi}\right)^{1/2} \int_{-\infty}^{\infty} \exp[-(x_r^2 + x_{r'}^2)/2]$$

$$\times H_m(x_r) H_{m'}(x_{r'}) \, dx_{r'} \, , \tag{4.8.3}$$

where $x_r = x_{r'}\sqrt{f} - (\Delta q/b_{r0})$. First of all we note that after the simultaneous replacements of Δq with $-\Delta q$ and of the integration variable $x_{r'}$ with $-x_{r'}$ the arguments of both Hermite polynomials in (4.8.3) change their signs. Calling on the known properties of the Hermite polynomials [4.164] we obtain

$$S_{rmr'm'}(-\Delta q) = (-1)^{m+m'} S_{rmr'm'}(\Delta q) \, , \tag{4.8.4}$$

whence it follows that the Franck–Condon factor (4.5.2) is independent of the sign of Δq. Now the overlap integral (4.8.3) is calculated for $\Delta q \geq 0$.

We define the generating function for the overlap integral as

$$G(z) = \sum_{m=0}^{\infty} \frac{(z-1)^m \xi^{m/2}}{\sqrt{m!}} S_{rmr'm'} \, . \tag{4.8.5}$$

Exploiting the well-known formula for the Hermite polynomials' generating function and performing the change of the integration variable in (4.8.3), $x_{r'} = x[2/(f+1)]^{1/2}$, we obtain

$$G(z) = \left(\frac{2\sqrt{f}}{2^{m'} m'! \, \pi (f+1)}\right)^{1/2} \exp\left(v^2 - \frac{\xi}{f-1}\right) \int_{-\infty}^{\infty} \exp[-(x-y)^2]$$

$$\times H_{m'}\left[x\left(\frac{2}{f+1}\right)^{1/2}\right] dx \, ,$$

$$v = \left(\frac{\xi(f-1)}{2(f+1)}\right)^{1/2} \left(\frac{f+1}{f-1} - z\right); \qquad y = z\left(\frac{\xi f}{f+1}\right)^{1/2} .$$

The integral can be found in common textbooks [2.213, 4.164]. After the integration, the generating function becomes

$$G(z) = \left(\frac{2\sqrt{f}}{2^{m'} m'! (f+1)}\right)^{1/2} \exp\left(v^2 - \frac{\xi}{f-1}\right) \left(\frac{f-1}{f+1}\right)^{m'/2}$$

$$\times H_{m'}\left[z\left(\frac{2\xi f}{f^2-1}\right)^{1/2}\right] . \tag{4.8.6}$$

This formula is valid for any values of $f \neq 1$; for $f = 1$ the generating function is easily obtained by means of taking the limit of $f \to 1$. For $f < 1$ (4.8.6) formally contains imaginary expressions since $(f-1)^{1/2} = \pm i(1-f)^{1/2}$. However, using the properties of the Hermite polynomials, it is easily verified that, as expected, expression (4.8.6) remains real for $f < 1$, too, so the choice of the above root sign is not important.

The definition (4.8.5) implies that the overlap integral is proportional to the m-th derivative of the generating function for $z = 1$,

$$S_{rmr'm'} = \frac{1}{\sqrt{\xi^m m!}} \left(\frac{d^m G(z)}{dz^m} \right)_{z=1} .$$

Differentiation of (4.8.6) yields

$$\frac{d^m G(z)}{dz^m} = B \sum_{k=0}^{m} C_m^k \left[\frac{d^{m-k}}{dz^{m-k}} \exp(v^2) \right] \left\{ \frac{d^k}{dz^k} H_{m'} \left[z \left(\frac{2\xi f}{f^2 - 1} \right)^{1/2} \right] \right\} ,$$

where B absorbs all factors independent of z. Here, the derivative in brackets is proportional to $H_{m-k}(iv)$ while that in braces to

$$H_{m'-k} \left[z \left(\frac{2\xi f}{f^2 - 1} \right)^{1/2} \right] ,$$

The case of $m' \leq m$ is treated in a similar way. The final expression [4.165, 166] is easy to transform in such a way that it would not involve any imaginary quantities [2.54]. For $f < 1$ it is

$$\begin{aligned} S_{rmr'm'} = & \left(\frac{2\sqrt{f}}{(1+f)m!m'!2^{m+m'}} \right)^{1/2} \exp[-\xi/(1+f)] \\ & \times \sum_{k=0}^{\min(m,m')} \frac{m!m'!}{(m-k)!k!(m'-k)!} \left(\frac{4\sqrt{f}}{1+f} \right)^k \left(\frac{1-f}{1+f} \right)^{(m+m')/2-k} \\ & \times H_{m-k} \left[\left(\frac{2\xi}{1-f^2} \right)^{1/2} \right] F_{m'-k} \left[\left(\frac{2\xi f}{1-f^2} \right)^{1/2} \right] , \end{aligned} \tag{4.8.7}$$

where $F_m(x) \equiv \mathrm{i}^m H_m(\mathrm{i}x)$ is real for real x. For $f > 1$ one has to replace, on the right-hand side, f with $1/f$, thereupon ξ with ξ/f, then exchange m and m', and multiply the right-hand side with $(-1)^{m+m'}$.

For a transition starting at the $m = 0$ ground vibrational state the overlap integral (4.8.7) reduces to [2.184, 4.167, 168]

$$\begin{aligned} S_{r0r'm'} = & \left(\frac{2\sqrt{f}}{(1+f)m'!2^{m'}} \right)^{1/2} \\ & \times \exp[-\xi/(1+f)] \left(\frac{1-f}{1+f} \right)^{m'/2} F_{m'} \left[\left(\frac{2\xi f}{1-f^2} \right)^{1/2} \right] . \end{aligned} \tag{4.8.8}$$

For a nontotally symmetric vibration the equilibrium-position displacement vanishes (see Sect. 4.1). On letting $\xi = 0$ in (4.8.8) one obtains $S_{r0r'(2n+1)} = 0$ and

$$S_{r0r'(2n)} = \left(\frac{2\sqrt{f}}{1+f} \right)^{1/2} \frac{\sqrt{(2n)!}}{n!2^n} \left(\frac{1-f}{1+f} \right)^n . \tag{4.8.9}$$

Equations (4.8.7, 8) are simplified if the oscillator has the same frequency in both electronic states. Taking the limit of $f \to 1$ in (4.8.7) one obtains for $m \leq m'$

$$S_{rmr'm'} = \left(\frac{m!}{m'!}\xi^{m'-m}\mathrm{e}^{-\xi}\right)^{1/2} L_m^{m'-m}(\xi)\,, \tag{4.8.10}$$

where $L_m^{\alpha}(x)$ is the associated Laguerre polynomial, while for $m \geq m'$ one has to transpose m and m' on the right-hand side of (4.8.10). For $m = 0$

$$S_{r0r'm'} = \left(\frac{1}{m'!}\xi^{m'}\mathrm{e}^{-\xi}\right)^{1/2}. \tag{4.8.11}$$

It is interesting to compare (4.8.9) and (4.8.11) for the overlap integral from the $m = 0$ state in the cases where either a frequency change or an equilibrium-position shift is present. The prominent difference between them is that the denominator of (4.8.11) involves $\sqrt{m'!}$, whereas only a power factor survives in (4.8.9) for large n. Therefore, for comparable values of the relative displacements of the equilibrium positions and of the relative frequency changes, the frequency effect must result in a higher Franck–Condon factor [1.5, 4.169].

For a Morse oscillator (4.6.19) the energy and the normalized wave function of the vibronic state rm are

$$E_{rm} = \varepsilon_r + \hbar\omega_r\left(m + \frac{1}{2}\right) - \frac{(\hbar\omega_r)^2}{4D_r}\left(m + \frac{1}{2}\right)^2, \tag{4.8.12}$$

$$\chi_{rm}(q) = \left(\alpha_r\frac{m!\,2(\sigma_r - m)}{\Gamma(2\sigma_r - m + 1)}\right)^{1/2}\exp(-x_r/2)\,x_r^{\sigma_r - m} L_m^{2\sigma_r - 2m}(x_r)\,,$$

$$\sigma_r = \frac{2D_r}{\hbar\omega_r} - \frac{1}{2}, \qquad x_r = (2\sigma_r + 1)\exp[-\alpha_r(q - q_r)]\,, \tag{4.8.13}$$

$$\omega_r = \alpha_r\sqrt{2D_r/M}\,,$$

where $\Gamma(x)$ is the gamma function and ω_r is the frequency of low-amplitude vibrations in the Morse potential. A precise analytical expression for the overlap integral is only known in the case $\alpha_r = \alpha_{r'} \equiv \alpha$. Writing the Laguerre polynomial of (4.8.13) as a sum in powers of x_r and integrating (4.6.12) term by term one obtains [4.170–172]

$$\begin{aligned} S_{rmr'm'} = {} & 2\left[m!\,m'!(\sigma_r - m)(\sigma_{r'} - m')\frac{\Gamma(2\sigma_r - m + 1)}{\Gamma(2\sigma_{r'} - m' + 1)}\right]^{1/2} \\ & \times(1 - z)^{\sigma_r - m}(1 + z)^{\sigma_{r'} - m'}\sum_{k=0}^{m}\frac{(-1)^{k+m'}\Gamma(\sigma_r + \sigma_{r'} - m - m' + k)}{k!(m - k)!\,\Gamma(2\sigma_r - 2m + k + 1)} \\ & \times(1 - z)^k P_{m'}(\sigma_r - \sigma_{r'} - m + k - 1, 2\sigma_{r'} - 2m'; 1 + 2z)\,, \end{aligned} \tag{4.8.14}$$

where again $f = \omega_r/\omega_{r'}$ and $\Delta q = q_r - q_{r'}$,

$$z = \frac{1-\lambda}{1+\lambda}, \qquad \lambda = \frac{2\sigma_r + 1}{2\sigma_{r'} + 1}\exp[\alpha(q_r - q_{r'})] \equiv f e^{\alpha \Delta q}, \tag{4.8.15}$$

and $P_n(\beta, \gamma; x) \equiv P_n^{(\beta,\gamma)}(x)$ is the Jacobi polynomial [4.164]. Some particular cases of (4.8.15) are investigated in [4.106, 107, 171].

The adiabatic potential for an out-of-plane vibration is sometimes fitted with the Rosen-Morse (modified Pöschl–Teller) potential $\mathscr{E}(q) = -\mathrm{const}/\cosh^2\alpha q$. An analytical form of the Rosen–Morse overlap integral was given in [4.173]. Arbitrary diatomic potentials were investigated in [4.174], where general recurrence relations were obtained for the overlap integral of the sort of Ansbacher's relations for the harmonic oscillator.

5. Quasiclassical Methods

Quasiclassical methods based of the JWKB approximation provide a powerful method for investigating the effect of nuclear motion on radiationless transitions. While being widely accepted in the theory of electronic transitions in atomic and molecular collisions, they have not yet been given proper attention in the theory of radiationless transitions because of their complexity and, which may be even more important, because the bound-state overlap integrals entering the theory are easily calculated quantum-mechanically, without invoking the quasiclassical approximation.

In fact, the utility of this approximation for the problems under study in this book is in no way associated with the easiness of numerical calculations, which in some cases may even be more complicated than direct quantum calculations. Rather, the quasiclassical results provide a new insight into the physical mechanism governing nonradiative and radiative transitions between electronic states. In particular, they reveal that some novel knowledge about the potential energy surfaces, which cannot be obtained from the available spectroscopic data and, hence, cannot be used in the quantum calculations, is needed to calculate the transition rate correctly.

The chapter opens with the introductory Sect. 5.1, where physical and mathematical foundations of the methods as well as their previous applications are briefly reviewed. Sections 5.2 and 5.3 describe in detail the derivation of the principal results, and their fundamental physical implications are formulated in Sect. 5.4.

The rest of the chapter is devoted to problems that seem, at first glance, to have nothing to do with radiationless transitions. Nevertheless, all of them have very much in common. Thus, a model of parallel terms considered in Sect. 5.5 relates to radiative and nonradiative transitions between different terms, as well as to radiative transitions within a single term. The latter case, being very important from the viewpoint of both verification of the applicability of the quasiclassical method and interpretation of the experimental data on overtone transitions, is individually discussed in Sect. 5.6. The collision model for radiationless transitions, described in Sect. 5.7, and the calculation of vibronic levels for two coupled electronic states in Sect. 5.8 give further examples of the application of the method to the problems involving intramolecular interstate coupling.

5.1 Introductory Remarks

In the calculations of the overlap integrals in Sect. 4.8 the specificity of nonradiative transitions was in no way taken into account. The formulae derived there are applicable for any values of the potential parameters and of the vibrational quantum numbers, and hence, are appropriate to describe both nonradiative and radiative transitions for an arbitrary change in quantum numbers.

A fundamental feature of the nonradiative transitions is that they are subject to definite limitations imposed by the energy conservation law (4.5.10) on the vibrational quantum numbers of the final electronic state. Representing the total energy of the molecule as a sum in (4.1.30) we obtain from (4.5.10) the vibrational energy in the final state as

$$\varepsilon_{r'm'} = \varepsilon_{rm} + \Delta E \ , \qquad (5.1.1)$$

where ΔE is a purely electronic energy gap of (4.5.18). In the absence of electronic (quasi)degeneration the electronic energy ΔE may greatly exceed the typical energy of the vibration quantum, and then the final state $r'm'$ corresponds to a large vibrational excitation at any values of the initial vibrational energy ε_{rm}. If, in addition, as is very typical for polyatomic molecules, the number of active modes in the nonradiative transition (see Sect. 2.10) is not large, condition (5.1.1) entails large values of the final vibrational numbers of these modes. Hence, the vibrational wave function of the final state is quasiclassical, enabling a calculation of the overlap integrals by means of the quasiclassical (JWKB) method.

This approach was elaborated by Landau as early as in 1932 [1.39, 40]. The integration path is moved from the real axis into the complex plane circumventing the turning points for the classical motion in both of the terms. Far away from the turning points the wave functions may be substituted by their quasiclassical expressions (for *any* values of the vibrational quantum numbers), and then the integral is performed by the saddle-point method. The criterion of applicability of the quasiclassical approximation consists in that the saddle points lie far enough from the turning points.

The Landau method was successfully applied in calculations of probabilities of tunneling transitions in a wide variety of physical systems. Specifically, *Landau* [1.39] calculated the probabilities of radiationless transitions in atomic second-kind collisions in both the overbarrier (classical) and underbarrier (tunneling) regions, and in *Landau–Lifshitz*'s book [1.40] this method was employed to solve a general problem how to calculate the quasiclassical matrix elements of an arbitrary operator. It is true that in the problems mentioned both the initial and final states are quasiclassical. In contrast, in the case of radiationless transitions in molecules the initial state is frequently not a quasiclassical state because the radiationless transitions usually start from the ground or a low-lying excited vibronic level m of the initial electronic state r. Nevertheless, as we shall see later, under very minor restrictions on the potential parameters, the *quasiclassical method applies for any value of m* including $m = 0$, provided that the condition

$$m' \gg 1 \tag{5.1.2}$$

is fulfilled; as follows from (5.1.1) it is equivalent to (4.6.17), where $\omega = \omega_{r'}$ is the frequency of low-amplitude vibrations in the final electronic state. The rationale behind this result is that under the conditions stated the dominant contribution to the integral along the contour comes from those regions of the complex q-plane which are far enough from the turning points for the classical motion in both terms r' and r; and far from the turning points the wave function $\chi_{rm}(q)$ is quasiclassical, even for small values of m.

As we shall see later, the saddle points for the overlap integral are real and complex points of intersection of the electronic terms. If the electronic terms cross on the real axis and the initial state energy is much higher than the potential energy at the crossing point (the overbarrier passage), the problem is greatly simplified since the saddle points are in the classically allowed region and there is no necessity in any analytical continuation of the JWKB wave functions into the complex plane. A straightforward calculation of the transition-rate constant in this case immediately leads to the Landau-Zener formula [1.39, 40, 5.1]. *Lax* [4.69] proposed a different version of the quasiclassical method and applied it to derive the quasiclassical Franck–Condon principle in the theory of optical transitions. *Rebane* [5.2] employed the Lax method to evaluate the radiationless transition rate and obtained the Landau-Zener expression valid only in the overbarrier region. The Landau-Zener probability was also recovered in the calculations of the wave-packet dynamics in the limit of strong coupling [5.3], and within a steady-state approach at high initial vibrational excitation [5.4]. Formosinho [5.5] calculated the transition rate in the tunneling region quasiclassically, but obtained an incorrect result because he disregarded the important role of the crossing points of the electronic terms. In the literature there are numerical calculations of the Franck–Condon factors for the Morse potential using the JWKB wave functions [2.53, 5.6, 7].

At small initial vibrational energies, when no real term-crossing points exist in the classically accessible region, the radiationless transition is of a tunneling nature. Applying the quasiclassical method in the tunneling case meets severe mathematical difficulties, which, nevertheless, can well be overcome, and simple-enough analytical expressions for the overlap integrals of various potentials can be deduced [5.8] (see Sect. 5.2, 3).

The most important result of the application of the quasiclassical method is the Franck–Condon principle for nonradiative transitions [1.39, 5.8, 9] formulated in Sect. 5.4. This principle has interesting implications with regard to the electronic and vibrational selection rules (see Sects. 5.3.3 and 5.4.4).

If the initial vibrational energy is high enough so that the nuclear motion in the upper electronic state may also be treated quasiclassically, then the collision model of Sect. 5.7 can be applied to calculate the transition rate. Of special interest is the analysis of the nonadiabatic transitions within a one-dimensional model since the quasiclassical method enables one to find a rigorous solution to this problem with no resort to the adiabatic-perturbation theory.

5.2 Overlap Integral for a Harmonic Oscillator

5.2.1 Basic Derivation

In this section we describe in detail the quasiclassical approach to the calculation of the overlap integral (4.6.12) for a harmonic oscillator, with an emphasis placed on this specific example to illustrate the basic ideas. Due to the property (4.8.4) of the harmonic-oscillator overlap we can assume $\Delta q > 0$. Note that notations of Sect. 4.8 are used below.

The analyticity of the vibrational wave functions within every arbitrary bounded domain of the complex plane permits to displace the integration path in (4.6.12) from the real axis. Thus, the overlap integral is written as (the arguments of the wave functions are omitted)

$$S_{rmr'm'} = \int_L \chi_{rm}\chi_{r'm'}\,dq\,, \tag{5.2.1}$$

where L is an arbitrary contour going from the left-hand to the right-hand half-plane, avoiding all turning points for the classical motion both in the term r,

$$q_{rm}^{\pm} = q_r \pm b_{rm} \equiv q_r \pm b_{r0}(2m+1)^{1/2}\,, \tag{5.2.2}$$

and similar turning points $q_{r'm'}^{\pm}$ in the term r'; b_{rm} denotes the amplitude of classical vibrations in the state rm. The contour passes by the turning points from above and satisfies the requirement for the integral to converge at both limits.

Far from the turning points $q_{rm}^{\pm}$ the wave function χ_{rm} may be replaced with its asymptotic expression (in the limit of either $\hbar\to 0$ or $|q|\to\infty$), which, generally, is a linear superposition of these two exponentials:

$$e_{rm}^{\pm} = \left(\frac{\omega_{rm}}{2\pi v_{rm}}\right)^{1/2}\exp\left(\pm\frac{\mathrm{i}}{\hbar}\int_{q_{rm}^{+}}^{q} p_{rm}\,dq \pm \frac{\mathrm{i}\pi}{4}\right), \tag{5.2.3}$$

where

$$p_{rm} = \{2M[E_{rm} - \mathscr{E}_r(q)]\}^{1/2} = M\omega_{rm}[b_{rm}^2 - (q-q_r)^2]^{1/2}\,, \tag{5.2.4}$$

$$v_{rm} = p_{rm}/M \text{ and} \tag{5.2.5}$$

$$\omega_{rm} = \pi\left(\int_{q_{rm}^{-}}^{q_{rm}^{+}}\frac{dq}{v_{rm}}\right)^{-1} \tag{5.2.6}$$

are momentum, velocity and frequency, respectively, of the classical motion in the term $\mathscr{E}_r(q)$ at a given energy E_{rm}. For the harmonic oscillator $\omega_{rm} = \omega_r$, but in general, ω_{rm} is a function of m.

Equation (5.2.4) is recast as

$$p_{rm} = M\omega_r[(q_{rm}^+ - q)(q - q_{rm}^-)]^{1/2} . \tag{5.2.7}$$

To specify the momentum branch we draw a cut in the lower half-plane linking the turning points $q_{rm}^{\pm}$ (shown in Fig. 5.1 by the wavy line; the displacement of the equilibrium positions is not large so that the turning points for the upper term $q_{rm}^{\pm}$ are located between those for the lower term $q_{r'm'}^{\pm}$). Let us adopt a convention to choose the arguments of each factor under the root sign in (5.2.7) in such a way that inequalities

$$\begin{aligned} &0 \leq \arg\{q - q_{rm}^-\} \leq \pi \text{ and} \\ &-\pi \leq \arg\{q_{rm}^+ - q\} \leq 0 \end{aligned} \tag{5.2.8}$$

would be valid for the points q in the upper half-plane. From (5.2.8) follows that in the upper half-plane

$$-\pi/2 \leq \arg\{p_{rm}\} \leq \pi/2 . \tag{5.2.9}$$

In particular, on the real axis

$$\arg\{p_{rm}\} = \begin{cases} \pi/2 & \text{for } q < q_{rm}^- , \\ 0 & \text{for } q_{rm}^- < q < q_{rm}^+ , \\ -\pi/2 & \text{for } q > q_{rm}^+ . \end{cases} \tag{5.2.10}$$

In the lower half-plane $\arg\{p_{rm}\}$ is defined by continuity. Thus, if a point q is

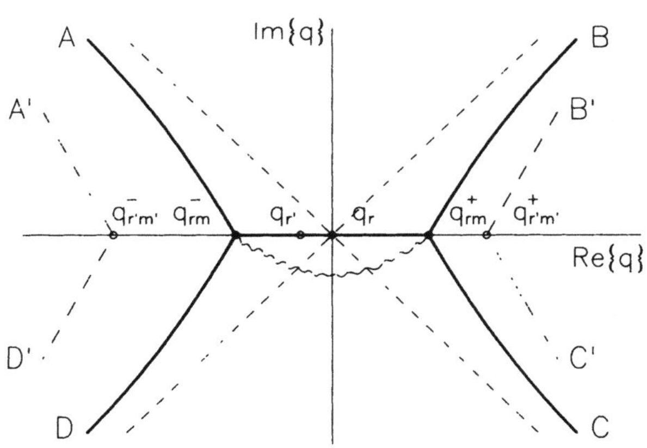

Fig. 5.1. Anti-Stokes lines (full lines) and the cut (*wavy* line) for a harmonic potential $\mathscr{E}_r(q)$

moving from the upper half-plane into the lower around the turning point q_{rm}^+ from the right, then $\arg\{q - q_{rm}^-\}$ decreases to zero and becomes negative, whereas $\arg\{q_{rm}^+ - q\}$ decreases to $-\pi$ and becomes lower than $-\pi$, so that $\arg\{p_{rm}\}$ becomes lower than $-\pi/2$, etc.

With these definitions the functions (5.2.3) are single-valued and analytical all over the complex plane with the cut specified. As an example, consider e_{rm}^+. When a point q makes a full clockwise turn around the turning points along any closed contour but not crossing the cut, the argument of the velocity acquires the increment of -2π, and that of the function e_{rm}^+, as follows from (5.2.3), the increment

$$\Delta \arg\{e_{rm}^+\} = \pi + \frac{1}{\hbar}\oint p_{rm}\,dq\,.$$

The integral over the closed contour can be recast as a sum of two equal integrals from q_{rm}^- to q_{rm}^+ along the upper and lower sides of the cut where the momenta differ by their signs only, and the integral over the upper side can be replaced by an integral over the real axis. Invoking the Bohr-Sommerfeld quantization rule

$$\frac{1}{\hbar}\int_{q_{rm}^-}^{q_{rm}^+} p_{rm}\,dq = \pi\left(m + \frac{1}{2}\right), \tag{5.2.11}$$

one obtains $\Delta\arg\{e_{rm}^+\} = 2\pi(m+1)$, whence the single-valuedness of e_{rm}^+ is inferred.

At any fixed value of q and $\hbar \to 0$ each of the functions (5.2.3) are either exponentially increasing or decreasing, or oscillating; the oscillation occurs on the anti-Stokes lines defined by the condition

$$\mathrm{Re}\left\{\frac{\mathrm{i}}{\hbar}\int_{q_{rm}^+}^{q} p_{rm}\,dq\right\} = 0\,. \tag{5.2.12}$$

The definitions (5.2.10) evidently show that the part of the real axis between the turning points is an anti-Stokes line. In the close vicinity of a turning point the potential may be approximated with a linear function, and in the case of a linear potential, as is easily verified, two more anti-Stokes lines emerge from each turning point making the 120° angles between them and with the line $q_{rm}^- - q_{rm}^+$. When moving away from the turning points these lines asymptotically approach the bisectrices of the coordinate angles, AC and BD (see Fig. 5.1), since in the limit of $|q| \to \infty$, (5.2.12) reduces to $\mathrm{Re}\{(q - q_r)^2\} = 0$. All the above formulae and definitions apply to the potential $\mathscr{E}_{r'}(q)$ after replacing subscripts rm with $r'm'$.

The anti-Stokes lines for the potential $\mathscr{E}_r(q)$ are shown by full lines in Fig. 5.1. Together with the cut-line they divide the complex plane into a number of domains, and in each domain the functions (5.2.3) are either increasing or decreasing. For instance, e_{rm}^+ decreases in sector AB as well as in sector CD below the cut, while e^-_{rm} decreases in sectors AD and BC, which can be easily shown with the help of the definitions (5.2.9, 10). When a q-point is moving over the complex plane, the increase changes into a decrease and vice versa at each crossing of either an anti-Stokes line or the cut.

For real $q \to \pm\infty$, the exact wave function χ_{rm} falls off exponentially, and hence it is asymptotically proportional to the exponential e^-_{rm} decreasing in sectors AD and BC:

$$\chi_{rm} = C_{rm} e^-_{rm} . \tag{5.2.13}$$

The coefficient C_{rm} is easy to find by comparing (5.2.13) with the exact expression (4.8.1) for $q \to \pm\infty$. Simple algebra yields

$$C_{rm} = \frac{\sqrt[4]{2\pi}}{\sqrt{m!}} \left(\frac{m + 1/2}{e} \right)^{(m+1/2)/2} . \tag{5.2.14}$$

This formula is valid for any $m \geq 0$. In particular, for $m \gg 1$ one has $C_{rm} = 1$.

To deduce an asymptotic expression for χ_{rm} valid along the whole contour L, it is necessary to analytically continue the right-hand side of (5.2.13) into the complex plane. The technique of analytical continuation in the present case is quite trivial. It is based on a few, nearly obvious, considerations to be used here; a detailed description of the method along with necessary rigorous proofs can be found in the initial chapters of *Heading*'s monograph [5.10].

In any case, since the function e^-_{rm} is defined at complex q's, (5.2.13) gives an analytical continuation of χ_{rm} into the complex domains bordering upon the positive and negative real half-axes, that is, into sectors BC and AD. After *Heading* [5.10], a decreasing solution can be extended beyond the anti-Stokes lines bordering the decrease area; thus, (5.2.13) gives an analytical continuation of χ_{rm} into sector AB. Hence it follows that it gives an asymptotic representation of the function χ_{rm} in the three sectors AD, AB and BC through which the contour L passes (for that same reason (5.2.13) also applies in sector CD below the cut).

Notice that an increasing solution cannot be continued toward the lines that are boundaries of the increase area, and all the more beyond these lines, since such a solution always contains a contribution from a damped exponential (with an unknown coefficient) which increases after an anti-Stokes line is crossed; in contrast to this, in the case of a decreasing solution, the coefficient of the second (increasing) exponential is known: it precisely equals zero.

Consider now $\chi_{r'm'}$. The anti-Stokes lines for the potential $\mathscr{E}_{r'}(q)$ look like those for $\mathscr{E}_r(q)$ (in part they are shown in Fig. 5.1 by dashed-dotted lines); the cut is also drawn in a similar way. The asymptotic representation of $\chi_{r'm'}$ within sectors A′D′, A′B′, B′C′ (and C′D′ below the cut) is given by

$$\chi_{r'm'} = e^-_{r'm'} , \tag{5.2.15}$$

where $C_{r'm'} = 1$ is set in accordance with (5.2.14) and (5.1.2).

As distinct from χ_{rm}, the function $\chi_{r'm'}$ behaves quasiclassically also on the real axis between the turning points, but there, (5.2.15) is not valid since, as follows from the above-stated Heading's rule, the analytical continuation is not allowed from the interior of the increase area toward the anti-Stokes boundary of that area. The wanted $\chi_{r'm'}$ asymptotic expression on the real axis between the turning points has the form [1.40, 5.10]

$$\chi_{r'm'} = e^{+}_{r'm'} + e^{-}_{r'm'}, \quad (q^{-}_{r'm'} < q < q^{+}_{r'm'}) . \tag{5.2.16}$$

A straightforward substitution of the asymptotic expressions (5.2.13, 15) into the integral (5.2.1) does not yet enable one to perform the integration by the saddle-point method. Indeed, the product $e^{-}_{rm}e^{-}_{r'm'}$ decreases in sectors A′D′ and B′C′, and increases in sectors A′B′ and C′D′ so that the pass leading from the western valley to the eastern one is located precisely on the real axis in the turning point region where (5.2.13, 15) do not hold. Therefore we first transform the integral (5.2.1) to a different form.

Let us write down the exact final state wave function as

$$\chi_{r'm'} = \chi^{+}_{r'm'} + \chi^{-}_{r'm'} , \tag{5.2.17}$$

where $\chi^{+}_{r'm'}$ and $\chi^{-}_{r'm'}$ are *exact solutions* to the Schrödinger equation decreasing in sectors A′B′ and C′D′, respectively. The above representation works all over the whole complex plane since the exact solutions are analytical everywhere (the cuts and anti-Stokes lines only pertain to the asymptotic forms of the solutions). The asymptotic expressions for those solutions are easily obtained from (5.2.16). For instance, on the real axis between the turning points we obtain

$$\chi^{+}_{r'm'} = e^{+}_{r'm'} . \tag{5.2.18}$$

Moving from the real axis up, one enters the decrease area from where solution (5.2.18) can be continued into the increase areas A′D′ and B′C′. Hence the asymptotic expression (5.2.18) is valid everywhere along the contour *L*.

Although we do not need the $\chi^{+}_{r'm'}$ asymptote in the lower half-plane, notice that (5.2.18) is also valid when moving from the real axis down to the cut line, and the continuation beyond the cut can be easily performed on the basis of the requirement for the asymptotic wave functions to be continuous across the cut, which is a consequence of the continuity of the exact wave functions. However, one may act in a simpler manner by noting that $\chi^{-}_{r'm'}$ decreases in the lower half-plane, and then employing (5.2.17, 15) one obtains, in sector C′D′ below the cut, merely

$$\chi^{+}_{r'm'} = \chi_{r'm'} = e^{-}_{r'm'} . \tag{5.2.19}$$

The asymptotic form for the solution $\chi^{-}_{r'm'}$ is found analogously. For example, according to the definition of this solution, it decreases in sector C′D′ below the cut, and consequently it is proportional to $e^{+}_{r'm'}$. By Heading's rule on the analytical continuation of a damped solution, this is valid in the adjacent sector

A'D', too (where the solution increases), with the same factor of proportionality. This factor is obtained by applying expansion (5.2.17) within sector A'D' to give $\chi_{r'm'}^{-} = -\chi_{r'm'}^{+}$ since there $\chi_{r'm'}$ decreases, while $\chi_{r'm'}^{\pm}$ increase. Inserting (5.2.18) one obtains the asymptotic solution

$$\chi_{r'm}^{-} = -e_{r'm'}^{+} \tag{5.2.20}$$

valid in sectors A'D', C'D' (below the cut), and B'C'.

The transformation of the integral (5.2.1) to be performed depends on the relation between the frequencies ω_r and $\omega_{r'}$. First consider the case of $\omega_r > \omega_{r'}$ ($f > 1$). For the moment, we turn the integration contour back onto the real axis and insert (5.2.17) into the integrand of (5.2.1),

$$S_{rmr'm'} = \int_{-\infty}^{+\infty} \chi_{rm}(\chi_{r'm'}^{+} + \chi_{r'm'}^{-})\, dq\ . \tag{5.2.21}$$

We split this integral into the sum of two integrals, each being convergent since, owing to the assumption of $f > 1$, the function χ_{rm} for $q \to \pm\infty$ decreases more rapidly than the functions $\chi_{r'm'}^{\pm}$ increase. Considering that the identity $\chi_{r'm'}^{-} \equiv \overline{\chi_{r'm'}^{+}}$ is valid along the real axis (the bar denotes a complex conjugate) one obtains

$$S_{rmr'm'} = 2\,\mathrm{Re}\left\{\int_{-\infty}^{+\infty} \chi_{rm}\chi_{r'm'}^{+}\, dq\right\} = 2\,\mathrm{Re}\left\{\int_{L} \chi_{rm}\chi_{r'm'}^{+}\, dq\right\}, \tag{5.2.22}$$

where the middle part of the contour L is again displaced from the real axis into the upper half-plane, while its tails are still kept on the real axis.

In the opposite case of $\omega_r < \omega_{r'}$ ($f < 1$), the splitting of the integral (5.2.21) is impossible because the individual integrals diverge. Therefore, we turn back to the integral (5.2.1) over the path L:

$$S_{rmr'm'} = \int_{L} \chi_{rm}(\chi_{r'm}^{+} + \chi_{r'm'}^{-})\, dq\ . \tag{5.2.23}$$

We shift the ends of the contour L into the lower half-plane and align them parallel to lines C' and D'. It is easily verified that along the lines parallel to the anti-Stokes lines the products $\chi_{rm}\chi_{r'm'}^{\pm}$ are essentially oscillating so that the integral (5.2.23) may be split into a sum of two convergent integrals. Now we note that the product $\chi_{rm}\chi_{r'm'}^{-}$ decreases in sector CD since now, owing to $f < 1$, $\chi_{r'm'}^{-}$ decreases faster than χ_{rm} increases. For this reason the integration contour of the second integral can be closed in the lower half-plane, and the integral vanishes since the exact solutions are analytical (here it may be emphasized once more that we are dealing with the exact solutions rather than their asymptotic representations). As for the first integral, the middle part of contour L may be infinitely displaced upwards since in sector AB the function $\chi_{r'm'}^{+}$ decreases faster than χ_{rm} increases. In this way for $f < 1$ one obtains

$$S_{rmr'm'} = \int_{L} \chi_{rm}\chi_{r'm'}^{+}\, dq\ . \tag{5.2.24}$$

The integrals (5.2.22, 24) are convenient in that they permit the implementation of the saddle-point method.

(i) The Case $f<1$. We insert the asymptotic wave functions (5.2.13, 18) and the explicit expressions (5.2.3) for the functions $e^{\pm}$ in the integral (5.2.24):

$$S_{rmr'm'} = \frac{1}{2\pi} C_{rm}(\omega_r\omega_{r'})^{1/2} \int_L \exp[\Phi(q)](v_{rm}v_{r'm'})^{-1/2}\, dq\,, \tag{5.2.25}$$

$$\Phi(q) = -\frac{\mathrm{i}}{\hbar}\int_{q_{rm}^+}^{q} p_{rm}\, dq + \frac{\mathrm{i}}{\hbar}\int_{q_{r'm'}^+}^{q} p_{r'm'}\, dq\,. \tag{5.2.26}$$

In the limit of $\hbar \to 0$ the integral (5.2.25) is calculated by the saddle-point method. Equating the first derivative of $\Phi(q)$ to zero we obtain the following equation for the saddle points q_*:

$$p_{rm} = p_{r'm'}\,, \tag{5.2.27}$$

i.e., at the saddle point the momenta of the initial and final states are the same. Equating their squared absolute values yields the equation

$$E_{rm} - \mathscr{E}_r(q) = E_{r'm'} - \mathscr{E}_{r'}(q) \tag{5.2.28}$$

which has two real roots,

$$q_*^{\pm} = q_r + \Delta q\,\frac{-1 \pm f(1+\eta)^{1/2}}{1-f^2}\,, \tag{5.2.29}$$

where

$$\eta = \frac{1-f^2}{\xi f}\left[m' + \frac{1}{2} - f\left(m + \frac{1}{2}\right)\right], \tag{5.2.30}$$

the rest of parameters having been defined in Sect. 4.8. The point q_*^+ is located to the right of $q_{r'm'}^+$, hence of q_{rm}^+, since small displacements Δq are supposed (see Fig. 5.1). Therefore it follows from (5.2.10) that the arguments of momenta in (5.2.27) at $q = q_*^+$ are the same and are equal to $-\pi/2$. Analogously, at the point q_*^- lying to the left of both $q_{r'm'}^-$ and q_{rm}^- the arguments are $\pi/2$. Hence $q_*^{\pm}$ are actual roots of (5.2.27).

Let $v_*^{\pm}$ be the values of the velocity at $q_*^{\pm}$ common to both states. The above argument enables us to write

$$v_*^{\pm} = |v_*^{\pm}|\exp(\mp \mathrm{i}\pi/2)\,, \tag{5.2.31}$$

and for the absolute value of the velocity the following expression is deduced:

$$|v_*^{\pm}| = \omega_r b_{rm}[(z_{rm}^{\pm})^2 - 1]^{1/2} \equiv \omega_{r'} b_{r'm'}[(z_{r'm'}^{\pm})^2 - 1]^{1/2}\,, \tag{5.2.32}$$

where $z^{\pm}$ are the relative distances from the equilibrium positions to the saddle points,

$$z_{rm}^{\pm} \equiv \frac{q_*^{\pm} - q_r}{b_{rm}} = \left(\frac{\xi}{m+1/2}\right)^{1/2} \frac{-1 \pm f(1+\eta)^{1/2}}{1-f^2},$$
$$z_{r'm'}^{\pm} \equiv \frac{q_*^{\pm} - q_{r'}}{b_{r'm'}} = \left(\frac{\xi f}{m'+1/2}\right)^{1/2} \frac{-1 \pm (1+\eta)^{1/2}}{1-f^2}. \quad (5.2.33)$$

Let us calculate the second derivative of $\Phi(q)$. Considering (5.2.27) we have at an arbitrary saddle point q_*

$$\Phi''(q_*) = -\frac{\mathrm{i}}{\hbar}\frac{d}{dq}(p_{rm} - p_{r'm'}) = -\mathrm{i}\frac{F_r - F_{r'}}{\hbar v_*}, \quad (5.2.34)$$

where $F_{r,r'} = -d\mathscr{E}_{r,r'}(q)/dq_{q=q_*}$ are the forces at q_*. Insertion of (5.2.31) into (5.2.34) yields

$$\Phi''(q_*^{\pm}) = \frac{\Delta F}{\hbar |v_*^{\pm}|} \quad (5.2.35)$$

since the force difference at these points is

$$F_r^{\pm} - F_{r'}^{\pm} = \pm\Delta F, \qquad \Delta F = M\omega_r\omega_{r'}b_{r0}[2\xi(1+\eta)]^{1/2}. \quad (5.2.36)$$

In order to find a correct integration path, one has to learn about the relief features of the surface $w = \mathrm{Re}\{\Phi(q)\}$ in the three-dimensional space of variables w, $\mathrm{Re}\{q\}$ and $\mathrm{Im}\{q\}$ [5.11]. In the present case the relief is very simple. There is a hill in the central domain involving the turning points where $w \sim 1$. The descent down hill to the North or South leads to valleys, while to the West or East it leads to the saddles $q_*^{\pm}$ behind which ascents start. Equations (5.2.35, 36) yield $\Phi''(q_*^{\pm}) > 0$, that is, the directions of the steepest descent from both saddles are the same and are perpendicular to the real axis. Thus, both passes connect the southern valley with the northern, and the contour L must be drawn in the way shown in Fig. 5.2a.

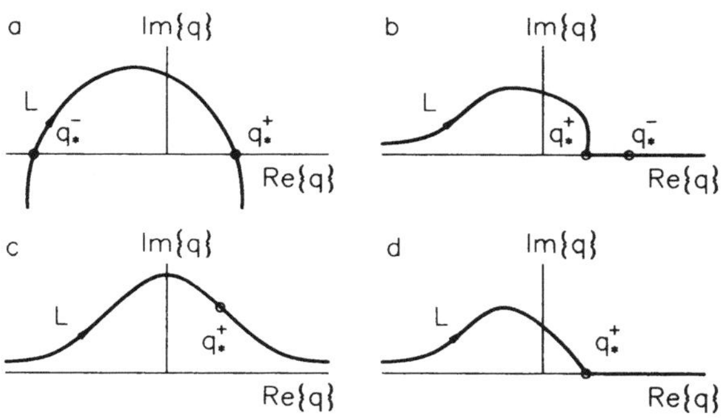

Fig. 5.2. Integration paths for $f < 1$ (**a**), $f_c > f > 1$ (**b**), $f > f_c$ (**c**), and $f = f_c$ (**d**)

We proceed with the calculation of the integral (5.2.25). It is equal to the sum of the contributions of two saddle points $q_*^\pm$:

$$S_{rmr'm'} = S^+_{rmr'm'} + S^-_{rmr'm'} . \tag{5.2.37}$$

Following the standard routine of the method of steepest descent we expand $\Phi(q)$ in the integrand of (5.2.25) in powers of $q - q_*^\pm$ and replace the velocities by their values (5.2.31) at the saddle points:

$$S^\pm_{rmr'm'} = \frac{C_{rm}}{2\pi|v_*^\pm|}(\omega_r\omega_{r'})^{1/2}\exp[\Phi(q_*^\pm)](\pm \mathrm{i}) \times \int\limits_{\pm \mathrm{i}\infty}^{\mp \mathrm{i}\infty} \exp[\Phi''(q_*^\pm)(q - q_*^\pm)^2/2]\,dq \tag{5.2.38}$$

(either upper or lower signs are taken simultaneously in all double-signed items). Integrating (5.2.38) and inserting (5.2.32, 35, 36) we obtain

$$S^\pm_{rmr'm'} = C_{rm}\left(\frac{\hbar\omega_r\omega_{r'}}{2\pi|v_*^\pm|\Delta F}\right)^{1/2}\exp[\Phi(q_*^\pm)] = \frac{C_{rm}}{2\sqrt{\pi}}\left\{\xi(1+\eta)\left(m+\frac{1}{2}\right)[(z^\pm_{rm})^2 - 1]\right\}^{-1/4}\exp[\Phi(q_*^\pm)] . \tag{5.2.39}$$

The final step is to calculate the function (5.2.26) at the saddle points. Notice that it takes on real values at both points because the integrals in (5.2.26) are along the real axis in the classically forbidden regions where, by definition (5.2.10), the momenta are purely imaginary. When calculating $\Phi(q_*^-)$ one has to utilize the quantization rule (5.2.11) and a similar rule for the potential $\mathscr{E}_{r'}(q)$. After a trivial integration one obtains

$$\Phi(q_*^\pm) = -\mathrm{i}\pi\zeta_\pm(m' - m) - (m + 1/2)J(\pm z^\pm_{rm}) + (m' + 1/2)J(\pm z^\pm_{r'm'}) , \tag{5.2.40}$$

where $\zeta_+ = 0$ and $\zeta_- = 1$. The function $J(z)$ for $z \geq 1$ is given by

$$J(z) \equiv 2\int\limits_1^z (z^2 - 1)^{1/2}dz = z(z^2 - 1)^{1/2} - \ln[z + (z^2 - 1)^{1/2}] . \tag{5.2.41}$$

It should be noted that the parameters (5.2.33), in absolute magnitude, are always greater than unity since the saddle points are located in the classically inaccessible region for both electronic terms (the values of $z^\pm = \pm 1$ correspond to the turning points).

(ii) The Case $f \geq 1$. For $f > 1$ we insert the asymptotic wave functions (5.2.13, 18) into the integral (5.2.22):

$$S_{rmr'm'} = \frac{1}{\pi}C_{rm}(\omega_r\omega_{r'})^{1/2}\,\mathrm{Re}\left\{\int\limits_L \exp[\Phi(q)](v_{rm}v_{r'm'})^{-1/2}\,dq\right\} , \tag{5.2.42}$$

where $\Phi(q)$ and the saddle points are again given by (5.2.26) and (5.2.29), respectively. Since, in the present case, the parameter η is negative, the points $q_*^{\pm}$ may be located either on the real axis (to the right of the turning points) or in the complex plane; the critical value of f, f_c, at which the two saddle points are united into one point is determined by equation $\eta = -1$.

First consider the case where both saddle points are on the real axis to the right of the turning points, and are well separated from each other so that their interference is negligible. The parameter η is in the interval $-1 < \eta \le 0$, and q_*^- is placed to the right of q_*^+.

We calculate the velocity and the second derivative of $\Phi(q)$ at $q = q_*^{\pm}$. By definition (5.2.10) one has $v_*^{\pm} = |v_*^{\pm}|\exp(-\mathrm{i}\pi/2)$, and the second derivative is found by (5.2.34, 36):

$$\Phi''(q_*^{\pm}) = \pm \frac{\Delta F}{\hbar |v_*^{\pm}|}, \tag{5.2.43}$$

where ΔF and $|v_*^{\pm}|$ are given by (5.2.36, 32). It follows that the direction of the steepest descent at the point q_*^+ is perpendicular, and at point q_*^- is parallel to the real axis.

The relief of the surface $w = \mathrm{Re}\{\Phi(q)\}$ looks as follows. There is still the hill at the center. His western slope leads to a valley while the eastern one to the saddle q_*^+. From the North and South the hill is girt with two gullies which are going, gradually rising, from the western valley to the East, and eventually meet at the saddle q_*^+. From the saddle q_*^+ to the East a slope rises up to the saddle q_*^- and then goes down towards the eastern valley. Since the contour L must go from the western valley to the eastern one [see the derivation of (5.2.22)] along the lines of stationary phase, $\mathrm{Im}\{\Phi(q)\} = \mathrm{const}$, it must be drawn as shown in Fig. 5.2b.

On the part of the contour from q_*^+ to $+\infty$ the function $\Phi(q)$ is real, whereas the velocities in both potentials are purely imaginary, so the integral over this part of the contour is purely imaginary and does not contribute to $S_{rmr'm'}$. For this reason one may integrate (5.2.42) until q_*^+. Expanding in (5.2.42) $\Phi(q)$ around q_*^+, taking the velocity at point q_*^+ off the integral sign, and inserting (5.2.43) one obtains

$$S_{rmr'm'} = \frac{C_{rm}}{\pi|v_*^+|}(\omega_r\omega_{r'})^{1/2}\exp[\Phi(q_*^+)]\mathrm{Re}\left\{\mathrm{i}\int_{\mathrm{i}\infty}^{0}\exp[\Phi''(q_*^+)(q-q_*^+)^2/2]\,dq\right\}.$$

A comparison with (5.2.38) shows that this expression is precisely equal to $S_{rmr'm'}^+$. Thus, for $-1 < \eta \le 0$ one has $S_{rmr'm'} = S_{rmr'm'}^+$, where $S_{rmr'm'}^+$ is given by (5.2.39). This result is quite natural since in (5.2.37) the second term vanishes for $f \to 1$ (point q_*^- moves to $-\infty$), while the first is continuous for $f = 1$ and thus yields the overlap integral for $f \ge 1$. The values of the parameters z_{rm}^+ and $z_{r'm'}^+$ for $f = 1$ are obtained from (5.2.33) by taking the limit $f \to 1$:

$$\begin{aligned} z_{rm}^+ &= (m' - m - \xi)[2\xi(2m+1)]^{-1/2},\\ z_{r'm'}^+ &= (m' - m + \xi)[2\xi(2m'+1)]^{-1/2}. \end{aligned} \tag{5.2.44}$$

Now we consider the case of $\eta < -1$ where the saddle points are in the complex plane. The point q_*^+ in the upper half-plane is defined as

$$q_*^+ = q_r + \frac{\Delta q}{f^2 - 1}(1 + \mathrm{i}f|1 + \eta|^{1/2}), \tag{5.2.45}$$

which is derived in a formal way from (5.2.29) using the substitution

$$(1 + \eta)^{1/2} \to -\mathrm{i}|1 + \eta|^{1/2}. \tag{5.2.46}$$

Now we draw a straight line through the point $q = q_r$ parallel to the imaginary axis. By definitions (5.2.7, 8) one has $\arg\{p_{rm}\} = 0$ on this line. Hence, if the velocity at point q_*^+ located to the right of this line is written as $v_*^+ = |v_*^+|\exp(-\mathrm{i}\gamma)$, then, due to (5.2.9), γ is enclosed into the interval $0 < \gamma < \pi/2$. Evaluation of the force difference at $q = q_*^+$ by (5.2.36), where substitution (5.2.46) is performed, and insertion of the result into (5.2.34) yields

$$\Phi''(q_*^+) = -\left|\frac{\Delta F}{\hbar v_*^+}\right| \mathrm{e}^{\mathrm{i}\gamma}. \tag{5.2.47}$$

The integration path is shown in Fig. 5.2c. It extends from the western into the eastern valley through the saddle q_*^+; the direction of the steepest descent from the saddle makes an angle of $-\gamma/2$ with the real axis. Calculating the integral in (5.2.42) in the vicinity of q_*^+ gives

$$\begin{aligned} S_{rmr'm'} &= C_{rm}\left(\frac{2\hbar\omega_r\omega_{r'}}{\pi|v_*^+ \Delta F|}\right)^{1/2} \mathrm{Re}\{\exp[\Phi(q_*^+) + \mathrm{i}\gamma/2]\} \\ &= \frac{C_{rm}}{\sqrt{\pi}}[\xi(m + 1/2)|(z_{rm}^+)^2 - 1|\,|1 + \eta|]^{-1/4}\mathrm{Re}\{\exp[\Phi(q_*^+) + \mathrm{i}\gamma/2]\}, \end{aligned} \tag{5.2.48}$$

where $\Phi(q_*^+)$ is given by (5.2.40); the parameters z_{rm}^+ and $z_{r'm'}^+$ are found from (5.2.33) with substitution of (5.2.46). The branches of the root and the logarithmic functions in (5.2.41) for complex z are specified by the condition for these functions to be real and positive for real $z \geq 1$.

Finally, we consider the case $\eta = -1$ when only one real saddle point exists, $q_*^+ = q_r + \Delta q/(f^2 - 1)$. At this point the second derivative of $\Phi(q)$ is zero. The third derivative is

$$\Phi'''(q_*^+) = -\frac{\omega_r(f^2 - 1)}{b_{r0}^2 f^2 |v_*^+|}.$$

The integration path is shown in Fig. 5.2d. The interval from q_*^+ to $+\infty$ again contributes nothing to the integral, so we integrate in (5.2.42) only until q_*^+. The direction of the steepest ascent toward the saddle makes an angle of $-\pi/3$ with the real axis. Evaluation of the integral in (5.2.42) in the vicinity of q_*^+ results in

$$S_{rmr'm'} = \frac{C_{rm}\Gamma(1/3)}{2\pi}\left(\frac{f}{3}\right)^{1/6}\{(f^2-1)(m+1/2)[(z_{rm}^+)^2-1]\}^{-1/3}\exp[\Phi(q_*^+)] ,$$

where $\Phi(q_*^+)$ is given by (5.2.40); z_{rm}^+ and $z_{r'm'}^+$ are found from (5.2.33) for $\eta = -1$.

The quasiclassical formulae derived above express the overlap integral in terms of elementary functions of the potential parameters and of the quantum numbers of the initial and final electronic states.

5.2.2 Conditions for Applicability of the Quasiclassical Approximation

The quasiclassical approach works provided that at each saddle point a standard quasiclassical condition is fulfilled for both the electronic terms; thus one has for the term r

$$\frac{\hbar M|F_r|}{|p_{rm}|^3} \ll 1 . \tag{5.2.49}$$

At the saddle points (5.2.29) this becomes

$$\frac{|z_{rm}^\pm|}{|(z_{rm}^\pm)^2-1|^{3/2}} \ll 2m+1 . \tag{5.2.50}$$

For the term r' the notations rm should be replaced with $r'm'$. These are the most general conditions for the applicability of the quasiclassical approximation.

For radiationless transitions between the lowest electronic states of polyatomic molecules the parameters and quantum numbers typically satisfy the conditions

$$\xi < 1, \quad f \sim 1, \quad m \geq 0, \quad m' \gg m+1 , \tag{5.2.51}$$

i.e., the shift of the equilibrium position does not exceed the amplitude of zero-point vibrations, and the frequencies in different electronic states are of the same order of magnitude (see definitions of the parameters in Sect. 4.8). Even the above quite minute restrictions prove to be sufficient to satisfy conditions (5.2.50) for both terms. To demonstrate this we evaluate the parameters z by (5.2.33). From (5.2.30) we have, approximately,

$$\eta \approx \frac{1-f^2}{\xi f} m' . \tag{5.2.52}$$

Since conditions (5.2.51) do not impose any significant limitations on the absolute magnitude of η, two cases are to be discussed.

For $|\eta| \lesssim 1$, the difference $1 - f^2 \approx \xi\eta/m'$ is small. Inserting this into (5.2.33) one obtains

$$|z_{rm}^{\pm}| \approx m'|\varrho_{\pm}|[2\xi(2m+1)]^{-1/2} \gg 1 ,$$
$$|z_{r'm'}^{\pm}| \approx \frac{1}{2}\left(\frac{m'}{\xi}\right)^{1/2} |\varrho_{\pm}| \gg 1 , \tag{5.2.53}$$

where the notation

$$\varrho_{\pm} = 2\frac{-1 \pm (1+\eta)^{1/2}}{\eta} \tag{5.2.54}$$

is introduced. In view of (5.2.53) the quasiclassical conditions (5.2.50) are evidently fulfilled for both electronic terms.

At $|\eta| \gg 1$ it follows from (5.2.33, 52) that

$$|z_{rm}^{\pm}| \approx \left[\frac{m'f}{(m+1/2)|1-f^2|}\right]^{1/2} \gg 1 , \tag{5.2.55}$$

$$z_{r'm'}^{\pm} \approx \pm(1-f^2)^{-1/2} \tag{5.2.56}$$

(for $f > 1$ the parameters $z_{r'm'}^{\pm}$ are purely imaginary). Inequality (5.2.50) is again met automatically, and the corresponding inequality for the second electronic term, after inserting (5.2.56), takes the form $|1-f^2|/f^3 \lessdot 2m' + 1$. This inequality establishes a precise meaning of the condition $f \sim 1$, and along with (5.2.55) imposes only an insignificant limitation on f,

$$\frac{m'}{m+1/2} \gg f \gg (2m'+1)^{-1/3} .$$

Thus we see that the quasiclassical approximation works in a wide range of ξ, f, m, and m', and hence can be applied in numerical and analytical calculations of the Franck–Condon factors for radiationless transitions.

5.2.3 Comparison of Frequency and Displacement Effects

Simplified models are often discussed in literature which allow for either a displacement of the equilibrium position of the oscillator [1.16, 22] or its frequency change [4.169, 5.12]. It is interesting to establish in which way the Franck–Condon factor changes when a small frequency change is introduced in the former case and a small displacement in the latter. A brief summary of the results obtained in the Russian edition of this book is outlined here.

First let ξ be fixed at a nonzero value while f tends to unity. Then only q_*^+ contributes to the overlap integral, and the Franck–Condon factor (4.5.2) is

$$F_{rmr'm'} = F_{rmr'm'}^{f=1} \exp\left[\frac{1-f}{2\xi}(m'-m)^2\right] . \tag{5.2.57}$$

It is seen that the frequency-change effect is negligible as compared to the displacement effect provided

$$|1-f| \ll \frac{2\xi}{(m'-m)^2}\,. \tag{5.2.58}$$

From (5.2.57) it is concluded that the Franck–Condon factor decreases if f increases in the vicinity of $f=1$. It is easy to account for such a result as follows. For a horizontal transition (4.5.10) the saddle point q_*^+ coincides with the electronic terms' crossing point, and with the frequency in the upper term increasing the crossing point is moving to the right (see Fig. 4.2), deeper into the classically forbidden region (and eventually into the complex plane at $f>1$), which causes a decrease in the Franck–Condon factor. And if f decreases, the crossing point approaches the turning points causing an increase of that factor.

We consider now the Franck–Condon factor as a function of ξ at fixed f. The overlap integral for $|\eta| \gg 1$ and $m' \gg m^2+1$ has the form

$$S_{rmr'm'} = \tfrac{1}{2}A\left[e^{B} + (-1)^{m'-m}e^{-B}\right] \tag{5.2.59}$$

for $f<1$, and

$$S_{rmr'm'} = A\cos\left[B-(m'-m)\tfrac{\pi}{2}\right] \tag{5.2.60}$$

for $f>1$, where

$$A = \left(\frac{\sqrt{2}}{m!\sqrt{\pi}}\right)^{1/2}\exp\left[-\frac{m'-m}{2}\ln\frac{1+f}{|1-f|}+\frac{1}{2}\left(m-\frac{1}{2}\right)\ln m' + \left(m+\frac{1}{2}\right)\ln\frac{2\sqrt{f}}{1+f}-\frac{\xi}{1-f^2}\right], \tag{5.2.61}$$

$$B = 2\left(\frac{\xi f m'}{|1-f^2|}\right)^{1/2}. \tag{5.2.62}$$

These expressions apply when both f and $|1-f|$ are on the order of unity.

At $\xi=0$ one has $B=0$, so the Franck–Condon factor vanishes at odd values of $m'-m$, as is expected from the exact equation (4.8.7). This is a result of interference between the two saddle points $q_*^{\pm}$ whose contributions to (5.2.59, 60) are equal in magnitude (since they are at equal distances from the corresponding turning points), but have different phases.

When ξ is non-zero and small so that $B \ll 1$, the displacement effect is obviously negligible. Thus the neglect of the displacement effect is justified at

$$\xi \ll \frac{|1-f^2|}{4fm'}\,. \tag{5.2.63}$$

At greater ξ values, when $B \gg 1$, the behavior depends on the sign of $1-f$. For

$f < 1$ the Franck–Condon factor deduced from (5.2.59) monotonically increases, and the difference between even and odd values of $m' - m$ disappears:

$$F^{f<1}_{rmr'm'} = \tfrac{1}{4} A^2_{\xi=0} e^{2B} , \tag{5.2.64}$$

where the slowly varying function A is replaced with its value at $\xi = 0$. In the case of $f > 1$ the cosine in (5.2.60) oscillates as a function of m'. Considering the fact that the transition probability involves summing up of expression (5.2.60) squared over the final states, we may replace the oscillating factor with a mean value to obtain (see Sect. 5.4.3)

$$F^{f>1}_{rmr'm'} = \tfrac{1}{2} A^2_{\xi=0} \tag{5.2.65}$$

for $m' - m$ both even and odd. In other words, with ξ increasing from zero to a high value the Franck–Condon factor is merely shared in equal portions between adjacent m' levels without any change in its m' dependence. At very large ξ its dependence upon the quantum numbers will certainly alter, too, but this will occur when the saddle points come close to the turning points, thus breaking down the quasiclassical approximation.

It is worthwhile to note that (5.2.64, 65) differ by a large factor, namely $\exp(2B)$. An implication of this fact is discussed in Sect. 5.3.3.

5.3 Overlap Integral for an Anharmonic Oscillator

5.3.1 Morse Oscillator with $\Delta\alpha = 0$

The second example to be considered is the Morse oscillator (4.6.19) with the same range-parameter α in both electronic states. The latter assumption drastically simplifies the saddle-point equation. We define new variables

$$Q = \frac{1}{\alpha}\{1 - \exp[-\alpha(q - q_r)]\}, \quad Q' = \frac{1}{\alpha}\{1 - \exp[-\alpha(q - q_{r'})]\} \tag{5.3.1}$$

and recast the potentials (4.6.19) as

$$\mathscr{E}_r(q) = \varepsilon_r + M\omega_r^2 Q^2/2, \qquad \mathscr{E}_{r'}(q) = \varepsilon_{r'} + M\omega_{r'}^2 Q'^2/2 , \tag{5.3.2}$$

where $\omega_{r,r'}$ are given by (4.8.13). The variables Q and Q' obey the relation

$$Q' = \frac{f}{\lambda}(Q + \Delta Q) , \tag{5.3.3}$$

where $\Delta Q = [\exp(\alpha \Delta q) - 1]/\alpha$ and λ [see (4.8.15)] are the displacement of the equilibrium position and the frequency ratio for the "harmonic oscillator" (5.3.2), whereas the Morse displacement Δq and the ratio f are still defined as in

Sect. 4.8. Additionally, we adopt the notations

$$b_r = (\hbar M\omega_r)^{1/2} \quad \text{and} \quad \xi = (\Delta Q/b_r)^2/2 .$$

The Morse energy levels are written as

$$E_{rm} = \varepsilon_r + \hbar\omega_r\mu_{rm} ,$$

where

$$\mu_{rm} = m + \tfrac{1}{2} - \tfrac{1}{2}\alpha^2 b_r^2 (m + \tfrac{1}{2})^2 \tag{5.3.4}$$

is the vibrational energy ε_{rm} in units of $\hbar\omega_r$. Only discrete levels will be considered. They are under the dissociation limit,

$$\varepsilon_{rm} < D_r , \tag{5.3.5}$$

and the quantum number is restricted from above by

$$m < \sigma_r \equiv \alpha^{-2} b_r^{-2} - \tfrac{1}{2} . \tag{5.3.6}$$

The same equations are valid for the electronic state r' on replacing rm with $r'm'$.

Compared to the harmonic oscillator, the Morse oscillator has an additional dimensionless parameter αb_r which governs the anharmonic level shift as well as the total number of discrete levels by (5.3.4, 6).

In calculating the overlap integral we shall assume that the parameters ξ, f, m, m' still satisfy conditions (5.2.51) which, as in the harmonic oscillator case, assure the fulfilment of the validity conditions for the quasiclassical approximation. As for the parameter α, on the order of magnitude, it is equal to the inverse of the interatomic separation, and from (4.2.6) one has $\alpha b_r \sim \kappa \ll 1$, where κ is the Born-Oppenheimer parameter (4.1.4). This inequality means that the anharmonic shift of the lowest levels with $m \sim 1$ is small as compared to the vibrational quantum, whereas the number of the discrete levels is large.

It is a simple matter to solve (5.2.28) for the saddle points in the present case of equal range-parameters, since, after (5.3.2–4) are inserted, it reduces to a quadratic equation in Q. Its two roots are

$$Q_*^{\pm} = \frac{-\Delta Q \pm \lambda |\Delta Q| (1 + \eta)^{1/2}}{1 - \lambda^2} . \tag{5.3.7}$$

The parameter η is defined in analogy with (5.2.30) as:

$$\eta = \frac{1 - \lambda^2}{\xi f} (\mu_{r'm'} - f\mu_{rm}) . \tag{5.3.8}$$

The involvement of the additional Morse parameter α significantly increases the variety of situations where particular formulae for the overlap integral are obtained. Therefore, the exhaustive study such as that of Sect. 5.2 for the harmonic oscillator will not be conducted here. Instead, the general method will be illustrated in the particular case of $\lambda < 1 (\eta > 0)$, where the roots of (5.3.7) are real and have opposite signs; namely, independently of the sign of ΔQ one has

$$Q_*^- < 0, \qquad Q_*^+ > 0\,, \tag{5.3.9}$$

which flows from inequality $\lambda(1+\eta)^{1/2} > 1$ derived directly from (5.3.8). Analysis of other cases (real roots of the same sign at $-1 < \eta < 0$, root degeneration at $\eta = -1$, and complex roots at $\eta < -1$) does not meet any additional difficulties and can be carried out in the same way as in Sect. 5.2.

Although the anharmonicity parameter αb_r is small, it implies by no means that the anharmonicity always provides only small corrections to the overlap integral of the harmonic oscillator. This would be the case on condition that the saddle points q_* which just govern the magnitude of the overlap integral are not too far from the equilibrium position q_r where the anharmonicity effect can still be neglected, i.e., on condition

$$\alpha|q_* - q_r| \approx \alpha|Q_*| \ll 1\,. \tag{5.3.10}$$

This inequality can be satisfied, provided the final vibrational energy $\varepsilon_{r'm'}$ is small enough, in which case the anharmonic contribution is small and is of no interest.

With an increase in $\varepsilon_{r'm'}$ the difference $|q_* - q_r|$ also increases, and the anharmonicity becomes of primary importance. Most interesting is the case of a strong anharmonicity,

$$\alpha Q_*^+ > 1\,, \tag{5.3.11}$$

which is easily implemented at high $\varepsilon_{r'm'}$ and low ε_{rm}. In fact, after simple algebra using (5.3.7, 8) condition (5.3.11) takes the form

$$\frac{\varepsilon_{r'm'}}{D_{r'}} > 1 - f^2 + f^2\frac{\varepsilon_{rm}}{D_r}\,, \tag{5.3.12}$$

where small terms on the order of $\alpha\Delta Q$ are dropped. Inequality (5.3.12) combined with (5.3.5) for the state r' specifies a wide range of variation of the final state vibrational energy $\varepsilon_{r'm'}$ where the condition of strong anharmonicity (5.3.11) is satisfied. Only this case is treated below.

To arrive at a better understanding of the physical situation in the limit of strong anharmonicity, it is appropriate to address the most important case of resonant states (4.5.10) when (5.2.28) for the saddle points reduces to $\mathscr{E}_r(q) = \mathscr{E}_{r'}(q)$. It is seen that the saddle points q_* are then term crossing points on the real axis in the complex plane. The negative root Q_*^- gives rise to a real saddle point

$$q_*^- = q_r - \frac{1}{\alpha}\ln(1 - \alpha Q_*^-) \tag{5.3.13}$$

located to the left of the classically accessible region. The positive root Q_*^+ produces, under condition (5.3.11), a complex saddle point in the upper half-plane,

$$q_*^+ = q_r - \frac{1}{\alpha}\ln|1 - \alpha Q_*^+| + \frac{\mathrm{i}\pi}{\alpha}, \tag{5.3.14}$$

and its complex conjugate $\overline{q_*^+}$ in the lower.

Figure 5.3 shows how the electronic terms are arranged with respect to one another in the case of a single real term-intersection point. In the opposite case of weak anharmonicity (5.3.10) there would be an additional real intersection point on the outer branch of the potential; the strong anharmonicity removes the outer intersection by shifting the intersection point into the complex plane.

Inequality (5.3.12) resulting from the strong anharmonicity criterion (5.3.11) has a transparent meaning. Figure 5.3 immediately reveals that with the lack of the outer potential branch intersection the vibrational energies of the isoenergetic initial and final states satisfy the inequality

$$D_r - \varepsilon_{rm} > D_{r'} - \varepsilon_{r'm'}, \tag{5.3.15}$$

which precisely coincides with (5.3.12).

The saddle points (5.3.13, 14) as well as anti-Stokes lines and cuts for the potential $\mathscr{E}_r(q)$ are shown in Fig. 5.4. Because the Morse potential is periodic with the period of $2\pi\mathrm{i}/\alpha$, the entire picture is repeated periodically along the imaginary axis.

The momentum argument in close vicinities of the turning points and on the real axis between them is determined by (5.2.8, 10), and the extension into the complex plane is implemented by continuity. For instance, on the real axis near the outer turning point q_{rm}^+ to the right of it one has from (5.2.10)

$$p_{rm} = \exp(-\mathrm{i}\pi/2)\{2M[\mathscr{E}_r(q) - E_{rm}]\}^{1/2}. \tag{5.3.16}$$

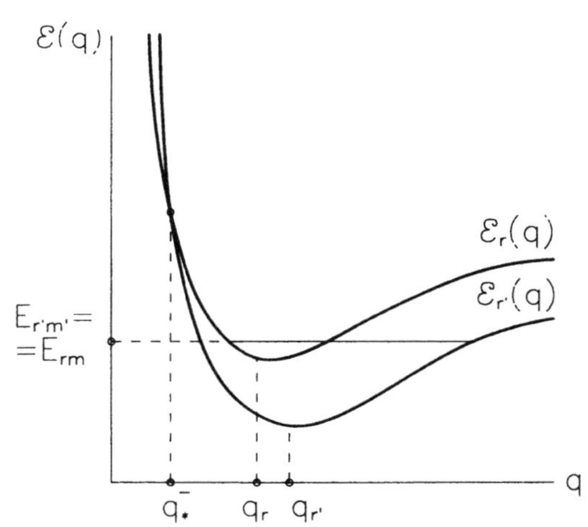

Fig. 5.3. Relative arrangement of the terms in the case of strong anharmonicity for $\lambda < 1$

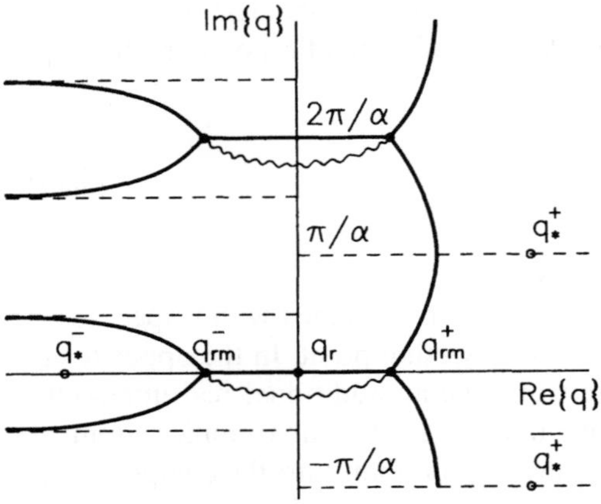

Fig. 5.4. Anti-Stokes lines (*full lines*) and cuts (*wavy lines*) for the Morse potential $\mathscr{E}_r(q)$. The saddle points nearest to the real axis are shown

When q moves from q_{rm}^+ to the right along the real axis, the expression under the root sign is kept real and positive, so one has

$$\arg\{p_{rm}\} = -\pi/2 \tag{5.3.17}$$

at any $q > q_{rm}^+$. Further, let us assume a real $q_0 > q_{rm}^+$ such that $\alpha q_0 \gg 1$. Then, at a real $q \geq q_0$, the momentum (5.3.16) is merely a constant independent of q,

$$p_{rm} = \exp(-i\pi/2)[2M(D_r - \varepsilon_{rm})]^{1/2} . \tag{5.3.18}$$

Hence, it is clear that (5.3.17) also holds for any complex q such that Re $\{q\} \geq q_0$. Moreover, on the lines Im $\{q\} = \pm\,\pi/\alpha$ the under-root expression in (5.3.16) also holds real and positive, and thus (5.3.17) is valid along these lines, too, where the saddle points q_*^+ and $\overline{q_*^+}$ are located. In a similar manner we obtain that the relation

$$\arg\{p_{rm}\} = \pi/2 \tag{5.3.19}$$

is valid on the real axis to the left of the inner turning point q_{rm}^-.

Further, we proceed as in Sect. 5.2. The overlap integral is decomposed into a sum of two integrals,

$$S_{rmr'm'} = \int\limits_{L_+} \chi_{rm}\chi_{r'm'}^+\,dq + \int\limits_{L_-} \chi_{rm}\chi_{r'm'}^-\,dq\ , \tag{5.3.20}$$

and the contours $L_\pm$ are displaced from the real axis in such a way that both integrals converge at the contours' ends, and that the functions χ_{rm} and $\chi_{r'm'}^\pm$ are allowed to be substituted with their simple asymptotic expressions; each contour must go from the left-hand half-plane into the right-hand one through the relevant saddles. Let us demonstrate that the contours pictured in Fig. 5.5 meet these demands.

It follows from the results of the previous section that χ_{rm} along both contours $L_\pm$ can be replaced with the simple asymptotic expression (5.2.13),

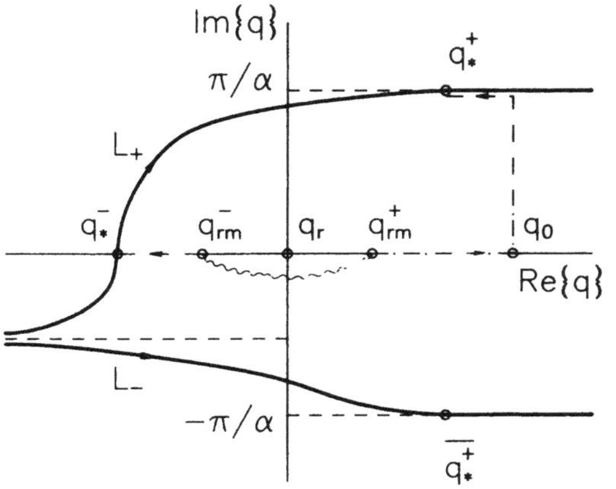

Fig. 5.5. Integration paths in (5.3.20) (*full lines*) and (5.3.30) (*dash-dotted lines*). The cut is shown by the wavy line

$\chi^+_{r'm'}$ along L_+ with (5.2.18), and $\chi^-_{r'm'}$ along L_- with (5.2.20). It should be stressed that except for the obligatory requirement of the integrals' convergence the choice of the integration paths in (5.3.20) is only dictated by practical considerations with regard to the possibility to apply the saddle-point method. For example, it would not be wrong to substitute L_- for L_+ in the first integral of (5.3.20) because the integral will be convergent anyway (see below). However, the function $\chi^+_{r'm'}$ cannot be now replaced all over that contour with a simple unique asymptotic expression of the kind of (5.2.18) since the contour crosses the anti-Stokes line separating two areas of increase of the function $\chi^+_{r'm'}$ that has different asymptotic representations (5.2.18, 19) within these areas, as well as at the boundary between them.

After the substitution of the asymptotic wave functions and of the definitions (5.2.3) into (5.3.20) one obtains in the integrand

$$\chi_{rm}\chi^{\pm}_{r'm'} \propto \exp[\Phi(q)] , \tag{5.3.21}$$

with $\Phi(q)$ from (5.2.26). For $\mathrm{Re}\{q\} \to +\infty$, the momentum p_{rm} tends to the constant (5.3.18) (and similarly behaves $p_{r'm'}$). Therefore,

$$\Phi(q) \approx \mathrm{const} - \frac{1}{\hbar}(|p_{rm}| - |p_{r'm'}|)q . \tag{5.3.22}$$

Considering (5.3.15, 18) one sees that the product (5.3.21) decreases for $\mathrm{Im}\{q\} = \mathrm{const}$, $\mathrm{Re}\{q\} \to +\infty$; hence, the integrals (5.3.20) converge at the right-hand tails of $L_\pm$. On the left-hand tails one has $\mathrm{Im}\{q\} = -\pi/2\alpha$, $\mathrm{Re}\{q\} \to -\infty$, and the dominant part of $\Phi(q)$ behaves as

$$\exp\left(\frac{2D_r(1-\lambda)}{\lambda\hbar\omega_r}\mathrm{e}^{-\alpha(q-q_r)}\right) = \exp\left(\mathrm{i}\frac{2D_r(1-\lambda)}{\lambda\hbar\omega_r}|\mathrm{e}^{-\alpha(q-q_r)}|\right) , \tag{5.3.23}$$

i.e., it oscillates, thus making the integrals of (5.3.20) convergent. Hence it is proved that both contours depicted in Fig. 5.5 satisfy all necessary requirements.

Let us calculate the velocity and the second derivative of $\Phi(q)$ at the saddle points. According to (5.3.17, 19), (5.2.31) holds at $q_*^{\pm}$, and the velocity at $\overline{q_*^+}$ is easily verified to coincide with its value at q_*^+, v_*^+. The absolute magnitude of the velocity is formally given by expression (5.2.32), where now $b_{rm} = b_r\sqrt{2\,\mu_{rm}}$ is the amplitude of the classical oscillations of the coordinate Q in the vibronic state rm (the amplitude of the Q' oscillations in the state $r'm'$, $b_{r'm'}$, is defined in the same way), and the parameters $z^{\pm}$ are

$$z_{rm}^{\pm} \equiv \frac{Q_*^{\pm}}{b_{rm}} = \left(\frac{\xi}{\mu_{rm}}\right)^{1/2} \frac{-\operatorname{sign}\Delta Q \pm \lambda(1+\eta)^{1/2}}{1-\lambda^2}\,,$$
$$z_{r'm'}^{\pm} \equiv \frac{Q'^{\pm}_*}{b_{r'm'}} = \left(\frac{\xi f}{\mu_{r'm'}}\right)^{1/2} \frac{-\lambda\operatorname{sign}\,\Delta Q \pm (1+\eta)^{1/2}}{1-\lambda^2}\,. \tag{5.3.24}$$

The second derivative of $\Phi(q)$ at the saddle points is given by (5.2.34). With the use of (5.3.9, 11) the force difference, $\Delta F^{\pm} = F_r^{\pm} - F_{r'}^{\pm}$, is written as

$$\Delta F^{\pm} = \pm\frac{1}{\lambda} M\omega_r^2 b_r [2\xi(1+\eta)]^{1/2}\,(1-\alpha Q_*^{\pm})\,. \tag{5.3.25}$$

Inserting this and (5.2.31) into (5.2.34) one obtains

$$\Phi''(q_*^{\pm}) = \mp\left|\frac{\Delta F^{\pm}}{\hbar v_*^{\pm}}\right|\,. \tag{5.3.26}$$

It is seen that the direction of steepest descent at q_*^- is perpendicular while at q_*^+ and $\overline{q_*^+}$ is parallel to the real axis.

Using (5.3.22, 23, 26) it is a simple matter to find out the relief of the surface $w = \operatorname{Re}\{\Phi(q)\}$. About the turning points there is a hill whose eastern slope leads to a valley, and the Western one to the saddle q_*^- beyond which an ascent starts. The saddle q_*^- joins two valleys located to the North-West and South-West of the hill in the domains $\pi/2\alpha < |\operatorname{Im}\{q\}| < 3\pi/2\alpha$, $\operatorname{Re}\{q\} \to -\infty$ (see Fig. 5.4), whereas q_*^+ and $\overline{q_*^+}$ connect the North-West and South-West valleys, respectively, to the East one. The contours $L_{\pm}$ pass from the South-West valley to the East one: L_+ through the saddles q_*^- and q_*^+, while L_- through $\overline{q_*^+}$.

Let $S_{rmr'm'}^{\pm}$ denote the contributions of $q_*^{\pm}$, respectively, to the integral over the contour L_+; then the contribution of $\overline{q_*^+}$ to the integral over L_- is equal to $\overline{S_{rmr'm'}^+}$, which is easily verified by a straightforward calculation. The calculation of the integrals in (5.3.20), by the method of steepest descent, yields

$$S_{rmr'm'} = S_{rmr'm'}^- + 2\operatorname{Re}\{S_{rmr'm'}^+\}\,, \tag{5.3.27}$$

$$S_{rmr'm'}^{\pm} = C_{rm}\left(\frac{\hbar\omega_{rm}\omega_{r'm'}}{2\pi|v_*^{\pm}\,\Delta F^{\pm}|}\right)^{1/2} \exp\left[\Phi(q_*^{\pm}) + (1-\zeta_{\pm})\frac{i\pi}{2}\right], \tag{5.3.28}$$

where $\zeta_+ = 0$ and $\zeta_- = 1$. The constant C_{rm} for the Morse oscillator only differs from that of the harmonic one by small terms on the order of $\alpha^2 b_r^2$ so that it may be calculated by (5.2.14). The frequency of classical vibrations in the state rm is found by (5.2.6):

$$\omega_{rm} = \omega_r[1 - \alpha^2 b_r^2(m + 1/2)] . \tag{5.3.29}$$

The same equation, after replacing rm with $r'm'$, applies to $\omega_{r'm'}$.

Finally, $\Phi(q_*^\pm)$ is calculated by (5.2.26). We introduce the notations

$$\delta_{rm}^\pm \equiv \frac{i}{\hbar} \int\limits_{q_{rm}^\pm}^{q_*^\pm} p_{rm}\, dq , \tag{5.3.30}$$

and similarly $\delta_{r'm'}^\pm$ (q_{rm}^+ and q_{rm}^- are the outer and inner turning points, respectively). Then, invoking the quantization rule (5.2.11) one has

$$\Phi(q_*^\pm) = i\pi\zeta_\pm(m - m') - \delta_{rm}^\pm + \delta_{r'm'}^\pm . \tag{5.3.31}$$

It is appropriate to draw the integration paths in the integrals (5.3.30) such as shown in Fig. 5.5 (dash-dotted line) because along those paths the $\arg\{p_{rm}\}$ keeps constant at a value of (5.3.17) or (5.3.19). Changing to the new integration variable Q by (5.3.1) one obtains

$$\delta_{rm}^\pm = \pm\frac{1}{\hbar} \int\limits_{\pm b_{rm}}^{Q_*^\pm} \frac{|p_{rm}|\, dQ}{1 - \alpha Q - i0} , \tag{5.3.32}$$

where the integral is taken over the real axis and the term $-i0$ in the denominator has appeared because the singular point $Q = 1/\alpha$ in the integral for δ_{rm}^+ is circumvented from above along a small hemicircle of radius $\alpha^{-1} \exp[-\alpha(q_0 - q_r)]$ vanishing in the limit of $q_0 \to \infty$. Equation (5.3.32) is recast as

$$\delta_{rm}^\pm = \mu_{rm} J(\pm z_{rm}^\pm, \pm \alpha b_{rm}) , \tag{5.3.33}$$

$$J(z, \beta) \equiv 2\int\limits_1^z \frac{(t^2 - 1)^{1/2}}{1 - \beta t - i0} dt . \tag{5.3.34}$$

In this integral the parameters β and z assume real values, $|\beta| < 1$ [this follows from (5.3.5)] and

$$z > 1, \qquad z|\beta| > 1 . \tag{5.3.35}$$

Here, the first inequality reflects the fact that, since the displacement of the equilibrium position is small, the points $Q_*^\pm$ are located off the line segment connecting the turning points of the "harmonic oscillator" (5.3.2); this is one of the necessary conditions for the present quasiclassical method to be applied. As for the second inequality resulting from the condition (5.3.11) of strong anharmonicity, it plays no role in the calculation of the integral.

Since we shall need the integral (5.3.34) for complex z values, too, condition $\mathrm{Im}\{z\} \geq 0$, more general than (5.3.35), will be assumed below. Elementary integration yields

$$J(z,\beta) = \frac{2}{\beta^2}\left\{(1-\beta^2)^{1/2}\ln\frac{z-\beta+[(1-\beta^2)(z^2-1)]^{1/2}}{\beta z - 1}\right.$$
$$\left. - \ln[z+(z^2-1)^{1/2}] - \beta(z^2-1)^{1/2} + \mathrm{i}\pi(1-\beta^2)^{1/2}\right\} . \quad (5.3.36)$$

The branches of the roots and logarithms are selected by the condition that these functions take on real values for real $z > 1$ and $\beta z > 1$; in particular, one should put $\arg\{\beta z - 1\} = 0$ for $\beta z > 1$ and π for $\beta z < 1$. In (5.3.31) the quantities $\delta^{\pm}_{r'm'}$ are also given by (5.3.33), with rm replaced by $r'm'$.

Thus, in the case considered, the overlap integral is given by (5.3.27, 28), where all the entries are expressed in terms of elementary functions of the potential parameters and quantum numbers of the initial and final electronic states.

Two remarks have to be made regarding the formulae obtained. First, the expression for $\delta^{\pm}_{rm}$ can be simplified under conditions (5.2.51) by using an asymptote of the function $J(z,\beta)$ for $z \gg 1 \gg |\beta|$; on the other hand, additional restrictions need to be imposed on the parameters and quantum numbers to simplify $\delta^{\pm}_{r'm'}$, and, in practice, such restrictions always emerge in calculations of the overlap integrals for specific molecules. Second, one should bear in mind that, since the barrier integrals $\delta^{\pm}_{rm}$ and $\delta^{\pm}_{r'm'}$ are large, the dominant contribution to the overlap integral comes from such a saddle point for which the barrier integrals are the smallest in absolute magnitude; hence in (5.3.27) one of the terms is, as a rule, always much larger than the other.

The case $\lambda \geq 1$ is treated in quite a similar way, and the corresponding results are given without detailed derivation. As in the case of the harmonic oscillator (see Sect. 5.2), a critical value of the parameter λ, λ_c, exists at which two roots $Q^{\pm}_{*}$ come together; this value is found from (5.3.8) with $\eta = -1$. For $1 \leq \lambda \leq \lambda_c$ both roots $Q^{\pm}_{*}$ are again real, but are of the same sign coinciding with that of ΔQ. The overlap integral is dominated by the root with the least absolute magnitude remaining finite for $\lambda \to 1$, and now instead of (5.3.27) we have

$$S_{rmr'm'} = \begin{cases} 2\,\mathrm{Re}\{S^{+}_{rmr'm'}\} & \text{for } \Delta Q > 0 \ , \\ S^{-}_{rmr'm'} & \text{for } \Delta Q < 0 \ , \end{cases} \quad (5.3.37)$$

while all the subsequent equations [(5.3.38) and on] remain unchanged. For $\lambda > \lambda_c$ the roots $Q^{\pm}_{c}$ are complex (they are complex conjugates to each other). Let Q^{+}_{*} stand for the root in the upper half-plane, and write $1 - \alpha Q^{+}_{*} = |1 - \alpha Q^{+}_{*}|\exp(\mathrm{i}\varphi)$, where $-\pi \leq \varphi \leq 0$. With such a definition of φ we have

$$q^{+}_{*} = q_r - \frac{1}{\alpha}\ln|1 - \alpha Q^{+}_{*}| - \frac{\mathrm{i}\varphi}{\alpha} \quad (5.3.38)$$

so that q_*^+ is in the upper half-plane as well, and q_*^- is its complex conjugate. In (5.3.20) the integral over L_+ is governed by q_*^+ while that over L_- by q_*^- so that these integrals are complex conjugates to one another. Then we obtain

$$S_{rmr'm'} = 2\,\mathrm{Re}\{S^+_{rmr'm'}\}\,, \tag{5.3.39}$$

$$S^+_{rmr'm'} = C_{rm}\left(\frac{\hbar\omega_{rm}\omega_{r'm'}}{2\pi|v_*^+\Delta F^+|}\right)^{1/2}\exp\left[\Phi(q_*^+)+\frac{\mathrm{i}}{2}\left(\frac{\pi}{2}-\varphi-\gamma\right)\right],$$

where

$$|v_*^+| = \omega_r b_{rm}|(z^+_{rm})^2-1|^{1/2}\,,$$

$$0 \le \gamma \equiv \arg\{[(z^+_{rm})^2-1]^{1/2}\} \le \pi$$

($\gamma = 0$ at real $z^+_{rm} > 1$); the force difference is found from (5.3.25), $\Phi(q_*^+)$ from (5.3.31, 33, 36), z^+_{rm} and $z^+_{r'm'}$ from (5.3.24), where the substitution (5.2.46) is performed.

5.3.2 Morse Oscillator with $\Delta\alpha \neq 0$ and Arbitrary Potentials

In conclusion, we discuss the general case of $\Delta\alpha \equiv \alpha_r - \alpha_{r'} \neq 0$. In the preceding section we were capable of expressing, by use of the quasiclassical approximation, the Morse overlap integral in terms of elementary functions of the parameters and quantum numbers because the saddle points q_* could be obtained in explicit analytical forms. For $\Delta\alpha \neq 0$ the saddle-point equation is transcendental and the quasiclassical overlap integral is no longer expressed via the elementary functions of the parameters and quantum numbers while it is still an elementary function of q_*. In connection with this it is interesting to note that at $\Delta\alpha = 0$ the exact overlap integral is expressed through higher transcendental functions (see Sect. 4.8), but no such kind of formula exists for $\Delta\alpha \neq 0$.

In the general case the Morse overlap integral can be expressed as a sum over the saddle points,

$$S_{rmr'm'} = \sum_{q_*} C_{rm}\left(\frac{\hbar\omega_{rm}\omega_{r'm'}}{2\pi|v_*\Delta F|}\right)^{1/2}\mathrm{Re}\{\exp[\Phi(q_*)+\mathrm{i}(\nu-\psi)]\}\,, \tag{5.3.40}$$

where ν is the steepest descent direction and ψ is the argument of the velocity at the saddle point which is calculated by the rule given before. Owing to the large values of the barrier integrals for deep tunneling transitions, either one real point or a pair of complex conjugate points will only contribute to the sum, as a rule.

The calculations by (5.3.40) are carried out in two stages. In the first stage the saddle points q_* are searched by solving the transcendental equation (5.2.28). To this end we let first $\Delta\alpha = 0$ and find the roots $q_*^\pm$ by (5.3.7, 13, 38); the roots displaced by the period of $2\pi\mathrm{i}/\alpha$ to the upper half-plane have to be incorporated, too, while those in the lower half-plane need not be considered explicitly as their contributions to (5.3.40) are complex conjugates of the contributions of the corresponding complex conjugate roots in the upper half-plane. Then we put $\Delta\alpha = \varepsilon$, where ε is small enough in absolute magnitude and has the same sign as the actual $\Delta\alpha$. The new values of $q_*^{\pm}$ are easily found by solving (5.2.28) numerically. Also, care should be taken of new roots in the left half-plane that may come from infinity; at small ε their approximate values are easily found analytically by solving (5.2.28) in the limit of $\mathrm{Re}\{q\} \to -\infty$. On letting then successively $\Delta\alpha = 2\varepsilon$, 3ε, etc., we calculate each time new values of all the roots by an appropriate iterative procedure using as a starting approximation the values calculated at the preceding step.

In the second stage the contribution of each particular saddle point q_* into the sum of (5.3.40) is calculated. To do this, the parameters

$$z_{rm} = \frac{1}{\alpha_r b_{rm}}\{1 - \exp[-\alpha_r(q_* - q_r)]\}$$

and $z_{r'm'}$ (by this same formula with rm replaced by $r'm'$) are found (perhaps, (5.2.28) is more suitable to be solved with the use of these variables), and

$$\Phi(q_*) = -\mu_{rm} J(z_{rm}, \alpha_r b_{rm}) + \mu_{r'm'} J(z_{r'm'}, \alpha_{r'} b_{r'm'})$$

is calculated, where $J(z, \beta)$ is given by (5.3.36) for $\mathrm{Im}\{z\} \geq 0$ and $\beta < 1$. Further, we calculate the velocity at the saddle point

$$v_* = \left\{\frac{2}{M}[E_{rm} - \mathscr{E}_r(q_*)]\right\}^{1/2} \equiv \left\{\frac{2}{M}[E_{r'm'} - \mathscr{E}_{r'}(q_*)]\right\}^{1/2}, \tag{5.3.41}$$

the force difference

$$\Delta F = \left(-\frac{d\mathscr{E}_r(q)}{dq} + \frac{d\mathscr{E}_{r'}(q)}{dq}\right)_{q=q_*}, \tag{5.3.42}$$

and the second derivative

$$\Phi''(q_*) = -\mathrm{i}\frac{\Delta F}{\hbar v_*}. \tag{5.3.43}$$

The direction of steepest descent is

$$\nu = [\pi - \arg\{\Phi''(q_*)\}]/2. \tag{5.3.44}$$

Notice that the argument of (5.3.43) is defined to within 2π only; hence, ν is defined by (5.3.44) to within π. Therefore, an additional limitation is imposed on ν to fall into the interval from $-\pi/2$ to $+\pi/2$ since the integration path at every point is directed from the left-hand into the right-hand half-plane (if $|\nu| = \pi/2$, then $\nu = +\pi/2$ in the left-hand and $\nu = -\pi/2$ in the right-hand half-plane).

In the case of arbitrary potentials $\mathscr{E}_r(q)$ and $\mathscr{E}_{r'}(q)$ the overlap integral may also involve contributions from the potentials' singularities in the complex plane. However, if these singularities are located more distant from the turning points than the saddle points are, (5.3.40) holds. The constant C_{rm} may still be calculated by (5.2.14), which is valid provided that the anharmonic shifts of the lower levels are small as compared to the vibrational quantum; further $\Phi(q_*)$ is found by (5.2.26), the classical vibrational frequency by (5.2.6), and the rest of the parameters by (5.3.41–44).

Equation (5.3.40) does not apply when two significant saddle points come close to one another so that their ranges of influence overlap. Such a situation is highly improbable, though the corresponding formula, if necessary, can be derived easily; an example of this kind (merging of two saddle points) was given in Sect. 5.2.1.

Let us touch upon possible applications of the quasiclassical formulae in numerical calculations. These formulae are simple enough and show high accuracy. In particular, they were used with actual values of parameters corresponding to C–C and C–H vibrations in aromatic hydrocarbons and some other molecules, and of the quantum numbers pertinent to radiationless transitions between lowest electronic states of these molecules [2.257, 5.8]. Preliminary calculations were carried out for $\Delta\alpha = 0$ (see also Sect. 5.4.3) since, in this case, the results can be checked against the exact formula (4.8.14). For $\lambda < 1$ the Franck–Condon factor is a monotonic function of the parameters and quantum numbers (the main contribution is from a real saddle point), and the error of (5.3.27) does not exceed 20%. For $\lambda > \lambda_c \approx 1$ the main contribution comes from a pair of complex conjugate points, and the Franck–Condon factor oscillates; as is seen from (5.3.39) it vanishes when the phase of the complex quantity $S^+_{rmr'm'}$ equals $(n + 1/2)\pi$, where n is an integer. The error of (5.3.39) is no more (actually much less) than 50% except for close vicinities of the Franck–Condon factor's zeros, where, because the leading term of the quasiclassical expansions in $\hbar$ vanishes, the higher terms come into play; however, allowing for these terms makes no sense because of a small value of the Franck–Condon factor. Finally, the formulae obtained do not apply in the vicinity of $\lambda = \lambda_c$ where closing of the saddle points occurs.

In early studies it was revealed that harmonic models often underestimate the transition rates as compared to measured values, and it was hoped that the anharmonicity might diminish this mismatch by increasing the transition rate [4.107, 5.13, 14]. However, it was then found that, in fact, the anharmonicity may both incrase and decrease Franck–Condon factor depending on the relevant parameters [2.300, 4.172, 173, 5.15]. The quasiclassical results enable us to furnish an explanation to this fact based on a consideration of the term crossing points.

As an example, consider Fig. 5.3 with $\Delta q < 0$. If we take the limit of $\alpha \to 0$ keeping the rest of the parameters unchanged, then the real intersection point q_*^- will shift to the left, and the Franck–Condon factor will decrease; the appearance of the second real point q_*^+ is of no importance since it is located too far away. Now imagine that $\Delta q > 0$ and the upper curve is displaced to the right so that no real intersection points exist. After the anharmonicity is removed for $\alpha \to 0$, a real intersection point q_*^+ appears and the Franck–Condon factor increases (see Sect. 5.3.3).

5.3.3 Selection Rule for the Morse Oscillator

The quasiclassical approximation reveals interesting features of the Franck–Condon factor as a function of the parameters and quantum numbers that permit the formulation of a peculiar selection rule for transitions between the electronic states separated by a large energy gap [5.9]. It turns out that when there is a real intersection point of the terms[1] (the range of monotony for $\lambda < 1$), it is, as a rule, just this point to provide the dominant contribution to the overlap integral, and the Franck–Condon factor takes on a relatively large value. On the other hand, in the absence of any real crossing points the overlap integral is controlled by a pair of complex conjugate points (the range of oscillation for $\lambda > 1$), and the Franck–Condon factor is small. Note that the same is true for the harmonic oscillator [see (5.2.64, 65)], but it has $\alpha = 0$ so that $\lambda = f$ by (4.8.15). Thus, as distinct from the harmonic Franck–Condon factor, the Morse one is quite sensitive to the sign of Δq [4.172, 5.16, 17], especially for

$$|1 - f| \ll \alpha |\Delta q| \,, \tag{5.3.45}$$

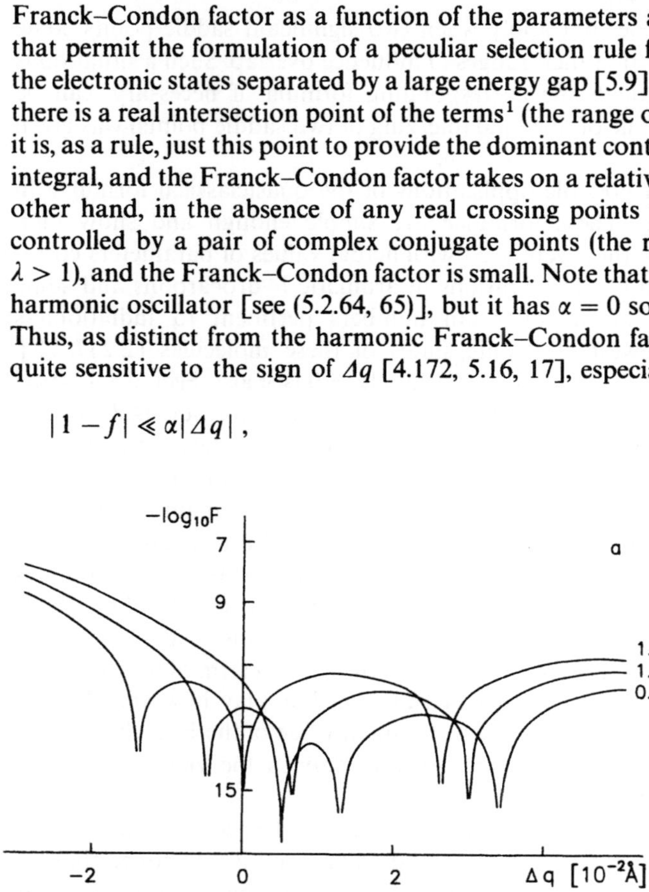

Fig. 5.6a–d. The Franck–Condon factor F of the Morse oscillator vs Δq for different values of the parameters and quantum numbers. (**a**) $m' = 10$, $\Delta\alpha = 0$ (numbers denote the f values); (**b**) $f = 1.01$, $\Delta\alpha = 0$ (numbers denote the m' values); (**c**) $m' = 8$, $f = 1.01$ (numbers denote the $\Delta\alpha$ values in Å^{-1}); (**d**) $m' = 7$, $f = 1.01$, $\Delta\alpha = 0.02\,\text{Å}^{-1}$; curve *1*: quantum Franck–Condon factor evaluated by direct numerical calculation of the overlap integral; curve *2*: quasiclassical approximation (5.3.46) allowing for the interference between the saddle points; curve *3*: classical approximation (5.4.4) with the interference effects averaged out

[1] For horizontal transitions (see (4.5.10) and Fig. 4.2) the saddle points are the points of term intersection [see (5.2.28)]

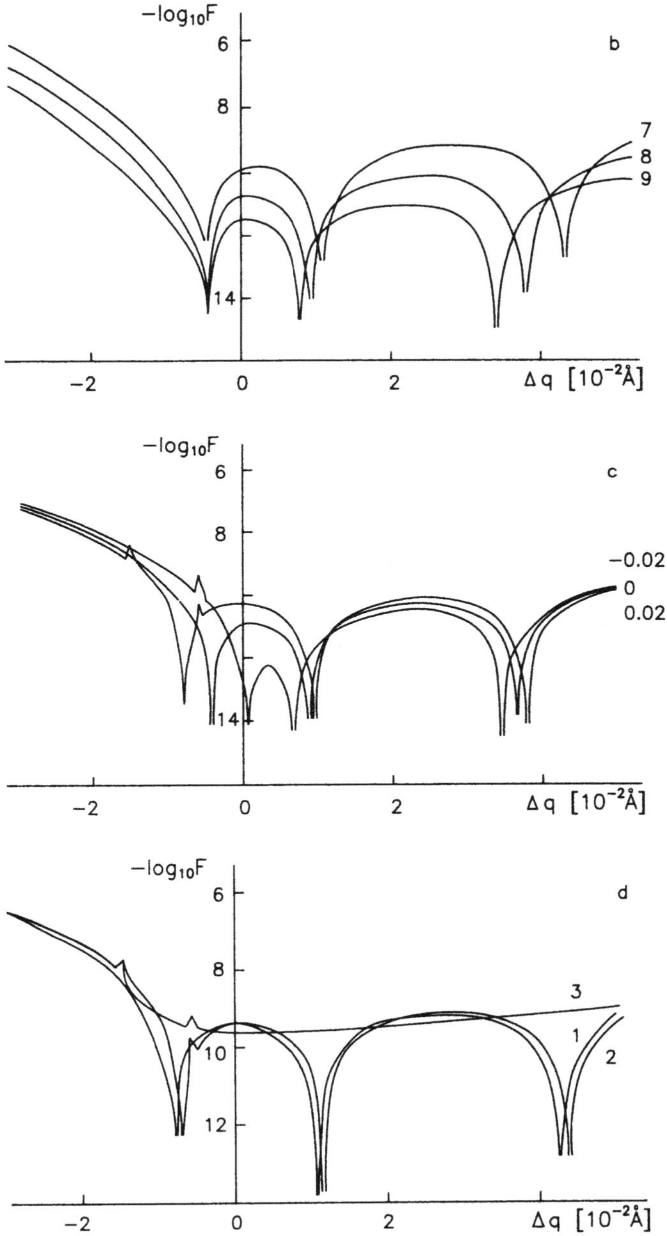

Fig. 5.6. Cont.

since then a change in the sign of Δq approximately replaces λ with $1/\lambda$ [see (4.8.15)], which causes a drastic alteration in the magnitude of the Franck–Condon factor due to the change in the type of term crossing.

This effect is illustrated in Fig. 5.6, where the Franck–Condon factor versus the term displacement Δq is depicted for the Morse potential for several values of the parameters and quantum numbers typical of the local C–H vibrations in aromatic hydrocarbons and of the $T_1 \rightsquigarrow S_0$ transitions. The initial quantum number is $m = 0$, the final one $m' = 7-10$ (this corresponds to a wide electronic gap of 2.5–3.4 eV), the dissociation limit and vibrational frequency in the final electronic state $D_{r'} = 44\,500\ \mathrm{cm}^{-1}$ and $\omega_{r'} = 3\,200\ \mathrm{cm}^{-1}$, respectively (giving the final state Morse range parameter $\alpha_{r'} = 1.775\ \text{Å}^{-1}$), and the frequency ratio $f = 0.99-1.03$. The reduced mass of the CH group was taken as the effective mass of the C–H local vibration. The shift of the equilibrium position was assumed to be equal to the change in the C–H bond length, $\Delta q = R_{T_1} - R_{S_0}$. Its absolute magnitude ranges from 0.01 Å to 0.03 Å while the sign is unknown [4.165]. The Franck–Condon factor is

$$F_{rmr'm'} = \left| \sum_{q_*} S_{rmr'm'}(q_*) \right|^2 , \tag{5.3.46}$$

where summation is over all significant saddle points q_* and the contribution of each point is calculated by (5.3.40), as described in Sect. 5.3.2.

Figures 5.6a, b show the results calculated at $\Delta\alpha = 0$ when the saddle-point equation (5.2.28) can be solved analytically (see Sect. 5.3.1). It is seen that the range of monotony is for $\Delta q < 0$ thus corresponding to a real term intersection while the oscillation range is mainly at $\Delta q > 0$ where the intersection is in the complex plane (the value of $\lambda = 1$ is reached next to the first minimum of each curve). The values of the Franck–Condon factor in the above two ranges differ by two to four orders of magnitude.

Numerical calculations for the Morse oscillator at $\Delta\alpha = 0$ revealed the Franck–Condon factor's oscillations versus ω_r and $\omega_{r'}$ [5.17, 18]. They are of the same nature.

Similar results are also obtained for $\Delta\alpha \neq 0$ (see Figs. 5.6c, d). The monotonic branches of the curves for $\Delta q < 0$ are again related to a real term crossing while the oscillations occur in the range of complex crossings, generally for $\Delta q > 0$. The spikes on the curves are false. Each spike signals a merging of two saddle points where the method of steepest descent does not apply in its standard form, and the exponent of the integrand should be expanded up to the third (rather than the second) derivative, as was done in Sect. 5.2.1 for the harmonic oscillator. Additionally, in the merging region the procedure for solving the transcendental equation (5.2.28) described in Sect. 5.3.2 does not work.

To assess the accuracy of the quasiclassical approximation the Franck–Condon factors were simultaneously calculated by straightforward numerical integration using the exact wave functions (4.8.13). At $\Delta\alpha = 0$ they were also calculated by the exact equation (4.8.14). Good agreement between the exact and quasiclassical Franck–Condon factors was obtained in all cases (a sample of this kind of comparison is given in Fig. 5.6d, curves 1, 2). Thus, the main conclusion is supported that the (real or complex) intersection points provided dominant contributions to the overlap integral.

Based on the results shown in Fig. 5.6 the following selection rule can be formulated: *At a large energy gap* (*on the order of* 2.5 eV *and more*) *the radiation-*

less transition in a molecule with C–H *bonds is allowed if the* C–H *bond length is shorter in the upper electronic state than in the lower, and is prohibited otherwise, with the prohibition factor being on the order of* 10^2–10^4.

The absolute magnitude of Δq in the aromatic molecules with a large number of benzene rings is lower than the above value typical of benzene and naphthalene [1.23, 5.19]. Simultaneously, a decrease in the quantity $|1-f|$ should be expected (for diatomic molecules this is established by the empirical *Badger*'s rule [5.20, 21]), which enhances the steepness of the curves in Fig. 5.6 in the vicinity of $\Delta q \approx 0$. Therefore, the indicated value of the prohibition factor for the $\Delta q > 0$ transitions holds for such molecules, too.

5.4 Franck–Condon Principle for Radiationless Transitions

5.4.1 General Formulation

It is well known [4.69] that the Franck–Condon principle for optical transitions is a direct consequence of the quasiclassical nature of the nuclear motion. Similarly, the quasiclassical nature of the nuclear wave function entails the Franck–Condon principle for radiationless transitions [1.39, 5.9].

The most important result of the application of the quasiclassical method outlined in this chapter is the conclusion that the main contribution to the overlap integral comes from such values of the nuclear coordinate where the classical momenta of the motion in both electronic terms are equal. For the resonant (horizontal) transition (4.5.10) these are the points of intersection of the potential curves on the real axis or in the complex plane.

Generally, at an arbitrary number of nuclear coordinates, the leading factor of the integrand in the overlap integral (4.2.18) is proportional to [1.39, 2.238]

$$\exp\left[\frac{\mathrm{i}}{\hbar}(S_r - S_{r'})\right], \tag{5.4.1}$$

where S_r and $S_{r'}$ are classical actions in the initial and final electronic states, respectively. In the limit of $\hbar \to 0$ the integral (4.2.18) is calculated by the saddle-point method. The equation for the saddle points q_* is obtained from the condition that the first derivatives of the exponent in (5.4.1), with respect to the coordinates q, vanish. Considering that the derivative of the action is the classical momentum one obtains

$$p_r = p_{r'}, \tag{5.4.2}$$

that is, at the configuration q_* the nuclear momenta of the states r and r', hence the kinetic energies, coincide. Owing to the conservation of the total energy in the radiationless transition the potential energies coincide, too,

$$\mathscr{E}_r(q_*) = \mathscr{E}_{r'}(q_*). \tag{5.4.3}$$

In other words, the q_* corresponds to a crossing of the potential surfaces of the initial and final electronic states.

Using the classical terminology one may say that the radiationless electronic transition occurs at the configuration q_* since just this configuration mainly contributes to the overlap integral. This result offers a generalization of the Franck–Condon principle for radiationless transitions: *A radiationless transition occurs at fixed nuclear coordinates and momenta at a configuration where the electronic terms cross.*

In the case of optical transitions the same argument leads, instead of (5.4.3), to the equation $\mathscr{E}_r(q_*) = \mathscr{E}_{r'}(q_*) + \hbar\omega$, where $\hbar\omega$ is the photon energy. Here, the configuration q_* corresponds to a crossing between the initial term $\mathscr{E}_r(q)$ and the final term $\mathscr{E}_{r'}(q)$ shifted up along the energy axis by $\hbar\omega$. Usually $\hbar\omega$ is such that q_* falls into the classically accessible region. In contrast, for the tunneling radiationless transitions q_* is in the classically inaccessible region or even in the complex region of the nuclear coordinates where the classical momenta (5.4.2) are complex.

At this juncture it is appropriate to cite the Landau paper [1.39], where the following is stated with regard to (5.4.2): "Physically this means: for the transitions, the points are essential where both q and p remain unchanged. This statement represents here something more than a derivation of the so-called Franck–Condon rule since it is valid not only for the real but also for the complex range of values of the variables". For the transitions between the resonant states (4.5.10) "essential are only the crossing points of the potential energy curves (or generally multidimensional surfaces)."

The quasiclassical approximation in the theory of radiationless transitions can also be formulated in the framework of the method of tunneling classical trajectories [1.13, 2.238, 5.22]. It turns out in this case, too, that the main contribution to the transition probability comes from an optimal trajectory passing through the point of electronic term intersection in the classically inaccessible region of the nuclear coordinates.

5.4.2 Real and Complex Term Intersections

From the preceding section the Franck–Condon principle reads that the radiationless transition occurs at a certain nuclear configuration where the electronic terms cross. If the term crossing takes place at real values of the nuclear coordinates, then there is a certain (exponentially low) probability to arrive at the crossing configuration q_* where the transition just occurs from one term to another in accord with the Franck–Condon principle's requirement about the constancy of the nuclear momenta.

This principle is also valid when no real intersection exists and the terms cross only at complex values of the nuclear coordinates. Here, however, one cannot speak of whatever probability to reach the complex "molecular configuration" q_* since this would contradict the basic principles of quantum mechanics. A real physical system is not capable of reaching such a configuration, and therefore, according to the Franck–Condon principle, the radiationless transition is in this case impossible. In fact, of course, there is a nonzero transition probability, but now the transition happens within a wide range of real nuclear coordinates nearby the complex "configuration" q_* (i.e., in the range of minimum term separation). In other words, the transition is not localized at a certain nuclear configuration, but, instead, is "spread" over a whole range of real nuclear coordinates. Such a delocalization greatly diminishes the transition probability, resulting in the selection rule of Sect. 5.3.3.

In the case of a complex term intersection, the radiationless transition can be regarded as a "double-quantum" process since it involves two pure quantum effects having no classical counterparts. These are a tunneling to the transition domain (the domain of the terms' approaching) and the subsequent sudden change in the nuclear coordinates and momenta caused by the change in the electronic state. From this point of view the transition via a real intersection is a "quantum-classical" process (the quantum effect of tunneling into the crossing region plus the classically allowed electronic transition at fixed coordinates and momenta).

It was demonstrated in Sect. 5.3.3 that for a light (hydrogen) atom the probability of the double-quantum process, where large electronic energy is transferred to vibrations, is only a factor of 10^2–10^4 as low as that of the quantum-classical one. Evidently this factor will be much larger for molecules without light atoms. Therefore, the double-quantum transitions are expected to be inefficient in such molecules.

In summary, the radiationless transition between two electronic states separated by a large energy gap occurs efficiently only in the presence of a real intersection of the potential surfaces, and is much less efficient when no such an intersection exists since in the latter case the exponentially small probability of tunneling into the transition region is additionally multiplied by a very low probability (because of the breakdown of the Franck–Condon principle) to change the electronic state. In connection with this it is interesting to note that in early studies on radiationless transitions [1.18, 2.32, 33] the existence of the potential surfaces' intersection *in the classically accessible region* was regarded as a necessary condition for the transition to occur; thus it was assumed that the radiationless transition is a classically allowed ("double-classical") process. Later it was established [2.36, 177] that the radiationless transitions between the lowest electronic states of organic molecules are classically forbidden (tunneling) processes. Nevertheless, as follows from the Franck–Condon principle, the tunneling transition also asks for a real term intersection, but this intersection is located *in the classically inaccessible region.*

5.4.3 Classical Franck–Condon Factor

For the values of the parameters corresponding to a complex electronic term intersection the curves in Fig. 5.6 show infinitely deep holes corresponding to zeros of the Franck–Condon factor. A formal reason for the zeros to appear was given at the end of Sect. 5.3.2, and a physical interpretation of this effect can be revealed with the help of the Franck–Condon principle.

Let q_* and $\bar{q}_*$ stand for a pair of complex conjugated points giving the main contribution to the sum of (5.3.46). The probability amplitudes for the respective transitions are complex conjugates to each other since the overlap integral is real. Writing them in the form $S_{rmr'm'}(q_*) = A\exp(\mathrm{i}\varphi)$ and $S_{rmr'm'}(\bar{q}_*) = A\exp(-\mathrm{i}\varphi)$ one obtains the Franck–Condon factor $F_{rmr'm'} = 4A^2\cos^2\varphi$, where φ is a large phase equal to the real part of the difference of the action integrals from q_* to the turning points in the initial and final electronic states. This phase strongly varies with Δq, and can assume values close to $(n + 1/2)\pi$, where n is an integer. Hence it follows that the zeros of the Franck–Condon factor result from an interference of the amplitudes of the transitions through the q_* and $\bar{q}_*$ configurations.

It is clear that the quasiclassical approximation (5.3.46) maintains the quantum superposition principle giving rise to the interference effects. However, it seems unlikely that in the course of a radiationless transition with a large energy transfer the nuclear subsystem would "remember" the precise phases of partial amplitudes of transitions via various configurations since those phases are large. Moreover, the positions of the interference minima cannot be localized because the molecular parameters (e.g., energy) are defined only approximately. Therefore it is appropriate, in the theory of transitions with a large energy transfer, to average the Franck–Condon factor over the phase of the probability amplitudes. Such averaging nullifies the cross-terms (and hence the interference effects) in (5.3.46) yielding the expression

$$F_{rmr'm'} = \sum_{q_*} |S_{rmr'm'}(q_*)|^2 , \tag{5.4.4}$$

which may be called *the classical Franck–Condon factor* since it embodies the classical principle of the additivity of the partial transition probabilities.

The classical Franck–Condon factor is a smooth function of Δq (see curve 3 in Fig. 5.6d). From the formulae of Sect. 5.3 it follows that expression (5.4.4) is also a smooth analytical function of noninteger quantum numbers. A different analytical extension of the overlap integral to the noninteger numbers was deduced by *Kühn* et al. [5.16] with the aid of purely formal mathematical methods.

The interference effect described above for complex intersections is also possible for real ones provided that contributions of a pair of real intersection points are equal in absolute magnitude. This is just the case for the undisplaced harmonic oscillator whose intersection points $q_*^\pm$ are located symmetrically on the real axis with respect to the equilibrium position (see (5.2.29, 30) in the limit of $\Delta q \to 0$ and $\xi \propto \Delta q^2 \to 0$). Equations (5.2.59, 60) for $B \to 0$ can be written as sums of equal contributions of those points that are in phase at $m' - m$ even and out of phase by π for $m' - m$ odd. Thus, the appearance of the Franck–Condon factor zeros for odd values of $m' - m$ is explained by the destructive interference of the transition amplitudes associated with the real intersection points.

5.4.4 Franck–Condon Principle and the Selection Rules

In the case of the Morse oscillator with a relatively small frequency change (5.3.45) the appearance of a real intersection of the potential curves, as required by the Franck–Condon principle, closely correlates with the sign of the displacement of the equilibrium position, which results in the selection rule given in Sect. 5.3.3. To date there is no experimental information regarding the sign of the change in the C–H bond lengths for the $T_1 \rightsquigarrow S_0$ transitions in aromatic hydrocarbons. If the theory were capable of calculating the absolute magnitudes of the transition rate constants at least to within a factor of 100, then a comparison of the calculated and measured constants would enable one to establish unambiguously, by this selections rule, the signs of the bond length changes, and thus to clear up the role of the double-quantum transitions in molecules with light atoms.

The Franck–Condon principle implies that the orbital selection rules are governed by the symmetry of the intersection configuration of the potential surfaces, q_*, rather than by the equilibrium configuration q_r (see (4.5.17, 19) and Sect. 4.7). For the internal conversion this difference is insignificant since the non-adiabaticity operator has nonzero transition matrix elements between any electronic states of similar multiplicity (see Sect. 4.7), i.e., such transitions are always orbitally allowed. But, in the case of the intersystem crossing, the electronic perturbation inducing the transition is the spin–orbit interaction. This operator may have no nonzero matrix elements for the transition between two given electronic states if the symmetry of the configuration q_* is high enough, and then the transition is orbitally forbidden. If only totally symmetric vibrations are active in the radiationless transition, the q_* configuration has the same symmetry as the equilibrium one, and hence, both configurations predict similar selection rules. However, generally, the symmetry of q_* is lower, and correspondingly the selection rules are less restrictive.

Previously *Freed* and *Lin* [1.1, 4.13] came to the same conclusion based on a Q-centroid approximation for the electronic matrix element [5.23]. The authors showed that this element should be calculated for a certain distorted nuclear configuration they called the Q-centroid (in analogy with the r-centroid in the theory of optical transitions [4.32]); the Q-centroid configuration is close to the configuration of the potential surface intersection along the coordinates of the dominant accepting modes.

The electronic matrix elements for different mechanisms of the $T_1 \rightsquigarrow S_0$ intersystem crossing in aromatic hydrocarbons are often estimated at the equilibrium nuclear configuration. However, in these molecules the T_1 and S_0 potential surfaces are not very strongly shifted and distorted relative to one another, and therefore, the term crossing may happen only in a configuration q_* severely distorted as compared to the equilibrium one, having severely distorted molecular orbitals. This circumstance must essentially affect the numerical values of the matrix elements and eventually the estimates of the relative importance of different mechanisms of the intersystem crossing. For this reason *a very interesting and actual problem of quantum chemistry is the calculation of the electronic wave functions, terms, and spin–orbit matrix elements in strongly distorted molecular configurations near term intersections.*

5.5 Transitions Between Parallel Terms

The electronic terms of low-lying excited states of polyatomic molecules are, as a rule, only slightly displaced and distorted with respect to the ground state. This is because they correspond to one-electron excitations, and a change in the state of one electron does not affect the vibration potential energy essentially due to a large number of atoms in the molecule. Of course, there are always exceptions but they mainly pertain to low-frequency (deformation) modes, whereas high-frequency (valence) modes are nearly insensitive to the electronic excitation. Just the latter ones play the main role as good acceptors of electronic energy in radiationless transitions between the terms separated by a large energy gap ΔE. The small displacement and distortion of the terms upon the electronic excitation correspond to *a weak vibronic coupling*. The weak coupling is also typical of rare-earth metal ions in solid and liquid media because the excitation of the electrons of inner shells does not affect the potentials of the local vibrations that are acceptors of the electronic energy.

According to the Franck–Condon principle the main contribution to the radiationless transition probability comes from real and/or complex points of intersection of the electronic terms where the classical momenta of the initial and final state coincide. At weak vibronic coupling the intersection points lie far from equilibrium in the classically inaccessible or even complex region of the nuclear coordinates. Therefore, the transition can be described in terms of tunneling (the so-called dynamical tunneling, see Sect. 5.5.2) through a long distance from the equilibrium to the region of an intersection or the closest approach of the terms. The weaker the vibronic coupling (i.e., the smaller the displacement and deformation of the terms), the farther this region moves away, and the tunneling depth increases.

The question is now if there are some restrictions upon the increase in the tunneling depth, and in particular by what factors it is controlled in the limit of *zero vibronic coupling* when the terms are exactly parallel to each other, i.e.,

$$\mathscr{E}_r(q) = \mathscr{E}_{r'}(q) + \Delta E \tag{5.5.1}$$

holds for any q. It proves that, when the vibronic coupling weakens, the tunneling depth rapidly tends to a finite limit that depends only upon ΔE and upon the shape of the potential curve of the ground electronic state $\mathscr{E}_{r'}(q)$.

We begin with a discussion of the physical assumptions underlying the model of parallel terms.

5.5.1 Model of Parallel Terms

The zero vibronic coupling corresponds to a small (but nonzero) interaction $U_{\text{el-nucl}}(x, q)$ in the Hamiltonian (4.1.1–3). Therefore, in the decomposition of the total Hamiltonian into the sum of (2.3.2) this interaction [along with $U_{\text{so}}(s, x, q)$] has to be assigned to the perturbation V. Then in the zeroth-order Hamiltonian H_0 the variables x and q are separated, and solutions to (4.1.7) are the functions $\varphi_r(x)$ independent of the nuclear coordinates q. As a result, (4.1.7) becomes purely electronic,

$$[T_{\text{el}}(x) + U_{\text{el}}(x)]\,\varphi_r(x) = \varepsilon_r \varphi_r(x)\,, \tag{5.5.2}$$

and the corresponding nuclear (diabatic) potential is

$$\mathscr{E}_r(q) = \varepsilon_r + U_{\text{nucl}}(q)\,. \tag{5.5.3}$$

The potentials of the electronic states r and r' are strictly parallel to each other, merely being displaced along the energy axis by (4.5.18) as in (5.5.1) (see Fig. 5.7a).

Equations (4.1.27–30) show that the vibrational wave functions $\chi_{rm}(q)$ and the energies ε_{rm} are, in fact, independent of the electronic state r, and the manifold of the vibronic levels (4.1.30) falls into two exactly identical vibrational progressions displaced by the electronic energy gap with respect to each other. This, in fact, corresponds to the zero vibronic coupling in the conventional

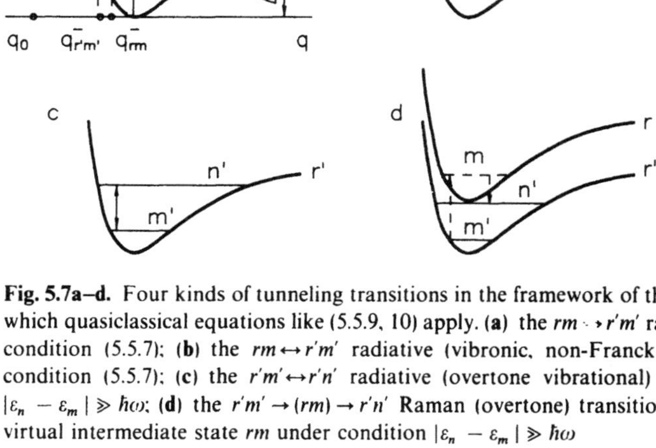

Fig. 5.7a–d. Four kinds of tunneling transitions in the framework of the model of parallel terms to which quasiclassical equations like (5.5.9, 10) apply. (**a**) the $rm \to r'm'$ radiationless transition under condition (5.5.7); (**b**) the $rm \leftrightarrow r'm'$ radiative (vibronic, non-Franck–Condon) transition under condition (5.5.7); (**c**) the $r'm' \leftrightarrow r'n'$ radiative (overtone vibrational) transition under condition $|\varepsilon_n - \varepsilon_m| \gg \hbar\omega$; (**d**) the $r'm' \to (rm) \to r'n'$ Raman (overtone) transition with a participation of a virtual intermediate state rm under condition $|\varepsilon_n - \varepsilon_m| \gg \hbar\omega$

meaning. Nevertheless, there still remains an electron-vibration interaction included into the perturbation V that induces transitions between the terms.

The perturbation matrix elements (4.1.38) do not involve any contributions from the non-adiabaticity operators L and K because the electronic wave functions $\varphi_r(x)$ are independent of q and expressions (4.1.17–19, 36) vanish identically. A more general solution to (4.1.7) may be written in the form $\varphi_r(x, q) = \varphi_r(x)\eta_r(q)$, where $\eta_r(q)$ is an arbitrary function of the nuclear coordinates; however, in this case, too, $L_{rr'}$ and $K_{rr'}$ vanish because of the orthogonality of $\varphi_r(x)$ and $\varphi_{r'}(x)$. Therefore, only the spin–orbit coupling, with the $U_{\text{el-nucl}}(x, q)$ term added, remain in (4.1.38). The resulting matrix element for a transition from the state $E_{rm} = \varepsilon_r + \varepsilon_m$ into $E_{r'm'} = \varepsilon_{r'} + \varepsilon_{m'}$ is written as

$$v_{rmr'm'} = \int \chi_m(q)\, V_{rr'}(q)\, \chi_{m'}(q)\, dq \,, \tag{5.5.4}$$

where the electronic index of the vibrational energies and wave functions is omitted. It is essential that in this expression $V_{rr'}(q)$ is merely a function of the nuclear coordinates,

$$V_{rr'}(q) = \int \varphi_r(x)\, U_{\text{el-nucl}}(x, q)\, \varphi_{r'}(x)\, dx + U_{rr'}(q) \,, \tag{5.5.5}$$

where the last term symbolizes a contribution from the spin–orbit coupling (its precise meaning is of no importance here). Thus, the necessity to treat the singularity of the non-adiabaticity operator noted in Sect. 4.3 (see also Sect. 5.7) no longer arises. It is worthwhile to note that the adiabatic approximation does not work at all in this model (see Sect. 6.7).

However, the model of parallel terms has its own difficulty originating from the inapplicability of the Condon approximation (4.5.1). The matter is that at $m' \neq m$ the overlap integral (4.2.18) vanishes because of the orthogonality of the wave functions $\chi_m(q)$ and $\chi_{m'}(q)$ while the transitions are still possible, because the matrix element (5.5.4) is generally nonzero. Hence, the explicit nuclear coordinate dependence of the electronic matrix element $V_{rr'}(q)$ needs to be allowed for. The quantum-mechanical theory of nonradiative transitions between parallel terms with application to rare-earth ions in condensed media was developed in [1.15, 5.24, 25]. The application of the quasiclassical approach asks for a more detailed specification of the validity conditions for the Franck–Condon principle, which was derived in Sect. 5.4 on the basis of the Condon approximation.

Basically, the matter is which limitations should be imposed on the function $V_{rr'}(q)$ to extend the conclusion of Sect. 5.4.1 to the present case. That conclusion was inferred from the fact that the overlap integral could be recast, with the use of decomposition (5.2.17), to the form of (5.2.22) [or to the equivalent form of (5.2.24)] where the integrand was proportional to expression (5.4.1). Precisely in the same way the matrix element (5.5.4) can be transformed. Substituting (5.2.17) and assuming the function $V_{rr'}(q)$ to be real for real q, one obtains

$$v_{rmr'm'} = 2\,\mathrm{Re}\left\{\int \chi_m(q)\, V_{rr'}(q)\, \chi^{+}_{m'}(q)\, dq\right\} \tag{5.5.6}$$

similar to (5.2.22) (the question regarding this integral's convergence will be discussed later). Here again the product $\chi_m(q)\chi^+_{m'}(q)$ is proportional to the factor (5.4.1). If $V_{rr'}(q)$ is a purely classical function independent of $\hbar$ (except for, perhaps, a constant multiplier that can be taken off the integral sign), then, in the quasiclassical limit of $\hbar \to 0$, the saddle point q_* for the integral of (5.5.6) is again determined by that factor, and hence, the ensuing Franck–Condon principle, i.e., (5.4.2) remains true. Hereafter this requirement is supposed to be met.[2]

5.5.2 Tunneling Nonradiative and Radiative Transitions

Consider the $rm \rightsquigarrow r'm'$ nonradiative transition between the parallel terms (see Fig. 5.7a). The electronic energy gap ΔE is supposed to be large as compared to the fundamental quantum $\hbar\omega$. Because the vibration energy change in the transition is equal to ΔE, due to the total energy conservation, one has

$$\varepsilon_{m'} - \varepsilon_m \gg \hbar\omega \, . \tag{5.5.7}$$

At the large vibrational energy change the nuclear motion is quasiclassical and the Franck–Condon principle can be applied. Let us find out at which nuclear coordinate value q_* the momentum difference between the final and initial states, $\Delta p = p_{r'm'} - p_{rm}$, vanishes. Invoking the explicit expression (5.2.4) for the classical momentum we obtain the following equation to define q_* in the model of parallel terms (5.5.3):

$$\Delta p = \frac{2M(\varepsilon_{m'} - \varepsilon_m)}{p_{r'm'} + p_{rm}} = 0 \, . \tag{5.5.8}$$

Since the numerator is a nonzero constant due to (5.5.7), this equation can be satisfied only at a singular point q_0 where the potential functions, hence the momenta, tend to infinity. For instance, the Morse potential (4.6.19) has $q_0 = -\infty$. A more realistic approximation to the true potential tends to infinity at a finite value of q_0 corresponding to an (infinitesimally) small interatomic separation.

In general, for typical potential curves, q_0 is on the left-hand side of the inner turning point $q^-_{r'm'}$ (see Fig. 5.7a) where the potential energy rapidly increases due to a strong interatomic repulsion. In as much as q_0 lies within the classically inaccessible region, it is inferred that the $rm \rightsquigarrow r'm'$ transition can be described in terms of tunneling from the inner turning point in the upper state q^-_{rm}, to q_0 with a subsequent return to the lower-state turning point $q^-_{r'm'}$. In the case of intersecting terms (as in Sects. 5.2, 3) the tunneling is up to the point of term intersection or closest approach (see Sect. 5.4.2).

---

[2] In the model by *Uzer* and *Child* [5.26] $V_{rr'}(q)$ depends on $\hbar$ in a manner similar to (5.4.1). For this reason the equation for q_* differs from (5.4.2) and the Franck–Condon principle is broken down. However, the use of essential quantum functions asks for a solid foundation.

To avoid confusion it should be stressed that no "actual" tunneling occurs since the probability to find the system far beyond the turning point is negligible. The language of "tunneling" is merely a fine means to represent the results of the mathematical analysis. In this connection it is worthwhile to refer once more to Miller's treatment of the tunneling nonadiabatic electronic transitions in reactive collisions [2.238]. There the probability amplitude, $A_{rr'}$, for the transition from the reagents' to the products' valley is expressed in the form of (5.4.1) where actions are calculated along a complex classical trajectory circumventing the term intersection point. Despite the fact that the entire trajectory enables one to determine $A_{rr'}$, no part of it can be connected with any physical quantity like the probability to fix the system at some middle point on the trajectory. To see this we express the exact stationary nuclear wave function as $\chi_r(q) + A_{rr'}\chi_{r'}(q)$, and then the probability to observe the system at a point q in the final electronic state is $P = |A_{rr'}\chi_{r'}(q)|^2$, where $A_{rr'}$ is still defined by the entire trajectory going from the reagents' to the products' valley, rather than by the part of it ending at point q. Obviously, no contradiction is in the statement that the trajectory defining the transition amplitude passes through such points q where P is negligible because the trajectory only appears in $A_{rr'}$, whereas the smallness of P originates from the rapid damping of $\chi_{r'}(q)$. This is in sharp contrast with the one-state one-dimensional underbarrier passage where $P = |\chi(q)|^2$ is the probability to fix the coordinate q and simultaneously the probability to penetrate through the barrier (provided that q is beyond the barrier). Here the trajectory enters the very $\chi(q)$, and the part of it ending at a given q defines the probability to find the system at that q.

The above ideas are very close to the concept of *dynamical tunneling* advanced by *Davis* and *Heller* [5.27] and *Gray* [5.28] and further developed by others [5.29–32]. This is illustrated by the following citation of [5.28]: "The sort of nonclassical path involved here is not the usual sort of tunneling path since it involves a complex angle variable, real axion and no obvious barrier penetration, nor does it connect in any obvious way the dynamically separated regions of classical phase space. Nonetheless, this generalized tunneling will be referred to as dynamical tunneling because the corresponding wave functions are, in fact, superpositions of states that are dynamically separated classically". As is clear from our previous discussion, no obvious barrier penetration is needed to make the transition between the states represented by the functions $\chi_r(q)$ and $\chi_{r'}(q)$; nevertheless, these states "are dynamically separated classically" (via the Franck–Condon principle of momentum conservation) resulting in the exponentially low transition probability.

Once the saddle point for the integral in (5.5.6) is found, a general quasiclassical expression for the matrix element can be immediately written because the magnitude of the integral is essentially governed by the value of expression (5.4.1) at q_* [1.40]:

$$v_{rmr'm'} = Q^{\mathrm{NRT}} \exp[\Phi(q_*)] \,. \tag{5.5.9}$$

For the parallel terms, q_* coincides with the singularity point q_0, while for the intersecting terms q_* is the intersection point. The above exponential factor is exactly the same as in (4.5.14) for the overlap integral [as well as in the corresponding expressions of Sects. 5.2, 3, e.g., in (5.2.39)]. This is nothing else as the product of the probability amplitude for the system to tunnel to the transition region defined by the Franck–Condon principle and that to return back into the classical region. Notice that while the former is vanishingly small (preventing the actual tunneling), the latter is huge, so their product takes on reasonable values.

The exponential factor of the quasiclassical transition matrix element (5.5.9) was originally found by *Landau* and *Lifshitz* [1.40]. It is the same for *any* admissible functions $V_{rr'}(q)$, being dependent of the potential alone. This factor controls both the absolute magnitude of the matrix element and its dependence of the system's parameters and vibrational quantum numbers.

As far as the preexponential factor Q^{NRT} is concerned, it depends upon both the potential and the perturbation $V_{rr'}(q)$ inducing the NonRadiative Transition (this is marked with the superscript NRT). It is a slow function of the parameters and quantum numbers as compared with the exponential factor. This fact is very significant. It implies that *the radiationless transition rate is principally governed by the shape of the potential curve in the region of interatomic repulsion and is nearly independent of the perturbation inducing the transition.* This statement needs a further discussion that will be given in Sect. 5.6.3.

A typical difficulty encountered in applications of (5.5.9) is that no general prescriptions can be given as to how to calculate Q^{NRT}. For example, from the *Landau-Lifshitz relation* (4.5.14) it is not obvious in which way the overlap integral for the parallel terms vanishes for $m' \neq m$. One may merely guess that this results from a cancellation of Q^{NRT} considering that the exponential factor does not vanish provided the repulsive potential rises more rapidly than the harmonic one.[3] Therefore, of special interest are the cases where Q^{NRT} is calculated accurately.

Two examples of such calculations were given in Sects. 5.2, 3 where complete quasiclassical formulae for the overlap integrals were deduced including the preexponential factors. As a simple yet useful exercise the reader can verify that in the limit of parallel terms (i.e., $f \to 1$, $\Delta q \to 0$, $\Delta\alpha \to 0$) the $m' \neq m$ quasiclassical overlap integrals obtained there tend to zero. One more example of the calculation of the preexponential factor will be given in Sect. 5.6.1, this time not for the overlap integral but for a general matrix element (5.5.4) in the parallel Morse terms. Notice that for nonparallel terms the problem is trivial since the Condon approximation applies and the factors Q^{NRT} for the matrix element (5.5.4) and for the overlap integral simply differ by a factor of $V_{rr'}(q_*)$ [as in

[3] Equation (5.5.9) does not apply to harmonic (and still slower) potentials, as $\Phi(q_*)$ diverges. Of course, the quasiclassical method works for such potentials, too, but the derivation of the corresponding expression is of little interest as actual molecular potentials rise quite rapidly in the repulsion domain.

(4.5.16)]. In contrast to this, a special analysis is required in the case of parallel terms because q_* coincides with the singular point of the potential, q_0, at which the function $V_{rr'}(q)$ may also have a singularity[4] so that the conventional saddle-point method does not apply.

Before we proceed to resolve the above-mentioned problem, we are going to emphasize a very important feature that provides a further justification to our heightened interest in this problem. The matter is that the quasiclassical relation (5.5.9) and the ensuing conclusion about the tunneling nature of the non-radiative transition can be immediately applied to radiative transitions as well, and in this way quite different physical phenomena are united in the framework of one and the same approach. In fact, if the dipole moment function of an $r \leftrightarrow r'$ electronic transition $d_{rr'}(q)$ is substituted for $V_{rr'}(q)$, then expression (5.5.4) will turn into the matrix element $d_{rmr'm'}$ of (4.5.3) determining *the intensities of vibronic transitions* in the electronic absorption and emission spectra (see Fig. 5.7b). For the $m \leftrightarrow m'$ transitions with a large detuning with regard to the $m = m'$ transition frequency the requirement of (5.5.7) is fulfilled and hence the matrix element (4.5.3) can be written in a quasiclassical form similar to (5.5.9),

$$d_{rmr'm'} = Q^{\mathrm{RT}} \exp[\Phi(q_*)] \,, \tag{5.5.10}$$

where the exponential factor is exactly same as in (5.5.9) and (4.5.14), and all the differences are incorporated into the preexponential Q^{RT} that now depends on the dipole moment of the Radiative Transition (which is marked by the superscript RT). Thus, at large detuning of the $rm \leftrightarrow r'm'$ from the $rm \leftrightarrow r'm$ transition the vibronic line intensities are governed by the potential curve in the repulsive region.

A similar conclusion also holds for the intensities of higher *overtone vibrational transitions* (see Fig. 5.7c) [5.32–37], in which case the electronic matrix element $d_{r'r'}(q) \equiv d(q)$ is merely the dipole moment function in the ground electronic state.

Finally, consider the overtone intensities in *Raman scattering* (Fig. 5.7d). The transition matrix element appears in the second order of the perturbation theory and contains terms of the form

$$\sum_{\substack{rm \\ r \neq r'}} \frac{d_{r'm'rm} d_{rmr'n'}}{E_{r'm'} - E_{rm}} \,.$$

For higher overtones one has $|\varepsilon_{n'} - \varepsilon_{m'}| \gg \hbar\omega$. Therefore, in each term either $|\varepsilon_{m'} - \varepsilon_m| \gg \hbar\omega$ or $|\varepsilon_{n'} - \varepsilon_m| \gg \hbar\omega$ is fulfilled, and the corresponding matrix elements are described by (5.5.10). Hence, the higher Raman overtones can also be described in terms of tunneling, and their intensities are governed by the repulsive branch of the molecular potential.

[4] Because the overlap integral vanishes, an uncertainty arises on the pattern of $\infty \cdot 0$.

In reality the terms r and r' are always slightly displaced and distorted. Nevertheless, the dominating (exponential) factor in (5.5.10) is the same independently of whether q_* is the term crossing or singularity point. Therefore, the transformation from the intersecting to the parallel terms affects the preexponential factor alone. Regarding the vibronic line intensities in optical spectra this transformation implies a change of the Franck–Condon to the Herzberg–Teller mechanism of the optical transitions [5.38].

Obviously, the location of q_* determines the tunneling depth. When the vibronic coupling weakens, the term crossing point q_* moves from the classically accessible region towards the singular point q_0, and at first glance it appears that in the limit of parallel terms the tunneling depth is determined by the q_0 position (for the Morse potential one would obtain an infinite tunneling depth, as $q_0 = -\infty$). However, it turns out that if the potential $\mathscr{E}_{r'}(q)$ near q_0 rises fast enough (e.g., exponentially), then $\Phi(q_*)$ eventually reaches a limiting value and no longer changes afterwards when q_* moves further towards q_0. This means that at given m' and m the tunneling depth is confined by a certain maximum distance from the boundary of the classical region (i.e., the turning point) to the boundary of the region of the $\Phi(q)$ constancy. From this argument it is natural to infer that the existence of term crossing affects the transition probability only unless the crossing point reaches the domain where $\Phi(q)$ is constant. Once q_* gets inside this domain, the transition matrix element (5.5.4) becomes independent of that crossing and is given by (5.5.9), for parallel terms.

5.6 Overtone Vibrational Transitions

The overtone vibrational transitions of Fig. 5.7c offer a unique opportunity to apply the model of parallel terms and to check the quasiclassical theory both against exact theoretical results and on the basis of an analysis of the wealth of experimental data on specific physical systems [5.32–38].

First, these transitions occur between the vibrational levels of the common (ground) electronic state. For this reason the initial and final terms are precisely identical. For other transitions in Fig. 5.7 the r and r' are two different terms. They are always somewhat displaced and distorted with respect to each other so that an additional examination is needed to justify the neglect of these differences for a given physical system. In the case of the Raman transition (Fig. 5.7d) it is necessary to make sure that the influence of other terms $r'' \neq r, r'$ is insignificant.

Second, in the problem of nonradiative transitions in polyatomic molecules one always has to take into account several vibrational modes with different vibronic couplings, which severely complicates the theory and makes any comparison with experiment more difficult. On the other hand, the overtone transitions may occur in diatomic molecules as well where no such problem is encountered at all. Moreover, it is well known that in a number of polyatomic molecules the overtone absorption is due to the so-called local vibrations, i.e., the vibrations of individual chemical bonds (see Sects. 2.10 and 4.6) that are described by the model of noninteracting linear anharmonic oscillators. This is true for the high-frequency X–H vibrations, whereas the overtone absorption of the molecules without hydrogen atoms, e.g., NF_3, SF_6, etc. [5.39] as well as the C–C absorption of organic molecules [5.40] are represented better with the model of normal vibrations, through some manifestations of the local-mode behavior are seen in these cases as well [5.41, 42].

Of course, in polyatomic molecules the coupling among the local modes, as well as between the local modes and other vibrational degrees of freedom, is essential as it gives rise to a broadening and splitting of the overtone absorption bands of the C–H [2.242–252, 255, 258, 5.43–68], N–H [5.68, 69], O–H [5.70–73], Si–H [5.74–77], etc., local vibrations. However, if only band positions and integral intensities are under study, one may disregard these effects and use the model of noninteracting linear oscillators.

Third, there are a good deal of experimental data on the intensities of overtone vibrational transitions in absorption and emission spectra generously cited throughout this section as well as theoretical calculations performed by different methods enabling a thorough check of quasiclassical theory. Data are also available on the probabilities of overtone vibrational transitions excited by fast electron impact [5.78].

5.6.1 Derivation of the Quasiclassical Formula

A general quasiclassical expression for the matrix element of the dipole moment function $d(q)$,

$$d_{mn} \equiv \langle m|d(q)|n\rangle = \int_{-\infty}^{+\infty} \chi_m(q)d(q)\chi_n(q)dq \,, \tag{5.6.1}$$

valid for a wide class of potentials was derived in [5.37, 79]. In this section the derivation will be given for an arbitrary potential and, as an example, for the Morse potential

$$\mathscr{E}(q) = D(1 - \mathrm{e}^{-\alpha q})^2 \tag{5.6.2}$$

under the condition that the vibrational energy difference between the upper (n) and lower (m) states generally exceeds the fundamental quantum,

$$y \equiv (\varepsilon_n - \varepsilon_m)/\hbar\omega \gg 1, \quad m \geq 0 \,. \tag{5.6.3}$$

As was shown in Sects. 5.1–3, no limitations need to be imposed on the lower state energy.[5]

[5] The case of $\varepsilon_m \gg \hbar\omega$ and $y \lesssim 1$ was treated in [5.79b, 80, 81].

In analogy with (5.5.6) the matrix element (5.6.1) is recast as

$$d_{mn} = 2\,\mathrm{Re}\left\{\int_{-\infty}^{+\infty} \chi_m(q)\, d(q)\, \chi_n^+(q)\, dq\right\}. \tag{5.6.4}$$

The further procedure is as in Sects. 5.2, 3 (we use the notations introduced there, with the omission of the electronic indices r, r' and the substitution of n for m'). We shift the integration path L_q from the real axis into the upper half-plane and replace the wave functions with their quasiclassical expressions (5.2.3, 15, 18):

$$d_{mn} = C_m C_n (\omega_m \omega_n)^{1/2} \pi^{-1}\, \mathrm{Re}\left\{\int_{L_q} d(q)(v_m v_n)^{-1/2} \exp[\Phi(q)]\, dq\right\}. \tag{5.6.5}$$

The definition of the quantities in this expression are given in (5.2.5, 6, 14, 26) and (5.3.29).

The most accurate way of treating (5.6.5) is, of course, to integrate it numerically along some contour L_q, but it is more appropriate to do this immediately in (5.6.1). A different method of [5.79b], outlined below, introduces a mean trajectory $q(t)$ (a trajectory with a mean energy ε between ε_n and ε_m) to obtain, without a significant loss in accuracy, a general-enough result valid in both the limit of (5.6.3) and that of footnote 5; in the latter case d_{mn} in the (n-m)-th Fourier component of the periodic function $d[q(t)]$ [1.40].

As discussed in Sect. 5.5.2, in the quasiclassical limit of $\hbar \to 0$ [this corresponds to (5.6.3)] the main contribution to the integral of (5.6.5) comes from the stationary-phase point q_0($q_0 = -\infty$ for the Morse potential) where $\Phi'(q) = 0$ and the Franck–Condon principle $p_n = p_m$ is fulfilled. Therefore, d_{mn} will be proportional to the factor

$$\exp[\Phi(q_0)] \equiv \exp[\mathrm{Re}\{\Phi(q_0)\}]\exp[\mathrm{i}\,\mathrm{Im}\{\Phi(q_0)\}]\,,$$

which in the case of the Morse potential having $\mathrm{Im}\{\Phi(q_0)\} = \pi(m-n)$ becomes

$$\exp[\Phi(-\infty)] = (-1)^{n-m} \exp(-\kappa_{mn})\,,$$

where

$$\kappa_{mn} = \lim_{q \to -\infty} \kappa_{mn}(q)\,, \tag{5.6.6}$$

$$\kappa_{mn}(q) = \frac{\sqrt{2M}}{\hbar}\left\{\int_{q_n^-}^{q} [\mathscr{E}(q) - \varepsilon_n]^{1/2}\, dq - \int_{q_m^-}^{q} [\mathscr{E}(q) - \varepsilon_m]^{1/2}\, dq\right\}. \tag{5.6.7}$$

These formulae are immediately deduced from (5.3.30, 31) and (5.2.4, 10) with a minor change in notations. In (5.6.7) the integrations are along the real axis from the inner turning points to the stationary-phase point at negative infinity, Notice that each of the integrals diverges when $q \to -\infty$, while κ_{mn} is finite since the momentum difference (5.5.8) tends fast enough to zero (cf. footnote 3). Thus the matrix element (5.6.5) can be represented in the form

$$d_{mn} = Q \exp[\mathrm{Re}\{\Phi(q_0)\}] , \tag{5.6.8a}$$

or, for the Morse potential,

$$d_{mn} = Q \exp(-\kappa_{mn}) , \tag{5.6.8b}$$

which is similar to (5.5.10). Condition (5.6.3) leads to the inequality $|\mathrm{Re}\{\Phi(q_0)\}|$, $\kappa_{mn} \gg 1$, which implies that the matrix element (5.6.8) is exponentially small. The preexponential factor is written as

$$Q = C_m C_n (\omega_m \omega_n)^{1/2} \pi^{-1} \mathrm{Re}\{Z \exp[\mathrm{i}\, \mathrm{Im}\{\Phi(q_0)\}]\} , \tag{5.6.9a}$$

where

$$Z = \int_{L_q} d(q) (v_m v_n)^{-1/2} \exp[\Phi(q) - \Phi(q_0)] dq . \tag{5.6.10}$$

For the Morse potential

$$\begin{aligned} Q &= (-1)^{n-m-1} C_m C_n (\omega_m \omega_n)^{1/2} (\omega y)^{-1} \varrho , \\ \varrho &= -(\omega y / \pi) \mathrm{Re}\{Z\} , \end{aligned} \tag{5.6.9b}$$

where y is given by (5.6.3). Since the exponent in the integrand of (5.6.10) tends to a constant when $q \to q_0$, the convergence of the integral is assured by the decrease of the preexponential function $(v_m v_n)^{-1/2}$ provided that $d(q)$ does not increase too rapidly (see below). We note that the integral cannot be computed by the saddle-point method since the stationary-phase point q_0 is an essential singular branch point for the exponent and at the same time an essential singular branch point for the preexponential function.[6] Still the stationary-phase method can be applied.

We transform the exponent in (5.6.10) by writing the momentum difference as in (5.5.8):

$$\Phi(q) - \Phi(q_0) = \frac{\mathrm{i}}{\hbar} \int_{q_0}^{q} (p_n - p_m) dq = 2\mathrm{i} M \omega y \int_{q_0}^{q} (p_n + p_m)^{-1} dq . \tag{5.6.11}$$

In the vicinity of the stationary-phase point the momenta p_n and p_m are large, but their difference (5.5.8) is small. Therefore, we can introduce a mean energy ε between ε_n and ε_m such that $p_n \approx p_m \approx M v_\varepsilon$, where v_ε is the velocity of the

---

[6] In the case of the essential singular point the saddle point method can be applied provided that the preexponential function is regular at this point [5.82].

"classical" motion (generally in the complex q-plane) with the energy ε along the corresponding mean trajectory implicitly given by equation

$$t(q) = \int_{q_\varepsilon^+}^{q} v_\varepsilon^{-1}\, dq \,. \tag{5.6.12}$$

The initial condition for this trajectory is $q(0) = q_\varepsilon^+$, where q_ε^+ stands for the turning point. The trajectory reaches q_0 at $t = t_0 = t(q_0)$. Inserting these relations into (5.6.10, 11) gives

$$Z = \int_{L_t} d[q(t)]\, \mathrm{e}^{i\Omega(t-t_0)}\, dt \,,$$

where

$$\Omega \equiv \omega y = (\varepsilon_n - \varepsilon_m)/\hbar$$

is the transition frequency and the integration path L_t is the image of L_q in the complex t-plane under transformation (5.6.12). In the limit of a large m this expression for Z, along with (5.6.8a, 9a), results in d_{mn} reducing to the Fourier component of $d[q(t)]$ [5.79b]).

As is seen from (5.6.12), t_0 is a branch point for $q(t)$ since near it one has

$$v_\varepsilon \propto [\mathscr{E}(q)]^{1/2} \,.$$

Let us draw a cut $\mathrm{Re}\,\{(t-t_0)\} = 0$, $\eta \equiv \mathrm{Im}\{(t-t_0)\} \geq 0$ and displace L_t up in the t-plane so that the displaced contour ascends towards the stationary-phase point t_0 along the left-hand side and then descends from t_0 along the right-hand side of the cut. If q_d and q_a stand for $q(t)$ on the descending and ascending branches of L_t, respectively, then [5.79c]

$$Z = \mathrm{i} \int_0^\infty [d(q_d) - d(q_\mathrm{a})]\, \mathrm{e}^{-\Omega\eta}\, d\eta \,. \tag{5.6.13}$$

Further, we consider only the Morse potential, for which (5.6.13) should be modified. Since the initial contour L_q in (5.6.1) starts at $q_0 = -\infty$, the transformed contour L_t must start at t_0. Hence, only the descending path contributes to Z, and only the first term in the integrand of (5.6.13) survives. If we define the dipole moment

$$V(x) = d(q) = d(x/\alpha)$$

as a function of the dimensionless displacement $x = \alpha q$, where α is the Morse range parameter of (5.6.2), then ϱ from (5.6.9b) becomes [5.79c, d]

$$\varrho[V(x); y] = y\, \mathrm{Re}\left\{(\mathrm{i}\pi)^{-1} \int_0^\infty V(x_d)\, \mathrm{e}^{-yz}\, dz\right\}. \tag{5.6.14}$$

The trajectory can be easily calculated from (5.6.12):

$$x_d = \alpha q_d = i\pi + \ln[\cosh(rz) + r\sinh(rz) - 1] - 2\ln r\,,$$
$$r \equiv \omega_\varepsilon/\omega = [1 - (\varepsilon/D)]^{1/2}, \qquad \varepsilon = (\varepsilon_n + \varepsilon_m)/2\,. \tag{5.6.15}$$

It runs along a straight line parallel to the real axis shifted upward the imaginary axis by π. Since no prescription has been obtained in the course of the derivation with regard to the choice of the mean energy ε, it is specified arbitrary as the arithmetic mean.

In (5.6.14) we indicated explicitly that ϱ is a functional of $V(x)$ and a function of y. It also depends on ε, but much more weakly. Thus, in the limit of (5.6.3), where only small values of z contribute to the integral of (5.6.14) the following approximate relation independent of ε is obtained by expanding (5.6.15) [5.37, 5.79a]:

$$\varrho[V(x);y] = y\,\mathrm{Re}\left\{(i\pi)^{-1}\int_0^\infty V(i\pi + \ln z)\,e^{-yz}\,dz\right\}. \tag{5.6.16}$$

The difference between (5.6.14) and (5.6.16) is significant when y is not large enough.

It should also be noted that (5.6.14, 16) diverge at the lower integration limit corresponding to the stationary-phase point provided that $V(x)$ increases too fast when $\mathrm{Re}\{x\} \to -\infty$, namely faster than $\exp(-\nu x)$ with $\nu \geq 1$ (see below).

The quantity κ_{mn} can be calculated accurately by (5.6.7) in the same way as δ^-_{rm} was calculated in Sect. 5.3. Using (5.3.34, 36), we obtain

$$\kappa_{mn} = (2x_e)^{-1}(A_n - A_m)\,,$$
$$x_e = \hbar\omega/4D = 1/(2\sigma + 1)\,,$$
$$A_n = \frac{1 + r_n}{2}\ln\frac{2}{1 + r_n} + \frac{1 - r_n}{2}\ln\frac{2}{1 - r_n}\,, \tag{5.6.17}$$
$$r_n = \omega_n/\omega = [1 - (\varepsilon_n/D)]^{1/2} = 1 - x_e(2n + 1)\,;$$

A_m is given by a similar relation.

The quasiclassical formulae (5.6.8) are easily proved to satisfy the necessary requirement that the vibrational overlap integral vanishes for $n > m$. The matrix element d_{mn} will be proportional to the overlap integral provided the dipole moment function is constant. On letting $d(q) \equiv V(x) \equiv d_0$, where d_0 is a *real* constant [see text preceding (5.5.6)], from (5.6.13, 14) one obtains $Z = 0$ and

$$\varrho[d_0;y] = d_0\,\mathrm{Re}\{(i\pi)^{-1}\} = 0\,,$$

hence $Q = 0$ according to (5.6.9). Thus, as was stated in Sect. 5.5.2, the condition of the wave functions' orthogonality is fulfilled due to the vanishing preexponential factor of (5.6.8).

For the linear dipole moment function $V(x) = x$ it follows from (5.6.14, 16) that $\varrho[x; y] = 1$, and the quasiclassical matrix element assumes the form

$$\langle m|x|n\rangle = (-1)^{n-m-1} C_m C_n (\omega_m \omega_n)^{1/2} (\omega y)^{-1} \mathrm{e}^{-\kappa_{mn}} . \tag{5.6.18}$$

Then, for an arbitrary $V(x)$ (5.6.8b, 9b) are recast as

$$\langle m|V(x)|n\rangle = \varrho[V(x); y] \langle m|x|n\rangle . \tag{5.6.19}$$

Thus, ϱ is nothing else as the ratio of the matrix element of the given function $V(x)$ to that of the linear function calculated in the quasiclassical approximation by (5.6.14, 16). Evidently, one can also define the exact matrix element ratio,

$$R[V(x); m, n, x_e] = \langle m|V(x)|n\rangle / \langle m|x|n\rangle , \tag{5.6.20}$$

calculated either by the available analytical formulae or numerically [5.83–86]. It is a functional of $V(x)$ and a function of three parameters, two quantum numbers and the anharmonicity constant x_e. The comparison of ϱ with R enables one to check the validity of (5.6.14, 16) as well as to assess the accuracy of the quasiclassical approximation. Inequality (5.6.3) implies that the accuracy is the higher, the larger y. As follows from the derivation of (5.6.16) in [5.37, 79a], the least necessary restriction on y from below is

$$y > 1 + (\varepsilon_n / D)^{1/2} , \tag{5.6.21}$$

where $\varepsilon_n \equiv \hbar\omega\mu_n$,

$$\mu_n = n + \tfrac{1}{2} - x_e (n + \tfrac{1}{2})^2 , \tag{5.6.22}$$

$$y = (n - m)[1 - x_e(n + m + 1)] . \tag{5.6.23}$$

5.6.2 Comparison with Exact Calculations

An exact expression has been derived for the matrix elements of the linear dipole moment function in the Morse potential [5.87, 88]. In the present notation it is

$$\langle m|x|n\rangle = (-1)^{n-m-1} (\omega_m \omega_n)^{1/2} (\omega y)^{-1} \left[\frac{n!\,\Gamma(2\sigma - n + 1)}{m!\,\Gamma(2\sigma - m + 1)} \right]^{1/2} , \tag{5.6.24}$$

where $\Gamma(x)$ is the gamma function. In the quasiclassical limit of (5.6.3), (5.6.24) must reduce to (5.6.18). Since the discrete levels obey the condition $n < \sigma$ and σ is large, (5.6.24) can be simplified using Stirling's formula for $n \gg 1$, $2\sigma - n \gg 1$ and $2\sigma - m \gg 1$. Then, one can easily verify that (5.6.24) reduces to (5.6.18), where κ_{mn} has to be replaced with the approximate relation following from (5.6.17) under the same conditions.

In the case of an arbitrary function $V(x)$, (5.6.16) should be compared with (5.6.20). Notice that the matrix element ratio is generally a function of three parameters, m, n and x_e, whereas the approximate quasiclassical formula (5.6.16) predicts a dependence upon the universal combination of the parameters (5.6.23) only.

For the exponential function $V(x) = \exp(-x)$ an exact formula was given in [5.88] which results in

$$R[\exp(-x); m, n, x_e] = -y\,. \tag{5.6.25}$$

On the other hand, computing ϱ with (5.6.16) for a general exponential function, $V(x) = \exp(-vx)$ with $v < 1$ yields

$$\varrho[\exp(-vx); y] = -y^v/\Gamma(v)\,. \tag{5.6.26}$$

Virtually, this relation is valid for any v because its right-hand side has no singularities for $v \geq 1$.[7] Evidently, at $v = 1$ it is identical with (5.6.25). Of course, such a literal coincidence is fortuitous, and, generally, ϱ and R are expected to agree only approximately. For example, the actual relative error of (5.6.26) as compared to the exact one of [5.88] is $\sim 1/(n-m)$. Thus, at $v = 2$ one has from (5.6.26) $\varrho[\exp(-2x); y] = -y^2$, while the exact formula of [5.88] yields $R[\exp(-2x); m, n, x_e] \approx -(n-m)(n-m+1)$ (the terms $\sim x_e$ are omitted in expressions for ϱ and R).

A better grasp of the errors of the quasiclassical formulae is permitted by numerical calculations of the matrix elements of the power functions in the Morse potential carried out in [5.37, 79a, c] for the C–H local vibrations in benzene. The benzene overtone absorption-band positions are given by the Birge-Sponer relation $\varepsilon_m - \varepsilon_0 = m(A + Bm)$. The parameters were taken to be $A = 3\,100$ and $B = -57.1\ \mathrm{cm}^{-1}$ [5.89], where $x_e = -B/(A - B) = 0.01809$. The CH radical reduced mass, $M = 0.930$ a.u. [5.90], was used as the oscillator effective mass, and the Morse parameter was $\alpha = \sqrt{-2MB/\hbar} = 1.77\ \text{Å}^{-1}$. Figure 5.8 shows the ratio θ of the matrix elements of the linear function $\langle m|x|n\rangle$, calculated with (5.6.17, 18) to the exact values calculated with (5.6.1) or (5.6.24) for $0 \leq m < n \leq 10$. As expected, the relative error $(\theta - 1)$ rapidly decreases with increasing y.

The extremely high accuracy of the quasiclassical approximation demonstrated in Fig. 5.8 for the Morse potential with the linear dipole moment function was also obtained for two symmetric potentials with nonlinear dipole moments studied in [5.79c], where (5.6.8a, 9a, 13) were used. In [5.79d] the matrix elements were calculated for the C–H vibration in HCN using (5.6.8b, 9b, 14, 15, 17) for the Morse potential with a 6th-power dipole moment function, and the error was within a few percent for all $n \leftarrow 0$ transitions, the largest error of 6% being for the $1 \leftarrow 0$ fundamental transition.

---

[7] For $v \geq 1$ the original integral (5.6.4) diverges, but this simply means that the integration path needs to be chosen differently, in full analogy with Sects. 5.2, 3. Instead, however, we utilize the fact that (5.6.26) deduced for $v < 1$ permits a straightforward analytical continuation to the region of $v \geq 1$.

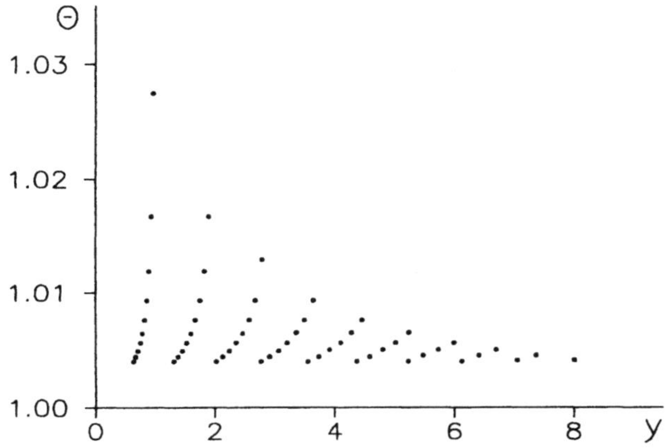

Fig. 5.8. The ratio of the quasiclassical to the exact matrix elements of the linear dipole moment function in the Morse potential for the C–H local vibrations in benzene

Figure 5.9 displays the results of calculations of ϱ with (5.6.16) and R with (5.6.20) for power dipole moment functions [the matrix elements in (5.6.20) were obtained by numerical integration of (5.6.1)]. Explicit expressions for $\varrho[x^k; y]$ are given in [5.37, 79a]; for $k = 2$ and 3 they are

$$\varrho[x^2; y] = -2\ln y - 1.1544\,,$$
$$\varrho[x^3; y] = 3\ln^2 y + 3.4633\ln y - 3.9353\,. \tag{5.6.27}$$

It is seen from Fig. 5.9 that the exact matrix element ratio R is actually a function of a single parameter y and is well fit with the quasiclassical curve (ϱ) down to y values as small as 1.2. A significant divergence between ϱ and R is only seen at $y < 1.2$, where even the least necessary condition (5.6.21) is not met.

The results of this section show that the accuracy of the quasiclassical approximation is high enough to analyze the higher overtone intensities in the molecular vibrational spectra.

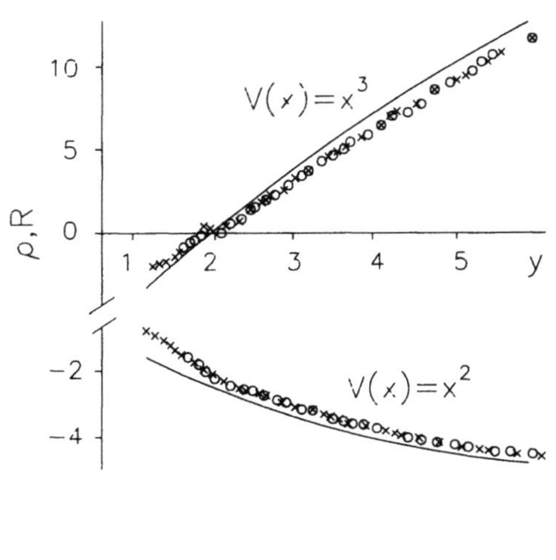

Fig. 5.9. Comparison of the quasiclassical (ϱ, *full lines*) and exact (R, *crosses* and *circles*) matrix elements ratios for the power functions in the Morse potential: ×, the OH radical ($x_e = 0.02201$) for $n \leq 9$, $n - m \leq 6$ [5.85]; ○, the C–H local vibrations in benzene ($x_e = 0.01809$) [5.37, 79]

5.6.3 Normal Intensity Distributions

The formulae obtained enable us to find out general trends of the intensity distributions in the overtone vibrational spectra. Recall that (5.6.7, 8) apply to any reasonable molecular potential given by an *analytical* function $\mathscr{E}(q)$ (the analyticity was essentially exploited in Sect. (5.6.1). As was mentioned in Sect. 5.5.2, the major dependence of d_{mn} on quantum numbers is provided by the exponential factor of (5.6.8), whereas Q is a slowly varying function. The latter is immediately obvious from (5.6.9) and Fig. 5.9 if it is born in mind that within the y range of that figure the matrix element $\langle m|x|n\rangle$ varies by several orders of magnitude. Consequently, *the intensity distribution is only very slightly affected by the dipole moment function* and is mainly governed by the potential, namely its repulsive branch in the region to the left of the inner turning points for the classical motion in the initial and final states.

The minor effect of the dipole moment function on the overtone intensities was first predicted on the basis of the Timm–Mecke approximation [5.83, 84, 91] and then discovered numerically by *Ferguson* and *Parkinson* [5.85] for the power functions. *Child* and *Halonen* [5.86] obtained similar results for power-exponential Mecke functions, but they additionally observed local low-intensity anomalies (see Sect. 5.6.5). Additionally, in recent calculations of the matrix elements (5.6.1) for the Morse and generalized Morse potentials *Lehmann* and *Smith* [5.92a] established that the repulsive branch of the potential about the inner turning point is the most important in determining the higher overtone intensities.

Nikitin et al. [5.92b] divided the integration axis in (5.6.1) into 5 regions and found that the $n \leftarrow 0$ transitions were "strongly quantum" for $n \geq 3$ (Table III in [5.92b]) in the sense that region I extending from minus infinity to the left turning point of the upper state provided the largest contribution to the calculated matrix element of the linear dipole moment in the Morse potential. They also derived a quasiclassical expression for this matrix element which differs insignificantly from (5.6.18) (see [5.79d]).

In an attempt to answer the basic question of "Where does the overtone intensity come from?" [5.92a] a confusion arises with regard to the meaning of (5.6.1) and (5.6.8) [5.58]. These equations embody two different methods to calculate one and the same quantity, d_{mn}. Equation (5.6.1) incorporated a segment of the real axis covering the classical region along with very narrow nonclassical vicinities of the turning points since the integrand decreases very rapidly beyond the turning points. On the other hand, (5.6.8) involves a wide nonclassical region extending from the inner turning point up to a point q_* where the repulsive branch of the potential reaches the values in excess of the dissociation limit (see Sect. 5.6.4). This seeming contradiction inherent to dynamical tunneling (see Sect. 5.5.2) is resolved by noting that, due to the properties of analytical functions, the potential defined at any restricted segment of the real axis is thereby immediately defined everywhere, so d_{mn} can be calculated on equal footing using different parts of the real axis.

Now if one deals with (5.6.1) alone, the statement that the repulsive branch is more important than the attractive branch simply means that the inner turning point contributes to the integral of (5.6.1) more than the outer. However, as soon as (5.6.8) is invoked, another very significant feature of the repulsive branch is revealed.

To see this explicitly, suppose first that the potential is known exactly for all q's. Then (5.6.1, 8) are numerically fully equivalent (apart from insignificant errors due to the quasiclassical approximation), as was demonstrated in Sect. 5.6.2. Now suppose that the potential is known only approximately, say, within 10% of the fundamental quantum, for all q's belonging to a given (either classical or nonclassical) segment. Then a striking difference arises between the two equations when the intensities of, say, the $n \leftarrow 0$ transitions are considered for increasing n. Namely, the accuracy of (5.6.1) falls off very rapidly so that the error in the intensity may reach 100% and more at moderate values of n. On the other hand, such an "enhancement" of the intensity error over the potential one does not occur in calculations of (5.6.8). In fact, apart from a numerical factor, the following estimate may be obtained using the equations derived later in this section:

$$\delta \ln d_{0n}^2 \propto (D/\hbar\omega n)^{1/2}(\delta\mathscr{E}/\hbar\omega) ,$$

where δ means the error in the corresponding quantity. It is seen that the intensity error is roughly on the same order of magnitude as the potential one, and, moreover, it decreases (rather than increases) with increasing n. In other words, in the limit of high values of n only the knowledge of the repulsive potential allows the calculation of the overtone intensities. In this sense the repulsive branch in the deep tunneling region may be spoken of as governing the overtone intensities. Concluding this discussion we emphasize once more, to avoid confusion, that the statement regarding the importance of this branch has only the aforementioned special meaning and pertains in no way to the integral of (5.6.1).

If we neglect a weak dependence of Q on the quantum numbers and regard Q as a nonzero constant, then (5.6.7, 8) enable us to extract the potential $\mathscr{E}(q)$ in the repulsion region from the intensities of the overtone vibrational transitions. To do this, we have to fit the repulsive branch of $\mathscr{E}(q)$ with a smooth function and, using this approximation, to deduce the dependence of the intensity on the upper and lower level energies. The $n \leftrightarrow m$ emission and absorption intensities are proportional to $f_{mn} = d_{mn}^2$. For brevity, the f_{mn} itself will be called the intensity.

First we derive some general relations. For any three quantum numbers $n > m_1 > m_2 \geq 0$ from (5.6.7, 8) one obtains

$$\ln(f_{m_1n}/f_{m_2n}) = 2\kappa_{m_2m_1} , \tag{5.6.28}$$

which shows that *the ratio of the emission intensities from a common upper level n is independent of n.*

Further, since we are interested in the intensity dependence on the upper level energy, we differentiate $\ln f_{mn}$ with respect to ε_n. From (5.6.7, 8) we have

$$\frac{\partial}{\partial \varepsilon_n} \ln f_{mn} = \frac{\sqrt{2M}}{\hbar} \int_{q_n^-}^{-\infty} \left[\mathscr{E}(q) - \varepsilon_n \right]^{-1/2} dq \tag{5.6.29}$$

(the derivative over the lower limit does not contribute since q_n^- is a turning point, $\mathscr{E}(q_n^-) = \varepsilon_n$). The change in the integration variable,

$$z = [\mathscr{E}(q) - \varepsilon_n]^{1/2}, \qquad 2z\,dz = \mathscr{E}'(q)\,dq \equiv -F(q)\,dq\,, \tag{5.6.30}$$

transforms the integral of (5.6.29) to the form

$$\frac{\partial}{\partial \varepsilon_n} \ln f_{mn} = -\frac{2\sqrt{2M}}{\hbar} \int_0^{\infty} \frac{dz}{F(q)}\,, \tag{5.6.31}$$

where q is a function of z and ε_n through (5.6.30). Equations (5.6.28, 31) are valid independently of the specific shape of $\mathscr{E}(q)$.

Now we assume that the repulsive branch of the potential at large negative displacements q can be approximated with a smooth function monotonically increasing for $q \to -\infty$. The approximation with the exponential function,

$$\mathscr{E}(q) = Ce^{-2\beta q}\,, \tag{5.6.32}$$

looks as the most appropriate. Such an approximation is well known in the theory of atomic collisions [4.4], and the constants C and 2β were obtained in experiment for a wealth of atomic pairs including at least one rare gas atom [5.90]. From (5.6.30, 32) one has $F(q) = 2\beta(z^2 + \varepsilon_n)$. Substituting this into (5.6.31) and performing elementary integration, one obtains

$$\log_{10} f_{mn} = \text{const} - a\sqrt{\mu_n}\,, \tag{5.6.33}$$

where const is a function of m only,

$$a = \frac{\pi}{\beta \ln 10}\left(\frac{2M\omega}{\hbar}\right)^{1/2} = \frac{1.364}{\beta r_e}\left(\frac{\hbar\omega}{B_e}\right)^{1/2}. \tag{5.6.34}$$

μ_n is defined in (5.6.22), r_e is the equilibrium interatomic separation and $B_e = \hbar^2/2Mr_e^2$ is a formal rotational constant. If the lower level m is high above the ground level and the approximation (5.6.32) is valid down to the point q_m^-, then the whole derivation can be repeated with respect to the ε_m dependence of f_{mn}. Then a well-known result by *Landau* and *Lifshitz* [1.40] is obtained (see [1.40], problem 1, Chap. 7, p. 189):

$$\log_{10} f_{mn} = \text{const} - a(\sqrt{\mu_n} - \sqrt{\mu_m})\,, \tag{5.6.35}$$

where const is now independent (or a very slowly varying function) of both

m and n. By the same method for the relative intensities we find from (5.6.28)

$$\log_{10}(f_{m_1n}/f_{m_2n}) = \text{const} + a\sqrt{\mu_{m_1}}\,, \tag{5.6.36}$$

where const is a function of m_2. Finally, if m_2 is large enough, from (5.6.35) we have

$$\log_{10}(f_{m_1n}/f_{m_2n}) = a(\sqrt{\mu_{m_1}} - \sqrt{\mu_{m_2}})\,. \tag{5.6.37}$$

Equation (5.6.34) predicts an isotope effect on the intensity distribution, $a \propto M^{1/4}$, since $\omega \propto M^{-1/2}$.

We shall call (5.6.33–37) *the normal intensity distributions* in the overtone vibrational spectra because they describe smooth intensity variations versus energy due to the exponential factor of the dipole moment matrix element (5.6.8). They are portrayed by straight lines on the plots of the intensity logarithm versus the square root of the reduced energy of the upper level or versus the difference of the square roots of the energies of two levels. It is seen from (5.6.34) that the values of the constant β can be found from the slopes of these straight lines.

Figure 5.10 shows experimental data on liquid benzene. The C–H overtone intensities are normalized to the first overtone transition. In accord with (5.6.33) the points are well fitted with a straight line having the slope $a = 4.67$. Invoking the data of [5.93–95] gives the average slope $a = 4.73$. Then the parameter β can be calculated by (5.6.34) which is recast in a more suitable form,

$$\beta = \frac{10.51}{a}(\overline{M}\,\overline{\omega})^{1/2}\,, \tag{5.6.38}$$

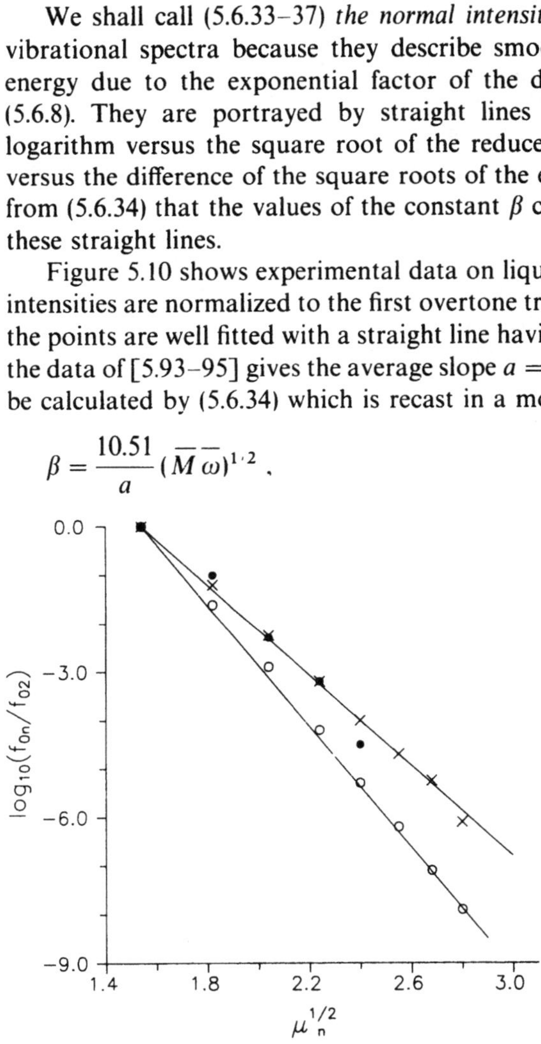

Fig. 5.10. Intensity distribution in the overtone absorption spectrum of the benzene C–H local vibrations. Experimental data from [5.93] (●) and [5.89] (×), and theoretical calculation by (5.6.24) for the Morse potential and the linear dipole moment function (○)

where β is in $Å^{-1}$, $\bar{M}$ is the reduced mass in atomic units, and $\bar{\omega}$ the fundamental frequency in 10^3 cm^{-1}. As a result one obtains $\beta_{exp}(C\text{–}H) = 3.81$ $Å^{-1}$.

The normal intensity distribution holds for an arbitrary smoothly rising repulsive potential that can be matched with the exponential function (5.6.32) over the range of displacements essential for overtone transitions, in particular for the Morse potential. For the sake of comparison, the relative intensities for the linear dipole moment function calculated with (5.6.24) for $x_e = 0.01809$ are also shown in Fig. 5.10. The points are fitted by a straight line with a higher slope, $a = 6.30$, than the experimental points, indicating that β_{exp} (C–H) > β_{Morse} (C–H). Hence the actual C–H bond repulsive potential rises steeper than the Morse potential. A similar treatment of the absorption data on the HF molecule [5.96] gives β_{exp} (HF) $\approx$ β_{Morse} (HF) [5.97].

In [5.35, 79a] it is shown that experimental data on the overtone absorption due to C–H vibrations in liquid halomethanes [5.98, 99] and $(CF_3)_3CH$ [5.91], as well as due to Si–H vibrations in gaseous SiH_4 [5.75] also confirm the normal distribution (5.6.33).

In halomethanes the C–H overtone intensities are insensitive to the halogen atom substitution. The C–H intensities of the $C_nF_mCl_nH$ compounds obey the so-called *convergence rule* [5.91], i.e., they converge to a common value with increasing of the overtone number up to the third overtone while the fundamental intensities vary quite significantly among those compounds. The universality and transferability of the higher C–H overtone intensities in large [5.100, 101] and small (HCN, C_2H_2) [5.102–104] polyatomic molecules were also discovered. All these findings are accounted for as follows [5.32, 92, 99]. The repulsive branch of $\mathscr{E}(q)$ for small interatomic separations contributing to the higher overtone intensities is much less sensitive to the specific molecular structure than both the attractive one for larger separations and the dipole function governing the fundamental transition intensity.

Figure 5.11 displays the overtone emission data on the HF and DF molecules. The $m_2 = 2$ level was supposed to be high enough to apply the normal intensity distribution of (5.6.37) for $2 \leq m_2 < m_1 < n$. The points are fitted with

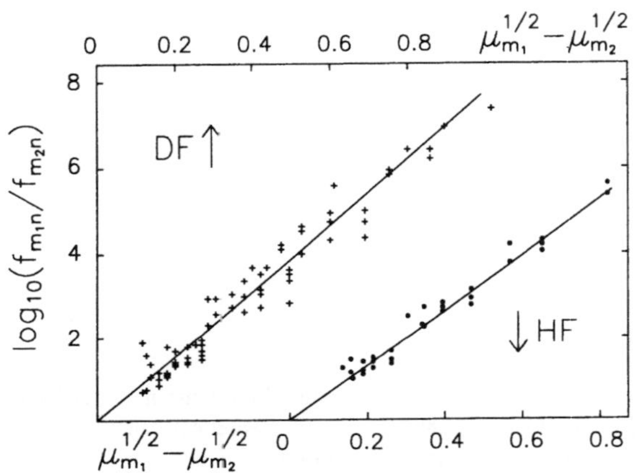

Fig. 5.11. Intensity distribution in the overtone emission spectrum of the HF and DF molecules: Experimental data from [5.105] for $m_2 \geq 2$

straight lines, and the DF/HF slope ratio is 1.17 in accord with the predicted $M^{1/4}$ dependence of (5.6.34). Experimental points with $m_2 = 0$ and 1 are shown to satisfy the intensity distribution of (5.6.36) [5.34].

These data can also be used to verify relation (5.6.28). In Fig. 5.11 one can see vertical sets of points corresponding to fixed values of m_1 and m_2 and varying values of n. Small scatter of the points within each group implies that the prediction of (5.6.28) is confirmed at $m_2 \geq 2$. This result holds true at $m_2 = 0$ and 1 as well. While the relative intensities $f_{m_1 n}/f_{m_2 n}$ vary by several orders of magnitude, they keep constant within a factor of two provided that m_1 and m_2 are fixed. Similar results were obtained in [5.34] from the analysis of the emission data on the OH radical [5.106].

5.6.4 Dynamical Tunneling Depth

The normal distributions (5.6.33–37) show that the $n \leftrightarrow m$ transition intensity decreases exponentially with increasing energy of the upper level n (for fixed m). This fact can be accounted for in a natural way based on the idea about the dynamical tunneling discussed in Sects. 5.5.2 and 5.6.3.

Consider the function $\kappa_{mn}(q)$ of (5.6.7). When q moves to the left of the inner turning point q_n^-, this function increases monotonically since for $\varepsilon_n > \varepsilon_m$ one has

$$\hbar\, \partial\kappa_{mn}(q)/\partial q = |p_n| - |p_m| < 0 \,. \tag{5.6.39}$$

If the potential rises more rapidly than q^2, the $\kappa_{mn}(q)$ tends to a finite limit, κ_{mn}, and a point q_1 can be defined such that

$$\kappa_{mn} - \kappa_{mn}(q_1) = 1 \,. \tag{5.6.40}$$

In view of $\kappa_{mn} \gg 1$ this implies that for $q \leq q_1$ the function $\kappa_{mn}(q)$ insignificantly differs from its limiting value, and hence $\kappa_{mn}(q_1)$ can be substituted for κ_{mn} in (5.6.8), the difference (5.6.40) being absorbed by the preexponential factor. Thus $|q_1|$ is nothing else as the tunneling depth (with its origin placed at the equilibrium position $q = 0$) we were talking about in Sect. 5.5.2. This quantity has not the conventional meaning of the barrier width since no actual barrier penetration occurs in the present case where the suppression of the transition stems from the dynamical restrictions imposed by the Franck–Condon principle (see Sect. 5.6.3).

Let us find out the energy dependence of the tunneling depth. Transforming the left-hand side of (5.6.40), as was done in the integrand of (5.6.11), and assuming $\mathscr{E}(q_1) \gg \varepsilon_{n,m}$, one obtains the equation for q_1 in the form

$$\omega y \left(\frac{M}{2}\right)^{1/2} \int\limits_{-\infty}^{q_1} [\mathscr{E}(q)]^{-1/2}\, dq = 1 \,, \tag{5.6.41}$$

where y is the energy difference of (5.6.3). Differentiating with respect to y gives $\partial q_1/\partial y < 0$, implying that when the energy difference grows q_1 moves to the left and, hence the tunneling depth increases. Since the probability of a tunneling process exponentially decreases with increasing tunneling depth [this is also evident from (5.6.39)], one obtains that the overtone transition intensity exponentially decreases with increasing overtone number. In the previous section this result was deduced for a potential whose repulsive branch permits the exponential approximation (5.6.32), and here it is shown to be valid for any potential growing for $q \to -\infty$ rapidly enough to ensure the convergence of the integral (5.6.41).

An explicit expression for q_1 can be obtained for the potential (5.6.32). Inserting (5.6.32) into (5.6.41) leads to

$$q_1 = -\frac{1}{\beta}\ln\left[\left(\frac{M}{2C}\right)^{1/2}\frac{\omega y}{\beta}\right].$$

The potential energy at q_1 is $\mathscr{E}(q_1) = D(\alpha y/\beta)^2$, where it was convenient to introduce the parameters of the Morse function (5.6.2). In the previous section it was obtained that an approximate relation, $\beta \approx 2\alpha$, holds for the C–H benzene vibrations, and the same takes place for other diatomic systems studied [5.32–34]. Therefore, for $y > 2$, that is, for the second and higher overtones, the transition probability feels the repulsive potential up to the energies in excess of the dissociation limit. Thus, the overtone intensities provide information on the potential in such a region of nuclear displacements where it cannot be obtained by conventional spectroscopic methods measuring the line positions and intensities of the classically allowed (fundamental vibrational or Franck–Condon vibronic) transitions.

5.6.5 Intensity Anomalies in Absorption Spectra

The normal intensity distributions of Sect. 5.6.3 were derived from the exponential factor of the dipole moment matrix element (5.6.8) which depends on the potential function alone. The dipole moment function enters the preexponential factor Q (through the factor ϱ) which is a slowly varying function of the energy, and, as a rule, affects insignificantly the overtone intensity distribution. However, one interesting exception exists to this rule.

Consider the y dependence of ϱ shown in Fig. 5.9. For the quadratic dipole function, ϱ does not vanish and conserves its sign within the entire range of the y variation, whereas for the cubic one ϱ passes through zero at $y \approx 2$. Obviously, in the latter case an anomaly would appear against the background of the smooth normal intensity distribution manifesting in a weakening of the intensities of one or two bands nearest in energy to this zero.

It is easy to see from (5.6.8, 9, 18, 19) that the low-intensity anomaly can be revealed by calculations of the exact matrix element ratio R of (5.6.20), as in the quasiclassical limit of (5.6.3) the exponential factor cancels and R merely converts to ϱ. Just in this way the anomaly was first discovered by *Meredith* and *Smith* [5.107] for a cubic dipole moment function. Numerical calculations for the Mecke functions [5.86] and for a dipole function allowing for nonadiabatic corrections [5.108] also predicted the appearance of such anomalies.

The above anomaly manifests itself not only in the absolute magnitude but also in the signs of the matrix elements d_{mn}, the latter being defined as soon as the signs of the wave functions $\chi_m(q)$ are fixed. The signs are conventionally specified by the condition that $\chi_m(q) > 0$ for $q \to +\infty$ [in the Morse potential this corresponds to the standard form of (4.8.13)]. Although only absolute magnitudes of the matrix elements can be found directly from the experimental vibrational transition intensities, there are indirect methods to determine their signs [5.107, 109–115], e.g., via the Herman–Wallis factors that describe the intensity distributions in vibrational spectra [5.113].

As is seen from (5.6.18, 19), for the linear dipole moment function $V(x) = x$ the signs are successively alternating[8], and precisely the same sign alternation must be observed for any other $V(x)$ provided that the corresponding ϱ factor does not pass through zero preserving its sign all over the entire range of the y variation. For instance, for $\varrho > 0$ and $n = 1, 2, \ldots, 10$ the matrix elements d_{0n} have the signs $(+ - + - + - + - + -)$. However, this succession brakes down if the anomaly emerges. Let ϱ go across zero between the $n = 6$ and 7 levels. Then the above succession is $(+ - + - + - - + - +)$. In summary, the anomaly can be revealed by both the local intensity weakening relative to the normal value and the violation of the successive alternation of the signs of the matrix elements.

Figure 5.12 shows the calculated HF matrix elements from [5.107]. For the quadratic dipole moment function the matrix elements monotonically decrease in absolute magnitude and have successively alternating signs, whereas for the cubic one there is a deep hole at $n = 6$ and a violation of the sign succession, as in the above example. Similar anomalies are also seen in the $8 \leftrightarrow 1$ and $9 \leftrightarrow 2$

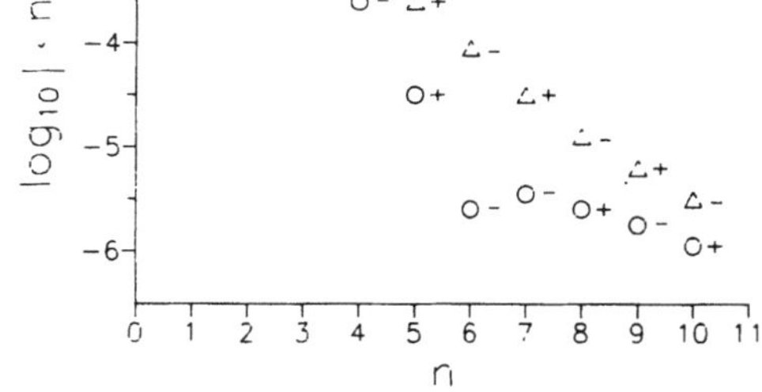

Fig. 5.12. Absolute values (in Debye) and signs of the matrix elements for the HF overtone absorption calculated with the RKR potential for quadratic (△) and cubic (○) dipole moment functions (from [5.107] Fig. 14)

[8] The existence of definite trends in the matrix element sign alternation was noted in [5.85, 116].

transition intensities (see [5.107], Appendix 2).[9] However, *Meredith* and *Smith* [5.107] concluded that the intensity distribution as a whole severely depends on the dipole moment function, and hence one can extract $d(q)$ from the intensities in a wide range of displacements. In fact, the anomaly originating from the $d(q)$ behavior is of local nature, the intensities of all other (normal) bands are again principally governed by the potential.

Calling on our example on the sign alternation we see that in the case under consideration the ϱ factor for the quadratic $d(q)$, ϱ_2, must be positive for all $n = 1$–10, while that for the cubic one, ϱ_3, must cross zero somewhere near $n = 6$. To verify these inferences the ϱ_2 and ϱ_3 values were calculated in [5.37, 79a] by (5.6.16) for the Morse potential since the dipole moment matrix elements do not alter significantly between the Morse and RKR potentials for HF [5.107, Fig. 7]. The results presented in Table 5.1 are in accordance with the above predictions. The HF d_{0n} absorption matrix elements were measured only for $n \leq 5$ [5.96, 107, 109, 113]; the analysis of the HF and DF d_{mn} emission matrix elements from [5.105] did not permit to discover the predicted anomaly, probably because either the y values were insufficiently high or the scatter of the experimental points was too large. However, this anomaly is easily revealed in the absorption spectra of other hydrogen halides. The ϱ factor for HCl, HBr, and HI passes through zero between the level pairs $n = 3$ and 4, 2 and 3, 1 and 2, respectively (see Table 5.1 for the HCl results).

Based on these results and on the assumption that the normal band intensities have still to obey the normal distribution (5.6.33), the decadic logarithms of

Table 5.1. The calculated ϱ factor (in Debye) for the HF quadratic and cubic and the HCl sixth-power polynomial dipole moment functions, and the HCl dipole moment matrix elements with the signs derived in [5.111, 115] from the overtone absorption spectrum

n	HF		HCl	
	ϱ_2	ϱ_3	ϱ_6	d_{on}
1	0.74	1.11	0.56	7.12×10^{-2}
2	0.83	0.84	0.42	-7.75×10^{-3}
3	0.88	0.57	0.16	5.15×10^{-4}
4	0.92	0.34	-0.13[a]	-3.06×10^{-5}[a]
5	0.95	0.14	-0.37	-8.42×10^{-6}
6	0.97	-0.04[a]	-0.56	6.61×10^{-6}
7	0.99	-0.19	-0.69	-3.27×10^{-6}[b]

[a] The change in the ϱ sign and the breakdown of the strictly alternating-sign succession both signal the appearance of an anomaly
[b] The sign is arbitrary

[9] In Appendix 2 of [5.107] the $6 \leftarrow 0$ matrix element has a lower value and the opposite sign as in [5.107, Fig. 14], but this does not influence the conclusion about the anomaly appearance.

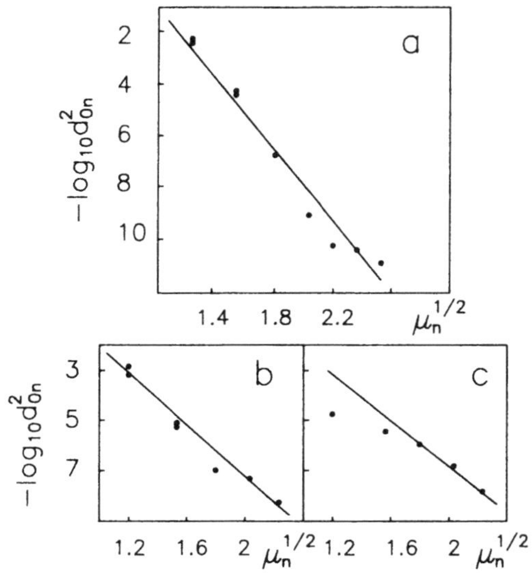

Fig. 5.13a–c. Experimental dipole moment matrix elements squared from the overtone absorption spectra of hydrogen halides. The straight lines of the normal distributions are drawn by the least-squares method disregarding the anomalous points, predicted by the ϱ factor calculations. **(a)** HCl: the anomalous bands $n = 4$ and 5 (Data from [5.111, 115, 117–119]); **(b)** HBr: the anomalous band $n = 3$ (Data from [5.114]); **(c)** HI: the anomalous bands $n = 1$ and 2 (Data from [5.112])

the experimental d_{0n}^2 values were plotted in Fig. 5.13 versus the square root of the upper-state reduced energy. It is seen that the anomalous bands actually have the intensities about an order of magnitude as small as the normal intensities on the straight lines. It is interesting to note that along the hydrogen halide series the anomaly position monotonically shifts to the lower n values from HF to HI.

The anomaly for the CO molecule was also predicted for $n = 5$ [5.32, 37, 79] while the experimental data are available only for $n \leq 4$ [5.118, 120, 121]. The experimental data on alkyl phenyl ketones [5.122] indicate an appearance of the anomaly around the sixth overtone of the C–H vibration. Calculations in [5.79d] for the C–H vibration in HCN predict the anomaly at $n = 9$ and 10.

5.7 Collision Model

In Sects. 5.1–3 the quasiclassical method was described for calculating the overlap integral of the vibrational wave functions, the most important quantity controlling the tunneling radiationless transition rate constant. However, in the case of nonadiabatic transitions, i.e., the transitions between adiabatic electronic terms induced by the non-adiabaticity operator L a specific difficulty arises originating from this operator's singularity (see Sects. 4.2–4) in that same region that provides the major contribution to the overlap integral, viz., the region of a (real or complex) electronic term intersection (see Sect. 5.4). This singularity does not affect the exponential factor of the transition probability arising from the Franck–Condon factor, but it must be taken into account when calculating the preexponential factor (see Sect. 4.5). In particular, an incorrect preexponential factor is obtained if the transition rate is calculated by the adiabatic perturbation theory in which the adiabatic states serve as nonperturbed states and the perturbation is the non-adiabaticity operator.

The above point deserves a more detailed discussion because it is a common practice to calculate the nonadiabatic transition rate just by perturbation theory. The idea of such an approach is that the nonadiabatic transition rate should be small due to the smallness of the Born-Oppenheimer parameter κ, entering the operator L. However, if the perturbation series in L is developed, the smallness of κ will result in that every member of that series will be exponentially small (because every member involves the Franck–Condon factor), but successive members will not be proportional to powers of κ, and virtually the problem will reduce to a summation of a numerical series. Thus, while the total sum of the series is indeed small in the tunneling energy region, it cannot be approximated with the first expansion term, hence perturbation theory is inapplicable.

This statement can be verified on the basis of a simple, yet exactly solvable model to be dealt with in this section. Let the nuclei possess quite a large vibrational energy in the initial electronic state for their motion to be treated quasiclassically. According to Sect. 5.4 the transitions between the molecular electronic states induced by the nuclear motion (or by other perturbations) are localized in those regions of the configurational nuclear-coordinate space that correspond to the closest approach or an intersection of the electronic terms. Therefore, the characteristics of the nuclear motion at a large-enough distance from the above regions insignificantly affect the transition rate constant, and the radiationless transition can be regarded as a cumulative effect of multiple "intramolecular second-kind collisions" [2.32, 5.123–128] occurring with the repetition rate equal to the nuclear vibration frequency ω. If the electronic transition probability in a single collision is P, the total radiationless transition rate constant is $k = \omega P/2\pi$. Thus, in the collision model the problem is to calculate the transition probability per collision.

The probability P can be derived in an asymptotically accurate form in a one-dimensional case of linear *diabatic* terms. This problem became classical due to the work by *Nikitin* et al. [5.129], where the energy region around the intersection point was studied and P was found in the overbarrier and tunneling regions. In Sect. 5.7.1 we solve this problem in a deep tunneling region by constructing the full transition matrix (see also [5.130]). *Stueckelberg* [5.131] investigated terms of a general form and the energy region above the intersection point, and he succeeded to construct the quasiclassical approximation for a set of two coupled second-order equations.

The probability P can also be found by the semiclassical method where the nuclear motion is specified by a trajectory $q(t)$ and the Schrödinger equation reduces to a set of two first-order equations for transition amplitudes [4.4, 5.22]. With this method the probability was obtained in [5.132] in the overbarrier energy range and in [5.133, 134] in the tunneling range.

5.7.1 Transition Matrix in the Linear Model

The model. In the region of the electronic terms' approach essential for the transition, the total wave function is no longer represented as a product (4.1.23) of the electronic and nuclear components, but instead should be written either in the form of expansion (4.1.9) over the adiabatic basis or the equivalent expansion (4.3.1) over the diabatic one. The coefficients of these expansions (the components of the nuclear wave function in different electronic states) obey the sets of coupled equations (4.1.21) or (4.3.2), respectively. We shall suppose that there are only two electronic states, 1 and 2, and both bases will be utilized as they are completely equivalent.

The transformation from the diabatic to the adiabatic basis in the two-state problem was discussed in Sect. 4.3, and we employ the notations introduced there. In the diabatic basis $\{\psi_1, \psi_2\}$ (the arguments of all functions are omitted) the nuclear Hamiltonian is

$$H = -\frac{\hbar^2}{2}\frac{d^2}{dq^2} + h\,, \tag{5.7.1}$$

where the nuclear mass is forced to be unity, and h represents the square matrix with elements (4.3.4). Its diagonal elements as functions of the nuclear coordinate q are diabatic terms, and the off-diagonal elements describe term coupling. The model of linear terms with a constant coupling independent of q,

$$\mathrm{h} = \begin{pmatrix} -f_1 q & v \\ v & -f_2 q \end{pmatrix}, \tag{5.7.2}$$

is to be considered. Here the q origin is placed at the term crossing, f_1 and f_2 are constant forces, and v is the coupling constant. For the sake of argument, $f_1 > f_2$ will be assumed.

The transformation to the adiabatic basis $\{\varphi_1, \varphi_2\}$ is accomplished by h-matrix diagonalization via the rotation transformation (4.3.5). In this way the adiabatic potentials (4.3.6) and the rotation angle (4.3.7) are obtained; for matrix (5.7.2) they are

$$\mathscr{E}_{1,2} = -Fq \mp (f^2q^2 + v^2)^{1/2}\,, \tag{5.7.3}$$

$$\alpha = -\frac{1}{2}\arctan\frac{v}{fq}\,, \tag{5.7.4}$$

where $2F = f_1 + f_2$ and $2f = f_1 - f_2 > 0$. The total wave function is written in the form of expansions over the diabatic and adiabatic bases,

$$\Psi = \psi_1 \tilde{\chi}_1 + \psi_2 \tilde{\chi}_2 \equiv \varphi_1 \chi_1 + \varphi_2 \chi_2 \,. \tag{5.7.5}$$

The expansion coefficients (the components of the diabatic and adiabatic nuclear wave functions) are also interconnected by transformation (4.3.5):

$$\begin{pmatrix} \chi_1 \\ \chi_2 \end{pmatrix} = \begin{pmatrix} \cos\alpha & \sin\alpha \\ -\sin\alpha & \cos\alpha \end{pmatrix} \begin{pmatrix} \tilde{\chi}_1 \\ \tilde{\chi}_2 \end{pmatrix} . \tag{5.7.6}$$

They obey the sets of coupled equations ensuing from the Schrödinger equation $(H - E)\Psi = 0$ with the Hamiltonian (5.7.1, 2). Substituting the first expansion of (5.7.5) for Ψ, left-multiplying this equation in turn by ψ_1 and ψ_2 and integrating over the electronic variables, we obtain a set of equations for the diabatic amplitudes, which is written in matrix form as

$$\begin{pmatrix} -\dfrac{\hbar^2}{2}\dfrac{d^2}{dq^2} - f_1 q - E & v \\ v & -\dfrac{\hbar^2}{2}\dfrac{d^2}{dq^2} - f_2 q - E \end{pmatrix} \begin{pmatrix} \tilde{\chi}_1 \\ \tilde{\chi}_2 \end{pmatrix} = 0 \,. \tag{5.7.7}$$

Similarly a set of equations is derived for the adiabatic amplitudes,

$$\begin{pmatrix} -\dfrac{\hbar^2}{2}\dfrac{d^2}{dq^2} + \mathscr{E}_1 + \dfrac{\hbar^2}{2}\alpha'^2 - E & \dfrac{\hbar^2}{2}\alpha'' + \hbar^2\alpha'\dfrac{d}{dq} \\ \dfrac{-\hbar^2}{2}\alpha'' - \hbar^2\alpha'\dfrac{d}{dq} & -\dfrac{\hbar^2}{2}\dfrac{d^2}{dq^2} + \mathscr{E}_2 + \dfrac{\hbar^2}{2}\alpha'^2 - E \end{pmatrix} \begin{pmatrix} \chi_1 \\ \chi_2 \end{pmatrix} = 0 \,, \tag{5.7.8}$$

where prime means differentiation with respect to q. The non-adiabaticity operator is

$$L = \begin{pmatrix} 0 & L_{12} \\ -L_{12} & 0 \end{pmatrix}, \qquad L_{12} = \hbar^2\left(\frac{1}{2}\alpha'' + \alpha'\frac{d}{dq}\right) . \tag{5.7.9}$$

Equation (5.7.3) shows that the adiabatic terms approach one another inside the region $|q| \lesssim v/f$. Outside this region they approach the corresponding diabatic terms, the adiabatic basis approaches the diabatic one (since by (5.7.4) $\alpha \sim 1/q \to 0$), and the non-adiabaticity operator vanishes (since $\alpha', \alpha'' \to 0$). Hence the difference between the adiabatic and diabatic amplitudes is only significant in the domain of the term approach.

Formulation of the Problem. The problem of the second-kind collision transition probability is formulated on the basis of (5.7.8) for the adiabatic amplitudes. Consider the amplitudes' behavior at $q \to +\infty$. Owing to the steep weakening of the nonadiabatic coupling the equations become uncoupled and two independent equations are obtained.

$$\left(-\frac{\hbar^2}{2}\frac{d^2}{dq^2} + \mathscr{E}_n - E\right)\chi_n = 0 \quad (n = 1, 2), \tag{5.7.10}$$

each describing the system's motion within one term without any transitions between the terms. Each of these equations has two linearly independent solutions that will be written in the quasiclassical (JWKB) approximation. A general solution to (5.7.10) is

$$\chi_n = A_n^- Z_n^- + A_n^+ Z_n^+ , \tag{5.7.11}$$

where

$$Z_n^{\pm} = \left(\frac{2\pi\hbar}{\mathscr{P}_n}\right)^{1/2} \exp\left(\pm\frac{i}{\hbar}S_n \mp \frac{i\pi}{4}\right), \tag{5.7.12}$$

$$\mathscr{P}_n = [2(E - \mathscr{E}_n)]^{1/2}, \quad S_n = \int_{q_n}^{q} \mathscr{P}_n \, dq ; \tag{5.7.13}$$

q_n are turning points and $A_n^{\pm}$ constants. Two terms in (5.7.11) represent the incoming (from $+\infty$) and outgoing (to $+\infty$) waves.

Physically, it is clear that we may, e.g., force the coefficient of the incoming wave in term 1, A_1^-, to an arbitrary value and impose the condition that there was no incoming wave in term 2, $A_2^- = 0$. Then the coefficients of the outgoing waves, $A_n^{\pm}$, are to be determined by solving the simultaneous equations (5.7.8). The absolute magnitudes of these coefficients squared give the reflection and transition probabilities. Thus, only two out of four constants $A_n^{\pm}$ are arbitrary. Mathematically this is because these four constants are subject to two constraints that the solutions decreased at $q \to -\infty$.

We introduce the transition matrix T by relation

$$\begin{pmatrix} A_1^+ \\ A_2^+ \end{pmatrix} = T \begin{pmatrix} A_1^- \\ A_2^- \end{pmatrix}. \tag{5.7.14}$$

The problem is to find asymptotic (for $q \to +\infty$) solutions to (5.7.8) of the form (5.7.11) that decrease for $q \to -\infty$, and then to derive the transition matrix without using the perturbation expansion in L.

In view of the fact that the difference between the adiabatic and diabatic nuclear functions disappears in the limit of $q \to +\infty$, the diabatic functions are bound to acquire the asymptotic form of (5.7.11) with the same coefficients $A_n^{\pm}$. For this reason the transition matrix can be sought by solving the more simple system of (5.7.7).

Transformation to a p-Representation. In the case of linear terms this transformation reduces the order of the system (5.7.7). On letting for $n = 1, 2$

$$\tilde{\chi}_n(q) = \left(\frac{f_1 f_2}{f_n}\right)^{1/2} \int_L g_n(p) \exp\left(\frac{\mathrm{i}}{\hbar} pq - \frac{\mathrm{i}F}{\not{h} f} \int_0^p U\, dp\right) dp \tag{5.7.15}$$

we obtain the system of equations for g_n,

$$\begin{pmatrix} -\mathrm{i}\not{h}\dfrac{d}{dp} - U & V \\ V & -\mathrm{i}\not{h}\dfrac{d}{dp} + U \end{pmatrix} \begin{pmatrix} g_1 \\ g_2 \end{pmatrix} = 0\,, \tag{5.7.16}$$

where the notations

$$U = \tfrac{1}{2}p^2 - E, \qquad \not{h} = \hbar f_1 f_2 / f, \qquad V = v(f_1 f_2)^{1/2}/f \tag{5.7.17}$$

are introduced. From the first equation g_2 is expressed through g_1,

$$V g_2 = \mathrm{i}\not{h} g_1' + U g_1\,, \tag{5.7.18}$$

where prime now means differentiation over p. This is inserted into the second equation to yield

$$\not{h}^2 g_1'' + (W - \mathrm{i}\not{h} p) g_1 = 0, \qquad W \equiv U^2 + V^2\,. \tag{5.7.19}$$

Thus, in the linear model the problem is reduced to a single second-order equation that is to be solved in the quasiclassical approximation.

The Quasiclassical Approximation for (5.7.19) is developed along the same lines as the well-known JWKB approximation for the conventional Schrödinger equation [1.40]. The solution is tried in the form

$$g_1 = \exp\left(\frac{\mathrm{i}}{\not{h}}\sigma\right), \qquad \sigma = \sigma_0 + \frac{\not{h}}{\mathrm{i}}\sigma_1 + \cdots.$$

Substituting this into (5.7.19) and equating the coefficients of successive powers of $\not{h}$ to zero, we obtain

$$\sigma_0 = \pm \int \sqrt{W}\, dp, \qquad \sigma_1' = \frac{p - \sigma_0''}{2\sigma_0'}\,.$$

Integrating of the second equation gives

$$\sigma_1 = \pm \frac{1}{2} \ln\left(U + \sqrt{W}\right) - \frac{1}{2} \ln \sqrt{W}\,,$$

where integration constants are omitted. It is seen that the quasiclassical approximation does not work near zeros of the W function, $p_k = \pm[2(E \pm iV]^{1/2}$ $(k = 1, \ldots, 4)$, since at these points σ_1 increases infinitely. The quasiclassical condition

$$|W^{3/2}/\hbar Up| \gg 1 \tag{5.7.20}$$

is obtained in the same way as in the case of the usual Schrödinger equation [1.40]. The W zeros are located symmetrically with respect to both coordinate axes, and with E changing from $+\infty$ to $-\infty$ they are moving along the trajectories shown in Fig. 5.14.

We further proceed as in Sect. 5.2. We introduce the two functions:

$$\begin{aligned} e_1^+ &= \frac{(U + \sqrt{W})^{1/2}}{2^{1/2} W^{1/4}} \exp\left(\frac{i}{\hbar}\int_0^p \sqrt{W}\,dp\right), \\ e_1^- &= \frac{V}{2^{1/2}(U + \sqrt{W})^{1/2} W^{1/4}} \exp\left(-\frac{i}{\hbar}\int_0^p \sqrt{W}\,dp\right), \end{aligned} \tag{5.7.21}$$

which are quasiclassical solutions to (5.7.19). At $|p| \to \infty$ each exponential either increases, decreases or oscillates. The latter occurs along anti-Stokes lines defined by the equation $\mathrm{Im}\{\int \sqrt{W}\,dp\} = 0$. For $|p| \to \infty$ this equation takes the form $\mathrm{Im}\{p^3\} = 0$, whence it follows that in infinity the anti-Stokes lines approach the real axis and the directions $\arg\{p\} = \pm\pi/3, \pm 2\pi/3$. Considering as well that all the W zeros are single, and that each single zero lets out three lines making 120° angles between each other [5.10], one can easily establish the arrangement of the anti-Stokes lines. They are depicted by full lines in Fig. 5.15.

Let us draw cuts connecting the zeros to pairs p_1, p_2 and p_3, p_4, and define the $\sqrt{W}$ branch such that $\sqrt{W} > 0$ would hold on the real axis. Nevertheless,

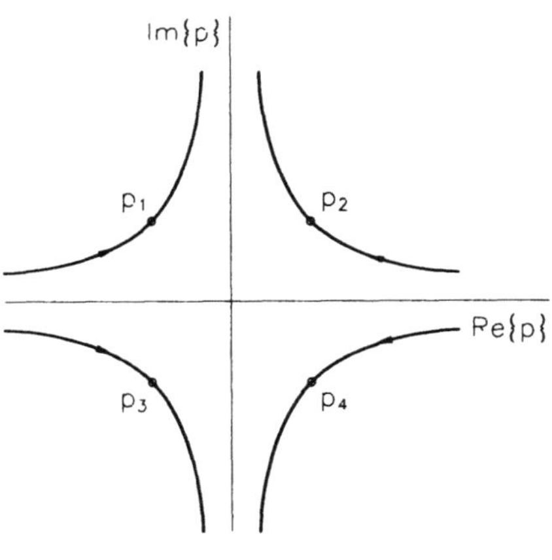

Fig. 5.14. The trajectories of the W zeros with E changing from $+\infty$ to $-\infty$

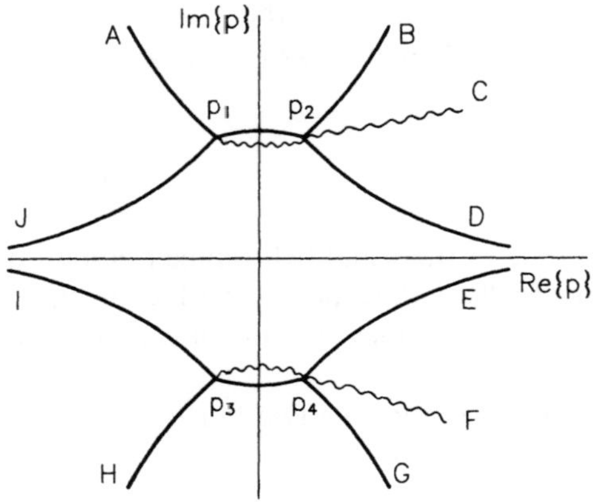

Fig. 5.15. Anti-Stokes lines (*full lines*) and cuts (*wavy lines*)

functions (5.7.21) still remain many-valued since the integral

$$\gamma = \frac{1}{\hbar} \oint_{p_1, p_2} \sqrt{W}\, dp \tag{5.7.22}$$

along a closed contour encircling the pair of the W zeros is not an integer multiple of 2π (in Sect. 5.2 the similar integral of (5.2.11) was a multiple of 2π due to the Bohr–Sommerfeld quantization rule). Therefore, we draw additional cuts from the p_2 and p_4 zeros (all the cuts are shown in Fig. 5.15). Then the functions (5.7.21) become single-valued and analytical all over the p-plane.

The anti-Stokes lines and cuts divide the complex plane into a number of areas. Within each of the outer sectors AB, BC, etc., the functions (5.7.21) either increase or decrease exponentially for $|p| \to \infty$. Owing to the above choice of the root branch we have $\sqrt{W} \propto p^2/2$ when p moves to infinity in any direction. Correspondingly,

$$e_1^{\pm} \sim \exp\left(\pm \frac{ip^3}{6\hbar}\right) \quad \text{for } p \to \infty \ . \tag{5.7.23}$$

It is seen that e_1^+ increases while e_1^- decreases along the positive imaginary axis, and the same is true all over the whole sector AB because the asymptotic behavior of these functions is changed only on the anti-Stokes lines. In the central area these functions also increase or decrease when moving away from the zeros, though both exponentials oscillate on the real axis.

Construction of Asymptotes of Exact Solutions all over the Complex Plane. Let g_1 be an exact solution to (5.7.19). Its asymptote inside any area in Fig. 5.15 is proportional to one of the functions (5.7.21), while on the anti-Stokes lines it is represented as a linear combination of these functions. The latter ones undergo hops across the cuts while the exact solution is analytical everywhere in the complex plane, which is immediately evident from (5.7.19). Therefore, the g_1 asymptote, which should also be analytical on the cuts, is determined by different expressions at the opposit cut sides.

Hence it follows that the g_1 asymptote all over the complex plane as a whole cannot be represented as a linear combination of functions (5.7.21) with constant coefficients. Nevertheless, one often writes down such combinations, but then the coefficients' hops are introduced on certain lines (the Stokes lines) located inside each area. The coefficients' hopping is called Stokes' phenomenon [5.10]. This formal method seems inappropriate here since in practice the hops' values can only be determined when the g_1 asymptotes in all areas are already known, i.e., actually after the principal problem has been resolved.

Let us construct the asymptotes of two linearly independent solutions to (5.7.19). As the first of the two, we take a solution g_1^{AB} decreasing in sector AB. In this sector its asymptote has to be proportional to the decreasing function e_1^-,

$$g_1^{AB} = Ce_1^- \,, \tag{5.7.24}$$

where C is an arbitrary constant. To find the asymptotes of this same solution in other areas, (5.7.24) is to be analytically continued beyond the anti-Stokes lines and cuts.

Similar problems have already been solved in Sects. 5.2, 3. The analytical continuation technique is based on the *Heading* theorem [5.10] stating that the decreasing solution can be continued across the anti-Stokes lines bordering the decrease area. In the present context this implies (5.7.24) to be valid, with the same value of C, in sectors BC and AJ where it increases. By the same theorem the increasing solution cannot be continued to an anti-Stokes line and in the region beyond the line. Hence, (5.7.24) loses its validity when approaching the p_1J line.

The Heading theorem has a simple meaning. It is easy to prove that the solution decreasing in a given area is unique (apart from a constant factor). Indeed the Wronskian of any two solutions is a constant, and if both solutions decrease within one and the same area, this constant is zero, that is, those solutions are linearly dependent. In view of the uniqueness of the decreasing solution, the problem of analytical continuation for such a solution makes sense. On the other hand, there are always two linearly independent increasing solutions, but their asymptotes are proportional to one and the same function, for instance, e_1^+ in sector AB. For this reason, the increasing asymptote does not yet specify a certain solution, so the question about the analytical continuation of such an asymptote into other areas loses its significance: two different solutions having the e_1^+ asymptote in AB will have different asymptotes in other sectors.

Let us continue the function (5.7.24) into the area CD across the long p_2C cut. On account of the exact solution being analytical all over the p-plane its asymptote has to be continuous on the cut in any case. Emphasize once more that the cuts define the analytical properties of functions (5.7.21) only, but not of the exact solutions and their asymptotes. In particular a different choice of the cuts would mean a redefinition of the functions (5.7.21), but it would not affect the asymptote's behavior that would remain unchanged in going across the cut.

The $\sqrt{W}$ function is continuous on the p_2C cut (this cut was exclusively drawn to make the $e_1^{\pm}$ phases single-valued), so g_1^{AB} in sector CD will be still proportional to e_1^- but this time with a different proportionality factor,

$$g_1^{AB} = Ce^{i\gamma} e_1^- . \tag{5.7.25}$$

It is easy to verify, using definitions (5.7.21, 22), that this expression coincides with (5.7.24) at any point on the cut. Therefore it gives an analytical continuation of the g_1^{AB} asymptote into sector CD; however, (5.7.25) loses its validity when approaching the p_2D anti-Stokes line.

Asymptote (5.7.24) can also be continued from sector AB into the central area across the p_1p_2 anti-Stokes line. If the points p_1 and p_2 are well separated from each other (the interaction V is large) so that the quasiclassical condition (5.7.20) is fulfilled in the region between them, the asymptote (5.7.24) also holds valid beyond that line as far as the short p_1p_2 cut. However, we are interested in a more general case permitting a violation of the quasiclassical approximation in the region between p_1 and p_2.

Now we shall consider deep tunneling transitions at large negative energies, $-E \gg V$. Then p_1 and p_2 are close to one another, and are remote of the second pair of zeros:

$$p_{1,2} = p_0\left(1 \pm \frac{iV}{2|E|}\right), \qquad p_0 = i\sqrt{2|E|} . \tag{5.7.26}$$

The region between them is $\Delta p \sim V\sqrt{|E|}$ wide. Inside this region one has $U \sim p_0 \Delta p \sim V$, $W \sim V^2$, and the ratio on the left-hand side of inequality (5.7.20) is proportional to the Landau–Zener parameter

$$\delta = \frac{V^2}{2\hbar\sqrt{2|E|}} = \frac{v^2}{\hbar(f_1 - f_2)\sqrt{2|E|}} . \tag{5.7.27}$$

Thus, at large negative energies the quasiclassical condition for the region between p_1 and p_2 reads $\delta \gg 1$. For arbitrary δ, when the quasiclassical approximation is not valid in this region, the asymptotical solution (5.7.24) can be continued into the central area only by solving exactly the original equation (5.7.19) in the vicinity of the p_1 and p_2 zeros.

The remoteness of p_3 and p_4 zeros enables one to expand U around p_0,

$$U \simeq p_0(p - p_0) . \tag{5.7.28}$$

Then (5.7.19) becomes

$$\hbar^2 g_1'' + [p_0^2(p - p_0)^2 + V^2 - i\hbar p_0]g_1 = 0 .$$

With the substitution

$$x = -i(p - p_0)(2\sqrt{2|E|}/\hbar)^{1/2} \tag{5.7.29}$$

it is reduced to the equation or the parabolic cylinder functions

$$\frac{d^2 g_1}{dx^2} - \left(\delta + \frac{1}{2} + \frac{x^2}{4}\right) g_1 = 0 \, .$$

The solution to this equation decreasing along the imaginary p-axis above p_0 (i.e., for $x \to +\infty$) has the form $g_1 = BD_{-\delta-1}(x)$, where B is a constant. Its asymptotes are [4.164]

$$g_1 = \begin{cases} Bx^{-\delta-1}\exp(-x^2/4) \text{ for } x \to +\infty \, , & (5.7.30) \\ B\dfrac{\sqrt{2\pi}}{\Gamma(1+\delta)}|x|^{\delta}\exp(x^2/4) \text{ for } x \to -\infty \, . & (5.7.31) \end{cases}$$

Let the $p_1 p_2$ cut cross the imaginary p-axis at p_0 (at $x = 0$). Then asymptote (5.7.30) decreasing for $x \to +\infty$ must coincide with (5.7.24) while asymptote (5.7.31) increasing for $x \to -\infty$ must be proportional to the function e_1^+ increasing toward the core of the central area.

Let us recast $e_1^{\pm}$ with the use of (5.7.28, 29). On the imaginary p-axis (i.e., for real x) one has $U = -Vx/2\sqrt{\delta}$ and $W = (V^2/4\delta)(x^2 + 4\delta)$. By the definition of the $\sqrt{W}$ branch one has to set

$$\sqrt{W} = \frac{V}{2\sqrt{\delta}} \begin{cases} -(x^2 + 4\delta)^{1/2} & \text{for } x > 0 \, , \\ (x^2 + 4\delta)^{1/2} & \text{for } x < 0 \end{cases} \tag{5.7.32}$$

(the roots on the right-hand side are now to be understood in the arithmetic sense). The integral in the exponent of e_1^- in the p-region above p_0 is represented in the form

$$\frac{1}{\hbar}\int_0^p \sqrt{W}\,dp = \frac{i\beta}{2} - \frac{\gamma}{2} + \frac{1}{\hbar}\int_{p_0+i0}^{p} \sqrt{W}\,dp \, , \tag{5.7.33}$$

$$\beta = -\frac{2i}{\hbar}\int_0^{p_0} \sqrt{W}\,dp \, . \tag{5.7.34}$$

On letting $t = p/p_0$ one obtains

$$\beta = \frac{1}{\nu}\int_0^1 [(1-t^2)^2 + (V/E)^2]^{1/2}\,dt \, , \tag{5.7.35}$$

where

$$\nu = \frac{\hbar}{(2|E|)^{3/2}} = \frac{2\hbar f_1 f_2}{(f_1 - f_2)(2|E|)^{3/2}} \ll 1 \tag{5.7.36}$$

is just the dimensionless small parameter virtually underlying the present quasiclassical approximation.

The second term of (5.7.33) is the contribution of the integral along the path winding clockwise around the cut from $p_0 - i0$ to $p_0 + i0$. Within the approximation of (5.7.28) it is half of the integral (5.7.22) with opposite sign; the latter integral is easily calculated to yield $\gamma = 2\pi\delta$. The last term in (5.7.33) is calculated with the use of (5.7.32, 29). In the limit of $x \to +\infty$ one has

$$\frac{1}{\hbar}\int_{p_0}^{p}\sqrt{W}\,dp = -\frac{i}{2}\int_{0}^{x}(x^2+4\delta)^{1/2}dx \simeq -i\left(\frac{x^2}{4}+\delta\ln x-\frac{1}{2}\delta\ln\delta+\frac{1}{2}\delta\right).$$

The integral of the e_1^+ exponent in the p-region below p_0 is evaluated in a similar way to give

$$e_1^- = -\sqrt{\delta}\exp\left[\frac{1}{2}\beta+i\pi\delta-\frac{x^2}{4}-(\delta+1)\ln x\right.$$
$$\left.+\frac{1}{2}\delta\ln\delta-\frac{1}{2}\delta\right] \quad \text{for } x\to+\infty\,, \tag{5.7.37}$$

$$e_1^+ = \exp\left(-\frac{1}{2}\beta+\frac{x^2}{4}+\delta\ln|x|\right.$$
$$\left.-\frac{1}{2}\delta\ln\delta+\frac{1}{2}\delta\right) \quad \text{for } x\to-\infty\,. \tag{5.7.38}$$

Now we are in a position to continue asymptote (5.7.24) into the central area. Equating (5.7.24) to (5.7.30) and substituting (5.7.37) we obtain

$$B = -C\sqrt{\delta}\exp\left(\tfrac{1}{2}\beta+i\pi\delta+\tfrac{1}{2}\delta\ln\delta-\tfrac{1}{2}\delta\right), \tag{5.7.39}$$

and a comparison of (5.7.31) with (5.7.38) shows that in the area below the cut

$$g_1^{AB} = -C\lambda e^{i\pi\delta}e_1^+\,, \tag{5.7.40}$$

$$\lambda = \frac{\sqrt{2\pi\delta}}{\Gamma(1+\delta)}\exp(\beta+\delta\ln\delta-\delta)\,. \tag{5.7.41}$$

Asymptote (5.7.40) loses validity near the p_1J and p_2D anti-Stokes lines, but there it can be determined as follows. Since g_1^{AB} oscillates on these lines, its asymptote has to be represented as a linear combination of functions (5.7.21) with the coefficients found under the condition that, when the p-point is displaced from the p_1J line into sector AJ or into the central area, the asymptote must coincide with (5.7.24) or (5.7.40), respectively. Hence it follows that on the p_1J line the g_1^{AB} asymptote is the sum of (5.7.24) and (5.7.40). Similarly it is the sum of (5.7.25) and (5.7.40) on the p_2D line.

Later we shall need the g_1^{AB} asymptotes on the real axis for $p\to\pm\infty$. The above argument shows that they may, in general, involve, additionally to (5.7.40), a term proportional to e_1^-. However, at this point it should be taken into account that $e_1^\pm$ behave differently on the real axis as well as on the anti-Stokes lines. As is seen from definition (5.7.21), both exponentials oscillate, but the modulus of e_1^+ tends to a constant, while that of e_1^- decreases as $1/p^2$. Hence this correction is insignificant, and asymptote (5.7.40) is valid on the real axis for $p \to \pm\infty$.

As the second solution which must be linearly independent of the first, we take g_1^{GH} which decreases in sector GH. According to (5.7.23) its asymptote in GH is of the form

$$g_1^{\mathrm{GH}} = De_1^+ , \tag{5.7.42}$$

where D is an arbitrary constant. By the Heading theorem this same expression holds in the adjacent sectors HI and FG, and the continuation across the p_4F cut is performed by the scheme described above. In sector EF one obtains

$$g_1^{\mathrm{GH}} = D\mathrm{e}^{-\mathrm{i}\gamma} e_1^+ \quad \text{with } \gamma = 2\pi\delta . \tag{5.7.43}$$

The analytical continuation of the above asymptotes to the corresponding anti-Stokes lines and to the real axis for $p \to \pm\infty$ is carried out in a trivial way. On these lines and beyond terms with e_1^- should be added to (5.7.42, 43), but they are insignificant for us as on the real axis for $p \to \pm\infty$ we have $|e_1^-| \sim 1/p^2$, while $|e_1^+| \to 1$. Therefore (5.7.42, 43) give the g_1^{GH} asymptotes for $p \to -\infty$ and $p \to +\infty$, respectively.

Notice that if g_1^{GH} is continued into the central area across the p_3p_4 line, as was done with g_1^{AB}, then we obtain an expression proportional to e_1^-. Here, the second term with e_1^+ is small, but becomes dominating for $p \to \pm\infty$ for the reason mentioned above. Therefore g_1^{GH} is represented by different functions on different tails of the real axis, and just this effect is responsible for the phase decrement of $2\pi\delta$ when p changes from $-\infty$ to $+\infty$ [cf.(5.7.42, 43)]. In contrast to this behavior, in the g_1^{AB} solution the exponential e_1^+ dominates all over the whole real axis, and for this reason its coefficient does not change. From the above argument it follows as well that g_1^{GH} and g_1^{AB} are linearly independent solutions.

The general solution, which involves the two arbitrary constants C and D is merely the sum of the two solutions obtained. Its asymptotes on the real axis are

$$g_1 = g_1^{\mathrm{AB}} + g_1^{\mathrm{GH}} = \begin{cases} (-C\lambda\mathrm{e}^{\mathrm{i}\pi\delta} + D)e_1^+ & \text{for } p \to -\infty , \\ (-C\lambda\mathrm{e}^{\mathrm{i}\pi\delta} + De^{-2\mathrm{i}\pi\delta})e_1^+ & \text{for } p \to +\infty . \end{cases} \tag{5.7.44}$$

The second component of the nuclear wave function, g_2, is connected with g_1 via (5.7.18). We define functions $e_2^\pm$ connected with $e_1^\pm$ through the same relation. Omitting small terms on the order of $\hbar$ we obtain them in explicit form:

$$\begin{aligned} e_2^+ &= \frac{V}{2^{1/2}(U+\sqrt{W})^{1/2}W^{1/4}} \exp\left(\frac{\mathrm{i}}{\hbar}\int_0^p \sqrt{W}\,dp\right) , \\ e_2^- &= \frac{(U+\sqrt{W})^{1/2}}{2^{1/2}W^{1/4}} \exp\left(-\frac{\mathrm{i}}{\hbar}\int_0^p \sqrt{W}\,dp\right) . \end{aligned} \tag{5.7.45}$$

The g_2^{AB} solution decreasing in sector AB is expressed through g_1^{AB} by relation (5.7.18). Since, by definition, e_2^- is expressed through e_1^- by the same relation, from (5.7.24) we obtain in sector AB

$$g_2^{AB} = Ce_2^- \tag{5.7.46}$$

with the same coefficient C. Analogously the solution decreasing in sector GH has there the asymptote

$$g_2^{GH} = De_2^+ . \tag{5.7.47}$$

Now we note that the $e_2^\pm$ behavior on the real axis is precisely opposite to the $e_1^\pm$ behavior. Namely, for $p \to \pm\infty$ we have $|e_2^+| \sim 1/p^2$, $|e_2^-| \to 1$. Therefore, the continuation of solutions (5.7.46, 47) to the real axis results in asymptotes proportional to e_2^-:

$$g_2 = g_2^{AB} + g_2^{GH} = \begin{cases} (C + D\lambda e^{-i\pi\delta})e_2^- & \text{for } p \to -\infty , \\ (Ce^{2i\pi\delta} + D\lambda e^{-i\pi\delta})e_2^- & \text{for } p \to +\infty . \end{cases} \tag{5.7.48}$$

Asymptotes (5.7.44, 48) are expressed via the two functions e_1^+ and e_2^-, which, on the real axis for $p \to \pm\infty$, tend to

$$e^\pm = \exp\left(\pm\frac{i}{\hbar}\int_0^p \sqrt{W}\,dp\right),$$

respectively. Therefore, (5.7.44, 48) can also be recast as

$$\begin{aligned} g_1 &= ae^+ , \quad g_2 = be^- \quad \text{for } p \to -\infty ; \\ g_1 &= ce^+ , \quad g_2 = de^- \quad \text{for } p \to +\infty . \end{aligned} \tag{5.7.49}$$

The coefficients obey the relations

$$\begin{pmatrix} c \\ d \end{pmatrix} = T\begin{pmatrix} a \\ b \end{pmatrix}, \tag{5.7.50}$$

where

$$T = \frac{1}{1+\lambda^2}\begin{pmatrix} \lambda^2 + e^{-2i\pi\delta} & -2i\lambda\sin\pi\delta \\ -2i\lambda\sin\pi\delta & \lambda^2 + e^{2i\pi\delta} \end{pmatrix} \tag{5.7.51}$$

is just the wanted transition matrix, as will be shown below.

Asymptotic Solutions to (5.7.7). To find the asymptotic diabatic nuclear wave functions one has to calculate the integrals of (5.7.15) for $q \to \pm\infty$. This can be carried out by the saddle-point method. As will be apparent below, the saddle points lie in the asymptotic regions of the complex p-plane so that, in order to find $\tilde{\chi}_n$, it is sufficient to calculate the integrals

$$\mathcal{I}e^{\pm} \equiv \int_L \exp\left(\frac{i}{\hbar}pq - \frac{iF}{\hbar f}\int_0^p U\,dp \pm \frac{i}{\hbar}\int_0^p \sqrt{W}\,dp\right)dp. \tag{5.7.52}$$

We equate to zero the first derivative of the exponent with respect to p:

$$FU - f_1 f_2 q = \pm f(U^2 + V^2)^{1/2}. \tag{5.7.53}$$

Squaring both sides and resolving the equation with respect to U we find two pairs of the saddle points: $p_{*1} = \pm\mathscr{P}_1(q)$ for $\mathcal{I}e^+$ and $p_{*2} = \pm\mathscr{P}_2(q)$ for $\mathcal{I}e^-$, where $\mathscr{P}_n(q)$ are classical momenta (5.7.13). For $q \to +\infty$ all the saddle points are on the real axis while for $q \to -\infty$ they are on the imaginary axis. In both cases $|p| \to \infty$, which justifies the use of the g_n asymptotes.

The steepest-descent directions at the saddle points are determined by the second derivative of the exponent in (5.7.52) that for the integrals with e^+ and e^- equals, respectively, to

$$-\frac{ip_{*1}}{\hbar f_1}, \quad -\frac{ip_{*2}}{\hbar f_2} \quad \text{for } q \to +\infty \text{ and}$$
$$-\frac{ip_{*1}}{\hbar f_2}, \quad -\frac{ip_{*2}}{\hbar f_1} \quad \text{for } q \to -\infty. \tag{5.7.54}$$

In Fig. 5.16 circles mark the saddle points (1 and 2 for $q \to +\infty$, 3 and 4 for $q \to -\infty$), and arrows indicate the directions of steepest descent,

$$\arg\{p\} = \frac{3\pi}{4} - \frac{1}{2}\arg\{p_{*n}\}. \tag{5.7.55}$$

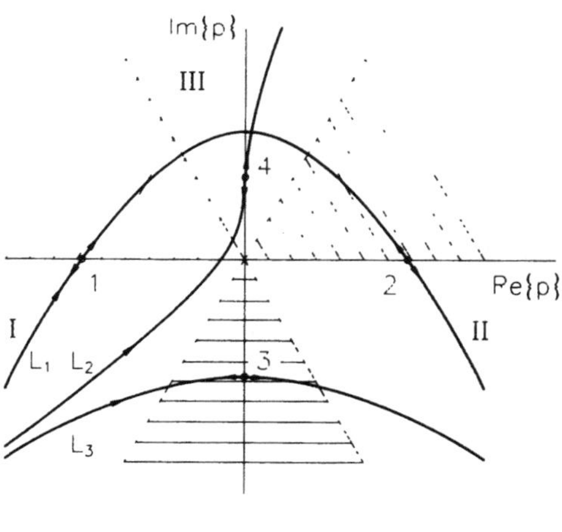

Fig. 5.16. Integration paths in (5.7.52) and the saddle points. The areas where the integrand grows are hatched

Now the integration path L can be drawn. Remind that the integrand has to decrease at its ends because this was exploited in going from (5.7.7) to (5.7.16). For $p \to \infty$ the leading term in the exponent of (5.7.52) has the form

$$\frac{ip^3}{6\hbar}\left(-\frac{F}{f} \pm 1\right) = -\frac{ip^3}{6\hbar f_{1,2}}\,.$$

The real part of this expression is positive inside the hatched areas of Fig. 5.16; in these areas the integrand increases. Therefore, the ends of L must be in sectors I, II, or III where it decreases. There are two independent paths on this pattern, L_1 and L_2. If the integral of (5.7.15) is taken along each of the two contours, and two linearly independent solutions to (5.7.19) are successively substituted into the integrand, then we obtain a full set of four linearly independent solutions to (5.7.7).

The exponent of (5.7.52) for e^+ at the saddle point $p_{*1} = +\mathscr{P}_1(q)$ is

$$\frac{i}{\hbar}\mathscr{P}_1(q)q - \frac{i}{\hbar f_1 f_2}\int\limits_0^{\mathscr{P}_1(q)}(FU - f\sqrt{W})dp\,. \tag{5.7.56}$$

If the integration variable is replaced with q' by equation $p = \mathscr{P}_1(q')$, the integrand will be equal to $f_1 f_2 q'$ because $\mathscr{P}_1(q')$ obeys (5.7.53) at $q = q'$ and with the plus sign on the right-hand side. The integration limits over q' are (q_1, q) where q_1 is the turning point. Integration by parts of (5.7.56) results in

$$\frac{i}{\hbar}\mathscr{P}_1(q)q - \frac{i}{\hbar}\left[q\mathscr{P}_1(q) - \int\limits_{q_1}^{q}\mathscr{P}_1(q')dq'\right] = \frac{i}{\hbar}S_1\,, \tag{5.7.57}$$

where S_1 is the classical action of (5.7.13). At the point $p_{*1} = -\mathscr{P}_1(q)$ one obtains the same result with the opposite sign, and at the points p_{*2} it is correspondingly $\pm iS_2/\hbar$.

Let $q \to -\infty$. At the saddle points of the upper half-plane we have

$$\mathrm{Im}\{\mathscr{P}_n\} > 0\,, \qquad \mathrm{Re}\{iS_n\} = -\int\limits_{q_1}^{q}\mathrm{Im}\{\mathscr{P}_n\}dq > 0\,. \tag{5.7.58}$$

Thus, while exponentials (5.7.52) decrease with fixed q and $p \to \infty$ in sector III, their values at the saddle points $p_{*n} = i|\mathscr{P}_n|$ increase unlimitedly for $q \to -\infty$, as is seen from (5.7.58); and all is vice versa in the hatched sector of the lower half-plane.

For $q \to -\infty$ the contour L_2 going from sector I into sector III passes through saddle 4, and hence, it corresponds to increasing solutions. As for decreasing ones, they are obtained by integration along the contour L_3 going from sector I into sector II through saddle 3. For $q \to +\infty$ this contour transforms to contour L_1 going through saddles 1 and 2.

Let $(\mathscr{L}e^+)_1$ denote the contribution of saddle point 1 to the integral $\mathscr{L}e^+$ for $q \to +\infty$, $(\mathscr{L}e^+)_2$ the contribution of point 2, etc. Using (5.7.12, 54, 55, 57) we obtain

$$\begin{aligned} (\mathscr{Y}e^+)_1 &= \sqrt{f_1}Z_1^- , \qquad (\mathscr{Y}e^-)_1 = \sqrt{f_2}Z_2^- , \\ (\mathscr{Y}e^+)_2 &= \sqrt{f_1}Z_1^+ , \qquad (\mathscr{Y}e^-)_2 = \sqrt{f_2}Z_2^+ . \end{aligned} \tag{5.7.59}$$

Now the asymptotes of the $\tilde{\chi}_n$ solutions for $q \to +\infty$ can be written. The integral of (5.7.15) is represented as the sum of the contributions from saddles 1 and 2:

$$\tilde{\chi}_n = \left(\frac{f_1 f_2}{f_n}\right)^{1/2} [(\mathscr{Y}g_n)_1 + (\mathscr{Y}g_n)_2] .$$

Then asymptotes of (5.7.49) for $p \to -\infty$ and $p \to +\infty$ are substituted into the first and second terms, respectively. Allowing for (5.7.59) we obtain

$$\tilde{\chi}_1 = \sqrt{f_1 f_2}(aZ_1^- + cZ_1^+) , \qquad \tilde{\chi}_2 = \sqrt{f_1 f_2}(bZ_2^- + dZ_2^+) .$$

Comparing these expressions with (5.7.11) and considering that the difference between χ_n and $\tilde{\chi}_n$ disappears in the limit of $q \to +\infty$, we see that the matrix of (5.7.51) is precisely the transition matrix defined in (5.7.14).

The Transition Probability is the off-diagonal T-matrix element squared. Owing to the smallness of the parameter (5.7.36) we have $\beta \gg 1$ and $\lambda \gg 1$. Hence, the transition probability can be written as

$$P = 4\sin^2 \pi\delta \ \lambda^2 .$$

Here, $\pi\delta$ plays the role of a scattering phase. When δ is an integer the transition probability vanishes, which is a consequence of the interference of the waves (5.7.12) in two electronic states. The limiting expressions are

$$P = 2\pi\delta e^{-2\beta} \quad \text{for } \delta \ll 1 , \tag{5.7.60}$$

$$P = 4e^{-2\beta}\sin^2 \pi\delta \quad \text{for } \delta \gg 1 . \tag{5.7.61}$$

In the latter equation the rapidly oscillating function $\sin^2 \pi\delta$ can be replaced with 1/2 (implying, e.g., an averaging of the transition probability over a narrow energy interval). Then

$$P = 2e^{-2\beta} . \tag{5.7.62}$$

The dependence of the transition probability on the Born-Oppenheimer parameter κ is provided only by the "nuclear velocity" $\sqrt{2|E|}$ which is proportional to κ^2. Therefore, from (5.7.27, 35) $\delta \sim \kappa^{-2}$ and $\beta \sim \kappa^{-6}$. Hence it follows that the transition probability cannot be obtained by any expansion in powers of κ, in particular with the use of perturbation theory in the non-adiabaticity operator. In this problem the correct small expansion parameter may only be either δ or, which is the same, the interaction matrix element v defining the minimum separation between the adiabatic terms within the region of their closest approach. Notice that if the terms do not approach each other (i.e., $\delta \gg 1$), the transition probability is nearly independent of v.

5.7.2 Perturbation Theory for Nonadiabatic Transitions

The limiting expression (5.7.60) for the transition probability is a result of perturbation theory in the static coupling of diabatic terms (v) for deep tunneling transitions. To prove this statement we shall calculate the transition amplitude

$$\mathscr{S}_{12} = \frac{v}{\hbar} \int_{-\infty}^{\infty} \tilde{\chi}_1(q)\tilde{\chi}_2(q)\,dq \tag{5.7.63}$$

which is proportional to the overlap integral of the zeroth-order (in v) diabatic wave functions. If the functions are normalized to unit flows, the nonadiabatic transition probability will be $P = |\mathscr{S}_{12}|^2$. The overlap in (5.7.63) for the linear terms can be found by a straightforward integration using exact solutions at $v = 0$ [1.40]. We will do this in a different way taking into account the quasiclassical nature of the motion in the transition region, aiming at a generalization of the result already obtained for nonlinear terms.

Like in Sect. 5.2, the integration path L is displaced from the real axis, as is shown in Fig. 5.17. The product $\tilde{\chi}_1\tilde{\chi}_2$ oscillates at both ends of L so that the integral converges. We decompose $\tilde{\chi}_1$ into a sum,

$$\tilde{\chi}_1 = \tilde{\chi}_1^+ + \tilde{\chi}_1^- , \tag{5.7.64}$$

where $\tilde{\chi}_1^\pm$ are exact wave functions decreasing in sectors A_1B and C_1B, respectively. Then $\mathscr{S}_{12}$ is decomposed into the sum of two convergent integrals, $\mathscr{S}_{12} = \mathscr{S}_{12}^+ + \mathscr{S}_{12}^-$. In sector $C_1\tilde{q}_1B$ the function $\tilde{\chi}_1^-$ decreases more rapidly than $\tilde{\chi}_2$ increases, so $\mathscr{S}_{12}^-$ can be supplemented with an integral over a large-

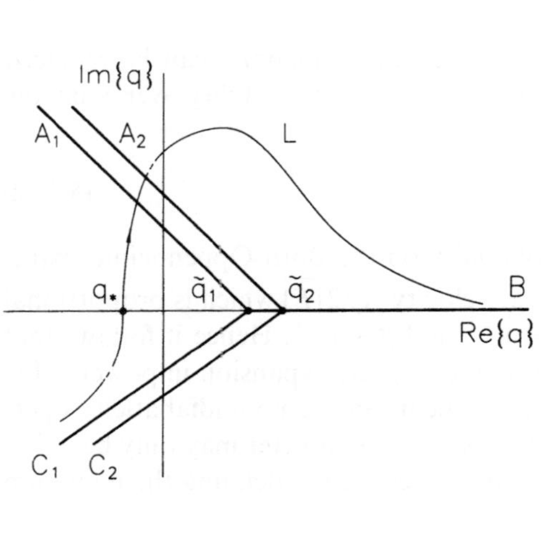

Fig. 5.17. Anti-Stokes lines (*full lines*) and the integration path in (5.7.65)

radius arc BC_1, after which it becomes evident that the integral equals zero because the exact solutions are analytical all over the complex plane. In the $\mathcal{S}^{+}_{12}$ integral the contour is shifted into the upper half-plane to move away from the turning points $\tilde{q}_1$ and $\tilde{q}_2$ (see Fig. 5.17). After L is shifted a long-enough distance, the quasiclassical representations of the form $\tilde{\chi}_2 = \tilde{e}_2^-$ and $\tilde{\chi}_1^+ = \tilde{e}_1^+$ become valid, where $\tilde{e}^{\pm}_{1,2}$ have the form of (5.2.3) to within a normalizing factor. Thus

$$\mathcal{S}_{12} = \frac{v}{\hbar} \int_L \tilde{e}_2^- \tilde{e}_1^+ \, dq \,. \tag{5.7.65}$$

Subsequent calculations involve the standard use of the saddle-point method along the lines of Sect. 5.2. The saddle is located at the term-crossing point q_* and is passed through by the steepest descent in the South-to-North direction. The result is (5.7.60) in which β and δ can be recast as

$$\beta = \frac{i}{\hbar}\left(\int_{\tilde{q}_1}^{q_*} \tilde{\mathcal{P}}_1 \, dq - \int_{\tilde{q}_2}^{q_*} \tilde{\mathcal{P}}_2 \, dq \right),$$
$$\delta = v^2 \left(\hbar \sqrt{2|E|} \left| \frac{d\tilde{\mathcal{E}}_1(q)}{dq} - \frac{d\tilde{\mathcal{E}}_2(q)}{dq} \right|_{q=q_*} \right)^{-1}, \tag{5.7.66}$$

where $\tilde{\mathcal{P}}_n$ are momenta (5.7.13) in the diabatic terms. These expressions are written in the form allowing an obvious generalization to nonlinear potentials.

For $\delta \gg 1$ the system behaves nearly adiabatically (see Sect. 5.7.1), and one can try to obtain (5.7.61) by perturbation theory in the nonadiabatic coupling [5.135]. To this end we shall first calculate, in the quasiclassical limit, the matrix element of the non-adiabaticity operator (5.7.9),

$$\mathcal{S}_{12} = \frac{1}{\hbar} \int_{-\infty}^{\infty} \chi_2(q) L_{12} \chi_1(q) \mathrm{d}q = \hbar \int_{-\infty}^{\infty} \chi_2(q) \alpha' \frac{d}{dq} \chi_1(q) \, dq$$

(the contribution of the second term of (5.7.9) will be considered later), where the zeroth-order adiabatic functions $\chi_{1,2}(q)$ are assumed to be normalized to unit flows. Invoking a decomposition $\chi_1 = \chi_1^+ + \chi_1^-$ similar to (5.7.64) we find

$$\mathcal{S}_{12} = 2\hbar \, \mathrm{Re} \left\{ \int_{-\infty}^{\infty} \chi_2 \alpha' \frac{d}{dq} \chi_1^+ \, dq \right\}. \tag{5.7.67}$$

We shift the integration path into the upper half-plane and draw it far from the turning points q_1 and q_2, circumventing the cut outgoing from the upper complex point of intersection of the adiabatic terms (5.7.3) (see Fig. 5.18a). Everywhere along the path we may use the quasiclassical asymptotes, $\chi_2 = e_2^-$ and $\chi_1^+ = e_1^+$, where $e^{\pm}_{1,2}$ are properly normalized functions of the form (5.2.3).

The function e_2^- decreases in sector $C_2 q_2 A_2$ and increases in $A_2 q_2 B$, while e_1^+ decreases in $A_1 q_1 B$ and increases in $A_1 q_1 C_1$. Therefore, everywhere on the contour L, except for its parts on the cut sides, the integrand of (5.7.67) is exponentially small (i.e., the decreasing functions turn out to be stronger than

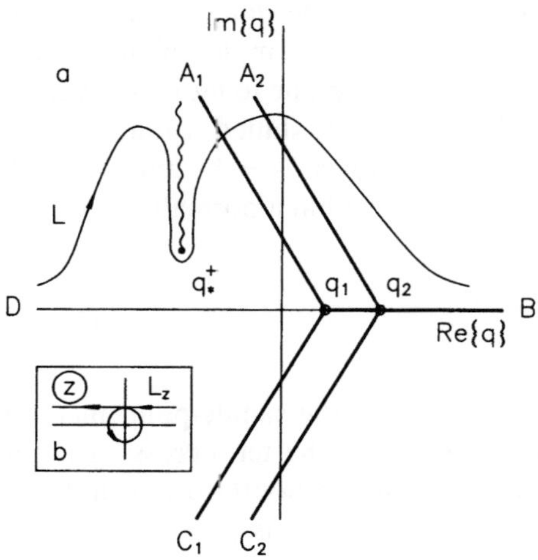

Fig. 5.18. Anti-Stokes lines (*full lines*), a cut (*wavy line*), the integration path in (5.7.68) (**a**), and the integration path in (5.7.70) (**b**)

the increasing ones). This permits to perform the integration in

$$\mathscr{S}_{12} = 2\hbar\,\mathrm{Re}\left\{\int_L e_2^- \,\alpha' \frac{d}{dq} e_1^+ \,dq\right\} \tag{5.7.68}$$

only along the cut. Proceeding in the spirit of the quasiclassical method, we differentiate in e_1^+ only the exponential factor to obtain $\mathscr{S}_{12} = 2\,\mathrm{Re}\,\{J_{12}\exp(-G)\}$, where G is given by the right-hand side of (5.7.66) in which the diabatic momenta and turning points should be replaced with the corresponding adiabatic quantities (see (5.7.72) below) and

$$J_{12} = \mathrm{i}\int_L dq \left(\frac{\mathscr{P}_2}{\mathscr{P}_1}\right)^{1/2} \alpha'(q)\,\exp\left[\frac{\mathrm{i}}{\hbar}\left(\int_{q_1^+}^{q} \mathscr{P}_1\,dq - \int_{q_*^+}^{q} \mathscr{P}_2\,dq\right)\right].$$

Expanding the integrand around q_*^+ in powers of $x = q - q_*^+$ yields, after simple algebra,

$$J_{12} = -\frac{1}{4}\int_{L_x} x^{-1}\exp(\mathrm{i}ax^{3/2})dx, \qquad a = \frac{2}{3\hbar}\exp(-\,\mathrm{i}\pi/4)\left(\frac{2fv}{|E|}\right)^{1/2}, \tag{5.7.69}$$

where the contour L_x circumvents the cut in the x-plane beginning at the point $x = 0$. It is verified by a straightforward calculation that for $\delta \gg 1$ the J_{12} integral converges inside the expansion region. Virtually, (5.7.69) does not involve any parameters and is completely evaluated with the use of the variable substitution $z = ax^{3/2}$ to yield

$$J_{12} = -\frac{1}{6}\int_{L_z} \frac{\mathrm{e}^{\mathrm{i}z}}{z}\,dz \tag{5.7.70}$$

(the contour L_z is depicted in Fig. 5.18b). Finally, $J_{12} = -\,\mathrm{i}\pi/3$.

At this juncture it is appropriate to look at the α'' term of the first-order adiabatic perturbation theory,

$$\Delta \mathscr{S}_{12} = \mathrm{Re}\left\{\hbar \int_L e_2^- \alpha'' e_1^+ \, dq\right\} .$$

Integration by parts gives

$$\Delta \mathscr{S}_{12} = -\hbar \int_L \left(e_2^- \alpha' \frac{d}{dq} e_1^+ + e_1^+ \alpha' \frac{d}{dq} e_2^- \right) dq = 0$$

since two members of the integrand exactly cancel within the x-expansion region. Thus, the final result is

$$\mathscr{S}_{12} = \frac{2\pi}{3} \mathrm{Im}\{\mathrm{e}^{-G}\} , \tag{5.7.71}$$

with

$$G = \frac{\mathrm{i}}{\hbar} \int_{q_1}^{q_*} \mathscr{P}_1 \, dq - \frac{\mathrm{i}}{\hbar} \int_{q_2}^{q_*} \mathscr{P}_2 \, dq \equiv \beta' - \frac{\mathrm{i}\gamma'}{2} . \tag{5.7.72}$$

For the nonadiabatic transition probability one obtains the following expression

$$P = \left(\frac{2\pi}{3}\right)^2 \mathrm{e}^{-2\beta'} \sin^2 \frac{\gamma'}{2} \tag{5.7.73}$$

which is to be compared with the corresponding exact expression (5.7.61) in the limit of $\delta \gg 1$. To this end we recast G via a contour integral in the q-plane with cuts outgoing from the complex adiabatic-term intersection points $q_*^{\pm}$ (see Fig. 5.19),

$$G = \frac{\mathrm{i}}{\hbar} \int_{C_1} \mathscr{P}(q) \, dq . \tag{5.7.74}$$

Each time when the q-point crosses a cut, the classical momentum of one adiabatic term is converted into that of the other. In the momentum expression

$$\mathscr{P}(q) = \left\{ 2\left(E - \frac{\tilde{\mathscr{E}}_1 + \tilde{\mathscr{E}}_2}{2} - \left[\left(\frac{\tilde{\mathscr{E}}_1 - \tilde{\mathscr{E}}_2}{2} \right)^2 + v^2 \right]^{1/2} \right) \right\}^{1/2}$$

the sign of the outer square root is specified by circumventing the q-point along a small hemi-circle in the lower half-plane. Introducing a new variable, $p = \mathscr{P}(q)$, and integrating (5.7.74) by parts we obtain

$$G = -\frac{\mathrm{i}}{\hbar} \int_{C_1'} q(p) \, dp$$

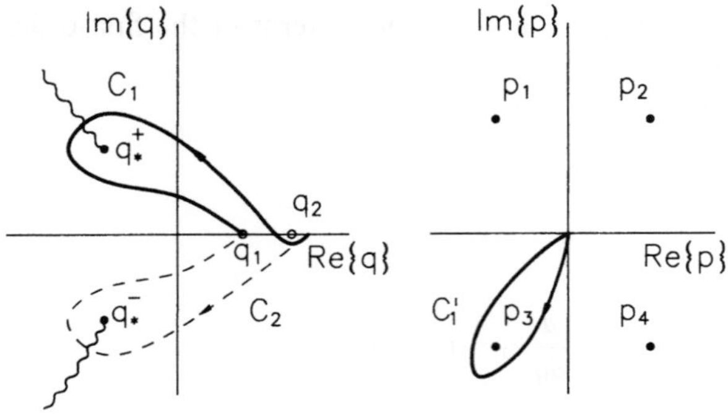

Fig. 5.19. Integration paths in the q- and p-planes

(the off-integral term vanishes since it is evaluated at the turning points where $\mathscr{P}(q) = 0$). Here

$$q(p) = \frac{1}{f_1 f_2}\left[F\left(\frac{p^2}{2} - E\right) + f\sqrt{W}\right].$$

Let us draw a contour C_1' in the p-plane corresponding to C_1 (see Fig. 5.19). Taking into account the single-valuedness of the first term in $q(p)$ we find

$$G = \frac{\mathrm{i}}{\hbar}\int_{C_1'} \sqrt{W}\,dp\,.$$

From definitions (5.7.22, 34, 72) it is seen that $\beta' = \beta$ and $\gamma' = \gamma$; as for the phase γ, it has been calculated before in the limit of $v/|E| \ll 1$ to be $\gamma = 2\pi\delta$. Thus the transition probability is

$$P = \left(\frac{2\pi}{3}\right)^2 \mathrm{e}^{-2\beta} \sin^2 \pi\delta\,. \tag{5.7.75}$$

The difference between this and the corresponding exact formula (5.7.61) for $\delta \gg 1$ signals that the adiabatic perturbation theory is only applicable to within the exponential accuracy. A lack of understanding of this fundamental fact, resulting from the singularity of the perturbation in the transition region, gave rise to a number of works whose authors tried to find explanations to some physical phenomena in preexponential factors of the transition probability by means of treating the non-adiabaticity operator as a small perturbation (e.g., identification of the promoting modes by the position dependent internal deuterium isotope effect [4.120] and the propensity rule of lowering the electronic energy gap by one promoting mode quantum [2.223, 4.163, 5.136]). Moreover, in the framework of such an approach, the diabatic (rather than adiabatic) functions were used, in whose space the L and K operators are not at all defined.

5.7.3 Method of the Classical $\mathcal{S}$-Matrix

The difficulties encountered in the construction of the correct theory of the transitions in the adiabatic limit can be overcome with the help of the (dating back to Landau's ideas) Miller method of classical $\mathcal{S}$-matrix invoking the quasiclassical propagator to effectuate a time continuation of the wave functions [1.40, 2.238, 5.137–141]. The calculation routine to find it is as follows. Consider the matrix element of the evolution operator,

$$K(q_1 t_1, q_2 t_2) = \langle q_2 | \exp[-\mathrm{i}H(t_2 - t_1)/\hbar] | q_1 \rangle ,$$

which represents the probability amplitude for a transition from a state with a coordinate q_1 at a moment t_1 into a state with a coordinate q_2 at a moment t_2. Following *Feynman* and *Hibbs* [5.142] we represent $K(q_1 t_1, q_2 t_2)$ in the form

$$K(q_1 t_1, q_2 t_2) \sim \sum \exp\left\{\frac{\mathrm{i}}{\hbar} S[q(t)]\right\} ,$$

where $S[q(t)]$ is the classical action corresponding to the trajectory $q(t)$ and the summation taken over all paths is only subject to the restrictions of $q(t_1) = q_1$ and $q(t_2) = q_2$. In the classical limit of $\hbar \to 0$, only those paths contribute to this sum that minimize the functional $S[q(t)]$, i.e., the classical trajectories, which enables one to introduce a notion of the quasiclassical propagator given by the same formula in which the summation is now over the classical paths only. Yet unspecified normalization factors can be found from the strict unitarity condition. The result of the corresponding, quite sophisticated, calculations is [5.143]:

$$K(q_1 t_1, q_2 t_2) = \sum_{\nu_0} \left[\left(\frac{\mathrm{i}}{2\pi\hbar}\right)^N \frac{\partial^2 S}{\partial q_2 \partial q_1}\right]^{1/2} \exp\left\{\frac{\mathrm{i}}{\hbar} S[q(t)] - \frac{1}{2} k\pi \mathrm{i}\right\} , \quad (5.7.76)$$

where ν_0 are classical paths connecting q_1 and q_2 for the time $t = t_2 - t_1$, k is the branching number of the normalization factor during the same time interval, and N is the number of degrees of freedom. The transition from (5.7.76) to an element of the classical $\mathcal{S}$-matrix is performed by a replacement of the total action with a short action S_0 and by going to the limit of $t \to +\infty$. Specifically, in the one-dimensional case

$$\mathcal{S}_{12} = \sum_{\nu_0} \left(\frac{\mathrm{i}}{2\pi\hbar} \frac{\partial^2 S_0}{\partial q_2 \partial q_1}\right)^{1/2} \exp\left\{\frac{\mathrm{i}}{\hbar} S_0[q(t)] - \frac{1}{2} k\pi \mathrm{i}\right\} . \quad (5.7.77)$$

This equation can be used to calculate the nonadiabatic transition probability if ν_0 is allowed to embrace classical paths connecting potential surfaces of different electronic states. Becoming necessarily complex, these paths can be continuously extended across the cuts encountered by them onto other sheets of the adiabatic potential treated as a united many-valued function of nuclear coordinates. Equation (5.7.77) is used in the simplest way, provided one assumes an insignificant limitation that far enough from the transition region in the classically allowed regions of the motion the adiabatic asymptotes $\chi_{1,2}$ have the form of plane waves

$$\chi_{1,2}(q) = (2\pi\hbar)^{-1/2}\exp(\mathrm{i}p_{1,2}q) .$$

Then, considering the element of the classical $\mathscr{S}$-matrix,

$$\langle q_2|\mathscr{S}|p_1\rangle = \sum_{\nu_0}\left(\frac{\mathrm{i}}{2\pi\hbar}\frac{\partial^2 S_0}{\partial q_2\partial p_1}\right)^{1/2}\exp\left[\frac{\mathrm{i}}{\hbar}S_0(q_2,p_1) - \frac{1}{2}k\pi\mathrm{i}\right],$$

and taking into account that equations

$$\frac{\partial^2 S_0}{\partial q_2\partial p_1} = \left(\frac{p_1}{p_2}\right)^{1/2}$$

and

$$S_0(q_2,p_1) = \int\limits_{q_1}^{q_2}\mathscr{P}(q,E)dq + p_1q_1$$

are applicable due to the quasiclassical nature of the motion, we find the following expression for the nonadiabatic transition probability:

$$P = \left|\sum_{\nu_0}\exp\left[\frac{\mathrm{i}}{\hbar}\int\limits_{q_1}^{q_2}\mathscr{P}(q,E)\,dq - \frac{1}{2}k\pi\mathrm{i}\right]\right|^2 . \tag{5.7.78}$$

Without loss of generality q_1 and q_2 can be moved to the turning points. It remains to count the number of zeros of $\partial p_2/\partial p_1$ along each of the contours connecting q_1 with q_2. This generally very complicated problem is resolved unambiguously in the case of two zeros of the adiabatic splitting. The terms are connected by two independent paths like C_1 and C_2 in Fig. 5.19 with the numerical phase difference of π. Then

$$\frac{\mathrm{i}}{\hbar}\int\limits_{C_{1,2}}\mathscr{P}(q,E)dq = \beta \mp \frac{\mathrm{i}\gamma}{2}, \tag{5.7.79}$$

and the final result is

$$P = 4\mathrm{e}^{-2\beta}\sin^2\frac{\gamma}{2} . \tag{5.7.80}$$

For the linear model it is identical with the exact expression (5.7.61), and it is easily extended to potentials not too much "curved". Equation (5.7.80) is only valid in the asymptotic limit of $\gamma \gg 1$. Therefore, actually, the adiabatic limit is described by the simple expression $P = 2\exp(-2\beta)$.

In conclusion, we would like to place emphasis on the decomposition of the basic contour integral into the real and complex parts in (5.7.80). As is seen, it can serve as a starting point in calculations of the preexponential factors of the nonradiative transition rates.

5.8 Two-State Vibronic Levels

The basic information about the molecular levels with the involvement of nonadiabatic coupling can be obtained with the use of the two-state model represented by the Hamiltonian of (5.7.1) with

$$\mathrm{h} = \begin{pmatrix} \tilde{\mathscr{E}}_1 & v \\ v & \tilde{\mathscr{E}}_2 \end{pmatrix}, \tag{5.8.1}$$

where $\tilde{\mathscr{E}}_1$ and $\tilde{\mathscr{E}}_2$ are diabatic terms with a single real intersection point q_* (see Fig. 5.20). Around q_* one has

$$\mathrm{h} = \tilde{\mathscr{E}}(q_*) + \begin{pmatrix} -f_1(q - q_*) & v(q_*) \\ v(q_*) & -f_2(q - q_*) \end{pmatrix}, \tag{5.8.2}$$

and the validity region for this expansion will be supposed to be large enough to incorporate all significant nonadiabatic effects. Under the additional assumption that the motion along the adiabatic terms $\mathscr{E}_{1,2}$ is quasiclassical, the problem specified by the Hamiltonian (5.8.1) can be treated by the method of matching of quantum domains [4.4]. To this end we define, for (5.8.1) for $q > q_*$ in close vicinity to the left turning points, linearly independent solutions $Z_n^{\pm}$ and

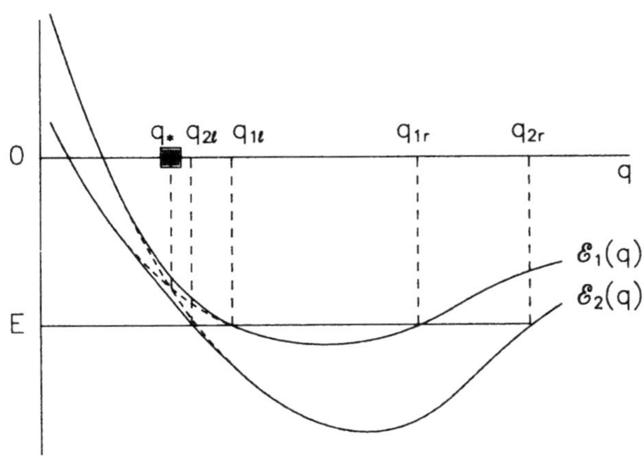

Fig. 5.20. Molecular terms with an avoided crossing. Dashed rectangle marks an interval around the diabatic terms intersection q_*, in which the terms may be considered as linear ones; q_{nl} and q_{nr} $(n = 1,2)$ are the turning points for $E < \tilde{\mathscr{E}}(q_*)$

arbitrary constants $A_n^{\pm}$ by analogy with (5.7.11–13). We introduce the amplitudes $B_{nl}^{\pm} = A_n^{\pm} \exp(\mp \mathrm{i}\pi/4)$ for which relation (5.7.14) can be recast as

$$\begin{pmatrix} B_{1l}^{+} \\ B_{2l}^{+} \end{pmatrix} = -\mathrm{i}T \begin{pmatrix} B_{1l}^{-} \\ B_{2l}^{-} \end{pmatrix} . \tag{5.8.3}$$

The quasiclassical continuation of the solutions into the q-region near the right turning points implies the transformation

$$\begin{pmatrix} B_{1r}^{\pm} \\ B_{2r}^{\pm} \end{pmatrix} = \begin{pmatrix} \exp(\pm \mathrm{i}S_1^{\mathrm{a},\alpha}/\hbar) & 0 \\ 0 & \exp(\pm \mathrm{i}S_2^{\mathrm{a},\alpha}/\hbar) \end{pmatrix} \begin{pmatrix} B_{1l}^{\pm} \\ B_{2l}^{\pm} \end{pmatrix}, \tag{5.8.4}$$

$$\begin{aligned} S_n^{\mathrm{a}} &= \int_{q_{nl}}^{q_{nr}} \mathscr{P}_n(q)dq\,, \quad E < \tilde{\mathscr{E}}(q_*)\,, \\ S_n^{\alpha} &= \int_{q_*}^{q_{nr}} \mathscr{P}_n(q)dq\,, \quad E > \tilde{\mathscr{E}}(q_*)\,, \end{aligned} \tag{5.8.5}$$

The condition for the solutions to decrease inside the nonclassical region of $q > q_{nr}$,

$$\begin{pmatrix} B_{nr}^{+} \\ B_{nr}^{-} \end{pmatrix} = b_n \begin{pmatrix} \exp(\mathrm{i}\pi/4) \\ \exp(-\mathrm{i}\pi/4) \end{pmatrix}, \tag{5.8.6}$$

(where b_n are new constants) together with (5.8.3, 4) lead to the equation

$$\left[1 + \begin{pmatrix} \exp(\mathrm{i}S_1^{\mathrm{a},\alpha}/\hbar) & 0 \\ 0 & \exp(\mathrm{i}S_2^{\mathrm{a},\alpha}/\hbar) \end{pmatrix} T \begin{pmatrix} \exp(\mathrm{i}S_1^{\mathrm{a},\alpha}/\hbar) & 0 \\ 0 & \exp(\mathrm{i}S_2^{\mathrm{a},\alpha}/\hbar) \end{pmatrix}\right] \begin{pmatrix} b_1 \\ b_2 \end{pmatrix} = 0 \tag{5.8.7}$$

which has a nontrivial solution under this quantization condition:

$$\begin{aligned} &1 + (T_{11}T_{22} - T_{12}T_{21}) \exp[2\mathrm{i}(S_1^{\mathrm{a},\alpha} + S_2^{\mathrm{a},\alpha})/\hbar] \\ &\quad + T_{11}\exp(2\mathrm{i}S_1^{\mathrm{a},\alpha}/\hbar) + T_{22}\exp(2\mathrm{i}S_2^{\mathrm{a},\alpha}/\hbar) = 0\,. \end{aligned} \tag{5.8.8}$$

The transition matrix T in the tunneling energy region is given by (5.7.51) with

$$\beta = \frac{\mathrm{i}}{\hbar}\left[\int_{q_{1l}}^{q_*} \mathscr{P}_1(q)dq - \int_{q_{2l}}^{q_*} \mathscr{P}_2(q)dq\right], \tag{5.8.9}$$

while for $E > \tilde{\mathscr{E}}(q_*)$ it is of the form (see, for instance, [4.4, 5])

$$\begin{aligned} T &= \exp[\mathrm{i}(S_1^{\beta} + S_2^{\beta})/\hbar] \\ &\times \begin{pmatrix} (1-\nu^2)e^{2\mathrm{i}\chi+\mathrm{i}\Delta} + \nu^2 e^{-\mathrm{i}\Delta} & -2\mathrm{i}\nu(1-\nu^2)^{1/2}\sin(\chi+\Delta) \\ -2\mathrm{i}\nu(1-\nu^2)^{1/2}\sin(\chi+\Delta) & (1-\nu^2)e^{-2\mathrm{i}\chi-\mathrm{i}\Delta} + \nu^2 e^{\mathrm{i}\Delta} \end{pmatrix}, \end{aligned} \tag{5.8.10}$$

where

$$S_n^\beta = \int_{q_*}^{q_{nl}} \mathscr{P}_n(q)dq\,, \quad \Delta = \frac{1}{\hbar}(S_1^\beta - S_2^\beta)\,, \tag{5.8.11}$$

$$v = \mathrm{e}^{-\pi\delta}\,, \quad \chi = \arg\{\Gamma(\mathrm{i}\delta)\} - \delta\ln\delta + \delta + \frac{\pi}{4}\,,$$

with δ being given by (5.7.27), as in the tunneling region.

Owing to the exact relations $\det\{T\} = 1$ and $T^+ = T^{-1}$ ensuring the conservation of flow and the reversibility in time, as well as to the obvious equality $S_n^\alpha + S_n^\beta = S_n^\mathrm{a}$, (5.8.8) can be transformed to

$$1 + \exp[2\mathrm{i}(S_1^\mathrm{a} + S_2^\mathrm{a})/\hbar] + (1-P)^{1/2}[\exp(2\mathrm{i}S_1^\mathrm{a}/\hbar + \mathrm{i}\psi) + \exp(2\mathrm{i}S_2^\mathrm{a}/\hbar - \mathrm{i}\psi)] = 0\,, \tag{5.8.12}$$

where $P = |T_{12}|^2$ is the nonadiabatic transition probability and

$$\psi = \begin{cases} \arctan\left[\dfrac{\sin 2\pi\delta}{\lambda^2 + \cos 2\pi\delta}\right], & E < \tilde{\mathscr{E}}(q_*)\,, \\ \arctan\dfrac{(1-v^2)\sin 2\chi + v^2\sin 2\Delta}{(1-v^2)\cos 2\chi + v^2\cos 2\Delta}\,, & E > \tilde{\mathscr{E}}(q_*)\,. \end{cases} \tag{5.8.13}$$

The quantization condition (5.8.12) can finally be written in a real-valued form as

$$\cos[(S_1^\mathrm{a} + S_2^\mathrm{a})/\hbar] + (1-P)^{1/2}\cos[(S_1^\mathrm{a} - S_2^\mathrm{a})/\hbar + \psi] = 0\,. \tag{5.8.14}$$

It is appropriate to carry out separate investigations of this equation for the tunneling and overbarrier energy ranges. In the former case with the help of (5.8.13) it is transformed to [5.124]

$$\cos(S_1^\mathrm{a}/\hbar)\cos(S_2^\mathrm{a}/\hbar) + \frac{1}{\lambda^2}\cos(S_1^\mathrm{a}/\hbar - \pi\delta)\cos(S_2^\mathrm{a}/\hbar + \pi\delta) = 0 \tag{5.8.15}$$

allowing the use of the perturbation method. The basic inequality $\lambda^2 \gg 1$ implies that the above equation in the zeroth-order approximation describes the quasiclassical quantization in adiabatic terms $\mathscr{E}_1(q)$ and $\mathscr{E}_2(q)$:

$$S_1^\mathrm{a}(E_n^{(0)}) = \frac{2n+1}{2}\pi\hbar\,, \qquad S_n^\mathrm{a}(E_m^{(0)}) = \frac{2m+1}{2}\pi\hbar\,. \tag{5.8.16}$$

With the lack of accidental resonances between the $E_n^{(0)}$ and $E_m^{(0)}$ levels, the first-order corrections to the levels have the following maximum (estimated) value:

$$|E_{n,m}^{(1)}| \sim \hbar\omega_{n,m}\mathrm{e}^{-2\beta}\,, \tag{5.8.17}$$

where $\omega_{n,m}$ are classical frequencies at the energies $E^{(0)}_{n,m}$. If, however, a resonance occurs, then the level splitting $\delta E_{nm} = |E_n - E_m|$ is given by

$$\delta E_{nm} = \frac{\hbar}{\pi}(\omega_n \omega_m)^{1/2} p^{1/2} . \tag{5.8.18}$$

For the overbarrier motion at $E > \tilde{\mathscr{E}}(q_*)$ the basic quantization condition has the form [5.123]

$$(1 - v^2)\cos(S_1^{\mathrm{a}}/\hbar + \chi)\cos(S_2^{\mathrm{a}}/\hbar - \chi) + v^2 \cos(S_1^{\mathrm{d}}/\hbar)\cos(S_2^{\mathrm{d}}/\hbar) = 0 , \tag{5.8.19}$$

where not only the "adiabatic", but also the "diabatic" actions,

$$S_1^{\mathrm{d}} = S_1^{\beta} + S_2^{\alpha} , \qquad S_2^{\mathrm{d}} = S_2^{\beta} + S_1^{\alpha} , \tag{5.8.20}$$

are involved that are formed from the adiabatic ones in a stepwise manner at the diabatic-term intersection point q_*.

Within the zeroth-order approximation in v^2 (for $\delta \to \infty$, $\chi \to 0$) the levels found from (5.8.19) are identical with those for the uncoupled adiabatic terms, and the basic corrections to them are on the order of

$$|E^{(1)}_{n,m}| \sim \hbar\omega_{n,m}\chi , \qquad (\chi \sim \delta^{-1}) . \tag{5.8.21}$$

For $v \to 1$ ($\delta \to 0$, $\chi \to \pi/4$) these are the diabatic actions (5.8.20) that are quantized, and the corresponding unperturbed levels $\tilde{E}^{(0)}_{n,m}$ are shifted by

$$|\tilde{E}^{(1)}_{n,m}| \sim \hbar\omega_{n,m}\delta . \tag{5.8.22}$$

A remarkable feature is that (5.8.18) for the accidental-resonance splitting holds true in these cases as well.

The limitations imposed by the matching method upon the quantization conditions (5.8.15, 19) can be analyzed in the framework of the model of displaced parabolic terms,

$$\tilde{\mathscr{E}}_{1,2} = \frac{m\omega^2 q^2}{2} + F_{1,2}q , \tag{5.8.23}$$

with a constant coupling v [5.144]. By means of the transformation

$$\tilde{\chi}(q) = \exp(-m\omega q^2/2\hbar)\,\zeta(q) \tag{5.8.24}$$

the corresponding set of equations is reduced to a problem of the Landau–Zener kind with the adiabatic term splitting $W(p)$ in the p-representation of the following form

$$W(p) = \frac{1}{s}\left[\left(\frac{p^2}{2} - E\right)^2 /s^2 + V^2\right]^{1/2} , \tag{5.8.25}$$

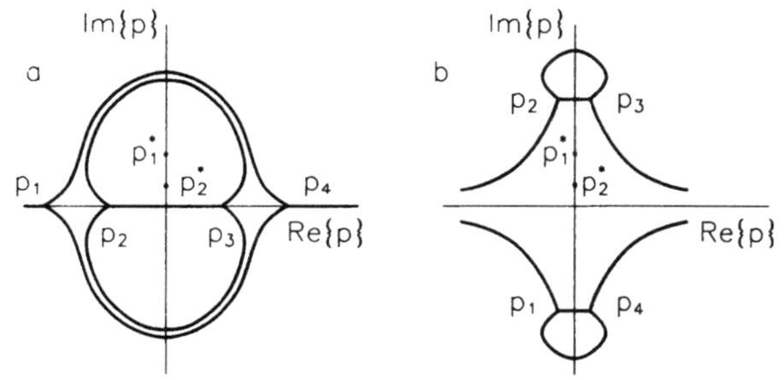

Fig. 5.21. Anti-Stokes lines for parabolic terms for $E > \tilde{\mathscr{E}}(q_*)$ (**a**) and $E < \tilde{\mathscr{E}}(q_*)$ (**b**)

$$s^2 = \left(1 + \frac{\mathrm{i}p\omega}{F_1}\right)\left(1 + \frac{\mathrm{i}p\omega}{F_2}\right). \tag{5.8.26}$$

The $W(p)$ has four zeros that are connected in pairs by anti-Stokes lines as shown in Fig. 5.21 for two energy ranges of $E \gtrless \tilde{\mathscr{E}}(q_*)$. Along these lines the quasiclassical approximation is applicable at a long-enough distance from both the $W(p)$ zeros and the $W(p)$ singularities p_i^* at $s = 0$. In the coordinate representation these singularities correspond to domains near the minima of the terms (5.8.23). Therefore, if the levels under study are high enough to be quasiclassical, the solution to the model can be obtained by matching of the quantum regions with the pairs of the $W(p)$ zeros along the anti-Stokes lines of Fig. 5.21, in full analogy with the linear term model. The ensuing solution has to be single-valued on the contour surrounding the s^2 zeros, which results in the quantization conditions of (5.8.15, 19) with the parameters δ, β and S^{d} defined by the integration across the cut of $\mathscr{P}(q)$,

$$\begin{aligned} \beta - \mathrm{i}\pi\delta &= \frac{\mathrm{i}}{\hbar}\int_{C_1} \mathscr{P}(q)dq\,, \quad E < \tilde{\mathscr{E}}(q_*)\,, \\ S^{\mathrm{d}} - \mathrm{i}\pi\delta &= \frac{1}{\hbar}\int_{C_2} \mathscr{P}(q)dq\,, \quad E > \tilde{\mathscr{E}}(q_*)\,. \end{aligned} \tag{5.8.27}$$

The C_1 contour connects the points q_{nl} with each other, while C_2 connects q_{nl} and q_{mr} for $n \neq m$.

Nonperturbative treatments of the nonadiabatic coupling in two- and three-state models with applications to optical and Raman spectra are developed in [5.145].

6. The Statistical Limit

The statistical limit deserves special consideration since it is typical of the nonradiative transitions in polyatomic molecules with large energy gaps. This chapter describes the methods developed to calculate the transition rate and its dependence upon the initial vibrational energy.

6.1 Accepting Modes, Effective States and the Transition Rate

The electronic relaxation processes in small and large moleculars are of very different nature. In small molecules they are due to molecule–medium interactions, and in a nonradiative transition the electronic energy is transferred not only to intramolecular vibrations but also to the medium's degrees of freedom, in particular to the relative motion of colliding particles or to the vibrational motion of the medium [1.15, 26, 6.1–11].

In large molecules the mechanism of the nonradiative transitions is quite different. Owing to the large number of vibrations it is always possible to pick out a relatively small set of accepting modes active in the nonradiative transition, i.e., the modes taking most of the electronic energy (see Sect. 2.10). The rest of the modes are passive and form, with respect to the former set, "an inert medium"; an interaction of the accepting modes with this medium broadens the acceptors' descrete levels to implement the energy conservation law in the radiationless transition via energy exchange with this intramolecular medium. The existence of a large set of passive vibrations significantly relaxes the influence of the external medium because the role of the latter is reduced to a small additional acceptor level broadening, and, in the statistical limit, this influence disappears at all as the nonradiative transition rate becomes independent of the level broadening of the accepting modes and is completely governed by the internal characteristics of these modes.

The statistical limit was outlined in detail in Sect. 2.11. Its most important feature is the existence of the effective states by which we mean such states (or energy levels) of the accepting modes that exhibit the following properties: i) the intramolecular coupling matrix element, v_{eff}, or the Franck–Condon factor between a given initial state and the final states is largest for the effective states and is nearly constant among states of a given energy equal to the initial energy; ii) the density of states, ϱ_{eff}, is large so that the mean state separation on the energy scale is much lower than the level width, γ, due to the interaction with the intramolecular medium (2.11.2). A similar definition of the effective states was also adopted in the theory of Intramolecular Vibrational Redistribution (IVR) [6.12]. It should be noted that the statistical limit does not occur in models where no effective states exist [6.13–15, 5.15].

With the effective states postulated to exist in the statistical limit, we can neglect the presence of other states with relatively small Franck–Condon factors when calculating the nonradiative transition rate by the general formula (3.4.9). Then, as was already discussed in Sect. 2.7, we arrive at the Bixon-Jortner problem of Sect. 2.6 and immediately obtain the transition rate constant in the form of (2.7.21) which only involves the parameters v_{eff} and ϱ_{eff} pertaining to the effective states of the accepting modes without any dependence on γ.

Such an approach to the problem of the statistical limit, only based on intuitive physical ideas on the nature of the intramolecular radiationless transitions, remains, however, barren unless one specifies a method to find the accepting modes and effective states as well as a method to calculate the parameters v_{eff} and ϱ_{eff} entering both the basic inequality (2.11.2) characteristic of the statistical limit and (2.7.21) for the statistical limit transition rate constant. Here, it will be shown how to solve this problem provided the dependence of the Franck–Condon factor upon the final vibrational quantum numbers is known, which places the effective-states method on a real solid foundation and makes it, in principle, suitable to actual calculations. It should be emphasized that this method provides the same statistical limit expression for the nonradiative transition rate as other methods, e.g., the method used in the classical papers by *Huang* and *Rhys* [4.163] and *Krivoglaz* [6.16] (see Sect. 6.5), or the commonly accepted saddle-point method described in Sect. 6.3. But, it is favorable disparate of them in that it allows to calculate an important parameter, ϱ_{eff}, enabling to decide whether the given electronic transition in the given molecule may be assigned to the statistical limit (2.11.2) or to the intermediate case (2.11.4).

Consider the simplest model of noninteracting harmonic oscillators with displaced but undistorted potential surfaces (4.6.2, 4, 6), i.e., without frequency-change and Duchinsky effects. We further suppose the transition to begin at the $m = 0$ ground vibrational level of the initial electronic state r. Then, the Franck–Condon factor of (4.5.2) and (4.8.11) may be written in the form

$$F_{r0r'm'} = \prod_{\mu} F_{r0r'm'_\mu}, \qquad F_{r0r'm'_\mu} = \frac{1}{m'_\mu!}\,\xi_\mu^{m'_\mu}\,\mathrm{e}^{-\xi_\mu}, \tag{6.1.1}$$

where m' denotes a set of vibrational quantum numbers $\{m'_\mu\}$.

In the Condon approximation the rate constant for the radiationless transition from the initial state $r0$ to all the final states $r'm'$ is proportional to the so-called Franck–Condon weighted density of states [1.1] (or the density-of-states weighted Franck–Condon factor [2.313, 5.14]),

$$k_{r0} = \frac{2\pi}{\hbar} v_{el}^2 \sum_{m'} F_{r0r'm'}\, \delta_\gamma(E_{r0} - E_{r'm'}) , \tag{6.1.2}$$

where v_{el} is the electronic matrix element of the intramolecular coupling operator inducing the transition. The subscript γ of the delta function indicates that the discrete vibrational levels have a width so that the energy is conserved only approximately to within an uncertainty on the order of γ. According to (3.4.9) the $\delta_\gamma(E)$ is the Lorentzian function whose width is a function of E, rapidly vanishing with increasing $|E|$ (see end of Sect. 3.4). The approximation to be used here is that an average width γ is introduced and the summation in (6.1.2) is subject to the constraint $|E_{r0} - E_{r'm'}| < \gamma$. This is justified because, in the statistical limit, the radiationless transition rate is independent of γ, hence of the specific shape of $\delta_\gamma(E)$.

Now we proceed to the determination of the accepting modes and effective states. Let us find a set of the final vibrational quantum numbers,

$$m'_0 \equiv \{m'_{0\mu}\} \tag{6.1.3}$$

for which the Franck–Condon factor reaches a maximum under the additional constraint that the energy of the state (6.1.3) is equal to the initial energy E_{r0}. To this end we have to solve the problem of finding the conditional extremum,

$$\ln F_{r0r'm'} = \max , \tag{6.1.4}$$

under the additional constraint of energy conservation which can be written, using (4.6.6, 8, 10), as

$$\sum_\mu \hbar\omega_\mu m'_\mu = \Delta E , \tag{6.1.5}$$

where ΔE is the electronic energy gap (4.5.18). This approach was used in [1.1, 2.314] to analyze the ΔE dependence of the radiationless transition rate as well as to evaluate the contributions of individual modes in order to define good acceptors (the so-called "maximum term method", or "Boltzmann statistics method"; the latter name stems from a formal analogy between the conditional-extremum problem under consideration and the problem of the determination of the equilibrium statistical distribution in a many-particle system). However, in those studies only the maximum Franck–Condon factor was calculated while the problem of finding the effective states and $\varrho_{\rm eff}$ was not considered.

Under the adiabatic approximation we use here the electronic energy has to be much larger than the energy of any vibrational quantum, $\Delta E \gg \hbar\omega_\mu$. Considering the energy conservation law (6.1.5) we see that, at least for some modes, the inequality

$$m'_\mu \gg 1 \tag{6.1.6}$$

has to be fulfilled. By the definition of Sect. 2.10 only those modes are the good acceptors. To reveal them one has to solve the conditional-extremum problem of (6.1.4, 5) and then to select the modes whose quantum numbers (6.1.3) satisfy inequality (6.1.6). We form the Lagrange function

$$L_0 = \ln F_{r0r'm'} + \lambda_0 \sum_\mu \omega_\mu m'_\mu \, ,$$

where λ_0 is the Lagrangian multiplier (as in (6.1.3), the subscript "0" indicates the $m = 0$ set of quantum numbers). We shall assume that from the very beginning only good acceptors μ satisfying inequality (6.1.6) are included in the expression for the transition rate, and transform (6.1.1) with the use of Stirling's formula to

$$F_{r0r'm'} = \prod_\mu (2\pi m'_\mu)^{-1/2} \exp\left(m'_\mu \ln \frac{\xi_\mu}{m'_\mu} + m'_\mu - \xi_\mu \right) . \tag{6.1.7}$$

Formally treating m'_μ as continuous variables, we find the first derivatives of L_0 over m'_μ and equate them to zero. The resulting equation defines the desirable quantum numbers (6.1.3). A trivial calculation using (6.1.6) yields

$$m'_{0\mu} = \xi_\mu \exp(\omega_\mu \lambda_0) \, . \tag{6.1.8}$$

Substituting (6.1.8) into (6.1.5) we obtain the equation to find the Lagrangian multiplier λ_0,

$$\sum_\mu \hbar \, \omega_\mu \xi_\mu \exp(\omega_\mu \lambda_0) = \Delta E \, . \tag{6.1.9}$$

The set of quantum numbers (6.1.8) specifies a final state with the largest Franck–Condon factor

$$F_{r0r'm'_0} = \left[\prod_\mu (2\pi m'_{0\mu})^{-1/2} \right] \exp\left[-\frac{1}{\hbar} \lambda_0 \Delta E + \sum_\mu (m'_{0\mu} - \xi_\mu) \right] , \tag{6.1.10}$$

under the additional restriction of (6.1.5) that embodies the energy conservation law for the radiationless transition.

Equations (6.1.6, 8) establish a criterion for the μ-th mode to be a good acceptor [2.237, 6.14]

$$\xi_\mu \exp(\omega_\mu \lambda_0) \gg 1 \, . \tag{6.1.11}$$

(*Freed* et al. [1.1, 2.56] suggested some different criterion.) It follows that the accepting properties of the μ-th mode are the better, the larger the displacement of the equilibrium position and the higher the frequency. Further, we notice from (6.1.9) that λ_0 monotonically increases with ΔE, so the accepting properties of all modes increase with increasing ΔE.

Now we investigate how the Franck–Condon factor varies at small deviations of m'_μ from $m'_{0\mu}$ that do not violate the energy conservation law (6.1.5).

Expanding the logarithm of (6.1.7) in powers of $m'_\mu - m'_{0\mu}$ to the second order yields [2.202]

$$F_{r0r'm'} = F_{r0r'm'_0} \exp\left(-\sum_\mu \frac{(m'_\mu - m'_{0\mu})^2}{2m'_{0\mu}}\right), \tag{6.1.12}$$

i.e., the Franck–Condon factor decreases by the Gaussian law with a large dispersion $(2m'_{0\mu})^{1/2}$. Thus we see that there exists a large number of the final states $r'm'$ isoenergetic with the initial state $r0$ and having the Franck–Condon factor values close to (6.1.10). These are just the effective states. Their quantum numbers m'_μ satisfy the inequalities

$$|m'_\mu - m'_{0\mu}| \lesssim (2m'_{0\mu})^{1/2}, \tag{6.1.13}$$

and the transition matrix elements for these states are all close to the maximum value

$$v_{\text{eff}} = v_{\text{el}}(F_{r0r'm'_0})^{1/2}. \tag{6.1.14}$$

Now we calculate the total rate constant for the transition from the initial state $r0$ to all the effective final states. Substituting (6.1.12, 14) into (6.1.2) we obtain

$$k_{r0} = \frac{2\pi}{\hbar} v_{\text{eff}}^2 \sum_{m'} \exp\left(-\sum_\mu \frac{(m'_\mu - m'_{0\mu})^2}{2m'_{0\mu}}\right) \delta_\gamma(E_{r0} - E_{r'm'}). \tag{6.1.15}$$

In the statistical limit the density of the effective states, ϱ_{eff}, obeys inequality (2.11.2) which implies that the average spacing between the levels $E_{r'm'}$, over which the summation in (6.1.15) is carried out, is small as compared to the width γ of the delta function. Therefore, the delta function slowly varies with m'_μ, and so does the exponential factor due to (6.1.6, 13). This allows the replacement of summation over m' with integration.

Usually the broadening is small as compared to the vibrational quantum $\gamma \ll \hbar\omega_\mu$. Combined with (6.1.6) this implies γ to be small in relation to the width (in energy units) of the exponential factor in (6.1.15), $\gamma \ll \hbar\omega_\mu(2m'_{0\mu})^{1/2}$. Therefore, in the integral, one may set $\gamma = 0$. It is easy to see that the resulting expression is the density of the effective states,

$$\varrho_{\text{eff}} = \int \exp\left(-\sum_\mu \frac{(m'_\mu - m'_{0\mu})^2}{2m'_{0\mu}}\right) \delta(E_{r0} - E_{r'm'}) \prod_\mu dm'_\mu, \tag{6.1.16}$$

and the transition rate (6.1.15) takes the form of (2.7.21).

The integral of (6.1.16) is simply calculated by using the integral representation for the delta function,

$$\delta(E) = \frac{1}{2\pi\hbar} \int_{-\infty}^{\infty} e^{-iEt/\hbar} dt. \tag{6.1.17}$$

The final expression for ϱ_{eff} is

$$\varrho_{\mathrm{eff}} = \frac{1}{\hbar}\left[\prod_{\mu}(2\pi m'_{0\mu})/2\pi\sum_{\mu}\omega_{\mu}^{2}m'_{0\mu}\right]^{1/2} . \tag{6.1.18}$$

Substituting v_{eff} and ϱ_{eff} in (2.7.21) we find the following expression for the transition rate

$$k_{r0} = \frac{v_{\mathrm{el}}^{2}}{\hbar^{2}}\left(2\pi/\sum_{\mu}\omega_{\mu}^{2}m'_{0\mu}\right)^{1/2}\exp\left[-\frac{1}{\hbar}\lambda_{0}\Delta E + \sum_{\mu}(m'_{0\mu} - \xi_{\mu})\right] . \tag{6.1.19}$$

As expected, the density of effective states rapidly increases with the number of accepting modes and their vibronic activity in the radiationless transition, i.e., with their final quantum numbers $m'_{0\mu}$. The latter statement means that ϱ_{eff} depends not only on the electronic gap ΔE but also on the frequencies of the active vibrations, ω_{μ}, and on the parameters ξ_{μ} characterizing the electronic-vibrational interaction, viz., the stronger this interaction for a given mode and the higher its frequency, the greater its activity in the radiationless transition and hence its contribution to ϱ_{eff}.

For the case of a single accepting mode μ, (6.1.18) yields $\varrho_{\mathrm{eff}} = 1/\hbar\omega_{\mu}$. If, as is mostly the case, the level broadening γ is much lower than $\hbar\omega_{\mu}$, the statistical limit criterion (2.11.2) is surely not fulfilled in the single-mode case. Nevertheless, if the molecule as a whole fulfils the Freed-Jortner irreversibility criterion (2.4.41), i.e., the total density of states is large enough, then such a molecule can still exhibit purely intramolecular nonradiative transitions but their rates are defined by the general expression (3.4.9) rather than (2.7.21); in particular, the rates will essentially depend on the broadening γ (intermediate case).

We suppose that there are N accepting modes of equal frequency ω. Resolving (6.1.9) one obtains $\lambda_{0} = \omega^{-1}\ln(\Delta E/\hbar\omega\xi)$, where

$$\xi = \sum_{\mu}\xi_{\mu} . \tag{6.1.20}$$

The quantum numbers (6.1.8) are $m'_{0\mu} = (\xi_{\mu}/\xi)(\Delta E/\hbar\omega)$; hence, the density of effective states and the transition rate from (6.1.18, 19) take the form

$$\varrho_{\mathrm{eff}} = \frac{1}{\hbar\omega}\left(\frac{1}{2\pi}\prod_{\mu}2\pi\frac{\xi_{\mu}}{\xi}\right)^{1/2}\left(\frac{\Delta E}{\hbar\omega}\right)^{(N-1)/2} ,$$

$$k_{r0} = \frac{v_{\mathrm{el}}^{2}}{\hbar}\left(\frac{2\pi}{\hbar\omega\Delta E}\right)^{1/2}\exp\left[-\left(\ln\frac{\Delta E}{\hbar\omega\xi} - 1\right)\frac{\Delta E}{\hbar\omega} - \xi\right] . \tag{6.1.21}$$

The latter equation was derived by a different method in [6.17] (see also [1.16, 2.272, 273, 314, 4.138]). Notice that at a large number of atoms (and hence a large N) the constants ξ_{μ} are on the order of $1/N$ so that both the parameter ξ and the transition rate constant (6.1.20) are independent of N in the limit of $N \to \infty$.

The statistical-limit transition-rate constant can be written in the form (2.11.5), where A and ω_* are positive constants (or, to be more specific, are slowly varying functions of ΔE). Thus, in the case under consideration, one obtains from (6.1.19)

$$\omega_*^{-1} = \lambda_0 - \frac{\hbar}{\Delta E}\sum_\mu (m'_{0\mu} - \xi_\mu) .$$

The verification that this expression is positive is easily carried out by substituting (6.1.8, 9) for $m'_{0\mu}$ and ΔE. Indeed, having rewritten this expression for ω_* in the form

$$\omega_*^{-1} = \frac{\hbar}{\Delta E}\sum_\mu \xi_\mu [x_\mu \exp(x_\mu) - \exp(x_\mu) + 1], \qquad x_\mu = \omega_\mu \lambda_0$$

one sees that the expression in square brackets is positive for any $x_\mu > 0$. On the order of magnitude $\hbar\omega_*$ is the energy of the vibrational quantum.

In the general case of arbitrary potential surfaces the statistical-limit transition rate is given by a similar equation as (2.11.5). Indeed, the exponential factor in (2.11.5) describes the ΔE dependence of the Franck–Condon factor, and just such a dependence is expected in the case of tunneling transitions between the arbitrarily-shaped potential surfaces. For this reason only the constants A and ω_* are affected by the specific shape of the potential surfaces.

The method to determine the accepting modes and effective states developed in this section for the simplest potential surfaces is quite general. It also applies to arbitrary surfaces, provided that an explicit dependence of the Franck–Condon factor and the energy $E_{r'm'}$ on the vibrational quantum numbers of the final electronic state is known.

The ΔE dependence of k on the pattern of (2.11.5) is conventionally called the energy-gap law [1.1–3]. This term was borrowed from the theory of electronic spectra of solids. If there is an energy gap ΔE in the electronic spectrum (as in the case, e.g., in semi- and superconductors), then the number of free charge carriers is given by an expression as (2.11.5), where the thermal energy $k_B T$ is substituted for $\hbar\omega_*$. The energy-gap law was observed in many diverse experimental systems [2.277, 278, 301, 6.18–20] and often discussed theoretically [2.254, 6.21]. Of interest are also violations of this law, e.g., in some solid-state systems [6.22, 23].

The energy-gap law is the principal experimental indication that a given group of similar compounds with various ΔE is assignable to the statistical limit. In the intermediate case the ΔE dependence of k is expected to be nonmonotonic since the smooth dependence of (2.11.5) is superimposed with a stronger nonmonotonic dependence of the function (3.4.17); the latter reaches a maximum at a resonance between the initial and final states, and rapidly decreases with an increase in energy mismatch.

To complete this section we shall deduce a general expression for ϱ_{eff}, valid for any potential surfaces and an arbitrary initial state rm. We begin with the general expression for the nonradiative transition-rate constant,

$$k_{rm} = \frac{2\pi}{\hbar} v_{\mathrm{el}}^2 \sum_{m'} F_{rmr'm'} \, \delta_\gamma (E_{rm} - E_{r'm'}) \,, \tag{6.1.22}$$

and look for a set of final vibrational quantum numbers,

$$m'_m \equiv \{m'_{m\mu}\} \,, \tag{6.1.23}$$

in order to maximize the Franck–Condon factor under the constraint of this state's energy being equal to the initial energy E_{rm},

$$\ln F_{rmr'm'} = \max, \qquad E_{r'm'} = E_{rm} \,. \tag{6.1.24}$$

We construct the Lagrange function

$$L_m = \ln F_{rmr'm'} + \hbar^{-1} \lambda_m E_{r'm'} \,, \tag{6.1.25}$$

where λ_m is the Lagrangian multiplier. For the determination of the quantum numbers (6.1.23) we obtain

$$\frac{\partial L_m}{\partial m'_\mu} \equiv \frac{\partial \ln F_{rmr'm'}}{\partial m'_\mu} + \lambda_m \Omega_{\mu r'} = 0 \,, \tag{6.1.26}$$

where $\Omega_{\mu r'} = \hbar^{-1} \partial E_{r'm'} / \partial m'_\mu$ is the classical vibration frequency of the μ-th oscillator in the r'-th electronic state (it is a function of m'_μ). Equations (6.1.24, 26) allow to find the extremum quantum numbers (6.1.23) and the Lagrangian multiplier.

Equation (6.1.22) is expanded in powers of $m'_\mu - m'_{m\mu}$:

$$k_{rm} = \frac{2\pi}{\hbar} v_{\mathrm{eff}}^2 \sum_{m'} \exp\left(- \sum_\mu \frac{(m'_\mu - m'_{m\mu})^2}{2D_\mu} \right) \delta_\gamma \left[\sum_\mu \hbar \Omega_{\mu r'} (m'_\mu - m'_{m\mu}) \right] ,$$

where $\Omega_{\mu r'}$ is taken at $m'_\mu = m'_{m\mu}$ and the notations

$$v_{\mathrm{eff}} = v_{\mathrm{el}} (F_{rmr'm'_m})^{1/2} \,, \tag{6.1.27}$$

$$D_\mu^{-1} = - \partial^2 \ln F_{rmr'm'} / \partial m'^2_\mu |_{m' = m'_m} > 0 \,, \tag{6.1.28}$$

are introduced.

The term in the exponent involving the first derivative of the logarithm of the Franck–Condon factor has vanished owing to (6.1.26) and to the energy conservation constraint imposed by the delta function. The second derivative is negative since the Franck–Condon factor has a maximum at $m' = m'_m$. Replacing the sum with an integral and setting $\gamma = 0$ we obtain

$$\varrho_{\mathrm{eff}} = \int \exp\left(- \sum_\mu \frac{(m'_\mu - m'_{m\mu})^2}{2D_\mu} \right) \delta \left[\sum_\mu \hbar \, \Omega_{\mu r'} (m'_\mu - m'_{m\mu}) \right] \prod_\mu dm'_\mu \,.$$

This integral is easily solved using (6.1.17) to yield the final result [2.57]

$$\varrho_{\rm eff} = \hbar^{-1}\left[\prod_{\mu}(2\pi D_\mu)/2\pi\sum_{\mu}\Omega^2_{\mu r'}D_\mu\right]^{1/2} . \tag{6.1.29}$$

Condition (6.1.28) implies that the Franck–Condon factor of the μ-th oscillator decreases faster than exponentially with increasing m'_μ. As follows from the quasiclassical equations of Chap. 5, this is always true of the harmonic oscillators, but in the case of anharmonic ones it may not be the case, especially at m'_μ near the dissociation limit. If expression (6.1.28) is negative, the quantum numbers $m'_{m\mu}$ correspond to a minimum rather than a maximum of the Franck–Condon factor, and hence, the set of $m'_{m\mu}$ does not relate to the effective states. In this case such final states will be effective where most of the energy is transferred to a single oscillator (any one of the group of good acceptors) [1.13, 2.302].

Some recent statistical-limit rate calculations [2.257, 6.24] have shown that "moderate acceptors" having middle-valued Franck–Condon factors contribute appreciably owing to the fact that they are present in a larger number than the best acceptors.

6.2 Generating-Function Method

The generating-function method was developed by *Lax* [4.69], *Kubo* [5.12] and *Kubo* and *Toyozawa* [6.17], *Pekar* and *Krivoglaz* [6.25] and *Davydov* [6.26] to investigate radiative and nonradiative transitions of localized electrons in crystals and molecules assuming that the Boltzmann distribution exists over the vibrational states in the initial electronic state. This method is also used to calculate the radiationless transition rates, k_{rm}, from the Single Vibronic Levels (SVL) of isolated molecules in the statistical limit [1.1, 2, 16, 2.272, 273, 4.138, 5.13, 6.27].

The SVL nonradiative transition rate is written in a form similar to (6.1.22)

$$k_{rm} = \frac{2\pi}{\hbar} v^2_{\rm el}\sum_{m'} F_{rmr'm'}\,\delta(E_{rm} - E_{r'm'}) , \tag{6.2.1}$$

where the subscript γ of the delta function is omitted since the statistical-limit rate constant is independent of γ, so that the vibrational level broadening can be disregarded (see Sect. 6.1). Inserting the integral representation of the delta function (6.1.17) in (6.2.1) and the electronic-vibrational energies of (4.1.30) the transition rate reads as

$$k_{rm} = \frac{v^2_{\rm el}}{\hbar^2}\int\limits_{-\infty}^{\infty}\mathscr{F}_{rm}(t)\exp(-{\rm i}\Delta Et/\hbar)\,dt , \tag{6.2.2}$$

where ΔE is the energy gap of (4.5.18) and

$$\mathscr{F}_{rm}(t) \equiv \sum_{m'} F_{rmr'm'} \exp[-\mathrm{i}(\varepsilon_{rm} - \varepsilon_{r'm'})t/\hbar] \tag{6.2.3}$$

is called the generating function for the radiationless transition from the *rm* state.

The generating function is closely related to the time-dependent vibrational Green's functions. The initial electronic state Green's function in the coordinate representation is defined by

$$G_r(q, \bar{q}; t) = \sum_m \chi_{rm}(q)\, \chi_{rm}(\bar{q}) \exp(-\mathrm{i}\varepsilon_{rm} t/\hbar) \, . \tag{6.2.4}$$

The final r'-state Green's function is written in a similar manner. Writing the Franck–Condon factor as a double-overlap integral we obtain the generating function (6.2.3) in the form

$$\mathscr{F}_{rm}(t) = \int \chi_{rm}(q)\chi_{rm}(\bar{q}) \exp(-\mathrm{i}\varepsilon_{rm} t/\hbar)\, G_{r'}(\bar{q}, q; -t)\, dq\, d\bar{q} \, . \tag{6.2.5}$$

We consider also the expression for an average nonradiative transition rate from the initial electronic state r where the Boltzmann distribution over the vibrational levels exists,

$$k_r(T) = \langle k_{rm} \rangle_T \equiv Z_r^{-1} \sum_m k_{rm} \exp(-\varepsilon_{rm}/k_\mathrm{B} T) \, . \tag{6.2.6}$$

Here, the angle brackets denote statistical averaging at temperature T, and

$$Z_r = \sum_m \exp(-\varepsilon_{rm}/k_\mathrm{B} T) \tag{6.2.7}$$

is the vibrational partition function. The generating function $\mathscr{F}_r(t, T)$ for the transition rate (6.2.6) is the statistical average of (6.2.3),

$$\mathscr{F}_r(t, T) = \langle \mathscr{F}_{rm}(t) \rangle_T \, . \tag{6.2.8}$$

Substitution of (6.2.4, 5) yields

$$\mathscr{F}_r(t, T) = Z_r^{-1} \int G_r\left(q, \bar{q}; t - \frac{\mathrm{i}\hbar}{k_\mathrm{B} T}\right) G_{r'}(\bar{q}, q; -t)\, dq\, d\bar{q} \, . \tag{6.2.9}$$

In the model of noninteracting vibrations the Franck–Condon factor is represented by a product, and the energies ε_{rm} and $\varepsilon_{r'm'}$ as sums of the contributions from the individual vibrations. Then, the generating function (6.2.3) factors into

$$\mathscr{F}_{rm}(t) = \prod_\mu \mathscr{F}_{rm_\mu}(t) \, , \tag{6.2.10}$$

$$\mathscr{F}_{rm_\mu}(t) = \sum_{m'_\mu} F_{rm_\mu r'm'_\mu} \exp\left[-\frac{\mathrm{i}}{\hbar}(\varepsilon_{rm_\mu} - \varepsilon_{r'm'_\mu})t\right] . \tag{6.2.11}$$

The Green's function shows a similar factorization

$$G_r(q, \bar{q}; t) = \prod_\mu G_{\mu r}(q_\mu, \bar{q}_\mu; t)\,, \tag{6.2.12}$$

$$G_{\mu r}(q_\mu, \bar{q}_\mu; t) = \sum_{m_\mu} \chi_{rm_\mu}(q_\mu)\,\chi_{rm_\mu}(\bar{q}_\mu) \exp\left(-\frac{\mathrm{i}}{\hbar}\varepsilon_{rm_\mu} t\right). \tag{6.2.13}$$

Equation (6.2.13) for the harmonic oscillator is frequently called Slater's sum [6.17].

Now we calculate the Green's function (6.2.13) and the generating function (6.2.11) for the harmonic oscillator. Suppressing the subscript μ, we substitute in (6.2.12) the explicit expressions for the vibrational energy and wave functions of (4.8.1):

$$G_r(q, \bar{q}; t) = \frac{1}{b_r\sqrt{\pi}} \exp\left[-\tfrac{1}{2}(x_r^2 + \bar{x}_r^2 + \mathrm{i}\omega_r t)\right]$$
$$\times \sum_{m=0} \frac{1}{2^m m!} H_m(x_r) H_m(\bar{x}_r) \exp(-\mathrm{i}m\,\omega_r t)\,.$$

The sum is calculated by Mehler's formula [4.164]. After some simplifications a well-known expression for the Slater's sum is obtained,

$$G_r(q, \bar{q}; t) = (2\pi \mathrm{i} b_r^2 \sin\omega_r t)^{-1/2}$$
$$\times \exp\left[\frac{\mathrm{i}}{4}(x_r - \bar{x}_r)^2 \cot\frac{\omega_r t}{2} - \frac{\mathrm{i}}{4}(x_r + \bar{x}_r)^2 \tan\frac{\omega_r t}{2}\right]. \tag{6.2.14}$$

Using this equation (after replacing $r \to r'$ and $t \to -t$) it is not difficult to deduce the one-dimensional harmonic-oscillator generating function since the double integral of (6.2.5) is solved easily. If there is no frequency change (the frequency ratio f is unity), then

$$\mathscr{F}_{rm}^{f=1}(t) = \exp[\xi(\mathrm{e}^{\mathrm{i}\omega t} - 1)]\, L_m\left(4\xi \sin^2\frac{\omega t}{2}\right), \tag{6.2.15}$$

where ω is the oscillator frequency, ξ the squared dimensionless equilibrium displacement of Sect. 4.8 and $L_m(x)$ the Laguerre polynomial. For a transition from the $m = 0$ ground vibrational state one obtains for any f

$$\mathscr{F}_{r0}(t) = \frac{2\sqrt{f}}{f+1}(1 - \theta \mathrm{e}^{2\mathrm{i}\omega_{r'} t})^{-1/2}$$
$$\times \exp\left[\frac{\mathrm{i}}{2}(\omega_{r'} - \omega_r)t + \frac{2\xi(\mathrm{e}^{\mathrm{i}\omega_{r'} t} - 1)}{f + 1 + (f-1)\mathrm{e}^{\mathrm{i}\omega_{r'} t}}\right], \tag{6.2.16}$$

where ω_r and $\omega_{r'}$ are the oscillator frequencies in the initial and final electronic states, respectively, $f = \omega_r/\omega_{r'}$ and $\theta = (1-f)^2/(1+f)^2 = (\omega_{r'} - \omega_r)^2/(\omega_{r'} + \omega_r)^2$. Finally, for a transition from the $m = 0$ state without frequency change we obtain from (6.2.15) (or from (6.2.16) for $f = 1$)

$$\mathscr{F}_{r0}^{f=1}(t) = \exp[\xi(e^{i\omega t} - 1)] . \tag{6.2.17}$$

In the general case of arbitrary m and f the harmonic-oscillator generating function has a very cumbersome form [4.21, 6.28].

The statistically averaged generating function (6.2.8) for $f = 1$ is easily calculated by means of substitution of (6.2.15) to yield

$$\mathscr{F}_r^{f=1}(t, T) = \exp[\xi(e^{i\omega t} - 1) - 2\xi\bar{m}(1 - \cos\omega t)] , \tag{6.2.18}$$

where

$$\bar{m} = [\exp(\hbar\omega/k_B T) - 1]^{-1} \tag{6.2.19}$$

is the mean number of quanta at temperature T. For $f \neq 1$ the generating function was given in [6.29].

In the general case when the frequency change and Duchinsky effects are present, the statistically averaged generating function can be evaluated using (6.2.9), provided that the Green's function (which assumes the form of the product (6.2.12) of the factors (6.2.14) for each of the r and r' electronic states having different sets of the normal coordinates, $\{q_\mu\}$ and $\{q'_\mu\}$) is written in invariant tensor form [1.16, 6.17]. Without Duchinsky effect (i.e., the sets $\{q_\mu\}$ and $\{q'_\mu\}$ are identical) the integration in (6.2.9) is easily performed, and after simple manipulations the generating function is reduced to a product of single-oscillator functions:

$$\mathscr{F}_r(t, T) = \left(\frac{2f(\cosh\beta - 1)}{(f-1)^2 \sin\alpha \sin\alpha' + 2f[\cosh(\beta + i\Delta\omega t) - 1]}\right)^{1/2}$$
$$\times \exp\left[2\xi i\left(f\cot\frac{\alpha'}{2} - \cot\frac{\alpha}{2}\right)^{-1}\right],$$

where $\alpha = \omega_r t - i\beta$, $\alpha' = \omega_{r'} t$, $\beta = \hbar\omega_r/k_B T$ and $\Delta\omega = \omega_r - \omega_{r'}$. In the limit of $T \to 0$ this expression transforms to (6.2.16).

If the electronic transition is induced by the non-adiabaticity operator (4.1.17), the generating function becomes very complicated. For the case without frequency and Duchinsky effects it was found in [2.316].

When the generating function is available, the transition rate calculation reduces to solving the integral in (6.2.2). Within the approximation of noninteracting harmonic vibrations the generating function is a product of oscillating functions (6.2.10), so, formally, the integral of (6.2.2) diverges. However, one should bear in mind that there always exists a vibrational-level damping that ensures the convergence of the integral.

6.3 Saddle-Point Method

6.3.1 First Saddle-Point Approximation (FSPA)

In order to calculate the integral in (6.2.2) it is natural to invoke the saddle-point method [2.202, 272, 6.27, 30–32] since there is a large parameter, viz., the ratio of the electronic energy ΔE to the vibrational quantum $\hbar\omega_\mu$. To illustrate the method we first consider, as in Sect. 6.1, the simplest model of non-interacting harmonic oscillators without frequency changes and with zero initial quantum numbers.

We substitute the generating function (6.2.10, 17) into the integral of (6.2.2),

$$k_{r0} = \frac{v_{\mathrm{el}}^2}{\hbar^2} \int_{-\infty}^{\infty} \exp\left[-\frac{\mathrm{i}}{\hbar} \Delta E t + \sum_\mu \xi_\mu (\mathrm{e}^{\mathrm{i}\omega_\mu t} - 1) \right] dt \, . \tag{6.3.1}$$

Equating to zero the first derivative of the exponent over t we arrive at the following equation for the saddle points:

$$\sum_\mu \hbar\omega_\mu \xi_\mu \exp(i\omega_\mu t) = \Delta E \, . \tag{6.3.2}$$

The solution is an infinite series of roots $t^{(n)}$ $(n = 0, \pm 1, \ldots)$, and, strictly speaking, the integral in (6.3.1) can be represented as an infinite sum over the contributions from the saddle points $t^{(n)}$. Moreover, as was already mentioned at the end of the previous section, the original integral (6.3.1) and, hence, this sum is formally divergent.

Nevertheless, it is usually supposed that it is sufficient to allow for a contribution from only one saddle point $t^{(0)}$ next to the imaginary axis, in order to calculate the statistical-limit transition rate. We call this the First Saddle-Point Approximation (FSPA) [2.237]. In the present case, the first saddle point is on the negative imaginary axis since in the asymptotic limit of $\Delta E \gg \sum_\mu \hbar\omega_\mu \xi_\mu$ (6.3.2) has a solution of the form

$$t^{(0)} = -\mathrm{i}\tau_0 \, , \tag{6.3.3}$$

where τ_0 is real and positive (in the general case of arbitrary potential surfaces the first saddle point is not bound to be on the imaginary axis). It is easy to see that τ_0 coincides with the Lagrangian multiplier λ_0 of Sect. 6.1, since, after the substitution of (6.3.3) into (6.3.2), one obtains an equation for τ_0 identical with (6.1.9) for λ_0 (again the subscript "0" signals that the initial vibrational quantum numbers m are equal to zero).

Since, for the model under consideration, the statistical-limit transition rate was already calculated in Sect. 6.1 by the effective-states method, the above statement is immediately verified by calculating the integral of (6.3.1) in the FSPA and comparing the result with (6.1.19). Expanding the exponent of the integrand in powers of $t - t^{(0)}$ to the second order and performing a trivial integration we obtain

$$k_{r0} = \frac{v_{\mathrm{el}}^2}{\hbar^2}\left(2\pi \Big/ \sum_{\mu} \xi_\mu \omega_\mu^2 \, \mathrm{e}^{\omega_\mu \tau_0}\right)^{1/2} \exp\left[-\frac{\Delta E \tau_0}{\hbar} + \sum_{\mu} \xi_\mu (\mathrm{e}^{\omega_\mu \tau_0} - 1)\right].$$

In view of $\tau_0 = \lambda_0$ mentioned above, this expression is identical with (6.1.19), which justifies in an indirect fashion the neglect of all the saddle points $t^{(n)}$ except $t^{(0)}$.

A more rigorous substantiation of the FSPA was given in [2.237] where a small vibrational level broadening, $\gamma \ll \hbar\omega_\mu$, was explicitly allowed for to ensure the convergence of the integral in (6.3.1), and the contribution from all the remaining saddle points $t^{(n)}$ $(n \neq 0)$ was evaluated. In this way the conditions were established that the transition rate is governed by a single point $t^{(0)}$ only and thus is independent of γ (these are in fact the conditions for the statistical limit defined in Sect. 2.11 in terms of the effective states). These results are discussed in Sect. 6.3.2 because they are interesting by themselves both from the point of view of the physical interpretation of the statistical limit and owing to their application in calculations of the nonradiative transition-rate constants in the intermediate case (see Chap. 7) when the statistical-limit conditions are not met and the contributions of all saddle points rather than $t^{(0)}$ alone must be taken into account [6.33].

Thus, it is shown that a purely formal mathematical action, viz., the calculation of the divergent integral by the saddle-point method in the FSPA, has quite a reasonable physical interpretation: it corresponds to allowing for the dominant contribution of the effective states and to the neglect of the remaining (ineffective) states.

Let us apply the saddle-point method to calculate the transition rate in the model of nontotally symmetric accepting modes that have no displacements of the equilibrium positions (see Sect. 4.1), and only frequencies are changed. There is much interest to this problem in literature because often the frequency changes for such modes are very large, and the Franck–Condon factor takes on very high values in this case (see Sect. 5.2.3). This is especially true of the out-of-plane vibrations in some aromatic hydrocarbons and carbonyl compounds, as some of them show a two- or even threefold frequency decrease in going from the ground to the first excited singlet state. The large frequency change may be caused by the *proximity effect* [4.31, 6.34, 35] when a higher excited electronic state is close to the initially excited state so that the potential surface of the latter is severely distorted because of the pseudo-Jahn-Teller effect. In numerous calculations the harmonic out-of-plane modes turned out to be the best acceptors of the electronic energy [2.298, 300, 6.24, 36–40]. However, when the anharmonicity is taken into account, they pass to the grade of moderate acceptors [2.257, 6.41]; therefore, the harmonic results seem to be unreliable. Nevertheless, it is believed that the out-of-plane bending modes with large frequency changes are responsible for: (i) the *saturation effect* revealed in the triplet state of some polyatomic molecules where the $T_1 \rightsquigarrow S_0$ nonradiative transition rate increasing with the vibrational energy E_{vib} saturates at a high E_{vib} [6.41–47]; (ii) an enhancement of the rate constants for InterSystem Crossing (ISC) [6.48], IVR [4.49], IVR-induced Internal Conversion (IC) [6.49] and predissociation [6.50] when exciting the out-of-plane vibrations in the initial state.

Let us now write the transition rate from the $m = 0$ initial state as the integral of (6.2.2) with the generating function $\mathscr{F}_{r0}(t)$ as a product of the factors (6.2.16) with $\xi_\mu = 0$:

$$k_{r0} = \frac{v_{el}^2}{\hbar^2}\left(\prod_\mu \frac{2\sqrt{f_\mu}}{1+f_\mu}\right)\int_{-\infty}^{\infty} \exp[\psi(t)]\,dt\ ,$$

$$\psi(t) = -\frac{i}{\hbar}\Delta Et + \frac{i}{2}\sum_\mu (\omega_{\mu r'} - \omega_{\mu r})t$$

$$-\frac{1}{2}\sum_\mu \ln[1 - \theta_\mu \exp(2i\omega_{\mu r'}t)]\ . \qquad (6.3.4)$$

Taking the derivative of $\psi(t)$ and setting $t = t^{(0)}$, we obtain

$$\sum_\mu \frac{\hbar\omega_{\mu r'}\theta_\mu \exp(2\omega_{\mu r'}\tau_0)}{1 - \theta_\mu \exp(2\omega_{\mu r'}\tau_0)} = \Delta E\ , \qquad (6.3.5)$$

where the zero-point energy change is absorbed in ΔE. It is seen that at large ΔE the best acceptors are those modes for which the denominator on the left-hand side of (6.3.5) is small. We define $s_\mu = (2\omega_{\mu r'})^{-1}\ln(1/\theta_\mu)$. Then, the best acceptors should fulfil the condition

$$2\omega_{\mu r'}(s_\mu - \tau_0) \ll 1\ . \qquad (6.3.6)$$

Expanding the left-hand side of (6.3.5) in powers of $s_\mu - \tau_0$ we obtain

$$\frac{1}{2}\sum_\mu \frac{1}{s_\mu - \tau_0} = \frac{\Delta E}{\hbar}\ . \qquad (6.3.7)$$

We assume that there are N good acceptors with $s_\mu \equiv s$. Then, (6.3.7) has a unique solution $\tau_0 = s - (N\hbar/2\Delta E)$, and condition (6.3.6) has the form $m'_\mu \equiv \Delta E/N\hbar\omega_{\mu r'} \gg 1$, implying that m'_μ, a mean number of quanta on a single oscillator in the final electronic state, should naturally be large for good acceptors.

Further, we calculate the second derivative $\psi(t)$ at $t = t^{(0)}$ and then, by a common procedure, the integral (6.3.4) in the vicinity of the saddle point (6.3.3).

The resulting transition rate is

$$k_{r0} = \frac{v_{\text{el}}^2}{\hbar^2} \frac{\sqrt{\pi N}}{\Delta E} \left(\prod_\mu \frac{2\sqrt{f_\mu}}{1+f_\mu} \theta_\mu^{f_\mu - 1} \right)$$

$$\times \exp\left(-\frac{1}{\hbar} \Delta E s + \frac{N}{2} \ln \frac{\Delta E}{N\hbar\omega_{r'}} - \frac{N}{2} \right), \tag{6.3.8}$$

where a mean frequency

$$\omega_{r'} = \left(\prod_\mu \omega_{\mu r'} \right)^{1/N}$$

was introduced. It is worthwhile to note that the rate constant exhibits the energy-gap law.

6.3.2 Validity Conditions for the FSPA

The FSPA was shown before to provide a correct expression for the statistical-limit transition rate identical with that derived by the effective-states method. A different argument in favor of the FSPA was proposed by *Fischer* [6.27] who pointed out that in the statistical limit the contribution of the rest of the saddle points should be small due to a phase cancellation unavoidable in the presence of many vibrational modes. In this subsection a phase-cancellation condition will be derived, as well as additional validity conditions for the FSPA, based on an estimate of the contributions of all saddle points [2.237]; these conditions will be compared with the statistical limit criterion (2.11.2). The saddle-point method may be of value in the intermediate case as well where these conditions are not fulfilled, and the FSPA is definitely invalid [2.237, 6.33, 51, 52].

Now, since the vibrational-state decay needs to be allowed for explicitly, we shall begin with the general expression (6.1.2) for the transition rate from the $m = 0$ initial state in the model of noninteracting harmonic oscillators with displaced equilibrium positions and undistorted frequencies. In (6.1.2) $\delta_\gamma(E)$ is the line shape function of the vibrational energy levels having a width γ (it was denoted as $r_\lambda(E)$ in Sect. 2.10). While the precise shape of this function is unknown, a good-enough approximation (if used with the precautions indicated later) can be provided by the Lorentzian function (2.7.34) with a constant width γ independent of quantum numbers (see Sect. 3.4).

As already discussed in Sect. 2.10, the level broadening in a polyatomic molecule is caused by intra- and intermolecular vibrational relaxation processes. The former processes are especially important since the intramolecular interactions are usually stronger than the intermolecular ones. The intramolecular relaxation arises owing to the fact that in the first stage of the radiationless transition when the system hops from one potential surface to another, a nonstationary, vibrationally hot state is produced. Similarly, nonstationary states are also prepared in high overtone vibrational transitions when the polyatomic molecule absorbs a photon with a large energy of 10000–20000 cm^{-1} so that the width γ and the level shape can be estimated from the experimental overtone absorption bands of different molecular vibrations. Thus, the measured C–H absorption bands have Lorentzian shapes whose widths amount to 100 cm^{-1} in benzene [2.250, 5.62] and other molecules [5.43, 51, 54] and 200 cm^{-1} in naphthalene [2.248], and those widths vary insignificantly for the C–H quantum numbers ≤ 6 to 8. As was assumed in Sect. 6.1, the level widths are smaller than the vibrational quanta of the intramolecular vibrations. More recent investigations have shown that the broadening is often inhomogeneous, and homogeneous widths are on the order of 3 cm^{-1} for benzene [2.258, 6.53] and other molecules [5.40, 48, 60], and even drop to 0.1 cm^{-1} [5.73, 6.54]. The existence of a nonzero level broadening in isolated polyatomic molecules is connected with a vibrational quasicontinuum [2.161, 186, 6.55] that arises due to a transition from regular to chaotic dynamics with an increase in E_{vib} [6.56–59]. It is worth mentioning that the broadening itself does not immediately imply a fast energy equilibration, especially in medium-sized molecules [6.60].

The use of the Lorentzian function (2.7.34) with a constant width encounters a difficulty pointed out by *Lukashin* and *Makshantsev* [3.4] and *Kühn* and *Metz* [6.14]. Namely, as follows from the general treatment of Sect. 3.4, the function $\delta_\gamma(E)$ has a Lorentzian shape only near its maximum, whereas far away from the peak it decays much faster than the function (2.7.34), thus assuring the fulfillment of the energy-conservation law in calculation of the sum over the final states m'. But, if the sum is calculated with (2.7.34), then, because of a slow decrease of the Lorentzian function and a rapid increase of the Franck-Condon factor with decreasing final state energy $E_{r'm'}$, the dominant contribution to that sum will come from the off-resonant states with small m' [5.15, 6.15], which is surely incorrect since the $E_{r'm'} = E_{rm}$ resonant states must contribute most. It will be seen later how this difficulty can be overcome in the saddle-point method by eliminating the contribution of the non-physical states.

Utilizing the integral delta function representation

$$\delta_\gamma(E) = \frac{1}{\pi\hbar}\,\mathrm{Re}\left\{\int_0^\infty \exp[-(\mathrm{i}E + \gamma)t/\hbar]\,dt\right\}, \tag{6.3.9}$$

we write the transition rate in the form

$$k_{r0} = \frac{2v_{el}^2}{\hbar^2}\,\mathrm{Re}\left\{\int_0^\infty \exp[\psi(t) - (\gamma t/\hbar)]\,dt\right\},$$

$$\psi(t) = -\frac{\mathrm{i}}{\hbar}\Delta E t + \sum_\mu \xi_\mu[\exp(\mathrm{i}\omega_\mu t) - 1]. \tag{6.3.10}$$

As compared to (6.3.1), the exponent of the integrand contains the additional term $-\gamma t/\hbar$ assuring the integral's convergence. If $\delta_\gamma(E)$ is calculated by some exactly solvable model instead of being approximated with the Lorentzian, then the integrand will show a more complicated decay pattern; an example will be given in Sect. 7.2.

Since γ is small relative to $\hbar\omega_\mu$ and ΔE, the saddle points for the integral of (6.3.10) can again be sought as solutions to (6.3.2). Additionally, the integrand has a singularity in the complex plane at $t = 0$, and the $t = 0$ point has generally to be incorporated together with the saddle points [5.11]. However, it is easily verified that the $t = 0$ contribution is nothing else but the contribution of the $E_{r'm'} \ll E_{r0}$ off-resonant states to the sum of (6.1.2), arising as consequence of the slow decreasing of the Lorentzian $\delta_\gamma(E)$ for large $|E|$. Indeed, calculating first the integral of (6.3.10) in the vicinity of $t = 0$ and then the off-resonant states' contribution to (6.1.2), we arrive at one and the same expression

$$2v_{el}^2\gamma/\hbar(\Delta E)^2 \,. \tag{6.3.11}$$

Notice that if the Gaussian rather than Lorentzian function were employed, which decays steeply enough and thus assuredly nullifies the off-resonant states' contribution, then the integrand of (6.3.10) would not show any $t = 0$ singularity at all, since the exponent would contain a function proportional to t^2 instead of t. Hence, in calculating the integral of (6.3.10) by the saddle-point method one has to take into account only the saddle points ignoring the $t = 0$ singular point.

In order to resolve (6.3.2) for the saddle points, we assume that among the good acceptors there is a group of N_M high-frequency vibrations (the maximum frequency modes) all having equal displacement parameters ξ_M and frequencies ω_M, and that the frequencies ω_μ of other vibrations are significantly lower, which reads as

$$\lambda_\mu \equiv \omega_\mu/\omega_M < 1/2 \,. \tag{6.3.12}$$

The above assumptions enable a determination of the saddle points $t^{(n)}$ by successive approximations after (6.3.2) is transformed to the form

$$1 - \frac{\hbar\omega_M\xi_M N_M}{\Delta E}\exp(i\omega_M t) = \sum_{\mu \neq M}\frac{\hbar\omega_\mu\xi_\mu}{\Delta E}\exp(i\omega_\mu t) \,. \tag{6.3.13}$$

For large ΔE the right-hand side of (6.3.13) can be regarded as a small perturbation. In zeroth order the right-hand side is dropped to give

$$t^{(n)} = -i\tau_0 + \frac{2\pi n}{\omega_M}, \quad (n = 0, 1, \ldots) \,, \tag{6.3.14}$$

$$\tau_0 = \frac{1}{\omega_M}\ln\frac{\Delta E}{\hbar\omega_M\xi_M N_M} \,. \tag{6.3.15}$$

In the next approximation we set

$$t^{(n)} = -i\tau_0 + \frac{2\pi n}{\omega_M} + \frac{i}{\omega_M}y_n \,. \tag{6.3.16}$$

Assuming that

$$|y_n| \ll 1 \tag{6.3.17}$$

and substituting $t^{(n)}$ from (6.3.14) into the right-hand side of (6.3.13), one obtains, after a simple algebra,

$$y_n = \sum_{\mu \neq \mathrm{M}} \frac{\omega_\mu m'_{0\mu}}{\Delta E} \exp(\mathrm{i}2\pi n \lambda_\mu) \,, \tag{6.3.18}$$

$$m'_{0\mu} = \xi_\mu \exp(\omega_\mu \tau_0) = \xi_\mu \left(\frac{\Delta E}{\hbar \omega_\mathrm{M} \xi_\mathrm{M} N_\mathrm{M}} \right)^{\lambda_\mu} . \tag{6.3.19}$$

From (6.3.18) it is seen that

$$|y_n| \sim \Delta E^{\lambda - 1} \,, \tag{6.3.20}$$

where λ is the mean value of λ_μ, and hence, due to (6.3.12), inequality (6.3.17) is satisfied for large ΔE.

Calculating $\psi(t)$ at the saddle points (6.3.16) one obtains

$$\psi(t^{(n)}) = \psi_0(t^{(n)}) - \frac{\Delta E}{2\hbar\omega_\mathrm{M}} y_n^2 + \mathrm{O}(y_n \Delta E^{\lambda}) \,.$$

The first term is the $\psi(t)$ value at the zeroth-order saddle points (6.3.14); as for the second and the third ones, they tend to zero for large ΔE, which follows from (6.3.12, 20). Therefore, it is sufficient to confine ourselves with the zeroth-order approximation (6.3.14).

The second derivative of $\psi(t)$ at the saddle points (6.3.14) is

$$\psi''(t^{(n)}) = -\frac{1}{\hbar}\omega_\mathrm{M} \Delta E + \mathrm{O}(\Delta E^{\lambda}) \,.$$

The second term is negligible for large ΔE. Since this expression is real and negative, the steepest-descent directions at all saddle points are aligned parallel to the real axis.

The integration path in (6.3.10) is distorted as shown in Fig. 6.1, and the integral is divided into two integrals over the two path segments. In the integral over the segment of the imaginary axis we make the substitution $t = -\mathrm{i}\theta$ and expand the exponential in powers of $\gamma\theta/\hbar$ [$\leq \gamma\tau_0/\hbar \ll 1$, see (6.3.15)]. The zeroth-order term of the expansion disappears, as the integral is purely imaginary, and in the next approximation the integral is

$$\frac{2v_{\mathrm{e}1}^2 \gamma}{\hbar^3} \int_0^{\tau_0} \theta d\theta \exp\left[-\frac{1}{\hbar} \Delta E \theta + \sum_\mu \xi_\mu (\mathrm{e}^{\omega_\mu \theta} - 1) \right] .$$

The exponent monotonically decreases with increasing θ (since its derivative is

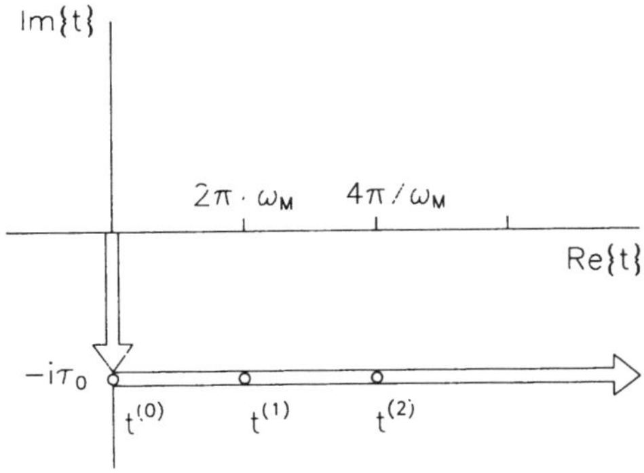

Fig. 6.1. Integration path in (6.3.10). Circles are the saddle points

negative for $\theta < \tau_0$), so at a large ΔE the main contribution comes from a small vicinity of the point $\theta = 0$. At small θ we can neglect the sum over μ and set the upper integration limit equal to infinity. As a result, (6.3.11) is obtained, as was already explained, due to the slow damping of the Lorentzian and corresponds to the off-resonant states' contribution. For this reason the integral over the imaginary axis must be omitted. We emphasize once more that if, instead of the Lorentzian, the Gaussian function was employed, which decays fast enough and, hence, completely excludes the non-physical contribution of the off-resonant states, the exponent in (6.3.9) would involve $(\gamma t/\hbar)^2$ instead of $\gamma t/\hbar$, and the imaginary-segment contribution would be exactly zero because of the integral being purely imaginary.

The integral over the horizontal part of the path is a sum of the contributions from the saddle points (6.3.14), the first saddle-point contribution being weighted with a factor of 1/2 since the path begins at this point. After simple manipulations the transition rate is obtained in the form

$$k_{r0} = k_{r0}^{(0)}\left(1 + 2\,\mathrm{Re}\left\{\sum_{n=1}^{\infty} R_n\right\}\right), \tag{6.3.21}$$

where

$$k_{r0}^{(0)} = \frac{v_{\mathrm{el}}^2}{\hbar}\left(\frac{2\pi}{\Delta E \hbar\omega_{\mathrm{M}}}\right)^{1/2} \exp\left[-\frac{1}{\hbar}\Delta E\tau_0 + \sum_{\mu}(m'_{0\mu} - \xi_\mu)\right]$$

is the contribution from the first saddle point $t^{(0)}$ and R_n is the desired relative contribution from the point $t^{(n)}$, $R_n = \exp(-\,\mathrm{i}n\varphi - \Gamma n + g_n - g_0)$, where $\varphi = 2\pi\Delta E/\hbar\omega_{\mathrm{M}}$ and

$$\Gamma = 2\pi\gamma/\hbar\omega_{\mathrm{M}} \ll 1\ , \tag{6.3.22}$$

$$g_n = \sum_{\mu \neq \mathrm{M}} m'_{0\mu}\exp(\mathrm{i}2\pi n\lambda_\mu)\ . \tag{6.3.23}$$

These equations enable us to find out the validity conditions for the FSPA.

Inequality (6.3.22), being a consequence of the smallness of the level broadening as compared to the vibrational quanta, implies that, generally, the contributions from a large number of the saddle points $t^{(n)}$ with $n \lesssim 1/\Gamma$ need to be included. Therefore, it is clear that the FSPA is a result of certain specific properties of g_n as a function of n, which, in the statistical limit, must fulfil the following three conditions.

Condition 1 is

$$g_0 = \sum_{\mu \neq M} m'_{0\mu} \gg 1 . \tag{6.3.24}$$

Indeed, if $g_0 \lesssim 1$, then the same is also valid for $\mathrm{Re}\{g_n\}$ for all n, since it is seen from (6.3.23) that $\mathrm{Re}\{g_n\} \leq g_0$. But, in such a case the $|R_n|$-versus-n decay rate is governed by the small Γ alone, and hence, all terms with $n \lesssim 1/\Gamma$ will contribute to (6.3.21).

The parameters $m'_{0\mu}$ are nothing else as occupation numbers of the accepting modes in the final electronic state, since, apart from notations, (6.3.19) is identical with (6.1.8). For this reason condition 1 is merely a criterion from the modes μ to be good acceptors [cf. (6.1.11)].

Condition 2 is that $\mathrm{Re}\{g_n\}$-versus-n must be a decreasing function,

$$\mathrm{Re}\{g_n\} \to 0 \quad \text{for } n \to \infty . \tag{6.3.25}$$

If the frequency ratios λ_μ are not too small, then already for n as low as 1 we have

$$\mathrm{Re}\{g_1\} = \sum_{\mu \neq M} m'_{0\mu} \cos(2\pi\lambda_\mu) < g_0 ,$$

and this is sufficient, owing to (6.3.24), to yield $|R_1| \ll 1$, i.e., the relative contribution from the point $t^{(1)}$ is small, and the same is true for the succeeding points $t^{(n)}$.

The property (6.3.25) is a consequence of the fact that g_n is a sum (if there are more than one good acceptor) of complex numbers with large absolute values whose phases are uncorrelated and increase with n. Due to the lack of phase coherence, individual members of the sum are partially cancelled; the cancellation being the deeper, the greater n. Thus (6.3.25) is *the phase cancellation condition.*

The phase cancellation condition needs a refined formulation in view of a possibility of the "theoretical recurrences" such as those discussed in Sect. 2.4.2. Indeed, g_n is the sum of a limited number of terms oscillating against n with frequencies $2\pi\lambda_\mu$, and, therefore, in reality (6.3.25) cannot be satisfied at arbitrarily large n's. Namely, there exists a large enough value of $n = n_T$ (an analogue of the Poincaré period) at which all the $n\lambda_\mu$ are integers, or are very close to integers, so that g_n is close to g_0 and $|R_n| \approx \exp(-\Gamma n)$. Hence, it follows that

Condition 3

$$\Gamma n_T \gg 1 \tag{6.3.26}$$

is necessary in order that the saddle points with $n \geq n_T$ would not contribute to the transition rate. The recurrence period n_T rapidly increases with the number of accepting modes, and hence, inequality (6.3.26) asks for a large-enough number of good acceptors.

Thus, the applicability conditions for the FSPA are inequalities (6.3.24) enabling to pick out, among the molecular vibrations, the good acceptors with large values of $m'_{0\mu}$, the phase cancellation condition (6.3.25), and condition (6.3.26) demanding that the number of good acceptors should be large. In fact, only inequalities (6.3.24, 26) impose limitations on the physical parameters of the molecule, whereas the phase cancellation condition is merely a mathematical property of g_n.

Because these same conditions simultaneously define the statistical limit, it is appropriate to make a comparison with the definition of the statistical limit given in Sect. 2.11 in the language of effective states. The effective states were defined in Chap. 2 as vibrational states of a quasimolecule, a molecular moiety involving the electronic degrees of freedom and good acceptors, while the poor acceptors form an "intramolecular medium" owing to which the quasimolecule levels have a significant width γ. Therefore, the statistical-limit condition (2.11.2) asking for a large density of the effective states has to be completed with the criterion (6.1.11) defining the good acceptors.

As was already said before, inequality (6.3.24) is equivalent to the criterion (6.1.11) for good acceptors. Now, in order to compare (6.3.26) with (2.11.2), consider the function

$$g(t) = \sum_{\mu \neq \mathrm{M}} \xi_\mu \exp(\mathrm{i}\omega_\mu t) ,$$

which takes on the values g_n at the saddle points (6.3.14). This function has a transparent meaning. In Sect. 7.2 it will be shown that the generating function for a radiationless transition coincides with the time-dependent correlation function of the intramolecular interaction inducing the electronic transition. The calculation of the correlation function in the model of displaced undistored harmonic oscillators gives an exponential whose exponent contains, in particular, a sum of correlation functions for molecular vibrational coordinates in the initial electronic state. In other words, it turns out that [6.51]

$$g(t) = \sum_{\mu \neq \mathrm{M}} c_\mu \langle r0 | x_\mu(t) x_\mu(0) | r0 \rangle , \tag{6.3.27}$$

where $x_\mu(t)$ is the Heisenberg operator of the μ-th nuclear coordinate and c_μ is a constant depending on the normalization of the nuclear coordinates.

Equation (6.3.27) shows that the variable t plays the role of time, and if a vibrationally excited initial state rm with a high-enough m was considered instead of the $r0$ state, then $x_\mu(t)$ could be interpreted as the classical trajectory of the nuclear motion (the fact that the saddle points (6.3.14) are displaced in the complex plane is explained by the tunneling nature of the radiationless transition: the classical trajectory cannot pass from one potential surface to another at real t [1.13, 2.238], see Sect. 5.7.3).

On the basis of this analogy one can expect that the behavior of $g(t)$ with increasing $\mathrm{Re}\{t\}$ (or, which is the same, of g_n with increasing n) is governed by the same laws as if the $x_\mu(t)$ in (6.3.27) were merely the classical nuclear coordinates. Then the Poincaré period T can be defined as such an increment in $\mathrm{Re}\{t\}$ that $g\{t\}$ returns to the initial value of $g(-\mathrm{i}\tau_0) = g_0$. Since $\mathrm{Re}\{t^{(n)}\} = 2\pi n/\omega_\mathrm{M}$, the recurrence time is connected with n_T by the relation $T = 2\pi n_\mathrm{T}/\omega_\mathrm{M}$. Recalling (6.3.22) one can recast (6.3.26) as

$$\gamma \gg \hbar/T\,. \tag{6.3.28}$$

In turn, T can be expressed in terms of the density of vibrational states, and precisely the effective states, since T is defined for the set of good acceptors only (the poor acceptors are not at all included to the sums of (6.3.23, 27) since their parameters $m'_{0\mu}$ are small). Inserting (2.4.40), with ϱ_eff substituted for ϱ_l, we obtain (2.11.2) to within an insignificant numerical factor. Thus, it is concluded that the FSPA validity conditions derived are completely equivalent to the statistical limit ones postulated in Sect. 2.11.

If the maximum frequency modes cannot be singled out in such a way that (6.3.12) would be met, we are not capable of solving (6.3.2), even approximately. Nevertheless, we shall show that also in this most general case the FSPA works under the above three conditions.

Let t be a saddle point different from (6.3.3),

$$t = x - \mathrm{i}(\tau_0 + y)\,, \tag{6.3.29}$$

where x and y are real and τ_0 is defined by

$$\sum_\mu \hbar\omega_\mu m'_{0\mu} = \Delta E\,, \tag{6.3.30}$$

which is obtained after the substitution of (6.3.3) into (6.3.2). Substituting (6.3.29) into (6.3.2) and separating the real and imaginary parts we obtain two simultaneous equations in x and y

$$\sum_\mu \hbar\omega_\mu m'_{0\mu} \exp(\omega_\mu y) \cos\omega_\mu x = \Delta E\,, \tag{6.3.31}$$

$$\sum_\mu \hbar\omega_\mu m'_{0\mu} \exp(\omega_\mu y) \sin\omega_\mu x = 0\,. \tag{6.3.32}$$

The following general conclusions can be drawn from these equations regarding the arrangement of the saddle points (6.3.29).

Equations (6.3.31, 32) are unaffected by the substitution of $-x$ for x, hence the saddle points are arranged in pairs symmetric with respect the imaginary axis. Since the integration path does not go into the left-hand half-plane, we may consider only points $x \geq 0$. Further, from (6.3.32) it follows that $x > \pi/\omega_M$, where ω_M is the maximum frequency. In fact, if this inequality is not satisfied, then all the members of the sum on the left-hand side of (6.3.32) are positive so that this equation is also not satisfied. Thus, the first saddle point (6.3.3) is separated from the others by a gap of $\mathrm{Re}\{t\} \leq \pi/\omega_M$. Now we subtract (6.3.31) from (6.3.30):

$$\sum_\mu \hbar\omega_\mu m'_{0\mu}(1 - e^{\omega_\mu y}\cos\omega_\mu x) = 0\,. \tag{6.3.33}$$

It is seen that $y \geq 0$ must hold since otherwise the factor in the parenthesis is positive for any μ and (6.3.33) is not fulfilled. Consequently, all the saddle points are shifted with respect to $t^{(0)}$ to the lower half-plane, $\mathrm{Im}\{t\} \leq \mathrm{Im}\{t^{(0)}\}$.

The relative contribution of the saddle point (6.3.29) to the transition rate within exponential accuracy is

$$|R| \approx \exp\left[\mathrm{Re}\{\psi(t) - \psi(t^{(0)})\} - \frac{1}{\hbar}\gamma\,\mathrm{Re}\{t\}\right]$$

$$= \exp\left[-\sum_\mu m'_{0\mu}(1 + \omega_\mu y - e^{\omega_\mu y}\cos\omega_\mu x) - \frac{1}{\hbar}\gamma x\right].$$

The following conclusions are inferred from this formula. The relative contribution of the saddle points with $x \ll \hbar/\gamma$ will be negligible, provided that both the quantum numbers $m'_{0\mu}$ are large (the good acceptors condition) and the phase cancellation occurs making the third term in the parenthesis smaller, after the sum over μ is done, than the sum of the first two. Further, a recurrence period T can be defined such that at $x = T$ all the $\omega_\mu x$ will be close to integer multiples of 2π. Then, as is seen from (6.3.33), y is close to zero and $|R| \approx \exp(-\gamma T/\hbar)$. The contribution of such a saddle point will be small, provided that (6.3.28) is met. Thus, the three above-given conditions for the FSPA applicability also hold in the general case of arbitrary relations between the frequencies of molecular vibrations.

Nitzan and *Jortner* [2.204] have considered the saddle points of the form $t^{(n)} = -\mathrm{i}\tau_0 + nT$. All the frequencies ω_μ were supposed to be integer multiples of some normalizing frequency, ω_N, so that the recurrence period is merely $T = 2\pi/\omega_N$. However, the existence of a multitude of other saddle points at $\mathrm{Re}\{t\} \ll T$ was disregarded, which was also noted by *Metz* [6.52]. Therefore the above authors obtained the FSPA validity condition in the form of a single inequality (6.3.28). Moreover, since no distinction was made between good and poor acceptors, the recurrence period was expressed via the total density of final states rather than the density of the effective states. But then, inequality (6.3.28) is merely the irreversibility criterion for the electronic transition in an isolated polyatomic molecule (see Sect. 2.4.2).

6.3.3 Saddle-Point Method Versus Effective-States Method

In summary, we have two methods to calculate the statistical-limit nonradiative transition rate, viz., the effective-states method and the saddle-point method based on the FSPA (two more methods are discussed at the end of this chapter). The former can be used when the Franck–Condon factor is available explicitly in the form of an analytically simple smooth function of the final vibrational quantum numbers, while the latter is of value when the generating function is known. Both methods yield identical results in the cases when they both apply. However, the effective-states method has the obvious advantage that it explicitly calculates the density of the effective states, ϱ_{eff}, and the knowledge of ϱ_{eff} along with a reasonable estimate for the vibrational level width, γ, makes it possible to directly check the fulfilment of the statistical-limit condition (2.11.2). If this condition is not met, but the total density of the final vibrational states with vibrational energy ΔE is large [see (2.11.4)], the nonradiative transition under study is still expected to occur in the isolated molecule, although its rate constant has to be calculated either from (3.4.9) or by the saddle-point method involving all the saddle points and level broadening, the resulting rate constant being significantly dependent upon broadening.

6.4 Single Vibronic Level (SVL) Transition-Rate Dependence upon Initial Vibrational Energy

Let us find out how the effective states and transition rates behave for Single Vibronic Levels (SVL) of the initial electronic state with nonzero quantum numbers m_μ. The initial vibrational energy will be assumed small as compared to the electronic gap,

$$E_{\text{vib}} \equiv \sum_\mu \hbar\omega_\mu m_\mu \ll \Delta E \,. \tag{6.4.1}$$

First, consider the model of displaced undistorted oscillators. From the energy-conservation law

$$\sum_\mu \hbar\omega_\mu (m'_\mu - m_\mu) = \Delta E \tag{6.4.2}$$

and (6.4.1) it follows that the quantum numbers of good acceptors are large, and hence a quasiclassical expression for the overlap integral can be used that is easily derived from the results of Sect. 5.2. The Franck–Condon factor is [5.8]

$$F_{rmr'm'} = \prod_\mu \frac{1}{m_\mu!(2\pi m'_\mu)^{1/2}} \exp[-2(m'_\mu - m_\mu)\ln[(m'_\mu - m_\mu)/\sqrt{\xi_\mu}] \\ + m'_\mu \ln m'_\mu + m'_\mu - 2m_\mu - \xi_\mu] \,. \tag{6.4.3}$$

We compose the Lagrange function (6.1.25) and, by solving (6.1.26), find the set of final vibrational quantum numbers (6.1.23) that maximize the Franck–Condon factor under the constraint of (6.4.2). After omission of the small term $1/2m'_\mu$ we arrive at quadratic equation in $m'_{m\mu}$ with the solution

$$m'_{m\mu} = m_\mu + \frac{1}{2} m'_{0\mu} e^{\omega_\mu \delta_m} \left[1 + \left(1 + \frac{4m_\mu}{m'_{0\mu}} e^{-\omega_\mu \delta_m} \right)^{1/2} \right], \tag{6.4.4}$$

where $m'_{0\mu}$ is given by (6.1.8) and $\delta_m = \lambda_m - \lambda_0$; the latter quantity obeys an equation that is obtained by the substitution of (6.4.4) into (6.4.2).

The transition rate is written in the form of (2.7.21), where $v_{\rm eff}$ and $\varrho_{\rm eff}$ are given by (6.1.27, 29), respectively, in which

$$D_\mu = m'_{m\mu} \frac{m'_{m\mu} - m_\mu}{m'_{m\mu} + m_\mu}, \qquad \Omega_{\mu r'} = \omega_\mu , \tag{6.4.5}$$

and the maximum Franck–Condon factor $F_{rmr'm'_m}$ is obtained from (6.4.3), where m'_μ has to be replaced with (6.4.4).

It is easily verified [see (6.4.11, 12)] that, owing to condition (5.2.51) for applicability of (6.4.3), both the increment of the Lagrangian multiplier (δ_m) and the relative increment of the final quantum numbers $m'_{m\mu}$ between $m = 0$ and $m \neq 0$ are small, on the order of $m_\mu/m'_{0\mu}$. The change of D_μ in (6.4.5), and hence that of the density of the effective states (6.1.29) will be on the same order of magnitude. Neglecting this small effect, and replacing (within the same approximation) $m'_{m\mu}$ with $m'_{0\mu}$ in the preexponential factor of (6.4.3), we obtain the following expression for the transition rate from the rm state

$$k_{rm} = k_{r0} \left(\prod_\mu \frac{1}{m_\mu!} \right) \exp\left[-\frac{1}{\hbar} \Delta E \delta_m + \sum_\mu (m'_{m\mu} - m'_{0\mu} - 2m_\mu) + \sum_\mu m_\mu \ln m'_{m\mu} \right], \tag{6.4.6}$$

where k_{r0} is given by (6.1.19).

Equation (6.4.6) can be simplified, provided that only the lowest vibronic levels are considered whose quantum numbers fulfil the stronger inequality

$$m'_{0\mu} \gg m_\mu^2 , 1 , \tag{6.4.7}$$

as compared to (5.2.51). Expanding the square root of (6.4.4) in powers of $m_\mu/m'_{0\mu}$, and keeping only the linear term, we obtain

$$m'_{m\mu} = m'_{0\mu} \exp(\omega_\mu \delta_m) + 2m_\mu . \tag{6.4.8}$$

The substitution of (6.4.8) into (6.4.2) yields an equation for δ_m:

$$\sum_\mu \hbar\, \omega_\mu [m'_{0\mu} \exp(\omega_\mu \delta_m) + m_\mu] = \Delta E , \tag{6.4.9}$$

which is transformed with the use of (6.1.9) to

$$\sum_{\mu} \omega_{\mu} m'_{0\mu} [\exp(\omega_{\mu} \delta_m) - 1] = - \sum_{\mu} \omega_{\mu} m_{\mu} \, . \tag{6.4.10}$$

Expanding the exponential on the left-hand of (6.4.10) results in

$$\delta_m = - \sum_{\mu} \omega_{\mu} m_{\mu} / \sum_{\mu} \omega_{\mu}^2 m'_{0\mu} \, , \tag{6.4.11}$$

whence it is seen that δ_m is a small quantity on the order of $m_{\mu}/m'_{0\mu}$. Therefore, we expand (6.4.8) to within the same accuracy to

$$m'_{m\mu} = m'_{0\mu} + \omega_{\mu} m'_{0\mu} \delta_m + 2 m_{\mu} \, , \tag{6.4.12}$$

and substitute this into (6.4.6). Neglecting small terms on the order of $m_{\mu}^2/m'_{0\mu}$ we finally obtain

$$k_{rm} = k_{r0} \prod_{\mu} (m'_{0\mu})^{m_{\mu}} / m_{\mu}! \, . \tag{6.4.13}$$

Equation (6.4.13) predicts a rapid increase in the transition rate with the initial quantum numbers of the good acceptors since $m'_{0\mu}$ is large. The better the acceptor (i.e., the larger the parameter $m'_{0\mu}$), the steeper is the increase; in particular it is the steeper, the larger ΔE. Hence, it is obvious that increasing the energy of the initial vibrational excitation of good acceptors may result in a violation of the Ermolaev-Sveshnikova rule (see Sect. 2.2) since the rate of the $S_1 \rightsquigarrow S_0$ internal conversion rises more rapidly than that of the $S_1 \rightsquigarrow T_1$ intersystem crossing [2.54, 63, 6.61–63] (however, the validity of the statistical limit for ISC needs a special consideration since at small ΔE the ISC may belong to the intermediate case and show a nonmonotonic dependence on E_{vib} [6.64]). Exceptions to this rule were found in [4.136, 6.65, 66] where transitions insensitive to E_{vib} were observed.

Equation (6.4.13) can also be deduced by the saddle-point method invoking the generating function (6.2.15). First of all we note that at the saddle point (6.3.3) the argument of the Laguerre polynomial is approximately $-\xi_{\mu}\exp(\omega_{\mu}\delta_0) \approx -m'_{0\mu}$, so the Laguerre polynomial of the generating function can be replaced with its asymptotic representation at large negative values of the argument,

$$L_m(x) \approx (-x)^m/m! \, . \tag{6.4.14}$$

Thus, the generating function and the saddle point equation are

$$\mathscr{F}_{rm}(t) = \prod_{\mu} \frac{1}{m_{\mu}!} \exp[\xi_{\mu}(\mathrm{e}^{\mathrm{i}\omega_{\mu}t} - 1) + \mathrm{i} m_{\mu}\omega_{\mu} t + m_{\mu} \ln \xi_{\mu}] \, , \tag{6.4.15}$$

$$\sum_{\mu} \hbar\omega_{\mu} \xi_{\mu} \exp(\mathrm{i}\omega_{\mu} t) = \Delta E - \sum_{\mu} \hbar\omega_{\mu} m_{\mu} \, . \tag{6.4.16}$$

Now we look for a saddle point of the form

$$t = -\mathrm{i}\tau_m \, . \tag{6.4.17}$$

Recall that, according to Sect. 6.3, the parameter $\tau_0 = \tau_{m=0}$ coincides with the Lagrangian multiplier λ_0 in the effective-states method. Similarly, τ_m is identical with λ_m, since after substitution of (6.4.17) into (6.4.16) and use of (6.1.8) we obtain (6.4.9).

The calculation of the integral in (6.2.2) with the generating function (6.4.15) at the saddle point (6.4.17) is trivial. Notice that, in fact, the difference between τ_m and τ_0 can be neglected provided condition (6.4.7) is met. In fact, in the expansion of the exponent in powers of $\delta_m = \tau_m - \tau_0$ the linear term does not appear since the first derivative vanishes at any saddle point, while the quadratic term provides small corrections on the order of $m_\mu^2/m'_{0\mu}$. As a result, (6.4.13) is easily obtained.

The replacement of τ_m with τ_0 has previously been proposed [4.21, 6.32] to evaluate the SVL $S_1 \rightsquigarrow T_1$ radiationless transition rates. However, one should bear in mind that at not large-enough values of the electronic energy gap both the statistical limit condition (2.11.2) (in other words, the applicability condition for the saddle-point or effective-states method as such) and inequality (6.4.7) allowing to disregard the difference between τ_m and τ_0 may be violated.

Let us see now in which way the transition rate varies when exciting the poor acceptors, which will be labeled with the subscript a. The effective-states method does not work here at all since it does not incorporate the poor acceptors, insignificantly affecting the transition rate as compared to the good ones. Therefore, we shall apply the saddle-point method in which the distinction between the poor and good acceptors is taken into account only at the stage of solving the saddle-point equation.

The total generating function is obtained by multiplying (6.4.15) with a product of terms like (6.2.15) corresponding to the poor acceptors,

$$\prod_a \exp[\xi_a(\mathrm{e}^{\mathrm{i}\omega_a t} - 1)]\, L_{m_a}\!\left(4\xi_a \sin^2 \frac{\omega_a t}{2}\right). \tag{6.4.18}$$

Recall that by criterion (6.1.11) of good acceptors the inequalities

$$\xi_\mu \exp(\mathrm{i}\omega_\mu t) \gg 1 \quad \text{and} \quad \xi_a \exp(\mathrm{i}\omega_a t) \ll 1 \tag{6.4.19}$$

are fulfilled for good and poor acceptors, respectively. The latter inequality permits a neglect of the poor acceptors' contribution to the saddle-point equation (see, e.g. (6.3.2) in the $m = 0$ case) so that the saddle point is controlled by the good acceptors only. Hence, when calculating the transition rate (6.2.2) we may take the factor (6.4.18) off the integral sign at $t = -\mathrm{i}\tau_m$. Thus, the dependence of the transition rate on the quantum numbers m_a is given by

$$k_{r\{m_\mu m_a\}} = k_{r\{m_\mu 0_a\}} \prod_a L_{m_a}\left(-4\xi_a \sinh^2 \frac{\omega_a \tau_m}{2} \right), \tag{6.4.20}$$

where $k_{r\{m_\mu 0_a\}}$ is given by (6.4.13) or by a more general formula (6.4.6).

Because of (6.4.19), the argument of the Laguerre polynomial in (6.4.20) is small, the approximate relation

$$\ln L_m(x) \approx -mx \tag{6.4.21}$$

can be employed to recast (6.4.20) in the form [2.237]

$$k_{r\{m_\mu m_a\}} = k_{r\{m_\mu 0_a\}} \exp\left[\sum_a 2\xi_a m_a (\cosh \omega_a \tau_m - 1) \right]. \tag{6.4.22}$$

This equation predicts a slow increase in the transition rate with the quantum numbers of the poor acceptors. However, the exponential dependence on the quantum numbers was also invoked in [6.37, 43, 45] to account for a rapid initial increase in the SVL transition rate with E_{vib}.

Brailsford and *Chang* [6.67] calculated the transition-rate dependence on the initial vibrational quantum numbers by the saddle-point method, and the generating function was found by a straightforward expansion of the Franck–Condon factors in powers of small displacement parameters ξ, which, virtually, corresponds to expansion (6.4.21); it is easily verified [2.237] that if expansion (6.4.21) is inserted into the exact generating function (6.4.18), one obtains just the Brailsford–Chang generating function. However, the authors utilized the above expansion not only for the poor, but also for the good acceptors, which is illegitimate, because in the latter case, as we saw above, the Laguerre polynomials' arguments are large and hence the approximation of (6.4.14) should be employed. Therefore [6.67], (III.17) gives an incorrect dependence on the quantum numbers of the good acceptors m_μ, while the dependence upon those of the poor ones, m_a, at $m_\mu = 0$ is reduced after some simplifications to the form of (6.4.22). Numerical investigations of the SVL transition-rate dependence upon the initial vibrational quantum numbers were carried out in [1.12, 2.54, 204, 313, 314, 316, 4.21, 6.68].

When the frequency changes of the good acceptors are taken into consideration, significant mathematical complications arise in the saddle-point method because of a complicated form of the generating function [4.21], and the effective-states method may fail because the Franck–Condon factor is no longer a smooth function of the final quantum numbers m'_μ. For instance, at $\xi_\mu = 0$ it is nonzero only for $m'_\mu - m_\mu$ even [see (4.8.7)], and for $\xi_\mu \neq 0$ and $f_\mu > 1$ it oscillates [see (5.2.60)]. However, in the case of

$$\xi_\mu \gg \frac{1 - f_\mu}{m'_\mu} > 0 \tag{6.4.23}$$

the Franck–Condon factor is a smooth function, and the transition rate from an $m \neq 0$ initial state can be calculated by the effective-states method. For the sake of simplicity we shall assume that there are N good acceptors with equal parameters, $\xi_\mu \equiv \xi$, $f_\mu \equiv f$ and $\omega_{\mu r'} \equiv \omega_{r'}$. Then, the one-oscillator Franck–Condon factor is given by (5.2.59), where the second term should be omitted since $B \gg 1$ in view of (6.4.23). The total Franck–Condon factor is

$$F_{rmr'm'} = \beta \prod_\mu \frac{C^{2m_\mu}}{m_\mu!}(m'_\mu)^{m_\mu - 1/2} \exp\left[-(m'_\mu - m_\mu)\omega_{r'} s + 4\left(\frac{\xi f m'_\mu}{1-f^2}\right)^{1/2}\right],$$

$$C = \frac{2\sqrt{f}}{f+1}, \tag{6.4.24}$$

$$\beta = \left[\frac{C}{2\sqrt{2\pi}} \exp\left(-\frac{2\xi}{1-f^2}\right)\right]^N, \qquad s = \frac{1}{\omega_{r'}} \ln \frac{1+f}{1-f}.$$

We compose the Lagrange function (6.1.25), omitting insignificant terms, as

$$L_m = \sum_\mu \left[m'_\mu \omega_{r'}(\lambda_m - s) + 4\left(\frac{\xi f m'_\mu}{1-f^2}\right)^{1/2}\right],$$

and find the quantum numbers (6.1.23) by solving (6.1.26),

$$m'_m = \frac{4\xi f}{\omega_{r'}^2 (s - \lambda_m)^2 (1-f^2)}. \tag{6.4.25}$$

As they are equal for all good acceptors, the subscript μ is omitted. This equation enables a determination of m'_m directly from the energy-conservation condition

$$\sum_\mu [\hbar\omega_{r'}(m'_\mu + 1/2) - \hbar\omega_r(m_\mu + 1/2)] = \Delta E.$$

On letting $m'_\mu \equiv m'_m$ one finds

$$m'_m = \frac{\Delta E}{N\hbar\omega_{r'}} + \frac{f}{N}\sum_\mu m_\mu - \frac{1-f}{2}. \tag{6.4.26}$$

Equations (6.4.25, 26) allow one to determine the Lagrangian multiplier, but it is of no use as the final quantum numbers are already known.

The parameters $D_\mu \equiv D$ and ϱ_{eff} from (6.1.28, 29) are

$$D = \left(\frac{1-f^2}{\xi f}\right)^{1/2} (m'_m)^{3/2},$$

$$\varrho_{\text{eff}} = \frac{1}{\hbar\omega_{r'}\sqrt{N}} (2\pi D)^{(N-1)/2}.$$

Since $D > 0$, the quantum numbers (6.4.26) correspond to the maximum Franck–Condon factor,

$$F_{rmr'm'_m} = \beta \prod_\mu \frac{C^{2m_\mu}}{m_\mu!} (m'_0)^{m_\mu - 1/2} \exp\left[-(m'_m - m_\mu)\omega_{r'} s + 4\left(\frac{\xi f m'_0}{1-f^2}\right)^{1/2} \right],$$

where $m'_0 \approx \Delta E / N\hbar\omega_{r'}$. This equation results from (6.4.24) after m'_m is substituted for m'_μ and then [wherever possible under condition (6.4.7)] m'_0 for m'_m; when calculating ϱ_{eff} one can set $m'_m = m'_0$ as well. After simple calculations the transition rate is obtained in the form

$$k_{rm} = k_{r0} \prod_\mu \frac{A^{m_\mu}}{m_\mu!}, \tag{6.4.27}$$

with

$$A = m'_0 C^2 \theta^{(f-1)/2},$$

$$k_{r0} = \frac{v_{\text{el}}^2 Q^N}{\hbar^2 \omega_{r'}} \left(\frac{2\pi}{DN}\right)^{1/2} \exp(-\Delta E s/\hbar) \tag{6.4.28}$$

and

$$Q = \frac{C}{2}\left(\frac{1-f^2}{\xi f} m'_0 \theta^{f-1}\right)^{1/4} \exp\left[-\frac{2\xi}{1-f^2} + 4\left(\frac{\xi f m'_0}{1-f^2}\right)^{1/2} \right].$$

Equation (6.4.28) can be also deduced by the saddle-point method using the generating function (6.2.16). The saddle point is near $t = -is$, and, despite the singularity the preexponential factor has at that point, a simple analysis shows that its influence on the saddle point is negligible under condition (6.4.23).

The increase in the radiationless transition rate with the initial vibrational excitation energy is experimentally observed in low-pressure gas phase and in free-jet expansions where the collisional relaxation rate of the excess vibrational energy in the initial electronic state is slow as compared to the radiative and nonradiative transition rates in molecules [1.12, 35, 2.55, 71, 229b, 280, 307, 315, 4.130, 146, 6.44, 49, 60, 64, 69–78] and in molecular ions [2.108]. This effect also explains the phenomenon, revealed by *Neporent* (see Sect. 2.2), of the fluorescence stabilization (enhancement) upon addition of foreign inert gases that increases the initial excess vibrational energy relaxation rate and thus decreases the radiationless transition rates.

6.5 Transition Rate from Statistically Equilibrated Initial States

The theory outlined in the preceding section pertains to such experimental situations where the excitation process (usually the photoexcitation) prepares the molecule in a certain Single Vibronic Level (SVL) *rm*, and the total lifetime of that state is short as compared to the characteristic times for energy redistribution and relaxation in the initial electronic state so that emission occurs from that very *rm* state. This situation is encountered in low-pressure vapors (pressure less than 1 Torr) and free jets for S_1 vibrational sublevels near the $m = 0$ ground vibrational state at the S_1 origin. With an increase in the initial vibrational excitation energy E_{vib} (which is initiated, for instance, by increasing the exciting radiation frequency ω_{ex}) both the vibronic-states density and the IVR rate rapidly increase. The former effect eventually leads to the impossibility to excite SVL in principle because of a finite spectral bandwidth of the exciting radiation (see Sect. 2.4.1); the latter to establishing the statistical initial vibrational energy distribution among all vibrational degrees of freedom which can be characterized by an effective temperature T, a function of E_{vib} (*Fischer*s communicating states model [6.79], see also [2.202, 6.32]). For radiationless transitions from the S_1 singlet state E_{vib} is the difference between $\hbar\omega_{\mathrm{ex}}$ and the $\Delta E_{S_1S_0}$ energy gap so that T is a function of ω_{ex}, while in the case of transitions from the T_1 triplet state prepared in the course of the $S_1 \rightsquigarrow T_1$ intersystem crossing E_{vib} is $\Delta E_{S_1T_1}$. The energy equilibration within the initial electronic state is conventionally discussed in terms of the communicating states model and vibrational temperature [2.51, 61, 6.37, 41, 43, 45, 49, 80, 81].

In high-pressure vapors and condensed media the molecule–environment interaction gives rise to an excess vibrational energy relaxation rapidly establishing the statistical equilibrium in the initial electronic state. In this case the effective temperature T coindices with the temperature of the medium, and is independent of the initial excess vibrational energy E_{vib}. There is a wealth of experimental data on the equilibrium-temperature dependence of the radiationless transition rates [1.1–3, 15, 35, 2.275, 4.119, 6.35, 82–84].

The thermally equilibrated transition rate (6.2.6) for $T \neq 0$ can be evaluated by the Langer method [6.85] employing classical trajectories in imaginary time [5.12, 6.17, 86]. Here it is calculated by the saddle-point generating-function method. The original integral for the average transition rate of (6.2.6, 18) is

$$k_r(T) = (v_{\mathrm{el}}^2/\hbar^2) \int_{-\infty}^{\infty} \exp[\psi(t)]\, dt \,,$$

$$\psi(t) = -\frac{\mathrm{i}}{\hbar}\Delta E t + \sum_{\mu} \xi_\mu[\exp(\mathrm{i}\omega_\mu t) - 1] - \sum_{\mu} 2\xi_\mu \bar{m}_\mu(1 - \cos\omega_\mu t)\,, \qquad (6.5.1)$$

where $\bar{m}_\mu$ are the mean occupation numbers of (6.2.19). On equating $\psi'(t) = 0$ we look for a saddle point of the form $t = -\mathrm{i}\tau_T$ [cf. (6.3.3)] and obtain the following equation for τ_T:

$$\sum_{\mu} \hbar\omega_\mu \xi_\mu[\exp(\omega_\mu\tau_T) + 2\bar{m}_\mu \sinh\omega_\mu\tau_T)] = \Delta E\,. \qquad (6.5.2)$$

The calculation of the integral in (6.5.1) in the first saddle-point approximation yields

$$k_r(T) = (v_{\text{el}}^2/\hbar^2)(2\pi/|\psi''|)^{1/2}\,\mathrm{e}^{\psi}\,, \tag{6.5.3}$$

$$\psi = -\Delta E\tau_T/\hbar - \sum_\mu \xi_\mu(2\bar{m}_\mu + 1) + \sum_\mu \xi_\mu[\exp(\omega_\mu\tau_T) + 2\bar{m}_\mu\cosh\omega_\mu\tau_T]\,, \tag{6.5.4}$$

$$\psi'' = -\sum_\mu \xi_\mu\omega_\mu^2[\exp(\omega_\mu\tau_T) + 2\bar{m}_\mu\cosh\omega_\mu\tau_T]\,. \tag{6.5.5}$$

In the limit of $T \to 0$ (6.5.3) reduces to k_{r0} of Sect. 6.3.1 [see (6.1.19)].

The above equations describe the radiationless transition-rate dependence upon the effective temperature T in frames of the simplest model of displaced undistorted potential surfaces with the perturbation operator independent of the nuclear coordinates. The only validity condition for these equations is the statistical limit criterion (2.11.2) in which ϱ_{eff} is the density of the effective states for the transition from the initial state with the statistically averaged quantum numbers $\bar{m}_\mu$. In fact, it is sufficient to fulfil this condition at $T = 0$ (or at $m_\mu = 0$) with ϱ_{eff} from (6.1.18) since ϱ_{eff} increases with the initial vibrational excitation (see Sect. 6.4).

In the above formulae the sums formally involve all the normal vibrations since no distinction was made between good and poor acceptors. In fact, however, the dominant contributions to (6.5.2, 5) come from the good acceptors (see Sect. 6.4). Therefore, if the subscript μ now labels as in Sect. 6.4, the good acceptors only, while the poor acceptors are labelled by the subscript a, then (6.5.2, 5) will remain unchanged, whereas the corresponding sums over a are to be added to (6.5.4).

Consider first the low-temperature limit when the mean occupation numbers of the good acceptors are small,

$$\bar{m}_\mu \ll 1\,. \tag{6.5.6}$$

Let τ_0 denote the value of τ_T at $T = 0$; it is equal to the Lagrangian multiplier λ_0 of Sect. 6.1 since at $T = 0$ (6.5.2) is identical with (6.1.9). Expanding the left-hand side of (6.5.2) in powers of $\delta_T = \tau_T - \tau_0$ one obtains

$$\delta_T = -\sum_\mu 2\omega_\mu\xi_\mu\bar{m}_\mu\sinh(\omega_\mu\tau_0)/\sum_\mu \omega_\mu^2 m'_{0\mu}\,, \tag{6.5.7}$$

where $m'_{0\mu}$ is given by (6.1.8) with $\lambda_0 = \tau_0$. A similar expansion is to be performed in (6.5.3) as well. In the preexponential factor one can set $T = 0$, while the exponent is expanded as

$$\psi = \psi_{T=0} + \sum_{\nu=\mu,a} 2\xi_\nu\bar{m}_\nu(\cosh\omega_\nu\tau_0 - 1) + \frac{1}{2}\delta_T^2\sum_\mu \omega_\mu^2 m'_{0\mu} \tag{6.5.8}$$

(the term linear in δ_T is absent since $\psi'(t)$ vanishes at the saddle point). Keeping only the terms linear in $\bar{m}_\mu$ we obtain the following transition-rate expression:

$$k_r(T) = k_r(0)\exp\left(\sum_{\nu=\mu,a} \zeta_\nu \bar{m}_\nu\right), \tag{6.5.9}$$

$$\zeta_\nu = 2\xi_\nu(\cosh \omega_\nu \tau_0 - 1), \tag{6.5.10}$$

where $k_r(0)$ coincides with k_{r0} of (6.1.19).

In the weak coupling limit of (5.2.51) it follows from (6.4.19) that for the good acceptors

$$\omega_\mu \tau_0 \gg 1 . \tag{6.5.11}$$

Using this inequality along with (6.5.10) and (6.4.19) we obtain

$$\zeta_\mu = m'_{0\mu} \gg 1 \quad \text{and} \quad \zeta_a \ll 1 \tag{6.5.12}$$

for the good and poor acceptors, respectively.

In fact, the τ_T dependence on temperature was neglected in the derivation of (6.5.9) since we set $\tau_T = \tau_0$ in the exact equation (6.5.4). The validity condition for this approximation is that the term omitted in (6.5.8) would be small as compared to that with δ_T^2; owing to (6.5.7,11) that condition is of the form

$$\left(\sum_\mu \omega_\mu m'_{0\mu} \bar{m}_\mu\right)^2 \ll 2\sum_\mu \omega_\mu^2 m'_{0\mu} ,$$

or approximately

$$\bar{m}_\mu \ll (1/N)(2/m'_{0\mu})^{1/2} , \tag{6.5.13}$$

which is a more severe limitation on $\bar{m}_\mu$ than was suggested in (6.5.6) as $m'_{0\mu}$ and the mean number of modes, N, are large.

We consider, in more detail, the effective temperature dependence of the transition rate predicted by (6.5.9). First suppose that the frequencies, hence the mean occupation numbers of the good and poor acceptors, are not very different. Then, in view of (6.5.12), the poor acceptor's contribution can be neglected at all, and the transition rate assumes the form

$$k_r(T) = k_r(0)\exp\left(\sum_\mu m'_{0\mu}\bar{m}_\mu\right). \tag{6.5.14}$$

Suppose that there are N good acceptors with equal frequencies ω and parameters ξ. Then $\bar{m}$ and m'_0 (the subscript μ is suppressed) can be expressed in terms of the vibrational (E_{vib}) and electronic (ΔE) energies:

$$\bar{m} = E_{\text{vib}}/N\hbar\omega , \qquad m'_0 = \Delta E/N\hbar\omega \tag{6.5.15}$$

(in the latter expression an insignificant addition of E_{vib} to ΔE is ignored), and

(6.5.14) takes the form

$$k_r(T) = k_r(0)\exp(\alpha E_{vib}) ,$$
$$\alpha = \Delta E/N(\hbar\omega)^2 . \qquad (6.5.16)$$

If the initial vibrational excitation of the molecule is predominantly due to the difference between the absorbed photon energy $\hbar\omega_{ex}$ and the purely electronic transition energy E_0, then $E_{vib} = \hbar\omega_{ex} - E_0$ is to be substituted into (6.5.16) to yield Fischer's formula [6.79] which describes the radiationless transition-rate dependence upon the exciting radiation frequency. The exponential dependence predicted was experimentally observed in some aromatic hydrocarbons at E_{vib} up to 20000 cm^{-1} [2.58–61, 6.87, 88]. The linear dependence of α on ΔE was also revealed in [2.61]; the experimental α^{-1} values are between 1000 and 2000 cm^{-1}.

If the vibrational excitation is exclusively of thermal origin and is controlled by the medium temperature T, then, according to (6.5.15,6) and (6.2.19), we have $E_{vib} = N\hbar\omega\exp(-\hbar\omega/k_BT)$. Inserting this into (6.5.16) we obtain the temperature behavior of the transition rate. An interesting implication of (6.5.16) is the prediction of an appreciable temperature dependence of the transition rate at such low temperatures that the excited vibrational state populations are as small as demanded by (6.5.13). Indeed, for identical acceptors, from (6.5.13) we obtain that the exponent in (6.5.16) is bound from above only by a weak inequality $\alpha E_{vib} \ll \sqrt{2m'_0}$, and may assume large values because m'_0 is large. Therefore, the variations of the transition rate in the temperature interval allowed by (6.5.13) may be significant, though this interval is quite narrow because of the steep temperature variation of the occupation number $\bar{m}$.

If there are poor acceptors with low frequencies, $\omega_a \ll \omega_\mu$, they appreciably affect the transition rate behavior in the temperature range $\hbar\omega_a \ll k_BT \ll \hbar\omega_\mu$, where the occupation numbers are

$$\bar{m}_a \approx k_BT/\hbar\omega_a \gg 1 , \qquad \bar{m}_\mu \approx \exp(-\hbar\omega_\mu/k_BT) \ll 1 . \qquad (6.5.17)$$

In particular, for temperatures $\hbar\omega_a \ll k_BT \ll \hbar\omega_\mu/\ln Nm'_{0\mu}$ the good acceptors are completely "frozen out" (the exponent of (6.5.14) becomes lower than unity), and the temperature dependence of the transition rate is only controlled by the low-frequency poor acceptors. By (6.5.9, 17) it is of the form

$$k_r(T) = k_r(0)\exp\left(\sum_a \zeta_a k_BT/\hbar\omega_a\right) . \qquad (6.5.18)$$

For $T \to 0$ the exponent of (6.5.9) becomes small, and only one member of the sum corresponding to a certain low-frequency vibration may be retained. Then the transition rate becomes [2.313, 6.32, 89–93]

$$k_r(T) = k_r(0)[1 + \zeta_v\exp(-\hbar\omega_v/k_BT)] , \qquad (6.5.19)$$

where ζ_v and ω_v are parameters of that vibration.

When the temperature rises and/or the electron-vibration coupling becomes stronger (the parameters ξ_μ increase), τ_T monotonically decreases, which immediately follows from (6.5.2). Consider the strong-coupling high-temperature limit when inequality

$$\omega_\mu \tau_T \ll 1 , \tag{6.5.20}$$

opposite to (6.5.11), is fulfilled. This case plays an especially important role in the theory of local electronic centers in solids [1.15, 16] since there low-frequency modes always exist. Expanding the left-hand side of (6.5.2) in powers of τ_T to the lowest order gives

$$\tau_T = \hbar(\Delta E - E_c)/D , \tag{6.5.21}$$

$$E_c = \sum_\mu \hbar\omega_\mu \xi_\mu , \qquad D = \sum_\mu \hbar^2 \omega_\mu^2 \xi_\mu (2\bar{m}_\mu + 1) . \tag{6.5.22}$$

Similar expansion should be performed in (6.5.3–5) to obtain

$$k_r(T) = \frac{v_{\rm el}^2}{\hbar} (2\pi/D)^{1/2} \exp[-(\Delta E - E_c)^2/2D] , \tag{6.5.23}$$

where E_c is the Stokes energy, i.e., the energy of vibrational excitation that would be created at a vertical $r \to r'$ optical transition from the equilibrium configuration of the initial electronic state. In the theory of optical transitions E_c coincides with the Stokes shift of the absorption and emission-band maxima under operation of the mirror symmetry law for optical spectra [1.15, 16, 22, 24, 25, 4.2].

The routine way to derive (6.5.23) is to expand $\psi(t)$ of (6.5.1) in powers of t up to the t^2 term inclusive [2.202, 273, 6.26]. The main validity condition for (6.5.23) is inequality (6.5.20), where τ_T is given by (6.5.21).

At high temperatures one has $\bar{m}_\mu = k_B T/\hbar\omega_\mu$, and the transition rate is represented by the Arrhenius equation

$$k_r(T) = \frac{v_{\rm el}^2}{\hbar} (\pi/k_B T E_c)^{1/2} \exp(-E_a/k_B T) \tag{6.5.24}$$

with the activation energy

$$E_a = (\Delta E - E_c)^2/4E_c . \tag{6.5.25}$$

E_a is the energy distance from the minimum of the potential surface of the initial electronic state to the lowest point of its intersection with the surface of the final state. This is made certain by solving the problem on the conditional extremum: $E_a = \min \mathscr{E}_r(q) - \varepsilon_r$ with the additional condition $\mathscr{E}_r(q) = \mathscr{E}_{r'}(q)$, where $\mathscr{E}_{r,r'}(q)$ are given by (4.6.2, 4, 6). The reader can easily verify that the solution to this problem obtained by the Lagrange method leads to (6.5.25).

Thus at high temperatures the transition is of the activation-type and is effectuated via the energy barrier E_a, whereas at low temperatures it proceeds via tunneling with a rate proportional to the overlap integral squared of the vibrational wave functions of the initial and final states. This is quite similar to the behavior of the transitions between different minima of one and the same potential surface [6.94].

Kato [2.64] has calculated the benzene S_0 and S_1 potential surfaces and found an intersection point near the saddle for the isomerization reaction (similar calculations on triplet states of benzene and pyrazine were performed in [6.40]). The $S_1 \rightsquigarrow S_0$ internal conversion-rate constant calculation was based on the concepts of reaction path and surface hopping. The result was that the internal conversion proceeds by the tunneling mechanism for $E_{\text{vib}} < 3700\ \text{cm}^{-1}$, while for $E_{\text{vib}} > 3700\ \text{cm}^{-1}$ the surface hopping followed by the isomerization comes into play. This offers a novel explanation to the benzene channel 3 [4.158]. A similar idea of a radiationless-transition-induced chemical reaction was developed in [2.52, 6.95].

In the pioneering works by *Huang* and *Rhys* [4.163] and *Krivoglaz* [6.16] the thermally equilibrated radiationless transition-rate constant of a local optical center in the crystal was calculated under the assumptions that all phonon frequencies are equal (the Einstein crystal) and the electronic energy gap is a multiple of the vibrational quantum:

$$\omega_\mu = \omega\,, \qquad \Delta E = \hbar\omega p\,, \tag{6.5.26}$$

where p is an integer. The calculation was performed by a replacement of the delta function with the quantity $1/\hbar\omega$ followed by a Franck–Condon factor summation over the final states conserving the initial energy (see the next section). The result obtained was very cumbersome since the perturbation operator inducing the transition was the non-adiabaticity operator (4.1.17) depending on derivatives over the nuclear coordinates. Therefore, for the sake of comparison, we shall employ the same method to calculate the transition rate in a simpler situation where the perturbation operator is independent of the nuclear coordinates, and we demonstrate that the result exactly coincides with (6.5.3) after relations (6.5.26) are substituted there. Replacement of the delta function with $1/\hbar\omega$ in the general equation (6.2.1) yields

$$k_{rm} = \frac{2\pi v_{\text{el}}^2}{\hbar^2\omega} \sum_{m'} F_{rmr'm'}\,, \tag{6.5.27}$$

where the summation over the final states is restricted by the energy conservation condition which, in view of (6.5.26), has the form $\sum_\mu (m'_\mu - m_\mu) = p$. We define a function $\Delta_{mm'}$ similar to the Kronecker symbol: $\Delta_{mm'} = 1$ at $E_{r'm'} = E_{rm}$ and 0 otherwise. Then (6.5.27) is recast as

$$k_{rm} = \frac{2\pi v_{\text{el}}^2}{\hbar^2\omega} \sum_{m'} F_{rmr'm'}\Delta_{mm'}\,, \tag{6.5.28}$$

where the summation is now over all final states. $\Delta_{mm'}$ is written as the integral

$$\Delta_{mm'} = \frac{\omega}{2\pi} \int\limits_{-\pi/\omega}^{\pi/\omega} \exp[\mathrm{i}(E_{r'm'} - E_{rm})t/\hbar]dt \,. \tag{6.5.29}$$

Substituting (6.5.29) into (6.5.28) and performing the Boltzmann averaging we find

$$k_r(T) = (v_{\mathrm{el}}/\hbar)^2 \int\limits_{-\pi/\omega}^{\pi/\omega} \mathscr{F}_r(t, T)\exp(-\mathrm{i}\Delta Et/\hbar)dt \,,$$

where $\mathscr{F}_r(t, T)$ is the generating function (6.2.8) equal to the product of the expressions (6.2.18). Thus

$$k_r(T) = \frac{v_{\mathrm{el}}^2}{\hbar^2\omega} \int\limits_{-\pi}^{\pi} \exp[-\mathrm{i}px + \xi(\mathrm{e}^{\mathrm{i}x} - 1) - 2\xi\bar{m}(1 - \cos x)]dx \,,$$

where ξ is given by (6.1.20). The integral is expressed through the Bessel function of imaginary argument after the integrand is expanded as a double series in powers of $\exp(\mathrm{i}x)$ and $\exp(-\mathrm{i}x)$. An elementary calculation yields

$$k_r(T) = \frac{2\pi v_{\mathrm{el}}^2}{\hbar^2\omega} \exp\left[\frac{p}{2k_{\mathrm{B}}T} - \xi(2\bar{m} + 1)\right] I_p(z) \,, \tag{6.5.30}$$

where $z = 2\xi[\bar{m}(\bar{m} + 1)]^{1/2}$ and $\bar{m}$ is the average quantum number (6.2.19). Equation (6.5.30) is similar to that of [4.163, 6.16] obtained with use of the non-adiabaticity operator.

It should be taken into account that, in fact, p is a large number, and hence, the Bessel function $I_p(z)$ has to be replaced with an appropriate asymptotic representation applicable at large values of the index and arbitrary values of the argument [4.164]. It is easily verified that the ensuing relation precisely coincides with (6.5.3) after (6.5.26) is inserted into (6.5.3). Omitting the details of the calculation we give the final result,

$$k_r(T) = \frac{v_{\mathrm{el}}^2}{\hbar^2\omega} \sqrt{2\pi}(p^2 + z^2)^{-1/4} \exp\left[\frac{\hbar\omega p}{2k_{\mathrm{B}}T} - \xi(2\bar{m} + 1) \right.$$
$$\left. + (p^2 + z^2)^{1/2} - p\ln\frac{p + (p^2 + z^2)^{1/2}}{z}\right] . \tag{6.5.31}$$

For $T \to 0$ this formula reduces to (6.1.21).

6.6 Summation of the Franck–Condon Factors

A straightforward way to calculate the statistical-limit transition-rate constant (6.1.22) consists of a direct summation of the Franck–Condon factors subject to the energy conservation restriction imposed by the delta function. This method was elaborated by *Metz* [6.52, 14] with the use of an algorithm similar to that proposed by *Stein* and *Rabinovitch* [6.96] for calculations of the density of vibrational states.

In the model of f noninteracting oscillators without Duchinsky mixing, the Franck–Condon factors and energies are expressed in terms of the corresponding single oscillator quantities as

$$F_{rmr'm'} = \prod_{\nu=1}^{f} F_{rm_\nu r'm'_\nu} ,$$

$$E_{rm} = \varepsilon_r + \sum_{\nu=1}^{f} \varepsilon_{rm_\nu} , \qquad E_{r'm'} = \varepsilon_{r'} + \sum_{\nu=1}^{f} \varepsilon_{r'm'_\nu} . \tag{6.6.1}$$

Let us define the following energy function for the oscillators with numbers ν from 1 to $\mu \leq f$:

$$W_m^{(\mu)}(E) \equiv \hbar\omega_N \sum_{m'_1, \ldots, m'_\mu} \left(\prod_{\nu=1}^{\mu} F_{rm_\nu r'm'_\nu} \right) \delta_\gamma \left(E - \sum_{\nu=1}^{\mu} \varepsilon_{r'm'_\nu} \right) , \tag{6.6.2}$$

where m is again a collective notation for the set of the initial quantum numbers, and ω_N is a normalization factor to be defined later. Then the starting expression (6.1.22) becomes

$$k_{rm} = \frac{2\pi}{\hbar^2 \omega_N} v_{el}^2 W_m^{(f)} \left(\Delta E + \sum_{\nu=1}^{f} \varepsilon_{rm_\nu} \right) , \tag{6.6.3}$$

where ΔE is the electronic energy gap (4.5.18). The function (6.6.2) can be calculated by the recurrence relation

$$W_m^{(\mu)}(E) = \sum_{m'_\mu} F_{rm_\mu r'm'_\mu} W_m^{(\mu-1)}(E - \varepsilon_{r'm'_\mu}) , \tag{6.6.4}$$

which is easily verified for $\mu > 1$ by the direct substitution of (6.6.2). If we also define

$$W_m^{(0)}(E) = \hbar\omega_N \delta_\gamma(E) , \tag{6.6.5}$$

then (6.6.4) becomes valid for all μ in the interval $1 \leq \mu \leq f$. The summations in (6.6.2, 4) are formally over all final quantum numbers, though only the states satisfying the energy conservation restriction really contribute to the sums.

Further, a common normalizing frequency ω_N is introduced such that all relevant energies, E and $\varepsilon_{r'm'_\mu}$, become integer numbers in units of $\hbar\omega_N$ [2.204]. Suppose that this has already been done, and calculate the functions $W_m^{(\mu)}(E)$ simultaneously for all the energy values $E = 0, 1, \ldots, E_{max}$, where E_{max} is a value high enough to exceed the really needed energy. Notice that these functions are not yet well-defined since the shape of the broadened delta function in (6.6.5) has not been specified until now. Since the rate constant is independent of γ in the statistical limit, the shape of the delta function is of little physical importance, so it is taken in the form of the step function of width $\gamma = \hbar\omega_N$,

$$\delta_\gamma(E) = (\hbar\omega_N)^{-1}\delta_{E,0} , \tag{6.6.6}$$

where $\delta_{E,0}$ is the Kronecker symbol. Now we have all the accessories to calculate k_{rm}. The actual calculation proceeds as follows.

(i) Initialize the array $W_m^{(0)}(E)$ such that $W_m^{(0)}(0) = 1$ and $W_m^{(0)}(E) = 0$ for $E = 1, 2, \ldots, E_{max}$.

(ii) For $\mu = 1$ perform the summation in (6.6.4) for each E from $E = E_{max}$ to $E = 0$ in steps of -1, and place the calculated $W_m^{(1)}(E)$ value at the memory site of the $W_m^{(0)}(E)$ arrary corresponding to the same E since this particular $W_m^{(0)}(E)$ will no longer be needed in the subsequent calculations for lower E values.

(iii) Repeat point (ii) for $\mu = 2, 3, \ldots, f$.

This algorithm works fast for any number of degrees of freedom, and it does not suffer from the necessity to round off the electronic and vibronic energies as was directly tested in our calculation [2.257]. In this way the transition rate is obtained as a function of E in the entire interval $0 \le E \le E_{max}$, but physically significant is only a narrow range of E around the actual E value entering (6.6.3).

The resulting k_{nr} versus E function fluctuates on the pattern shown in Fig. 7.1 of the following chapter, the fluctuations being more intense for lower values of ω_N. ω_N is selected such as to reduce the fluctuation amplitude to a reasonable level, enabling one to perform an averaging procedure to obtain the final k_{nr} value at the desired energy [2.198, 4.83, 6.37, 40].

6.7 Inductive-Resonant-Transfer Mechanism

The radiationless transition in the statistical limit can be regarded as an irreversible energy transfer from the electronic subsystem to vibrations, which was already recognized in early works [2.33, 6.97, 98] (see [1.26], Chap. 8). If, additionally, the electronic and vibrational subsystems are separated spatially, then one can speak of a spatial energy transfer through a distance R between the localization domains of the two subsystems. In this case a natural analogy arises with the purely electronic energy transfer, of which the theory was developed by *Förster* [6.99], *Dexter* [6.100], and *Galanin* and *Frank* [6.101]. *Sveshnikova*, *Ermolaev* and coworkers [6.1–6] applied this theory to quantitatively describe the radiationless transitions of rare-earth and transition-metal ion complexes in liquid and solid media as well as of organic molecules. In the comprehensive studies [1.26, 2.296, 6.23, 6.102–104] they developed a fruitful semiempirical approach based on the following assumptions.

The vibrational modes are divided in two nonoverlapping groups. The first group involves the modes coupled with the electronic subsystem. They have nonzero displacements of the equilibrium positions and manifest themselves in the optical (vibronic) spectra. The second group is composed of optically inactive modes that are, however, active in the overtone vibrational absorption due to high frequencies and anharmonicities. Examples of the first- and second-group vibrations in aromatic hydrocarbons are C–C and C–H modes, respectively. In the metal-ion complexes the vibronic spectrum is formed by low-frequency vibrations of the ion and the neighboring atoms, while the best acceptors of the electronic energy are the highest-frequency vibrations of the host. The interaction of the electronic subsystem with the second-group vibrations leads to a direct energy transfer, of which the probability, in full analogy with the Förster formula, turns out to be proportional to the overlap Z between the emission spectrum of the system including the optical electrons and first-group vibrations and the overtone absorption spectrum of the second-group modes.

This result is most directly obtained in the case of a very weak electron–vibration coupling characteristic of the rare-earth ions in condensed media. In this limit the adiabatic approximation is no longer valid since, if the driving force acting on the electrons from the side of the nuclei is weaker than the inertial force due to the nonzero electron mass, the electrons, despite their small masses, do not follow the nuclei adiabatically. Then we have to return to the original Hamiltonian (4.1.1) and to take, as a perturbation inducing the energy transfer, the part of $U_{\text{el-nucl}}(x, q)$ responsible for the interaction between the optical electrons and the second-group modes. If the separation between the two systems (R) is large enough, the multipole expansion can be invoked, and in the dipole–dipole approximation the nonradiative transition rate (k_{nr} in s^{-1}) is given by the Förster formula [1.26]:

$$k_{nr} = 8.8 \times 10^{-25} k_{\text{rad}} \Phi^2 n^{-4} R^{-6} Z, \qquad (6.7.1)$$

$$Z = \int I(\nu) \varepsilon(\nu) \nu^{-4} d\nu , \qquad (6.7.2)$$

where k_{rad} is the radiative rate in s^{-1}, $I(\nu)d\nu$ the dimensionless intensity distribution in the emission spectrum (normalized to the integral over the spectrum), $\varepsilon(\nu)$ the molar decimal absorption coefficient for the high-frequency vibrations in cm^2/mole, $R = |\boldsymbol{R}|$ their distance from the location of the emitting electronic dipole in cm, ν the frequency in cm^{-1}, n the refraction index, and $\Phi = \boldsymbol{n}_1 \cdot \boldsymbol{n}_2 - 3(\boldsymbol{n}_1 \cdot \boldsymbol{n}_3)(\boldsymbol{n}_2 \cdot \boldsymbol{n}_3)$ the orientational factor ($\boldsymbol{n}_1, \boldsymbol{n}_2$ and $\boldsymbol{n}_3$ are unit vectors directed along the electronic dipole, the vibrational dipole, and $\boldsymbol{R}$, respectively). If there are several high-frequency vibrations, the summation over them is implied in (6.7.1).

Equation (6.7.1) describes the resonant energy transfer due to the interaction of the induced dipoles. The energy transfer may also occur via higher multipole interactions, in which case the Dexter generalization of the theory should be applied. In all cases the most important feature, already mentioned, is that

$$k_{nr} \propto Z , \qquad (6.7.3)$$

the result that is also predicted by the Galanin–Frank formula obtained from classical considerations. The dependence on the pattern of (6.7.3) is a certain sign of the occurence of the energy transfer, while the R^{-6} dependence as well as the proportionality to k_{rad} are characteristic of the inductive-resonant dipole–dipole mechanism of (6.7.1).

If the interaction between the electrons and the first-group vibrations is weak and the emission spectrum is dominated by the zero-phonon line, the spectral overlap Z, and hence the nonradiative rate, are merely proportional to the overtone absorption coefficient at the electronic transition frequency [6.7, 9], as was also predicted by the parallel term model of Sect. 5.5.

The prominent feature of (6.7.1) is that it expresses the nonradiative transition-rate constant in terms of experimentally measurable quantities, thus providing a basis for a semiempirical description of the nonradiative transitions.

In the case of intramolecular transitions the spatial separation between the donor and acceptor systems is lower, and even may not be uniquely specified at all, in which case the multipolar expansion loses its significance and (6.7.1) becomes invalid. The electron–vibration interaction is stronger, so the adiabatic approximation applies, and the non-adiabaticity operator now serves as a perturbation inducing the internal conversion. However, if the vibrations can still be divided into the groups described before, then it can be shown that (6.7.3) holds [1.26, 6.105]. Generally, it applies in the framework of the Condon approximation to both the internal conversion and intersystem crossing. Since the nonradiative-transition rate is proportional to the Franck–Condon factor, this factor can be calculated semiempirically via the spectral overlap integral of (6.7.2). Such a calculation enables one to allow for both the anharmonicity of the second-group vibrations and the potential displacements and distortions for the first-group vibrations. The calculation of the spectral overlap integral provides a clear assessment of how much electronic energy is transferred to the high-frequency high-anharmonicity C–H, O–H, N–H vibrations and to the vibrations responsible for the vibronic spectrum of the luminescence.

The calculations along these lines correctly predict the k_{nr} dependence on the energy gap [6.3, 5] (Rare-Earth ions, RE^{3+}, in glasses), on the vibration frequency [6.102] (Er^{3+} in solutions) and anharmonicity [6.104] (Dy^{3+} in solutions), and on the Stokes-shift (Huang–Rhys) parameter [6.6] (Mn^{2+} in solutions and crystals). The data on the ISC rates in organic molecules shown in Table 6.1 demonstrate that k_{nr} is approximately proportional to the product $k_{rad}Z$, thus confirming the validity of the inductive-resonant-transfer mechanism in these systems.

The assumption of $R \gg a, b$ underlying the multipolar expansion that leads to the R^{-6} dependence in the dipole–dipole approximation is a priori justified for RE^{3+} and transition-metal ions. Thus, the electronic oscillator has the dimension of the d- or f-shell, $a \simeq 0.5$ Å, and the vibrational oscillator has the atom displacements of $b \simeq 0.1$ Å, while R in the metal complexes varies between 2 Å and 10 Å. The R^{-6} dependence was directly observed in VO^{2+} ions in solutions where the electronic energy is transferred to the C–H and O–H vibrations of the solvent as well as to the V=O vibration [6.106].

Table 6.1. The data on the $T_1 \to S_0$ phosphorescence and $T_1 \rightsquigarrow S_0$ intersystem crossing rates of several organic molecules and the calculated spectral overlap integral [6.5]

Compound	ΔE [cm^{-1}]	k_{rad} [s^{-1}]	k_{nr} [s^{-1}]	$k_{rad}Z \times 10^{24}$ [s^{-1} $mole^{-1}$ cm^6]
Phenanthrene	21 700	0.05	0.25	2.5
Naphthalene	21 250	0.02	0.48	3.0
Crysene	19 900	0.045	0.45	5.2
Pyrene	16 900	0.005	2.56	15.0
Acriflavine	17 100	0.13	0.28	2.7
Ethyoporphyrine	14 550	1.28	16.2	210

Table 6.2. The k_{nr} dependence for the $^4F_{9/2}$ level of the Dy^{3+} ion upon the size of the sphere, R_m^3, containing no C–H and O–H groups

Solvent	R_m^3 [Å^3]	k_{nr} [10^3 s^{-1}]	$k_{nr}R_m^3$ [10^6 s^{-1} Å^3]
H_2O	15.5	330	5.1
CH_3OD	42.5	50	2.1
C_2H_5OD	42.5	42	1.8
$(CH_3)_2SO$	125	25	3.6
$(C_4H_9O)_3PO$	275	15	4.1

The possibility to vary R in the RE^{3+} complexes from 2.5 Å to 6.5 Å without any change of the rest of the parameters enables one to check on the validity of the dipole–dipole approximation in these systems. The investigation was done for Yb^{3+}, Sm^{3+} and Dy^{3+} ions in solutions where R was varied by changing the solvent. Since there are different atomic groups accepting the electronic energy and located at different distances from the ion, the Förster equation (6.7.1) was integrated over R^2dR from R_m to infinity, where R_m is the largest radius of the sphere inside which no absorbing atomic groups are present. The corresponding Galanin–Frank classical equation is

$$k_{nr} = \frac{k_{rad}}{16\pi^4 n^4 R_m^3} \int I(\nu)\alpha(\nu)\nu^{-4}d\nu \,, \tag{6.7.4}$$

where $\alpha(\nu)$ is the linear vibrational-absorption coefficient of the medium in cm^{-1}. The data of Table 6.2 demonstrate the R_m^{-3} dependence for an RE^{3+} ion.

There is much similarity between the ions' and organic molecules' behavior despite the fact that the energy transfer in the ions occurs to the medium vibrations while in the molecules to the intramolecular ones. In fact, as was observed in the Cr^{3+} [6.107] and RE^{3+} [6.108] complexes, the transfer in the ions essentially occurs to the first coordination sphere while in the molecules no medium effects were observed. This simply means that in the latter case the medium's absorbing groups are located at a longer distance R, and hence, their influence is largely relaxed.

Another indication of the existence of the k_{nr} versus R dependence in large organic molecules is provided by the data on partial deuteration of naphtylamines, naphtols and naphtylammonium [2.296]. It was shown that k_{nr} was affected by both the N–H, O–H substituent groups and the C–H ring groups, and, when the electronic transition was localized near the substituted groups, the substituted groups' contribution to k_{nr} was twice as large as that of the C–H ring groups. On the other hand, the delocalization of the electronic excitation over the ring makes both contributions commensurable. Nevertheless, the applicability of the dipole–dipole approximation to k_{nr} calculations in organic molecules remains questionable since no reliable data confirming the R^{-6} dependence are available.

7. The Intermediate Case

The intermediate case displays a rich spectrum of physical effects, which are conventionally treated in the framework of the phenomenological effective Hamiltonian method described in Chap. 2. The present chapter continues the discussion of the most interesting effects, and then presents an attempt to go beyond the phenomenological approach using the correlation function formalism. In the last section we again return to the phenomenological method and describe a very simple kinetic model that accounts for the biexponential decay effect by the presence of reverse electronic transitions.

7.1 Physical Effects

The definition of the accepting modes given in Sect. 2.10 holds valid in the intermediate case as well. Therefore the concepts of the quasimolecule and of the inert intramolecular medium introduced there are still applicable. Remind that the electronic-vibrational levels of the quasimolecule involve the initial and final electronic states of the molecule as well as the vibrational states of the accepting modes that are active in the given electronic transition. The poor acceptors (the passive modes) form the inert medium, and the coupling with this medium gives rise to the quasimolecule level broadening; an additional broadening arises due to the coupling of the molecule with its environment. Thus, in the general expression (3.4.9) we can put $s = rm$ and $\lambda = r'm'$, where m and m' denote the sets of initial and final quantum numbers of the accepting modes, and $\Gamma_\lambda(E_s)$ is determined by (3.4.5) involving the interaction of the accepting modes with the inert medium.

The effective states can also be introduced in the intermediate case, but now they do not produce the strongest effect on the radiationless transition rate, as was the case in the statistical limit, because they are not numerous enough. More specific, additional final states of the quasimolecule for which the Franck–Condon factor does not have the largest value at the given total energy can contribute since the loss in the Franck–Condon factor may be compensated by a gain from the "level shape function" (3.4.17) at small energy mismatch. For this reason all the final states have to be incorporated in the intermediate case.

Obviously, the intermediate case in large molecules is similar to the case of small molecules in inert media. The only difference is that the quasimolecule level broadening in the large-molecule intermediate case is caused not only by the external medium, but also by the "internal medium" of the passive intramolecular vibrations, and owing to this fact, the large molecule displays irreversible transitions even when being completely isolated.

Based on this analogy, the intermediate case is expected to show the effects typical of the small molecules [1.1–3, 7, 2.123, 189, 206, 224, 3.8, 7.1–3]. Those effects have already been discussed in Sects. 2.5–7 and 3.5. However, in view of the importance of this point it is appropriate to list the principal experimentally observed implications of the intermediate case with special emphasis on those allowing a discrimination between the intermediate case and the statistical limit. It should be noted that a consistent classification of the intermediate-case and statistical-limit effects is only possible on the basis of the definitions of the limiting cases given in Sect. 2.11 with the use of the concept of effective states. According to inequality (2.11.4) the intermediate case is characterized by a large density of all final states ϱ_l, but a low density of the effective states ϱ_{eff}, at the given electronic energy ΔE. Both these quantities increase with ΔE much more rapidly than the level width γ. Therefore the appearance of the effects described below is expected at low ΔE values.

7.1.1 Absorption Spectra and Luminescence Kinetics

Unlike the statistical limit, the strong and weak coupling limits should be discriminated in the intermediate case. For weak coupling, only one strong resonance (3.5.31) originating from the initial bright s-state should be observed as well as a number of weak resonances associated with the dark λ-states. On excitation of the dominant resonance the luminescence intensity falls off exponentially, the emission quantum yield is lower than unity, and the transition rate constant is given by (3.4.9). Hence, if the possibility to observe the weak resonances is negligible, the absorption spectrum and the emission decay law in the weak coupling limit are identical for the intermediate case and the statistical limit, and these two limits can be distinguished only by the dependence of the radiationless transition rate upon ΔE and pressure in the gas phase, as well as by the deuteration effect.

Under the strong coupling condition a large number of absorption resonances should be observed around the initial bright s-state, and the luminescence decay kinetics must depend on the excitation conditions. If the total spectral width of the exciting radiation is lower than the separations between the resonances, then only one resonant state is excited, and the luminescence intensity decays by the exponential law (3.5.20) with the lower-than-unity quantum yield (3.5.22) (in the formulae of Sect. 3.5.1 the indices n and l related to the molecule should be replaced with the quasimolecule indices ν and λ, respectively). The difference with the statistical limit is that the radiative lifetime of the individual resonance (3.5.21) is longer than the radiative lifetime of the bright state since the $E_s \leftarrow E_0$ oscillator strength is distributed among many $E_\nu \leftarrow E_0$ transitions (E_0 is the ground-state energy). But, if the excitation covers more than one resonance, then the emission intensity decays nonexponentially: for incoherent excitation it is described by a sum of two or more exponentials with different decay times that are, however, of similar order of magnitude, while at coherent excitation intensity quantum beats occur.

7.1.2 Pressure Dependence of the Transition Rate

The statistical limit transition-rate constant k_s is independent of pressure in the gas phase because it is independent of the molecular level widths. In contrast, the intermediate-case constant k_{st} depends on the total level width, hence on pressure, though not so strongly as in the case of small molecules due to the presence of an additional intramolecular level broadening. This dependence enters the widths $\gamma_{\mathrm{rel}(s)}$ and $\gamma_{\mathrm{rel}(l)}$ of (2.5.8) via the vibrational-relaxation rate constant

$$\tau_{\mathrm{VR}}^{-1} = \tau_{\mathrm{IVR}}^{-1} + \sum_{i=a,b} k^{(i)} p_i \,,$$

where τ_{IVR}^{-1} is the monomolecular rate constant for the intramolecular vibrational redistribution in large polyatomic molecules ($\tau_{\mathrm{IVR}}^{-1} = 0$ in small molecules), $k^{(i)}$ the bimolecular rate constant for the vibrational relaxation induced by collisions with the molecules of the i-th kind, and p_i is the partial pressure ($i = a$ and b correspond to the compound under study and the buffer gas, respectively).

Considering the collisions within the Rigid-Sphere (RS) model and assuming that every collision event results in the relaxational transition with unit probability, one can deduce the following equation for the bimolecular constant

$$k_{\mathrm{RS}}^{(i)} = 4\sigma^2 \left(\frac{\pi}{M k_{\mathrm{B}} T}\right)^{1/2}, \qquad M = \frac{2M_a M_i}{M_a + M_i}, \qquad \sigma = \frac{1}{2}(\sigma_a + \sigma_i)\,,$$

where σ_a, σ_i, M_a and M_i are the diameters and masses of the colliding partners. At $i = a$ (self-relaxation), σ and M are the diameter and mass of the molecule under study, respectively, and the equation for the bimolecular constant reduces to that given in [7.4]. Expressing σ in Å, M in atomic units (the molecular weight) and T in Kelvin, one obtains $k_{\mathrm{RS}}^{(i)}$ in $\mathrm{Torr}^{-1}\,\mathrm{s}^{-1}$ as

$$k_{\mathrm{RS}}^{(i)} = 6.24 \times 10^7 \,\sigma^2 (MT)^{-1/2}\,.$$

Generally, one sets $k^{(i)} = w_i k_{\mathrm{RS}}^{(i)}$, where w_i is a correction factor allowing for a breakdown of this simplified model. The w_b values are close to unity provided that rare-gas atoms or nonpolar molecules serve as colliding partners. In the case of self-relaxation, w_a may take on large values [2.81].

For weak coupling, k_{st} increases with pressure provided that off-resonant states with

$$|E_s - E_\lambda| \gg \gamma_\lambda \tag{7.1.1}$$

give major contributions to the transition rate (3.4.9). But, if the sum in (3.4.9) is dominated by a λ-state in close resonance with the initial s-state, then the rate increase is changed to a decrease for states which do not obey inequality (7.1.1) [5.124, 7.5]. In any case, with γ_λ increasing, k_{st} eventually saturates to a constant k_s when γ_λ becomes larger than the level separations since this situation pertains to the statistical limit (see Sect. 2.5.2).

For strong coupling, k_{st} monotonically increases with pressure [see (3.5.24)] and then saturates to the statistical-limit rate constant. In a number of works (see Sect. 2.11) the pressure dependence was observed of the $S_1 \rightsquigarrow T_1$ intersystem crossing rate and of the quantum yield of formation of the triplet-state aromatic hydrocarbons in vapors.

The above considerations show that the most prominent feature of the pressure effect in the intermediate case is its non-Stern-Volmer behavior [2.76].

The interpretation of the pressure effects may become complicated by the presence of electronic intermolecular interactions (noninert media). In this case the transition rate is pressure dependent in both the intermediate case and the statistical limit [1.12].

7.1.3 Energy-Gap Dependence

The ΔE dependence of the statistical-limit transition-rate constant k_s is essentially governed by the Franck–Condon factor, and is represented by the energy-gap law of (2.11.5). In the intermediate case an additional ΔE dependence of the function (3.4.17) is superimposed on this smooth behavior. For strong coupling, the effect is not pronounced, as is seen from (3.5.24), so that the energy-gap law is still valid. In contrast, for weak coupling the ΔE dependence of the function (3.4.17) is strongly nonmonotonic since with changing ΔE a new final state comes in resonance with the initial state, and the Franck–Condon factors for the ineffective states fluctuate strongly between different states. It is thus concluded that for weak coupling in the intermediate case the energy-gap law is violated and the transition-rate constant is a nonmonotonic function of the electronic energy.

The calculation of the intermediate-case radiationless transition-rate constant is a complicated problem in view of the explicit rate dependence on the quantity γ_λ, which, in turn, is a strong function of the final-state energy (it diminishes with an increase in $|E_s - E_\lambda|$ very rapidly[1]). One of the conceivable ways to resolve this problem consists in studies of particular, exactly solvable models enabling one to obtain, in explicit form, either the function γ_λ [3.4] or the damping correlation function [5.124, 7.6–9], and eventually the transition rate constant (Sect. 7.2). A different approach is of phenomenological nature, and it consists in the replacement of the level-shape function $r_\lambda(E_s)$ of (3.4.17) with a simpler function depending on one phenomenological parameter only. Thus,

[1] Recall that γ_λ can be interpreted as a level width only at $E_\lambda = E_s$ (see the end of Sect. 3.4)

in Sect. 6.6 the level-shape function was approximated with a step function. In [2.237, 6.33], γ_λ is replaced with a constant and the generating function method is employed. In fact, after such a replacement the integral over t involves nonphysical contributions from off-resonant transitions violating the energy-conservation law, which is a consequence of the slow decrease of the Lorentzian function as compared to the Franck–Condon factor. However, when this integral is calculated by the saddle-point method and all the relevant saddle points are taken into account, the nonphysical transitions turn out to appear via the integrand's singularity at $t = 0$ so that they are easily eliminated by omission of the singular-point contribution (see Sect. 6.3.2).

Figure 7.1 shows the results of such a calculation of the $S_1 \rightsquigarrow T_1$ intersystem crossing in benzene. The calculation was performed by the saddle-point method allowing for many saddle points and using the Lorentzian shape function of constant width. The method was described in detail in Sect. 6.3.2 for the model of noninteracting harmonic oscillators with displaced but undistorted potential surfaces. However, the calculation was done with frequency changes (curves 1 and 3 in Fig. 7.1). There are six maximum frequency C–H modes for which the frequency values were taken at 3060 cm^{-1} and 3120 cm^{-1} in T_1 and S_1, respectively. The T_1 frequencies for the rest of modes were set to the S_0 values [4.21]. The displacement parameter for the v_1(C–C) totally symmetric mode was set at $\xi_1 = 0.125$, while that for the v_2(C–H) totally symmetric mode is small, and we assumed $\xi_2 = 0$, as well as $\xi_\mu = 0$ for all the nontotally symmetric modes. The electronic matrix element $v_{el} = 1$ cm^{-1} was employed.

For comparison, the results of a calculation in the first saddle-point approximation are also shown (curves 2 and 4 in Fig. 7.1). In this approximation the transition rate monotonically decreases with increasing ΔE by the exponential law [the energy-gap law of (2.11.5)], while the allowance for all the saddle points results in the nonmonotonic behavior. Similar results were obtained in [2.198, 6.52] by the method of Sect. 6.6. It should be emphasized that these results are

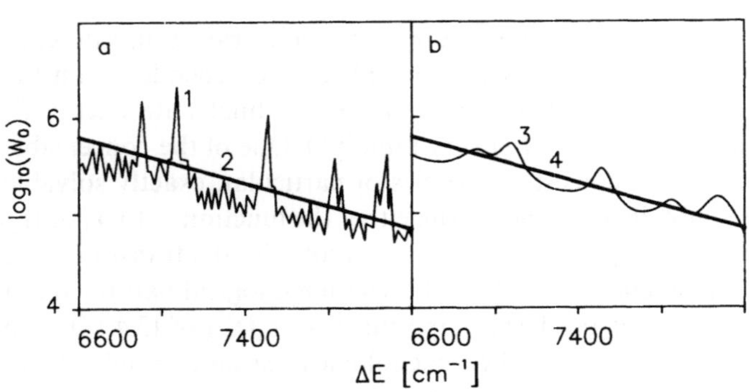

Fig. 7.1. The calculated $S_1 \rightsquigarrow T_1$ intersystem crossing rate in benzene vs the energy gap ΔE [6.33] for $\gamma = 10$ cm^{-1} (**a**) and 50 cm^{-1} (**b**)

valid in the weak coupling limit only, whereas in the strong coupling limit the energy-gap law applies in both the intermediate case and the statistical limit.

In view of a rapid increase in the density of effective states with ΔE it is expected that in polyatomic organic molecules the nonradiative transitions to the ground state will typically belong to the statistical limit, whereas those between the excited states to the intermediate case. This is confirmed by the aromatic hydrocarbon data analyzed by *Siebrand* and *Henry* [2.184, 274, 276]. As is seen from Fig. 2.5 and Table 2.1 of Sect. 2.11, the $T_1 \to S_0$ transition obeys the energy-gap law, while for the $S_1 \to T_1$ transition the ΔE dependence is nonmonotonic. Other examples of the intermediate case behavior are described in Sect. 2.2.

7.1.4 Vibrational and Rotational Energy Dependence

From the results of Sects. 6.4, 5 it follows that the statistical-limit k_s increases with the initial vibrational energy E_{vib}, which again stems from the Franck–Condon factor behavior. The same behavior is predicted in the intermediate case for strong coupling, but for weak coupling the dependence is not monotonic. Indeed, the difference between the initial and final state energies as a sum of the electronic, vibrational and rotational energy changes is

$$E_s - E_\lambda = \Delta E + \Delta E_{vib} + \Delta E_{rot} \,. \tag{7.1.2}$$

Then, with an increase in the initial and, correspondingly, final vibrational quantum numbers ΔE_{vib} undergoes variations on account of both the frequency differences in the two electronic states and the anharmonic level shifts. The resultant variations in the energy mismatch between the initial and final vibronic states lead, as was already pointed out in Sect. 7.1.3, to a nonmonotonic change in the transition rate. *Nitzan* et al. [7.1] predicted a monotonic *decrease* with increasing E_{vib}; such a behavior is possible, provided that the energy mismatch monotonically increases while the Franck–Condon factor does not change significantly.

The ΔE_{rot} effect was discussed in [2.237] in connection with the experimental results by *Howard* and *Schlag* [2.197] on naphthalene, and was treated in detail by those and other authors [2.197, 198, 4.34, 37, 38]. Although the naphthalene results were not confirmed in a further study [4.40], it is of interest to discuss possible sources of the rotational effects in the intermediate case.

Let us first assume that the matrix element $v_{s\lambda}$ is independent of the rotational quantum numbers so that these numbers are conserved in the electronic transition. Yet, the rotational energy does change due to a change in the moments of inertia. For instance, in the simplest case of the spherical top the energy change in the electronic transition is

$$\Delta E_{rot} = \Delta B J(J+1) \,, \tag{7.1.3}$$

where ΔB is a change in the rotational constant and J the rotational quantum number. Since $v_{s\lambda}$ is independent of J by assumption, the transition rate allowing for rotation is still given by (3.4.9), where the energy change has to be modified to include the rotational contribution, i.e., the replacement of

$$\Delta E \rightarrow \Delta E + \Delta E_{\text{rot}} \tag{7.1.4}$$

has to be performed. Equations (7.1.3, 4) show that the effect of ΔE_{rot} variations versus the initial rotational quantum numbers is adequately described as the corresponding ΔE variation effect. Since the average rotational energy is on the order of $k_B T$, ΔE_{rot} under the selective rotational excitation within a given vibronic band is estimated by (7.1.3) as $\Delta E_{\text{rot}} \sim (\Delta B/B) k_B T$, where B is the rotational constant. The relative change in the rotational constant between the S_1 and S_0 states amounts to $\Delta B/B = 0.05$ in benzene [7.10] and 0.03 in naphthalene [7.11]. Taking these values for the $S_1 \rightsquigarrow T_1$ transition we obtain, at room temperature, $\Delta E_{\text{rot}} \sim 10\ \text{cm}^{-1}$. Turning to Fig. 7.1 we see that ΔE changes of $10\ \text{cm}^{-1}$ for $\gamma = 10\ \text{cm}^{-1}$ result in a smooth (but generally nonmonotonic) transition-rate variation, because the mean separations between the maxima of curve 1 (Fig. 7.1) are just about 10–50 cm^{-1}. However, at lower γ a great deal of additional peaks appear. Thus, for $\gamma < 10\ \text{cm}^{-1}$ the transition rate will show several maxima and minima when the excitation wavelength spans a given vibronic band.

In general, the behavior is further modified by the rotational dependence of $v_{s\lambda}$. It should be emphasized, however, that no rotational effects are predicted in the statistical limit since the J dependence of ϱ_{eff} in (2.7.21) is negligible, and that of v_{eff}^2 vanishes after the summation over all the final rotational levels is performed, due to the rotational sum rule (see Sect. 4.4).

There are a lot of experimental data confirming the existence of the vibrational and rotational effects described above in the $S_1 \rightsquigarrow T_1$ intersystem crossing of polyatomic organic molecules in low-pressure vapors and free jets. Thus, in glyoxal $(CHO)_2$ the transition rate has a pronounced maximum in the range of $E_{\text{vib}} = 0$–$1000\ \text{cm}^{-1}$ that corresponds to the excitation of one ν_8 quantum ($E_{\text{vib}} = 800\ \text{cm}^{-1}$); it is interesting that this maximum disappears upon deuteration [2.199]. In the framework of the above-said this behavior is accounted for by the existence of a nearly resonant effective final state for the given initial state at $E_{\text{vib}} = 800\ \text{cm}^{-1}$; the deuteration shifts the vibronic levels, and the resonance disappears. *Lim* et al. [2.306] and *Stockburger* et al. [2.308] observed large-amplitude chaotic variations in the fluorescence quantum yield and in the intersystem-crossing rate from the naphthalene S_1 single vibronic levels. These data give evidence that the $S_1 \rightsquigarrow T_1$ intersystem crossing in the molecules mentioned is assigned to an intermediate-case weak-coupling-limit transition.

A strong rotational effect on the fluorescence quantum yield of formaldehyde was first observed by *Tang* et al. [7.12]. It was then followed by a striking discovery of the rotational effects in the molecules as large as pyrazine and related compounds (see Sect. 2.2). The experimental data were believed to be typical of an intermediate-case strong-coupling-limit molecule. As such, pyrazine displays the two-component fluorescence decay with the fast-component quantum yield being independent of rotation, while the slow component quantum yield showing the dependence of $\varphi_{slow} = 0.124/(2J + 1)$ for $J = 5 - 22$ [2.148]. As follows from Sects. 2.5, 6, in the strong-coupling limit φ_{slow} is proportional to the number of strongly coupled states N, so that from the above-mentioned data the relation $N \propto 2J + 1$ is deduced, i.e., N is proportional to the total number of the final rotational sublevels. This obviously contradicts to the selection rules derived in Sect. 4.7, since these rules connect each initial rotational state to a limited (no more than three) number of final states independently of the total number of the final rotational sublevels. To resolve this inconsistency it was proposed that the selection rules are violated, and the so-called "K-scrambling" (i.e., a strong mixing between the K-sublevels of a given J-manifold) occurs [2.123, 165], while the detailed mechanism for the K-scrambling has not yet been established. The analysis of the wealth of experimental data on the intermediate-case pyrazine properties goes beyond the scope of this monograph, so we refer the reader to the above reviews and to our review article [2.205].

7.1.5 Deuteration Effect

In Chap. 2, Siebrand's rule was formulated stating that the deuteration diminishes the statistical-limit transition rate. This rule is based on the energy-gap law (2.11.5) reflecting the exponential dependence of the Franck–Condon factor on the energy gap. The vibration-frequency lowering upon the deuteration increases the number of quanta of the accepting modes needed to 'exchange' the electronic energy; hence it diminishes the vibrational overlap and the Franck–Condon factor. In fact, the effective-states density entering the pre-exponential factor of (2.11.5) also increases, but the exponential factor is of dominant importance at large ΔE so that the deuteration lowers the transition rate. A similar behavior is also expected in the intermediate case for strong coupling.

On the other hand, for weak coupling the decrease in the Franck–Condon factor does not necessarily lower the intermediate-case transition rate since the change of the function (3.4.17) caused by a vibronic level shift may turn out to be more essential. Therefore, Siebrand's rule does not apply in the case under consideration, and the transition rate may both decrease and increase upon deuteration, allowing the occurrence of the inverse deuteration effect [1.42, 53]. This effect was observed in the $S_1 \rightsquigarrow T_1$ intersystem crossing from the naphthalene single vibronic levels [2.306, 308].

7.1.6 Comparative Description of the Intermediate Case and the Statistical Limit

The above effects, characteristic of the statistical limit and the intermediate case, are listed in Table 7.1 (see also Fig. 2.3). It is interesting to note that none of

Table 7.1. Physical effects in the statistical limit and the intermediate case

Effect	Statistical limit	Intermediate case	
		Strong coupling	Weak coupling
Light absorption	Single resonance	Several resonances	Single resonance
Emission decay kinetics	Exponential	a) Biexponential, with chaotic quantum beats on the slow component b) Exponential, with a lengthened radiative lifetime	a) Exponential b) Biexponential, with periodic quantum beats on the fast component
Transition rate dependence upon pressure	No	Yes	Yes
Dependence upon the electronic, vibrational, and rotational energies	Monotonic	Monotonic	Non-monotonic
Deuteration effect	Normal	Normal	Either normal or reverse

these effects taken separately does permit to tell the intermediate case from the statistical limit. Therefore a comprehensive study of all the effects is needed to assure a reliable assignment of the radiationless transitions in polyatomic molecules.

The results of the experimental investigations cited above in this section and in Chap. 2 suggest that in the isolated organic polyatomic molecules the $S_1 \rightsquigarrow T_1$ intersystem crossing can be classified as the intermediate case (either in the strong or the weak coupling limit), while the transitions to the ground state are assigned to the statistical limit.

7.2 Correlation-Function Method

The intermediate-case radiationless transition-rate constant is sensitive to the specific form of the quasimolecule–medium interaction, so of special interest are investigations of exactly solvable models. Of course, the choice of an appropriate model is to a large extent a decisive factor for such an approach to be successful.

For any model, the problem of the relaxation of a dynamical system interacting with a thermally equilibrated medium has to be solved with the use of the density matrix formalism [6.86, 7.13, 14] in which the radiationless transition correlation function is expressed (as will be seen below) in terms of the time-dependent correlation function for the intramolecular interaction [1.30, 6.27, 7.6].

The total Hamiltonian is written in the form

$$\mathcal{H} = \mathcal{H}_0 + V_M , \tag{7.2.1}$$

where V_M is an intramolecular coupling that mixes two electronic states under consideration, and the Hamiltonian $\mathcal{H}_0 = H_0 + H_c + V_{Mc}$ involves the quasimolecule zeroth-order Hamiltonian H_0 being diagonal in the electronic quantum number, the inert-medium Hamiltonian H_c and the quasimolecule–medium interaction V_{Mc}. Since only inert media are considered, the operators H_c and V_{Mc}, hence $\mathcal{H}_0$, do not mix different electronic states. In the model of two electronic states r and r' the operators H_0 and $\mathcal{H}_0$ are written as diagonal two-row matrices

$$H_0 = \begin{pmatrix} H_{0r} & 0 \\ 0 & H_{0r'} \end{pmatrix}, \quad \mathcal{H}_0 = \begin{pmatrix} \mathcal{H}_{0r} & 0 \\ 0 & \mathcal{H}_{0r'} \end{pmatrix}, \tag{7.2.2}$$

and the intramolecular coupling operator as

$$V_M = \begin{pmatrix} 0 & V_{rr'} \\ V_{r'r} & 0 \end{pmatrix}. \tag{7.2.3}$$

The elements of these matrices are operators acting on the nuclear coordinates of the quasimolecule and of the medium.

The separation of the total Hamiltonian into the sum of (7.2.1) corresponds to a physical situation where the electronic transition is a slow process as compared to the vibrational relaxation. For this reason V_M will be considered as a small perturbation, while V_{Mc} ought to be treated as accurately as possible.

The density matrix of the entire system in the interaction representation obeys the equation ($\hbar = 1$)

$$\mathrm{i}\frac{\partial R(t)}{\partial t} = [V(t), R(t)] , \tag{7.2.4}$$

$$V(t) = \exp(\mathrm{i}\mathcal{H}_0 t)\, V_M \exp(-\mathrm{i}\mathcal{H}_0 t) . \tag{7.2.5}$$

Applying time-dependent perturbation theory to (7.2.4) we can write the increment of the density matrix for the time interval Δt in the form

$$R(t+\Delta t) - R(t) = -\mathrm{i}\int_t^{t+\Delta t} dt_1 [V(t_1), R(t)]$$

$$-\int_t^{t+\Delta t} dt_1 \int_t^{t_1} dt_2 [V(t_1), [V(t_2), R(t)]] , \tag{7.2.6}$$

where only terms up to the second order in $V(t)$ are included.

Let ω be a characteristic intramolecular vibration frequency governing the $V(t)$ versus time behavior, β be a constant on the order of the off-diagonal matrix element of the intramolecular interaction, and let the following inequalities be fulfilled:

$$\omega \Delta t \gg 1 , \tag{7.2.7}$$

$$\beta \Delta t \gg 1 . \tag{7.2.8}$$

The former implies that in the linear term of expansion (7.2.6) the integrand oscillates in the interval of $(t, t + \Delta t)$, hence, on the order of magnitude, the linear term is β/ω. Below we shall see that in the quadratic term of expansion (7.2.6) the integrand oscillates versus $t_1 - t_2$ and is a slowly varying function of $t_1 + t_2$, so the order of magnitude of the quadratic term is $\beta^2 \Delta t/\omega$, and hence the linear term can be neglected according to inequality (7.2.8). Disentangling the double commutator yields

$$\begin{aligned} R(t + \Delta t) - R(t) = & - \int_t^{t+\Delta t} dt_1 \int_t^{t_1} dt_2 \\ & \times [V(t_1) V(t_2) R(t) - V(t_1) R(t) V(t_2)] \\ & + \text{H.C.} \, . \end{aligned} \tag{7.2.9}$$

From (7.2.9) and operators (7.2.2, 3) it follows that if the matrix $R(t)$ is diagonal in the electronic index at time t, the same holds true at $t + \Delta t$ so that we can set, for any t,

$$R(t) = \begin{pmatrix} R_r(t) & 0 \\ 0 & R_{r'}(t) \end{pmatrix},$$

where $R_{r,r'}(t)$ have the meaning as the quasimolecule density matrices in the initial and final electronic states, respectively. Indeed, according to (7.2.2, 3) operator (7.2.5) has no diagonal elements while the product $V(t_1) V(t_2)$ has no off-diagonal ones. As a result, (7.2.9) reduces to a set of two coupled equations for $R_{r,r'}(t)$.

Let the system be prepared in the electronic state r at $t = 0$. We shall be interested in the rate constant k_r from the initial r-state to the final r'-state. Then it is sufficient to deal with the following 'truncated' equation for $R_r(t)$:

$$\begin{aligned} R_r(t + \Delta t) - R_r(t) = & - \int_t^{t+\Delta t} dt_1 \int_t^{t_1} dt_2 \langle r | V(t_1) V(t_2) | r \rangle R_r(t) \\ & + \text{H.C.} \, . \end{aligned} \tag{7.2.10}$$

The term omitted, which is proportional to $R_{r'}(t)$, describes the reverse transitions. If the reverse transition rate is low, $k_{r'} \ll k_r$, then these transitions can be disregarded at times of interest,

$$t \lesssim 1/k_r \, . \tag{7.2.11}$$

Let R_c denote the thermally equilibrated density matrix of the medium independent of the quasimolecule's variables and τ_c the characteristic time for relaxation of the perturbations arising in the medium due to its interaction with the quasimolecule (it is the medium thermal equilibration time). If τ_c is short enough, one can impose the limitation of

$$\Delta t \gg \tau_c \,. \tag{7.2.12}$$

Since we do not consider the evolution of the system at time intervals below Δt, we can assume the medium to be thermally equilibrated for any t and Δt, and write the total density matrix as the product

$$R_r(t) = R_{rM}(t) R_c \,, \tag{7.2.13}$$

where $R_{rM}(t)$ depends upon the quasimolecule variables only. On substituting (7.2.13) into (7.2.10) and taking the trace over the medium variables we obtain

$$R_r(t+\Delta t) - R_r(t) = -\int_t^{t+\Delta t} dt_1 \int_t^{t_1} dt_2 \langle r|Q(t_1,t_2)|r\rangle R_r(t) + \text{H.C.} \,, \tag{7.2.14}$$

where by $R_r(t)$ we now mean the quasimolecule density matrix in the r-state, $R_{rM}(t)$, and

$$Q(t_1, t_2) \equiv \langle V(t_1)\, V(t_2)\rangle_c \tag{7.2.15}$$

is the time-dependent correlation function of the intramolecular interaction; $\langle \cdots \rangle_c$ stands for averaging over the medium variables performed with the equilibrium density matrix R_c.

Equation (7.2.14) describes the dynamical subsystem behavior in a definite external field created by the medium. The field is not stationary, but fluctuates in time. Therefore, either a dynamical or a statistical description of the medium is needed to calculate the correlation function (7.2.15); and the interaction V_{Mc} cannot be treated by perturbation theory since the latter is valid at times shorter than τ_c, whereas we are interested in the correlation-function decay at times according to (7.2.12). An example of solving such a kind of problem will be given at the end of this section.

Let $|rm\rangle$ denote the eigenfunctions of the Hamiltonian H_{0r} where m is a set of the quasimolecule vibrational quantum numbers. The quasimolecule–medium interaction results in nonstationary rm-states, which can be characterized by the vibrational relaxation rate γ. We shall assume that the quasimolecule level broadening (equal to γ) is small as compared to ω,

$$\gamma \ll \omega \,, \tag{7.2.16}$$

since otherwise it makes no sense to speak of any transitions between the rm

levels, and a new separation of the system into a quasimolecule and a medium has to be sought for.

In the rm representation the correlation function (7.2.15) is not diagonal in the m number since the $|rm\rangle$ functions are not eigenfunctions of $\mathscr{H}_0$; nevertheless, condition (7.2.16) permits to neglect its off-diagonal elements. To see this, consider the behavior of the diagonal and off-diagonal elements,

$$F_{rm}(\theta, \tau) \equiv \langle rm | Q(t_1, t_2) | rm \rangle , \tag{7.2.17}$$

$$F_{rmm'}(\theta, \tau) \equiv \langle rm | Q(t_1 t_2) | rm' \rangle , \tag{7.2.18}$$

as functions of the sum and difference of t_1 and t_2,

$$\theta = t_1 - t_2 , \qquad \tau = \frac{1}{2}(t_1 + t_2) . \tag{7.2.19}$$

If we set $\gamma = 0$ (i.e., $V_{Mc} = 0$), then, from definitions (7.2.15, 2, 3, 5) we obtain the correlation function in the form

$$Q(t_1, t_2) = \exp(\mathrm{i} H_{0r} t_1) V_{rr'} \exp(-\mathrm{i} H_{0r'} \theta) V_{r'r} \exp(-\mathrm{i} H_{0r} t_2) .$$

Now one sees that for $\gamma = 0$ the diagonal elements (7.2.17) only depend upon θ because $|rm\rangle$ are eigenfunctions of H_{0r}, while the off-diagonal elements (7.2.18) are functions of both variables (7.2.19) and oscillate with frequency ω.

For $\gamma \neq 0$ the time behavior of the off-diagonal elements does not change significantly due to condition (7.2.16): they predominantly oscillate with frequency ω. Therefore, after integration over dt_1 and dt_2, the off-diagonal elements on the right-hand side of (7.2.14) will be on the order of $1/\omega^2$. The diagonal elements for $\gamma \neq 0$ become dependent upon τ, their variation rate versus τ being just the vibrational relaxation rate γ. Indeed, at $\theta = 0$ one has $F_{rm}(0, \tau) = \langle\!\langle rm | V(\tau) V(\tau) | rm \rangle\!\rangle_c$, whence it is seen that the diagonal-element variation versus τ is associated with the variation of the mean square of the intramolecular interaction, which tends to its equilibrium value, at a given temperature of the medium, with the vibrational-relaxation rate.

We assume Δt such that $\gamma \Delta t \ll 1$. Then, in calculating the double integral of the diagonal element of the correlation function, one may set

$$F_{rm}(\theta, \tau) \simeq F_{rm}(\theta, t) , \tag{7.2.20}$$

after which the integral over $d\tau$ merely yields Δt. The resultant diagonal element on the right-hand side of (7.2.14) is on the order of $\Delta t/\omega$, and hence, is much larger, due to (7.2.7), than the off-diagonal one.

After omission of the off-diagonal terms in (7.2.14) and substitution of (7.2.17, 20) we obtain the following equation for the diagonal elements of the density matrix:

$$R_{rm}(t + \Delta t) - R_{rm}(t) = -2\,\mathrm{Re}\left\{ \int_0^{\Delta t} (\Delta t - \theta) F_{rm}(\theta, t) \mathrm{d}\theta \right\} R_{rm}(t) . \tag{7.2.21}$$

At this point we note that $F_{rm}(\theta, t)$ as a function of θ not only oscillates, but also decays with a certain rate Γ, which may be interpreted as the erasing rate of the system's "phase memory" due to the interaction with the medium. Usually the phase relaxes much faster than the energy,

$$\Gamma \gg \gamma\,, \tag{7.2.22}$$

and Δt can be chosen such that $\Gamma \Delta t \gg 1$. Then, in the integrand of (7.2.21) one can neglect θ as compared to Δt and shift the upper integration limit to infinity. On dividing both sides of (7.2.21) by Δt one obtains the kinetic equation

$$\frac{\partial R(t)}{\partial t} = -k(t)R(t)\,, \tag{7.2.23}$$

where $R(t)$ and $k(t)$ are matrices diagonal in $|rm\rangle$, and the diagonal element

$$k_{rm}(t) = 2\,\mathrm{Re}\,\{\int_0^\infty F_{rm}(\theta, t)\,d\theta\} \tag{7.2.24}$$

is the transition rate from the rm-state to all final $r'm'$-states.

Equation (7.2.24) shows that the diagonal matrix element of the correlation function of the intramolecular interaction coincides, to within a phase factor, with the generating function for the radiationless transition defined in Sect. 6.2. The convergence of the integral in (7.2.24) is ensured by the correlation-function decay caused by the quasimolecule-medium interaction. Hence it follows that this interaction must be treated as accurately as possible, with no recourse to perturbation theory.

This interaction also leads to nonexponential transition kinetics at times

$$t \lesssim 1/\gamma \tag{7.2.25}$$

when the rate constant is time-dependent. For

$$t \gg 1/\gamma \tag{7.2.26}$$

this dependence disappears and the rate constant is $k_{rm}(\infty)$.

A comparison of inequalities (7.2.26, 11) shows that the exponential kinetics occurs at times $1/\gamma \ll t \lesssim 1/k_{rm}$, i.e., it is observable provided that the electronic relaxation is slow as compared to the vibrational one (cf. Sects. 2.1, 6, 7), $k_{rm} \ll \gamma$. But even in this case the kinetics is not exponential for short times (7.2.25), as well as at very long times $t \gtrsim 1/k_{r'}$ due to the reverse transitions (see Sect. 7.3).

Combining all the inequalities and eliminating Δt we obtain the following validity conditions for (7.2.23, 24):

$$k_{rm} \ll \gamma \ll \tau_c^{-1}, \omega, \beta, \Gamma\,. \tag{7.2.27}$$

Strictly speaking, (7.2.23) is not the kinetic equation in the conventional sense, i.e., the equation for the rm-state populations, since $R(t)$ is the density matrix in the interaction representation. Indeed, let $\tilde{R}(t)$ be the density matrix in the Schrödinger representation to which $R(t)$ is related by

$$R(t) = \exp(\mathrm{i}\mathscr{H}_{0r}t)\tilde{R}(t)\exp(-\mathrm{i}\mathscr{H}_{0r}t)\,. \tag{7.2.28}$$

The rm-state populations are given by the diagonal elements of the $\tilde{R}(t)$ matrix, and since the $|rm\rangle$ functions are not eigenfunctions of the Hamiltonian $\mathscr{H}_{0r}$, the diagonal elements $R_{rm}(t)$ of the matrix (7.2.28) are not equal to the populations $\tilde{R}_{rm}(t)$.

The kinetic equation for the populations has to simultaneously describe both the electronic and vibrational relaxation. To derive such an equation it is necessary to transform (7.2.23) to the Schrödinger representation by (7.2.28), and then to the interaction representation with the Hamiltonian H_{0r} by

$$\tilde{R}(t) = \exp(-\mathrm{i}H_{0r}t)\bar{R}(t)\exp(\mathrm{i}H_{0r}t)\,.$$

Diagonal elements of the new matrix $\bar{R}(t)$ are the rm-state populations since $|rm\rangle$ are eigenfunctions of H_{0r}. In the derivation of the kinetic equation for $\bar{R}(t)$ the quasimolecule–medium interaction can now be treated as a small perturbation (in particular, the non-commutativity of the $k(t)$ and $\mathscr{H}_{0r}$ operators can be disregarded) because the principal effect of this interaction has already been accurately taken into account in the derivation of (7.2.24) for the transition rate. As a result, one obtains the kinetic equation for the populations, of which the right-hand side includes a sum of terms describing the vibrational relaxation and the term $-k(t)\bar{R}(t)$ responsible for the electronic relaxation, with $k(t)$ from (7.2.24) [7.6, 15].

As an example, consider the model in which the quasimolecule is represented by a single linear harmonic oscillator, the medium by a set of linear harmonic oscillators with a continuous frequency spectrum, the quasimolecule–medium interaction is a linear function of both the quasimolecule and medium coordinates and the intramolecular coupling is independent of the nuclear coordinates [7.6]. For this model we set

$$H_{0r} = \tfrac{1}{2}(p^2 + \omega^2 x^2) + \tfrac{1}{2}\Delta H\,, \tag{7.2.29}$$

$$H_{0r'} = \tfrac{1}{2}(p^2 + \omega^2 x^2) - \tfrac{1}{2}\Delta H\,,$$

$$\Delta H = \Delta E + Fx, \qquad V_{rr'} = V_{r'r} = \beta\,, \tag{7.2.30}$$

where x, p and ω are the coordinate, the momentum and the frequency of the intramolecular oscillator, respectively, ΔE is the electronic energy gap, F the parameter of the equilibrium position displacement and β is a real constant; further

$$H_c = \sum_k \tfrac{1}{2}(p_k^2 + \omega_k^2 q_k^2), \qquad V_{Mc} = \alpha x \sum_k q_k\,,$$

where p_k and q_k are the variables of the medium oscillators with the frequencies ω_k, and α is an interaction parameter independent of k. This model was also treated by *Nitzan* and *Jortner* [7.16].

Even in the framework of such a simplified model the calculation of the correlation function (7.2.15) is complicated. Therefore we omit the mathematical framework and only cite the final expression for the correlation function in the electronic state r:

$$\langle r|Q(t_1,t_2)|r\rangle = G\exp[\mathrm{i}\Delta E\theta - \tfrac{1}{2}D(|\mu_1|^2 + |\mu_2|^2 - 2\mu_1\mu_2^*)]$$

$$\times\exp[-\sqrt{D}(\mu_1^* - \mu_2^*)a^+]\exp[\sqrt{D}(\mu_1 - \mu_2)a]\,, \qquad (7.2.31)$$

where a^+ and a are the creation and annihilation operators for a fictitious oscillator with the Hamiltonian $(p^2 + \omega^2 x^2)/2$, i.e.,

$$a = \sqrt{\frac{\omega}{2}}x + \frac{\mathrm{i}}{\sqrt{2\omega}}p\,, \qquad (7.2.32)$$

$$\tfrac{1}{2}(p^2 + \omega^2 x^2) = \omega(a^+ a + \tfrac{1}{2})\,, \qquad (7.2.33)$$

μ_1 and μ_2 are the values of the function

$$\mu(t) = 1 - \exp(-\mathrm{i}\omega t - \gamma t) \qquad (7.2.34)$$

at times t_1 and t_2, respectively, and $D = F^2/2\omega^3$ is the dimensionless equilibrium position displacement squared (the asterisk means complex conjugation). The vibrational relaxation rate γ appearing in (7.2.34) is given by $\gamma = (\pi\alpha^2/4\omega^2)\varrho(\omega)$, where $\varrho(\omega)$ is the frequency spectrum density of the medium at the intramolecular oscillator frequency. The G factor in (7.2.31) is

$$G = \exp\{-\Gamma\theta - D[1 - \exp(-2\gamma t)][2(2\bar{n}+1)\sin^2(\omega\theta/2) + \mathrm{i}\sin(\omega\theta)]\}$$

(in accordance with (7.2.20) we substitute t for τ here and below), where

$$\Gamma = 2D(2\bar{n}+1)\gamma, \qquad \bar{n} = \left[\exp\left(\frac{\omega}{k_\mathrm{B}T_c}\right) - 1\right]^{-1} \qquad (7.2.35)$$

and T_c is the temperature of the medium.

To calculate the diagonal matrix element (7.2.17) we shall make a transformation from the a^+, a operators to new α^+, α operators diagonalizing the initial-state Hamiltonian H_{0r}. On setting

$$a = \alpha - \sqrt{D/2} \qquad (7.2.36)$$

and using (7.2.29, 30, 32, 33) one obtains $H_{0r} = \omega\alpha^+\alpha + \text{const}$. It follows that α^+ and α are the creation and annihilation operators for the intramolecular oscillator in the initial electronic state r. Inserting (7.2.36) into (7.2.31) we calculate (7.2.17), as in paper [7.6]. Neglecting small terms on the order of $\gamma\Delta t$

we obtain

$$F_{rm}(\theta,t)=\exp\left\{\mathrm{i}\Delta E\theta-\Gamma\theta-D[1-\exp(-\mathrm{i}\omega\theta)]-4D\bar{n}[1-\exp(-2\gamma t)]\right.$$
$$\left.\times\sin^2\frac{\omega\theta}{2}\right\}L_m\left[4D\exp(-2\gamma t)\sin^2\frac{\omega\theta}{2}\right], \tag{7.2.37}$$

where $L_m(z)$ is the Laguerre polynomial. At $\gamma=0$ this expression is reduced to (6.2.15) apart from a phase factor and notations.

With the use of (7.2.35) condition (7.2.22) takes the form $2D(2\bar{n}+1)\gg 1$. Hence, the phase relaxation is fast as compared to the vibrational energy relaxation at high-enough medium temperature and a large equilibrium-position displacement of the intramolecular oscillator.

The generating function (7.2.37) describes the radiationless transition from the initially prepared *rm* state. If the Boltzmann distribution over the quasimolecule *rm* levels at a temperature T_M, different from the medium temperature T_c, was prepared initially, the radiationless transition is described by the averaged correlation (generating) function

$$F(\theta,t;T_M)=\exp\left\{\mathrm{i}\Delta E\theta-\Gamma\theta-D[1-\exp(-\mathrm{i}\omega\theta)]-4D\lambda(t)\sin^2\frac{\omega\theta}{2}\right\}, \tag{7.2.38}$$

$$\lambda(t)=\bar{n}+(\bar{m}-\bar{n})\exp(-2\gamma t),\qquad \bar{m}=[\exp(\omega/k_{\mathrm{B}}T_M)-1]^{-1}. \tag{7.2.39}$$

Equation (7.2.38) shows that γ is the rate of equalizing of the quasimolecule and medium temperatures. If $T_M=T_c$ at the initial time, the radiationless transition rate is independent of time, and the kinetics is purely exponential for all t.

We now calculate the transition rate (7.2.24) with the correlation function (7.2.38). It will be assumed that inequality

$$4D\lambda(t)\gg 1 \tag{7.2.40}$$

is fulfilled for all t. Then, the correlation function is sharply peaked at $\theta_n=2\pi n/\omega$ $(n=0,\pm1,\dots)$, and the integral is represented as a sum of the integrals taken over small vicinities of these points. Expanding the exponent of (7.2.38) around θ_n in powers of $x=\omega(\theta-\theta_n)$ up to the second-order terms inclusive, we obtain the transition rate

$$k_r=\frac{\beta^2}{\omega}\left(B_0+2\sum_{n=1}^{\infty}B_n\right), \tag{7.2.41}$$

$$B_n=\mathrm{Re}\left\{\int_{-\infty}^{\infty}dx\exp\left[\frac{\mathrm{i}\Delta E-\Gamma}{\omega}2\pi n+\frac{\mathrm{i}(\Delta E-D\omega)-\Gamma}{\omega}x\right.\right.$$
$$\left.\left.-\tfrac{1}{2}D[2\lambda(t)+1]x^2\right]\right\},\quad n\geq 0. \tag{7.2.42}$$

The expansion in x is valid under condition

$$\delta x \equiv \max\left\{[D\lambda(t)]^{-1/2},\ \frac{\Delta E - D\omega}{D\omega\lambda(t)}\right\} \ll 1 ,$$

where δx denotes the convergence interval of the integral in (7.2.42). Therefore, the condition restricting the electronic energy, $(\Delta E/\omega) - D \ll D\lambda(t)$, should be fulfilled in addition to (7.2.40).

Neglecting the term $\Gamma x/\omega$ in the exponent of (7.2.42) we find

$$B_n = B_0 \operatorname{Re}\left\{\exp\left(2\pi n \frac{\mathrm{i}\Delta E - \Gamma}{\omega}\right)\right\}, \tag{7.2.43}$$

$$B_0 = \left(\frac{2\pi}{D[2\lambda(t) + 1]}\right)^{1/2} \exp\left(-\frac{(\Delta E - D\omega)^2}{2D\omega^2[2\lambda(t) + 1]}\right). \tag{7.2.44}$$

Insertion of (7.2.43) into (7.2.41) yields the transition rate

$$k_r = (\beta^2/\omega) B_0 P , \tag{7.2.45}$$

where

$$P = \frac{1 - \varepsilon^2}{(1 - \varepsilon)^2 + 4\varepsilon \sin^2(\pi\Delta E/\omega)}, \qquad \varepsilon = \exp(-2\pi\Gamma/\omega) .$$

The quantity P characterizes the "overlap" of the broadened initial and final electronic-vibrational levels [7.5]. For $\Gamma \gg \omega$ the levels overlap significantly so that $P = 1$, while for $\Gamma \ll \omega$ the overlap is sharply peaked at integer values of $\Delta E/\omega$ because the initial and final levels come into resonance.

Equations (7.2.44, 39) suggest that for short times (7.2.25) the transition rate decreases in time if $T_M > T_c$, and increases if $T_M < T_c$.

The condition of smallness of V_M resulting from (7.2.27) can be relaxed, provided that $\Gamma \ll \omega$ and the resonance picture applies. To escape the framework of perturbation theory in V_M, we introduce the total Hamiltonian of the system $\hat{\mathscr{H}}$, involving the electronic coordinates that reduces to $\mathscr{H}$ in the diabatic basis $(\psi_r, \psi_{r'})$. For the model under consideration this means in particular that

$$\langle \psi_r | \hat{V}_{Mc} | \psi_{r'} \rangle = V_{Mc} \delta_{rr'} . \tag{7.2.46}$$

We introduce the decomposition

$$\hat{\mathscr{H}} = \hat{H}_M + \hat{H}_c + \hat{V}_{Mc} \tag{7.2.47}$$

that separates the molecular Hamiltonian with the internal coupling,

$$\hat{H}_M = \hat{H}_0 + \hat{V}_M , \tag{7.2.48}$$

from the terms associated with the medium, and prognosticates the perturbation expansion in the molecule–medium coupling $\hat{V}_{Mc}$. The standard routine now is to solve the vibronic problem

$$\hat{H}_M|\Psi_m\rangle = E_m|\Psi_m\rangle \,, \tag{7.2.49}$$

which for the two-electronic-state model assumed was done in Sect. 5.8 in the quasiclassical approximation without essentially invoking the smallness of V_M.

We consider the picture of quasiresonances between some pairs of vibronic levels,

$$\Delta = |E_m - E_n| \ll \omega \,, \tag{7.2.50}$$

when the perturbation matrix $\langle\Psi_m|\hat{V}_{Mc}|\Psi_{m'}\rangle$is divided into two-by-two blocks corresponding to such pairs. For the model of (7.2.46) the assumptions about Γ and Δ, already adopted, permit us to consider the diagonal elements of these blocks as being equal. Actually, one has

$$\langle\Psi_{m,n}|\hat{V}_{Mc}|\Psi_{m,n}\rangle = \alpha(\langle\chi_r^{m,n}|x|\chi_{r'}^{m,n}\rangle + \langle\chi_r^{m,n}|x|\chi_{r'}^{m,n}\rangle)\sum_k q_k \,, \tag{7.2.51}$$

where $|\chi_{r,r'}^{m,n}\rangle$ are nuclear molecular functions in the adiabatic basis $(\varphi_r, \varphi_{r'})$. In view of (7.2.50) and (5.8.18) one has $|T_{12}| \ll 1$, and it follows from (5.8.3–8) that $|\chi_{r,r'}^{m,n}\rangle$ are the adiabatic functions and satisfy the condition

$$(\chi_r^n, \chi_{r'}^n) = (\chi_r^m, -\chi_{r'}^m) \tag{7.2.52}$$

to within $|T_{12}|$. Thus,

$$\langle\Psi_{m,n}|\hat{V}_{Mc}|\Psi_{m,n}\rangle \equiv \bar{V} = \alpha(x_{rr} + x_{r'r'})\sum_k q_k \tag{7.2.53}$$

to within the second-order terms in $\bar{V}\Delta/\omega$.

As a result, the individual (m, n) vibronic block of $\mathscr{H}$ takes the form

$$H = \bar{E} + H_c + \tfrac{1}{2}(\sigma_z \Delta + \sigma_x V) \,, \tag{7.2.54}$$

$$\bar{E} = \frac{E_m + E_n}{2} + V, \qquad V = 2\alpha(x_{rr} - x_{r'r'})\sum_k q_k \,. \tag{7.2.55}$$

In the interaction representation, eliminating H_c from the components of the density matrix of the system in the vibronic states Ψ_m, Ψ_n,

$$\varrho = \tfrac{1}{2}(1 + \boldsymbol{R}\cdot\boldsymbol{\sigma})R_c \quad (\boldsymbol{R} = (X, Y, Z)) \tag{7.2.56}$$

we obtain the equations

$$\begin{aligned} \frac{\partial X}{\partial t} &= -\Delta Y \,, \\ \frac{\partial Y}{\partial t} &= \Delta X - V(t) Z \,, \\ \frac{\partial Z}{\partial t} &= V(t) Y \,, \end{aligned} \tag{7.2.57}$$

where

$$V(t) = V[q_k(t)] \,. \tag{7.2.58}$$

Now we can use time-dependent perturbation theory in $V(t)$ assuming τ_c to be short enough, as in (7.2.12). The calculations lead to the master kinetic system

$$\frac{\partial \langle X \rangle_c}{\partial t} = -\Delta \langle Y \rangle_c \,,$$

$$\frac{\partial \langle Y \rangle_c}{\partial t} = \Delta \langle X \rangle_c - \Gamma \langle Z \rangle_c \,, \tag{7.2.59}$$

$$\frac{\partial \langle Z \rangle_c}{\partial t} = -\Gamma \langle Z \rangle_c \,,$$

where

$$\Gamma = 2\,\mathrm{Re}\left\{ \int_0^\infty \left\langle V\left(\tau + \frac{\theta}{2}\right) V\left(\tau - \frac{\theta}{2}\right) \right\rangle \cos \Delta\theta \, d\theta \right\} . \tag{7.2.60}$$

For this system, there exist three characteristic constants,

$$k_1 = \Gamma, \qquad k_{2,3} = \frac{\Gamma}{2} \pm \left(\frac{\Gamma^2}{4} - \Delta^2 \right)^{1/2} . \tag{7.2.61}$$

For $\Gamma \ll \Delta$ they describe the levelling-off of the vibronic-level populations and the damping of the beats that may be interpreted as successive forward and backward electronic transitions. If, however, $\Delta \ll \Gamma$, then it is the minimum root,

$$k_3 = k_{mn} = \frac{\Delta^2}{\Gamma} \,, \tag{7.2.62}$$

that should be identified as the radiationless transition rate for the level $E_m \simeq E_n$. In view of (5.8.18) one has

$$k_{mn} = \frac{\omega_m \omega_n}{\Gamma} |T_{12}|^2 \,. \tag{7.2.63}$$

This equation is valid in a wide interval of values of the parameter δ.

7.3 Kinetic Model

The kinetic model is used to account for the biexponential fluorescence decay of some organic molecules [1.3, 2.76, 81, 85, 90]. Within this model the biexponential kinetics is ascribed to the occurrence of the reverse $r \leftarrow r'$ electronic transition, which is considered to be just as irreversible as the direct $r \to r'$ transition. This implies that the balance (kinetic) equation of Sect. 2.1 is equally valid for both the direct and reverse transitions.

Consider two isoenergetic states S_1^* and T_1^* (the asterisk means vibrational excitation), and write down the following kinetic equations for the populations:

$$\frac{d}{dt}[S_1^*] = -\kappa_S[S_1^*] + k_{TS}[T_1^*] ,$$

$$\frac{d}{dt}[T_1^*] = -\kappa_T[T_1^*] + k_{ST}[S_1^*] , \qquad (7.3.1)$$

$$\kappa_S = k_f + k_{ST} + k_S, \qquad \kappa_T = k_{TS} + k_T ,$$

where k_f is the emission rate from S_1^*, k_{ST} the direct $S_1^* \rightsquigarrow T_1^*$ radiationless transition rate, k_S the sum of the rates of other S_1^* deactivating processes (the vibrational relaxation of S_1^* to the S_1 ground vibrational state and the $S_1^* \rightsquigarrow S_0$ internal conversion), k_{TS} the rate of the reverse $S_1^* \leftarrow T_1^*$ radiationless transition, k_T the sum of the T_1^* deactivation rates due to vibrational relaxation to the T_1 ground vibrational state and the $T_1 \rightsquigarrow S_0$ intersystem crossing. The solution to (7.3.1) with the initial conditions $[S_1^*]_{t=0} = [S_1^*]_0$ and $[T_1^*]_{t=0} = 0$ is

$$[S_1^*] = [S_1^*]_0(k_+ - k_-)^{-1}[(\kappa_S - k_-)\exp(-k_+ t)$$

$$+ (k_+ - \kappa_S)\exp(-k_- t)] , \qquad (7.3.2)$$

$$k_\pm = \tfrac{1}{2}(\kappa_S + \kappa_T) \pm [\tfrac{1}{4}(\kappa_S - \kappa_T)^2 + k_{ST}k_{TS}]^{1/2} .$$

Thus, the fluorescence intensity equal to $k_f[S_1^*]$ decays by the biexponential law. This is due to the inclusion of the $S_1^* \leftsquigarrow T_1^*$ reverse transition, since, if $k_{TS} = 0$, either k_+ or k_- is equal to κ_S and one of the exponentials in (7.3.2) disappears.

The application of the kinetic model is justified provided that both the direct and reverse transitions are irreversible. According to the Freed-Jortner criterion, the irreversibility demands a high density of the final states. Hence S_1^* (as well as T_1^*) must be a state with a high vibrational excitation. The highly excited vibrational states are mixtures of a large number of states with different sets of quantum numbers. Therefore, the quantum-mechanical calculation of the direct and reverse transition-rate constants implies the summation of the partial rates over all the initial states at a given energy.

At fast vibrational relaxation in the initial state, S_1^* rapidly relaxes to the S_1 ground vibrational state. In this case, the fluorescence decay in the kinetic model is purely exponential because the reverse $S_1^* \leftsquigarrow T_1^*$ transition is impossible. The same is valid for fast relaxation in the final state when the reverse transition is insignificant in view of a low T_1^* population.

The kinetic model is also invoked to explain the pressure dependence of the fluorescence quantum yield in vapors. For example, it was observed in [2.41] that the S_2 fluorescence intensity (normalized to the S_1 one) of pyrene and 3, 4-benzpyrene decreased with increasing pressure. The reverse $S_2^* \leftarrow S_1^*$ transitions enhancing the S_2 population become insignificant at high pressures due to the fast $S_1^* \rightarrow S_1$ relaxation, which results in lowering of the S_2 relative fluorescence intensity.

It should be pointed out that the existence of the reverse transitions still remains very questionable since the density of states in the upper electronic state is much lower than in the lower state at the same total energy. The experimental evidence for the reverse transitions is based on the observation of the fluorescence emission following the multiphoton IR excitation of molecules in the S_0 ground electronic state [2.185, 186]. The onset of the emission is seen when the S_0 vibrational excitation reaches the S_1 origin. However, the duration of the emission does not exceed the IR-laser pulse duration, whereas the fluorescence emission must have the S_1 lifetime when irreversible $S_1^* \leftarrow S_0^*$ transitions really occur. In fact, the fluorescence observed is merely due to mixing of the excited S_0^* states with the S_1 ground vibrational state.

8. Conclusion

In this book we intended to give a systematical account of the modern understanding of the nature of radiationless transitions, and to ensure a complete-enough and up-to-date coverage of the mathematical techniques employed in calculations of the rate constants for these processes. The material exposed affords an opportunity to draw conclusions concerning the calculation methods, and to assess some promising directions for future studies. Leaving aside fine details of the intermediate-case transitions we shall make concluding remarks concerning the most abundant class of polyatomic molecules in which the radiationless transitions occur in the statistical limit.

In this limit the dynamics of the radiationless transition is simple enough: in the course of oscillations the nuclei tunnel into the transition region where the molecular electronic shell is rearranged. In accordance with this picture the transition rate is controlled by three principal factors: the mean vibration frequency, the probability of the nuclear-configuration change, and the probability of the electronic transition itself. An additional, important parameter is the recurrence frequency, i.e., the frequency with which the system in the final electronic state returns to the region where the barrier width for the nuclear transformation is of least width. It is expressed in terms of the density of the effective final states that plays a fundamental role in the classification of radiationless transitions; in particular, the large values of the latter parameter (the low recurrence frequency) correspond to the statistical limit. In turn, the effective-states density is governed by the number of active modes and by the change of the potential's surface shape in the electronic transition.

The change of the nuclear configuration is a result of tunneling along the active-acceptor coordinates across that domain of the configurational space where the total energies of the electronic states participating in the transition become equal. The respective probability is controlled by the overlap of the nuclear wave functions associated with the electronic states under consideration, and essentially depends on their shapes.

The electronic-transition probability is a function of the Landau-Zener parameter and is governed by the coupling strength of the participating electronic states. The shape of the adiabatic terms and the interaction between them are closely related to each other. Particularly, the quasiclassical approximation shows that the shape of the electronic terms also specifies their coupling. If the terms' splitting is small within the domain of their closest approach (as is the case for the intersystem crossing induced by a weak spin–orbit coupling), then the electronic transition probability is small, being proportional to this splitting squared. If it is large (as for the internal conversion), then the probability is essentially independent of the splitting and is close to unity.

Thus, the nuclear tunneling turns out to be a dominant factor for the radiationless transition rate. The tunneling probability is exponentially small and, in practice, can be calculated in the quasiclassical limit only. It is worthwhile to emphasize that "exact" computer calculations often are misleading, especially in the case of large electronic energy gaps, since the tunneling probability is highly sensitive to mostly unknown details of the potential behavior in nonclassical regions of coordinate space, and the specificity of the computer methods is that they strongly enhance the errors primarily introduced in the calculation via the uncertainties in the potential parameters, while the quasiclassical methods are free from such shortcomings. Though, the quasiclassical calculations are quite sophisticated, and possible prospects to perform such calculations for multidimensional systems rely on the tunneling-trajectory method that regretfully has not been developed enough yet. For one-dimensional potentials the quasiclassical overlap integrals are known and represented in suitable analytical forms.

The above-mentioned sensitivity of the transition rate to the shape of the potential surfaces may give rise to a new line in experimental and theoretical studies, viz., the determination of the potential energy surfaces from the data on transition rates for the classes of molecules with similar electronic structures.

Appendix: Commutation Rules for Angular Momentum Operators in the Laboratory and Molecule Frame

Let $\boldsymbol{Z}$ stand for any one of the rotation ($\boldsymbol{N}$), spin ($\boldsymbol{S}$), and total ($\boldsymbol{J} = \boldsymbol{N} + \boldsymbol{S}$) angular momenta, $\tilde{Z}_i$ for its projections onto the axes of the Laboratory Coordinate Frame (LCF) with the unit directional vectors $\tilde{\boldsymbol{e}}_i$ ($i = x, y, z$):

$$\tilde{Z}_i = \boldsymbol{Z} \cdot \tilde{\boldsymbol{e}}_i \,. \tag{A.1}$$

The projections obey the convential commutation relations

$$[\tilde{Z}_i, \tilde{Z}_j] = \mathrm{i} e_{ijk} \tilde{Z}_k \,, \tag{A.2}$$

where e_{ijk} is the completely antisymmetric third-rank unit tensor and summation over the twice repeated index is implied here and below (all the momenta are in units of $\hbar$).

The operators $\tilde{S}_i$ act only on spin variables while $\tilde{N}_i$ only on rotation ones, i.e., three Euler angles ϑ establishing the orientation of the three unit vectors of the Molecule's Coordinate Frame (MCF) $\boldsymbol{e}_i$ relative to LCF. Hence

$$[\tilde{S}_i, \tilde{N}_j] = 0 \,. \tag{A.3}$$

One more relation results from a general commutation rule for components of an arbitrary vector with angular momentum projections. If one considers a rotation momentum associated with the operator of an infinitesimal rotation of MCF with respect to LCF, then by a vector one should mean the quantity of the type $\boldsymbol{b} = b_j \boldsymbol{e}_j$, where b_j are constants independent of ϑ. In particular, the unit vectors $\boldsymbol{e}_j$ are true vectors in the above sense, whereas the $\tilde{\boldsymbol{e}}_i$'s are not since they are unchanged under rotations of MCF. Developing $\boldsymbol{b}$ in the LCF unit vectors we obtain three quantities, $\tilde{b}_i = \boldsymbol{b} \cdot \tilde{\boldsymbol{e}}_i$, that transform through each other like vector components under MCF rotations. They are related to the constants b_j (projections of $\boldsymbol{b}$ in MCF) via the rotation matrix $C_{ij} = \tilde{\boldsymbol{e}}_i \cdot \boldsymbol{e}_j$ by

$$\tilde{b}_i = C_{ij} b_j, \quad b_j = \tilde{b}_i C_{ij}; \quad C_{ij} C_{i'j} = \delta_{ii'} \,. \tag{A.4}$$

The commutation rule mentioned is [1.40]

$$[\tilde{b}_i, \tilde{N}_j] = \mathrm{i} e_{ijk} \tilde{b}_k \,. \tag{A.5}$$

Inserting here (A.4) and recalling that the constants b_j are arbitrary we obtain commutation relations for components of the rotation matrix and the LCF projections of the rotation angular momentum [4.36]:

$$[C_{ij}, \tilde{N}_k] = \mathrm{i} e_{ikl} C_{lj} . \tag{A.6}$$

A useful relation can be derived for the rotation matrix from the conditions of mutual orthogonality of the unit vectors,

$$\boldsymbol{e}_i \times \boldsymbol{e}_j = e_{ijk} \boldsymbol{e}_k, \qquad \tilde{\boldsymbol{e}}_i \times \tilde{\boldsymbol{e}}_j = e_{ijk} \tilde{\boldsymbol{e}}_k .$$

Inserting the identity $\tilde{\boldsymbol{e}}_i = C_{ij} \boldsymbol{e}_j$ into the latter and exploiting the former one finds after simple algebra and renaming the indices

$$e_{ijk} C_{jj'} C_{kk'} = C_{ii'} e_{i'j'k'} . \tag{A.7}$$

Further, let Z_j be projections of $\boldsymbol{Z}$ in MCF:

$$Z_j = \boldsymbol{Z} \cdot \boldsymbol{e}_j = \tilde{Z}_i C_{ij} . \tag{A.8}$$

For $\boldsymbol{Z} = \boldsymbol{N}$ we note that the right-hand side of (A.6) vanishes at $i = k$. This signifies that C_{ij} commutes with $\tilde{N}_i$ at any i and j so that the order of multiplication in the definition (A.8) is irrelevant.

Let us right-multiply (A.6) by $C_{kk'}$ and sum it up over k with the help of (A.7). Inserting the definition (A.8) we obtain the commutation rule for the rotation matrix components with the MCF projections of the rotation angular momentum,

$$[C_{ij}, N_k] = -\mathrm{i} C_{il} e_{ljk} . \tag{A.9}$$

For applications the knowledge of the commutation relations similar to (A.2, 3) between the MCF projection of the angular momentum operators as well as between the MCF and LCF projections is necessary. First of all, of relevance here are the well-known relations between the MCF projections of the rotational momentum [1.40, 4.36],

$$[N_i, N_j] = -\mathrm{i} e_{ijk} N_k , \tag{A.10}$$

which differs from (A.2) by the sign of the right-hand side, and the relations

$$[N_i, \tilde{N}_j] = 0, \qquad [J_i, \tilde{J}_j] = 0 , \tag{A.11}$$

showing that the MCF projections of the rotational or total momentum can assume a definite value simultaneously with the LCF projection of the same momentum. In contrast, this is not true of the spin angular momentum since

$$[S_i, \tilde{S}_j] \neq 0 . \tag{A.12}$$

We also notice that, as distinct from (A.10), the MCF components of the spin momentum satisfy the relation

$$[S_i, S_j] = \mathrm{i}e_{ijk}S_k \tag{A.13}$$

similar to that of (A.2) for the LCF components.

These and other relations are easy to be deduced with the aid of formulae (A.2, 3, 6–9). As an example, we give the proof of the second relation in (A.11). After (A.8) is substituted, the relation considered is recast in the form

$$[J_i, \tilde{J}_j] = [\tilde{J}_l C_{li}, \tilde{J}_j] = [\tilde{J}_l, \tilde{J}_j]C_{li} + \tilde{J}_l[C_{li}, \tilde{J}_j] .$$

The first commutator is calculated by (A.2), and in the second one we write $\tilde{J}_j = \tilde{N}_j + \tilde{S}_j$. Remembering that $\tilde{S}_j$ commutes with C_{li} and calling on (A.6) we find

$$[J_i, \tilde{J}_j] = \mathrm{i}e_{jkl}(\tilde{J}_k C_{li} + \tilde{J}_l C_{ki}) = 0 .$$

It is here taken into account that the expression in braces is symmetric with respect to permutation of indices k and l, and therefore its convolution over k and l with the antisymmetric tensor vanishes.

For reference, several projection's commutators should be quoted:

$$[J_i, S_j] = [N_i, \tilde{Z}_j] = [Z_i, \tilde{J}_j] = 0 , \tag{A.14}$$

$$[N_i, Z_j] = -\mathrm{i}e_{ijk}Z_k, \quad [J_i, J_j] = -\mathrm{i}e_{ijk}J_k . \tag{A.15}$$

The operators J_i and S_i do not commute with $\tilde{N}_j$ and $\tilde{S}_j$, and no simple relations such as (A.2, 10, 15) exist for the corresponding four commutators.

The operator of the angular momentum squared may be written in any one of two equivalent forms: $\boldsymbol{Z}^2 = Z_i^2 = \tilde{Z}_j^2$. Let $\boldsymbol{Y}$, just as $\boldsymbol{Z}$, stand for any of the three momenta under study. Employing the above-given commutation relations we obtain

$$[\boldsymbol{Y}^2, \boldsymbol{Z}^2] = [\tilde{Z}_i, \boldsymbol{N}^2] = [\tilde{Z}_i, \boldsymbol{S}^2] = [\tilde{J}_i, \boldsymbol{J}^2] = 0 ,$$

$$[N_i, \boldsymbol{N}^2] = [Z_i, \boldsymbol{S}^2] = [Z_i, \boldsymbol{J}^2] = 0 ; \tag{A.16}$$

the operators S_i and J_i do not commute with $\boldsymbol{N}^2$; $\tilde{S}_i$ and $\tilde{N}_i$ do not commute with $\boldsymbol{J}^2$.

Now consider the operator U_{so} of the spin-orbit coupling in a nonlinear polyatomic molecule [see the definition in (4.4.8)]. The operator of the total spin is $\boldsymbol{S} = \sum_a \boldsymbol{s}_a$. U_{so} commutes with none of the operators $\boldsymbol{S}^2$, S_i, $\boldsymbol{N}^2$, N_i. On the other hand, the following relations hold:

$$[U_{\mathrm{so}}, \tilde{J}_i] = [U_{\mathrm{so}}, J_i] = [U_{\mathrm{so}}, \boldsymbol{J}^2] = 0 . \tag{A.17}$$

To prove this we write

$$[U_{so}, \tilde{N}_i] = \sum_a [\tilde{b}_{aj}, \tilde{N}_i]\tilde{s}_{aj} = \mathrm{i}e_{jik}\sum_a \tilde{b}_{ak}\tilde{s}_{aj} ,$$

$$[U_{so}, \tilde{S}_i] = \sum_{aa'} \tilde{b}_{ak}[\tilde{s}_{ak}, \tilde{s}_{a'i}] = \mathrm{i}e_{kij}\sum_a \tilde{b}_{ak}\tilde{s}_{aj} . \qquad \text{(A.18)}$$

Adding up these two identities we obtain on the left-hand side the first commutator of (A.17) while the right-hand one disappears. The second commutator in (A.17) is obtained from the first one by using (A.8) and bearing in mind that C_{ij} commutes with U_{so} since $\boldsymbol{b}_a$ is merely a function of ϑ. The third commutator in (A.17) is evidently also equal to zero, because U_{so} commutes with the projections of the total momentum, and hence, with the sum of those projections squared.

References

Chapter 1

1.1 K.F. Freed: Energy dependence of electronic relaxation processes in polyatomic molecules, in *Radiationless Processes in Molecules and Condensed Phases*, ed. by F.K. Fong, Topics Appl. Phys., Vol. 15 (Springer, Berlin, Heidelberg 1976) pp. 23–168
K.F. Freed: The theory of radiationless processes in polyatomic moleeules, in *Stereo- and Theoretical Chemistry*, Topics Curr. Chem., Vol. 31 (Springer, Berlin, Heidelberg 1972) pp. 105–139
K.F. Freed: J. Chem. Phys. **52**, 1345 (1970)

1.2 R. Englman: *Non-Radiative Decay of Ions and Molecules in Solids* (North-Holland, New York 1979)
B. Di Bartolo (ed.): *Radiationless Processes*, NATO Adv. Study Inst. Ser., Series B: Physics, Vol. 62 (Plenum, New York 1980)

1.3 P. Avouris, W.M. Gelbart, M.A. El-Sayed: Chem. Rev. **77**, 793 (1977)

1.4 W. Siebrand: Nonradiative processes in molecular systems, in *Dynamics of Molecular Collisions*, ed. by W.H. Miller, Modern Theoretical Chemistry, Vol. 1, Part A (Plenum, New York 1976) pp. 249–302

1.5 B.R. Henry, W. Siebrand: Radiationless transitions, in *Organic Molecular Photophysics*, ed. by J.B. Birks, Vol. 1 (Wiley, London 1973) pp. 153–237
B.R. Henry, W. Siebrand: Radiationless transitions – a postscript, in *Organic Molecular Photophysics*, ed. by J.B. Birks, Vol. 2 (Wiley, London 1975) pp. 303–311

1.6 J. Jortner, S. Mukamel: Molecular radiationless processes, in *Molecular Energy Transfer*, ed. by R. Levine, J. Jortner (Wiley, New York 1976) pp. 178–211

1.7 J. Jortner, S. Mukamel: Preparation and decay of exited molecular states, in *The World of Quantum Chemistry*, ed. by R. Daudel, B. Pullman (Reidel, Boston 1974) pp. 145–209
S. Mukamel, J. Jortner: Time evolution of excited molecular states, in *Excited States*, ed. by E.C. Lim, Vol. 3 (Academic, New York 1977) pp. 57–107

1.8 J. Jortner: Pure Appl. Chem. **24**, 165 (1970)

1.9 J. Jortner: Theoretical studies of radiationless transitions, in *Transitions Non-Radiatives dans les Molécules*, J. Chim. Phys. Phys.-Chim. Biol. (Suppl.) **10**, 9 (1970)

1.10 J. Jortner, S.A. Rice, R.M. Hochstrasser: Radiationless transitions in photochemistry, in *Adv. Photochem.*, ed. by J.N. Pitts, Jr., G.S. Hammond, W.A. Noyes, Jr., Vol. 7 (Interscience, New York 1969) pp. 149–309

1.11 G.W. Robinson: Molecular electronic radiationless transitions, in *Excited States*, ed. by E.C. Lim, Vol. 1 (Academic, New York 1974) pp. 1–34

1.12 K.F. Freed: Collisional effects on electronic relaxation processes, in *Adv. Chem. Phys.*, ed. by K.P. Lawley, Vol. 42 (Wiley, New York 1980) pp. 207–269

1.13 A.A. Ovchinnikov, M. Ya. Ovchinnikova: Problems of nonlinear radiationless transitions in chemistry, in *Adv. Quantum Chem.*, ed. by P.-O. Löwdin, Vol. 16 (Academic, New York 1982) pp. 161–227

1.14 R. Voltz: Theory of molecular decay processes, in *Org. Mol. Photophys.*, ed. by J.B. Birks, Vol. 2 (Wiley, London 1975) pp. 217–302

1.15 Yu.E. Perlin, B.S. Tsukerblat: *Effects of Electronic-Vibrational Coupling in Optical Centers of*

Paramagnetic Impurity Ions (Stiintsa, Kishinev 1974) (in Russian)
Yu.E. Perlin, A.A. Kaminskiĭ: Phys. Status Solidi (b) **132**, 11 (1985)
1.16 Yu.E. Perlin: Usp. Fiz. Nauk **80**, 553 (1963) [English transl.: Sov. Phys.- Usp. **80**, 542 (1963)]
1.17 J.T. Yardley: Dynamic properties of electronically excited molecules, in *Chem. Biochem. Appl. Lasers*, ed. by C.B. Moore, Vol. 1 (Academic, New York 1974) p. 231–279
1.18 B.R. Henry, M. Kasha: Annu. Rev. Phys. Chem. **19**, 161 (1968)
1.19 V.A. Kovarskiĭ: *Many-Quantum Transitions* (Stiintsa, Kishinev 1974) (in Russian)
1.20 V.A. Kovarskiĭ: *Kinetics of Radiationless Transitions* (Acad. Sci. Moldavian SSR, Kishinev 1968) (in Russian)
1.21 V.A. Kovarskiĭ, N.F. Perel'man, I.S. Averbukh, S.A. Baranov, S.S. Todirashku: *Nonadiabatic Transitions in Strong Electromagnetic Field* (Stiintsa, Kishinev 1980) (in Russian)
1.22 M.D. Frank-Kamenetskiĭ, A.V. Lukashin: Usp. Fiz. Nauk **116**, 193–229 (1975) [English transl.: Sov. Phys.- Usp. **18**, 391 (1975)]
1.23 V.G. Plotnikov: Usp. Khim. **49**, 327 (1980) [English transl.: Russian Chem. Rev. **49**, 172 (1980)]
V.G. Plotnikov, A.A. Ovchinnikov: Usp. Khim. **47**, 444 (1978) [English transl.: Russian Chem. Rev. **47**, 247 (1978)]
1.24 N.M. Kristoffel: *Theory of Low Radius Impurity Centers in Ionic Crystals* (Nauka, Moscow 1974) (in Russian)
1.25 A.F. Lubchenko: *Quantum Transitions in Impurity Centers in Solids* (Naukova Dumka, Kiev 1978) (in Russian)
1.26 V.L. Ermolaev, E.N. Bodunov, E.B. Sveshnikova, T.A. Shakhverdov: *Radiationless Electronic Excitation Energy Transfer* (Nauka, Leningrad 1977) (in Russian)
1.27 V.L. Ermolaev, E.B. Sveshnikova: Acta Phys. Pol. **34**, 771 (1968) (in Russian)
1.28 J.B. Birks: *Photophysics of Aromatic Molecules* (Wiley, London 1970)
1.29 J.B. Birks: The spectroscopy of the π-electronic states of aromatic hydrocarbons, in *Org. Mol. Photophys.*, ed. by J.B. Birks, Vol. 1 (Wiley, London 1973) pp. 1–15
1.30 E.W. Schlag, S. Schneider, S.F. Fischer: Annu. Rev. Phys. Chem. **22**, 465 (1971)
1.31 M.A. El-Sayed: Annu. Rev. Phys. Chem. **26**, 235 (1975)
1.32 S.K. Lower, M.A. El-Sayed: Chem. Rev. **66**, 199 (1966)
1.33 G.B. Tolstorozhev: *Radiationless Transitions in Isolated Molecules* (Acad. Sci. Byelorussian SSR, Minsk 1971) (in Russian)
1.34 R.N. Nurmukhamedov: *Absorption and Luminescence of Aromatic Compounds* (Khimiya, Moscow 1971) (in Russian)
1.35 N.A. Borisevich: *Excited States of Complex Molecules in Gas Phase* (Nauka i Tekhnika, Minsk 1967) (in Russian)
1.36 P. Seybold, M. Gouterman: Chem. Rev. **65**, 413 (1965)
1.37 M. Kasha: Disc. Faraday Soc. **9**, 14 (1950)
1.38 B.I. Stepanov, V.P. Gribkovskiĭ: *Introduction to the Theory of Luminescence* (Acad. Sci. Byelorussian SSR, Minsk 1963)
1.39 L.D. Landau: Phys. Zs. Sovjet. **1**, 88 (1932)
1.40 L.D. Landau, E.M. Lifshitz: *Quantum Mechanics*, 3rd rev. edn. (Pergamon, Oxford 1977)
1.41 M. Bixon, J. Jortner: J. Chem. Phys. **48**, 715 (1968)
M. Bixon, J. Jortner: J. Chem. Phys. **50**, 4061 (1969)
1.42 B. Sharf, R. Silbey: J. Chem. Phys. **53**, 2626 (1970)
B. Sharf, R. Silbey: Chem. Phys. Lett. **5**, 314 (1970)
1.43 R.G. Bennett, P.J. McCartin: J. Chem. Phys. **44**, 1969 (1966)
R.E. Kellogg: J. Chem. Phys. **44**, 411 (1966)
U. Laor, P.K. Ludwig: J. Chem. Phys. **54**, 1054 (1971)
U. Laor, J.C. Hsieh, P.K. Ludwig: Chem. Phys. Lett. **22**, 150 (1973)
A. Amirav, J. Jortner: Chem. Phys. Lett. **132**, 335 (1986)
V.Z. Kresin, W.A. Lester, Jr.: Chem. Phys. Lett. **138**, 59 (1987)
N. Ohmori, T. Suzuki, M. Ito: J. Phys. Chem. **92**, 1086 (1988)
1.44 C.S. Parmenter, H.M. Poland: J. Chem. Phys. **51**, 1551 (1969)

1.45 A. Amirav, C. Horwitz, J. Jortner: J. Chem. Phys. **88**, 3092 (1988)
1.46 D.C. McClure: J. Chem. Phys. **17**, 665 (1949)
D.C. McClure: J. Chem. Phys. **20**, 682 (1952)
M. Mizushima, S. Koide: J. Chem. Phys. **20**, 765 (1952)
1.47 H.F. Hameka: Spin-orbit interactions in organic molecules, in *The Triplet State*, ed. by A.B. Zahlan (Cambridge Univ. Press, London 1967) pp. 1–27
1.48 H.F. Hameka, L.J. Oosterhoff: Mol. Phys. **1**, 358 (1958)
1.49 E. Clementi: J. Mol. Spectrosc. **6**, 497 (1961)
1.50 A.C. Albrecht: J. Chem. Phys. **38**, 354 (1963)
1.51 M.A. El-Sayed: J. Chem. Phys. **38**, 2834 (1963)
1.52 F. Metz, S. Friedrich, G. Hohlneicher: Chem. Phys. Lett. **16**, 353 (1972)
1.53 F. Metz: Chem. Phys. Lett. **22**, 186 (1973)

Chapter 2

2.1 J.H. Callomon, J.E. Parkin, R. Lopez-Delgado: Chem. Phys. Lett. **13**, 125 (1972)
2.2 B. Dick, B. Nickel: Chem. Phys. **110**, 131 (1986)
2.3 A. Amirav, M. Sonnenschein, J. Jortner: J. Phys. Chem. **88**, 5593 (1984)
2.4 V.L. Ermolaev, V.A. Lyubimtsev: Acta Phys. Pol. **A71**, 731 (1987)
V.L. Ermolaev, V.A. Lyubimtsev: Izv. Akad. Nauk SSSR, Ser. Fiz. **52**, 770 (1988) [English transl.: Bull. Acad. Sci. SSSR Phys. Ser. **52**, 133 (1988)]
V.L. Ermolaev, V.A. Lyubimtsev: Opt. Spektrosk. **56**, 1026 (1984) [English transl.: Opt. Spectrosc. (USSR) **56**, 630 (1984)]
2.5 K.-J. Choi, H.-B. Lin, M. Topp: Fluorescence spectroscopy of subpicosecond states in liquids, in *Picosecond Phenomena*, ed. by C.V. Shank, E.P. Ippen, S.L. Shapiro, Springer Ser. Chem. Phys., Vol. 4 (Springer, Berlin, Heidelberg 1978) pp. 27–33
K.-J. Choi, L.A. Hallidy, H.-B. Lin, M.R. Topp: Fluorescence spectroscopy of ultrashort-lived states of aromatic compounds, in *Advances in Laser Chemistry*, ed. by A.H. Zewail, Springer Ser. Chem. Phys., Vol. 3 (Springer, Berlin, Heidelberg 1978) pp. 155–162
B.F. Plummer, Z.Y. Al-Saigh, M. Arfan: Chem. Phys. Lett. **104**, 389 (1984)
2.6 T. Elsaesser, F. Lärmer, W. Kaiser, B. Dick, N. Niemeyer, W. Lüttke: Chem. Phys. **126**, 405 (1988)
2.7 M. Mahaney, J.R. Huber: Chem. Phys. Lett. **105**, 395 (1984)
2.8 E.A. Mangle, M.R. Topp: J. Phys. Chem. **90**, 802 (1986)
2.9 W.D. Lawrance, A.E.W. Knight: J. Phys. Chem. **94**, 1249 (1990)
2.10 W. Leupin, S.J. Berens, D. Magde, J. Wirz: J. Phys. Chem. **88**, 1376 (1984)
2.11 A. Maciejewski: Chem. Phys. Lett. **164**, 166 (1989)
M. Szymanski, R.P. Steer, A. Maciejewski: Chem. Phys. Lett. **135**, 243 (1987)
2.12 C. Rulliere, A. Declemy, Ph. Kottis: Chem. Phys. Lett. **110**, 308 (1984)
2.13 S.L. Shapiro (ed.): *Ultrashort Light Pulses: Picosecond Techniques and Applications*, Topics Appl. Phys., Vol. 18 (Springer, Berlin, Heidelberg 1977)
2.14 C.V. Shank, E.P. Ippen, S.L. Shapiro (eds.): *Picosecond Phenomena*, Springer Ser. Chem. Phys., Vol. 4 (Springer, Berlin, Heidelberg 1978)
2.15 A.H. Zewail (ed.): *Advances in Laser Chemistry*, Springer Ser. Chem. Phys., Vol. 3 (Springer, Berlin, Heidelberg 1978)
2.16 T. Yajima, K. Yoshihara, C.B. Harris, S. Shionoya (eds.): *Ultrafast Phenomena VI*, Springer Ser. Chem. Phys., Vol 48 (Springer, Berlin, Heidelberg 1988)
2.17 F.W. Wise, M.J. Rosker, C.L. Tang: J. Chem. Phys. **86**, 2827 (1987)
N.A. Borisevich, I.V. Kryukov, P.G. Kryukov, G.B. Tolstorozhev, E.V. Khoroshilov, M.Yu. Shakhbazyan, A.V. Sharkov: Dokl. Akad. Nauk SSSR **308**, 350 (1989) [English transl.: Sov. Phys.-Dokl. **34**, 815 (1989)]
R.M. Bowman, M. Dantus, A.H. Zewail: Chem. Phys. Lett. **161**, 297 (1989)
H.L. Fragnito, J.-Y. Bigot, P.C. Becker, C.V. Shank: Chem. Phys. Lett. **160**, 101 (1990)

2.18 W. Rhodes: Chem. Phys. **4**, 259 (1974)

2.19 A. Inoue, N. Ebara: Chem. Phys. Lett. **109**, 27 (1984)
A. Inoue, N. Ebara: J. Phys. Chem. **89**, 4654 (1985)

2.20 R.A. Coveleskie, D.A. Dolson, C.S. Parmenter: J. Chem. Phys. **72**, 5774 (1980)
R.A. Coveleskie, D.A. Dolson, C.S. Parmenter: J. Phys. Chem. **89**, 645 (1985)
R.A. Coveleskie, D.A. Dolson, C.S. Parmenter: J. Phys. Chem. **89**, 655 (1985)

2.21 M.A. Dunkan, T.G. Dietz, M.G. Liverman, R.E. Smalley: J. Phys. Chem. **81**, 7 (1981)

2.22 J.W. Perry, N.F. Scherer, A. H. Zewail: Chem. Phys. Lett. **103**, 1 (1983)

2.23 E. Sekreta, J.P. Reilly: Chem. Phys. Lett. **149**, 482 (1988)

2.24 D.A. Hansen, E.K.C. Lee: J. Chem. Phys. **62**, 183 (1975)

2.25 V.P. Klochkov, E.G. Korsakova: Opt. Spektrosk. **67**, 110 (1989) [English transl.: Opt. Spectrosc. (USSR) **67**, 61 (1989)]
L.E. Brus, J.R. McDonald: J. Chem. Phys. **58**, 4223 (1973)

2.26 D.H. Levy: Sci. Am. **250**, 96 (1984)

2.27 A. Terenin: Acta Physicochim. URSS **18**, 210 (1943)
G.N. Lewis, M. Kasha: J. Am. Chem. Soc. **66**, 2100 (1944)
G.N. Lewis, M. Kahsa: J. Am. Chem. Soc. **67**, 994 (1945)

2.28 A. Jabłoński: Nature (London) **131**, 839 (1933)
A. Jabłoński: Z. Phys. **94**, 38 (1935)
S.P. McGlynn, T. Azumi, M. Kinoshita: *Molecular Spectroscopy of the Trilet State* (Prentice-Hall, Englewood Cliffs 1969)

2.29 K. Itoh: Chem. Phys. Lett. **1**, 235 (1967)
R. Huber, M. Schwoerer: Chem. Phys. Lett. **72**, 10 (1980)
M. Schwoerer, R.A. Huber, W. Hartl: Chem. Phys. **55**, 97 (1981)

2.30 M. Sonnenschein, A. Amirav, J. Jortner: J. Phys. Chem. **88**, 4214 (1984)

2.31 Y. Hirata, I. Tanaka: Chem. Phys. **25**, 381 (1977)
Y. Hirata, N. Mataga, Y. Mukai, Y. Koyama: Chem. Phys. Lett. **134**, 166 (1987)
G.W. Scott, L.D. Talley, R.W. Anderson, Jr.: The excited state absorption kinetics of quinoxaline in room temperature solution, in *Picosecond Phenomena*, ed. by C.V. Shank, E.P. Ippen, S.L. Shapiro, Springer Ser. Chem. Phys., Vol. 4 (Springer, Berlin, Heidelberg 1978) pp. 19–22
H. Schröder, H.J. Neusser, E.W. Schlag: Chem. Phys. Lett. **54**, 4 (1978)
N.A. Borisevich, Yu.I. Bubekov, G.B. Tolstorozhev, Yu. Vishchakas, V. Kabelka, M. Milyauskas: Chem. Phys. Lett. **103**, 115 (1983)
K. Hamanoue, T. Nakayama, M. Shiozaki, Y. Funazaki, N. Nakajima, H. Teranishi: J. Chem. Phys. **85**, 5698 (1986)
T. Kobayashi, S. Koshihara: Chem. Phys. Lett. **104**, 174 (1984)

2.32 J. Franck, R. Livingston: J. Chem. Phys. **9**, 184 (1941)

2.33 H. Sponer: Rad. Res. (Suppl.) **1**, 558 (1959)

2.34 P. Pringsheim: *Fluorescenz und Phosphorescenz im Lichte der neuern Atom Theorie* (Springer, Berlin 1928)

2.35 G.W. Robinson: J. Mol. Spectrosc. **6**, 58 (1961)

2.36 G.R. Hunt, E.F. McCoy, I.R. Ross: Austr. J. Chem. **15**, 591 (1962)

2.37 E.F. McCoy, I.G. Ross: Austr. J. Chem. **15**, 573 (1962)

2.38 D. Huppert, P.M. Rentzepis: Picosecond kinetics, in *Molecular Energy Transfer*, ed. by R. Levine, J. Jortner (Wiley, New York 1976) pp. 270–291
P.M. Rentzepis: Chem. Phys. Lett. **3**, 717 (1969)
A. Amirav, J. Jortner: J. Chem. Phys. **81**, 4200 (1984)

2.39 T.M. Woundenberg, S.K. Kulkarni, J.E. Kenny: J. Chem. Phys. **89**, 2789 (1988)
S.K. Kulkarni, J.E. Kenny: J. Chem. Phys. **89**, 4441 (1988)

2.40 Z.S. Ruzevich: Opt. Spektrosk. **15**, 357 (1963) [English transl.: Opt. Spectrosc. (USSR) **15**, 191 (1963)]
G. Binsch, E. Heilbronner, R. Jankow, D. Schmidt: Chem. Phys. Lett. **1**, 135 (1967)
R.C. Dhingra, J.A. Poole: Chem. Phys. Lett. **2**, 108 (1968)

2.41 P.A. Geldof, R.P.H. Rettschnick, G.J. Hoytink: Chem. Phys. Lett. **4**, 59 (1969)

2.42 N. Ohta, H. Baba, G. Marconi: Chem. Phys. Lett. **133**, 222 (1987)
H. Baba, A. Nakajima, M. Aoi, K. Chihara: J. Chem. Phys. **55**, 2433 (1971)
2.43 H. Zuckerman, B. Schmitz, Y. Haas: Chem. Phys. Lett. **151**, 323 (1988)
H. Zuckerman, B. Schmitz, Y. Haas: J. Phys. Chem. **93**, 4083 (1989)
2.44 A.L. Motyka, M.R. Topp: Chem. Phys. **121**, 405 (1988)
K.-M. Bark, R.K. Forcé: J. Phys. Chem. **93**, 7985 (1989)
V.V. Ryl'kov, E.A. Cheshev: Opt. Spektrosk. **63**, 778 (1987) [English transl.: Opt. Spectrosc. (USSR) **63**, 462 (1987)]
V.V. Ryl'kov, E.A. Cheshev: Opt. Spektrosk. **63**, 1030 (1987) [English transl.: Opt. Spectrosc. (USSR) **63**, 608 (1987)]
2.45 T. Kobayashi, S. Nagakura: Chem. Phys. **23**, 153 (1977)
S. Kobayashi, K. Kikuchi, H. Kokubun: Chem. Phys. **27**, 399 (1978)
2.46 N. Ohta, H. Baba: Mol. Phys. **59**, 921 (1986)
2.47 V.P. Klochkov, E.G. Korsakova, E.B. Verkhovskii: Opt. Spektrosk. **62**, 360 (1987) [English Transl.: Opt. Spectrosc. (USSR) **62**, 216 (1987)]
V.P. Klochkov, E.G. Korsakova, E.B. Verkhovskii: Opt. Spektrosk. **65**, 584 (1988) [English transl.: Opt. Spectrosc. (USSR) **65**, 347 (1988)]
2.48 V.A. Lyubimtsev, V.L. Ermolaev: Opt. Spektrosk. **61**, 511 (1986) [English transl.: Opt. Spectrosc. (USSR) **61**, 320 (1986)]
V.A. Lyubimtsev, V.L. Ermolaev: Opt. Spektrosk. **61**, 766 (1986) [English transl.: Opt. Spectrosc. (USSR) **61**, 481 (1986)]
V.A. Lyubimtsev, V.L. Ermolaev: Opt. Spektrosk. **64**, 1030 (1988) [English transl.: Opt. Spectrosc. (USSR) **64**, 614 (1988)]
2.49 B. Nickel, J. Hertzberg: Chem. Phys. **132**, 219 (1989)
2.50 U. Brühlmann, J.R. Huber: Chem. Phys. **68**, 405 (1982)
U. Brühlmann, P. Russegger, J.R. Huber: Chem. Phys. Lett. **84**, 479 (1981)
2.51 A. Sur, P.M. Johnson: J. Chem. Phys. **84**, 1206 (1986)
R.J. Lipert, S.D. Colson: J. Phys. Chem. **93**, 135 (1989)
2.52 A. Kearvell, F. Wilkinson: Chem. Phys. Lett. **11**, 472 (1971)
N.A. Borisevich, L.M. Bolot'ko, T.F. Raichyonok: Spectrosc. Lett. **15**, 125 (1982)
N.A. Borisevich, L.M. Bolot'ko, R.F. Raichyonok: Opt. Spektrosk. **50**, 137 (1981) [English transl.: Opt. Spectrosc. (USSR) **50**, 71 (1981)]
S. Hirayama, T. Inoue, Y. Ito, T. Matsuura: Chem. Phys. Lett. **115**, 79 (1985)
P.L. Holt, J.I. Selco, R.B. Weisman: J. Chem. Phys. **84**, 1996 (1986)
A. Kiermeier, K. Dietrich, E. Riedle, H.J. Neusser: J. Chem. Phys. **85**, 6983 (1986)
R. Duguid, B.D. Maxwell, Y. Munoz-Sola, K. Muthuramu, V. Rasbury, T.-V. Singh, H. Morrison, P.K. Das, G.L. Hug: Chem. Phys. Lett. **139**, 475 (1987)
P.M. Borell, H.-G. Löhmannsröben, K. Luther: Chem. Phys. Lett. **136**, 371 (1987)
2.53 V. Lawetz, W. Siebrand, G. Orlandi: Chem. Phys. Lett. **16**, 448 (1972)
2.54 G.S. Beddard, G.R. Fleming, O.L.J. Gijzeman, G. Porter: Chem. Phys. Lett. **18**, 481 (1973)
G.S. Beddard, G.R. Fleming, O.L.J. Gijzeman, G. Porter: Proc. R. Soc. London, Ser. A, **340**, 519 (1974)
G.S. Beddard, S.J. Formosinho, G. Porter: Chem. Phys. Lett. **22**, 235 (1973)
2.55 G.R. Fleming, C. Lewis, G. Porter: Chem. Phys. Lett. **31**, 33 (1975)
2.56 M.G. Prais, D.F. Heller, K.F. Freed: Chem. Phys. **6**, 331 (1974)
2.57 S.G. Fischer, A.L. Stanford, E.C. Lim: J. Chem. Phys. **61**, 582 (1974)
2.58 J.C. Hsieh, C.-S. Huang, E.C. Lim: J. Chem. Phys. **60**, 4345 (1974)
2.59 J.C. Hsieh, E.C. Lim: J. Chem. Phys. **61**, 736 (1974)
2.60 C.-S. Huang, J.C. Hsieh, E.C. Lim: Chem. Phys. Lett. **28**, 130 (1974)
2.61 C.-S. Huang, E.C. Lim: J. Chem. Phys. **62**, 3826 (1975)
G.D. Gillespie, E.C. Lim: Chem. Phys. Lett. **34**, 513 (1975)
Y. Hirata, E.C. Lim: J. Chem. Phys. **73**, 3804 (1980)
2.62 O. Sneh, A. Amirav, O. Cheshnovsky: J. Chem. Phys. **91**, 3532 (1990)
2.63 I. Yamazaki, K. Sushida, H. Baba: J. Chem. Phys. **71**, 381 (1979)

2.64 S. Kato: J. Chem. Phys. **88**, 3045 (1988)

2.65 B.I. Stepanov: Usp. Fiz. Nauk **58**, 3 (1956)

2.66 B.S. Neporent, B.I. Stepanov: Usp. Fiz. Nauk **43**, 380 (1951)

2.67 B.S. Neporent: Zh. Fiz. Khim. **21**, 1111 (1947)

2.68 G.A. Kundzich, A.A. Shishlovskiĭ: Dokl. Akad. Nauk SSSR **97**, 429 (1954)

2.69 N.A. Borisevich, G.B. Tolstorozhev: Dokl. Akad. Nauk SSSR **188**, 308 (1969) [English transl.: Sov. Phys.-Dokl. **14**, 878 (1970)]
N.A. Borisevich, G.B. Tolstorozhev: Izv. Akad. Nauk SSSR, Ser. Fiz. **34**, 654 (1970)

2.70 A. Terenin, A. Vartanian, B.S. Neporent: Trans. Faraday Soc. **35**, 39 (1939)

2.71 B.S. Neporent: The nature of diffuse vibronic spectra of complex molecules, in *Molecular Photonics*, ed. by A.A. Krasnowskiĭ (Nauka, Leningrad 1970) pp. 18–44 (in Russian)
B.S. Neporent: Opt. Spektrosk. **32**, 38 (1972) [English transl.: Opt. Spectrosc. (USSR) **32**, 19 (1972)]
B.S. Neporent: Opt. Spektrosk. **32**, 252 (1972) [English transl.: Opt. Spectrosc. (USSR) **32**, 133 (1972)]
B.S. Neporent: Opt. Spektrosk. **32**, 458 (1972) [English transl.: Opt. Spectrosc. (USSR) **32**, 240 (1972)]
B.S. Neporent: Opt. Spektrosk. **32**, 670 (1972) [English transl.: Opt. Spectrosc. (USSR) **32**, 355 (1972)]
B.S. Neporent: Opt. Spektrosk. **32**, 880 (1972) [English transl.: Opt. Spectrosc. (USSR) **32**, 468 (1972)]
B.S. Neporent: Pure Appl. Chem. **37**, 111 (1974)

2.72 W.R. Ware, P.T. Cunningham: J. Chem. Phys. **44**, 4364 (1966)

2.73 R. Williams, G.J. Goldsmith: J. Chem. Phys. **39**, 2008 (1963)
R.J. Watts, S.J. Strickler: J. Chem. Phys. **44**, 2423 (1966)

2.74 R.J. Watts, S.J. Strickler: J. Chem. Phys. **49**, 3867 (1968)

2.75 G.B. Kistiakovsky, C.S. Parmenter: J. Chem. Phys. **42**, 2942 (1965)

2.76 A. Frad, F. Lahmani, A. Tramer, C. Tric: J. Chem. Phys. **60**, 4419 (1974)
F. Lahmani, A. Tramer, C. Tric: J. Chem. Phys. **60**, 4431 (1974)

2.77 A.E.W. Knight, C.S. Parmenter: Chem. Phys. **15**, 85 (1976)
A.E.W. Knight, C.S. Parmenter: J. Lumin, **12/13**, 151 (1976)

2.78 K. Uchida, I. Yamazaki, H. Baba: Chem. Phys. Lett. **38**, 333 (1976)
K. Uchida, I. Yamazaki, H. Baba: Chem. Phys. **35**, 91 (1978)

2.79 I. Yamazaki, M. Fujita, H. Baba: Chem. Phys. **57**, 431 (1981)

2.80 K.G. Spears, M. El-Manguch: Chem. Phys. **24**, 65 (1977)

2.81 H. Reinessius, H. von Weyssenhoff: Chem. Phys. Lett. **52**, 34 (1977)

2.82 N. Ohta, H. Baba: Chem. Phys. Lett. **84**, 308 (1981)
N. Ohta, H. Baba: Chem. Phys. Lett. **106**, 382 (1984)
N. Ohta, H. Baba: Chem. Phys. Lett. **112**, 212 (1984)
N. Ohta, H. Baba: Chem. Phys. **82**, 41 (1983)
N. Ohta: Laser Chem. **10**, 109 (1989)

2.83 J.R. McDonald, L.E. Brus: J. Chem. Phys. **61**, 3895 (1974)

2.84 B. Soep, A. Tramer: Chem. Phys. **7**, 52 (1975)

2.85 R.P. Van der Werf, D. Zevenhuijzen, J. Kommandeur: Chem. Phys. Lett. **27**, 325 (1974)

2.86 R.P. van der Werf, J. Kommandeur: Chem. Phys. **16**, 125 (1976)

2.87 J. Chaiken, T. Benson, M. Gurnick, J.D. McDonald: Chem. Phys. Lett. **61**, 195 (1979)

2.88 J. Chaiken, M. Gurnick, J.D. McDonald: J. Chem. Phys. **74**, 106 (1981)
J. Chaiken, M. Gurnick, J.D. McDonald: J. Chem. Phys. **74**, 117 (1981)

2.89 R. van der Werf, E. Schutten, J. Kommandeur: Chem. Phys. **11**, 281 (1975)
R. van der Werf, E. Schutten, J. Kommandeur: Chem. Phys. **16**, 151 (1976)

2.90 R.A. Coveleskie, J.T. Yardley: Chem. Phys. **9**, 275 (1975)
R.A. Coveleskie, J.T. Yardley: Chem. Phys. **13**, 441 (1976)

2.91 N. Ohta, H. Baba: J. Phys. Chem. **90**, 2654 (1986)

2.92 G.D. Greenblatt, S. Ruhman, Y. Haas: Chem. Phys. Lett. **112**, 200 (1984)
2.93 O. Anner, H. Zuckermann, Y. Haas: J. Phys. Chem. **89**, 1336 (1985)
2.94 C.J. Werkhoven, T. Deinum, J. Langelaar, R.P.H. Rettschnick, J.D.W. van Voorst: Chem. Phys. Lett. **11**, 478 (1971)
2.95 D. Zevenhuijzen, R. van der Werf: Chem. Phys. **26**, 279 (1977)
2.96 N.F. Scherer, J.F. Shepanski, A.H. Zewail: J. Chem. Phys. **81**, 2181 (1984)
2.97 B.A. Jacobson, J.A. Guest, F.A. Novak, S.A. Rice: J. Chem. Phys. **87**, 269 (1987)
2.98 N. Ohta, O. Sekiguchi, H. Baba: J. Phys. Chem. **91**, 1076 (1987)
2.99 H. Bitto, H. Stafast, P. Russegger, J.R. Huber: Chem. Phys. **84**, 249 (1984)
2.100 H. Stafast, H. Bitto, J.R. Huber: J. Chem. Phys. **79**, 3660 (1983)
2.101 J. Mühlbach, J.R. Huber: J. Chem. Phys. **85**, 4411 (1986)
2.102 M. Dubs, J. Mühlbach, H. Bitto, P. Schmidt, J.R. Huber: J. Chem. Phys. **83**, 3755 (1985)
2.103 M. Dubs, J. Mühlbach, J.R. Huber: J. Chem. Phys. **85**, 1288 (1986)
2.104 M. Dubs, P. Schmidt, J.R. Huber: J. Chem. Phys. **85**, 6335 (1986)
2.105 H. Bitto, S. Derler, J.R. Huber: Chem. Phys. Lett. **162**, 291 (1989)
2.106 J. Jortner, R.D. Levine, S.A. Rice: Adv. Chem. Phys. **70**, 1 (1988)
2.107 D.S. Tinti, J.S. Baskin, A.H. Zewail: Chem. Phys. Lett. **155**, 243 (1989)
2.108 O. Braitbart, E. Castellucci, G. Dujardin, S. Leach: J. Phys. Chem. **89**, 3252 (1985)
D. Winkoun, D. Chapoulard, G. Dujardin, S. Leach: Can. J. Phys. **62**, 1361 (1985)
S. Leach: Acta Phys. Pol. **71A**, 671 (1987)
2.109 J.P. Maier, M. Ochsner, F. Thommen: Faraday Disc. Chem. Soc. No. 75, 77 (1983)
K. Qian, A. Shukla, J. Futrell: J. Chem. Phys. **92**, 5988 (1990)
2.110 H. Watanabe, S. Tsuchiya, S. Koda: J. Chem. Phys. **75**, 365 (1983)
2.111 P.J. Brucat, R.N. Zare: J. Chem. Phys. **78**, 100 (1983)
P.J. Brucat, R.N. Zare: J. Chem. Phys. **81**, 2562 (1984)
2.112 X.Z. Zhao, T.X. Lu, Z.F. Cui: Chem. Phys. Lett. **162**, 140 (1989)
2.113 G.W. Loge, J.J. Tiee, F.B. Wampler: J. Chem. Phys. **84**, 3624 (1986)
2.114 B.J. van der Meer, H.T. Jonkman, G.M. ter Horst, J. Kommandeur: J. Chem. Phys. **76**, 2099 (1982)
2.115 B.J. van der Meer, H.T. Jonkman, J. Kommandeur, W.L. Meerts, W.A. Majewski: Chem. Phys. Lett. **92**, 565 (1982)
2.116 B.J. Meer, H.T. Jonkman, J. Kommandeur: Laser Chem. **2**, 77 (1983)
J. Kommandeur, B.J. van der Meer, H.T. Jonkman: In *Intramolecular Dynamics*, ed. by J. Jortner, B. Pullman (Reidel, Dordrecht 1982) pp. 259–268
2.117 K.W. Holtzclaw, D.W. Pratt: Chem. Phys. Lett. **118**, 375 (1985)
2.118 W.M. van Herpen, K.E. Drabe, W.L. Meerts: Chem. Phys. Lett. **145**, 305 (1988)
2.119 W.M. van Herpen, W.L. Meerts, K.E. Drabe, J. Kommandeur: J. Chem. Phys. **86**, 4396 (1987)
2.120 W.M. van Herpen, P.A.M. Uijt de Haag, W.L. Meerts: J. Chem. Phys. **89**, 3939 (1988)
2.121 W. Siebrand, W.L. Meerts, D.W. Pratt: J. Chem. Phys. **90**, 1313 (1989)
2.122 W.L. Meerts, W.M. van Herpen: J. Mol. Struct. **173**, 201 (1988)
2.123 K.E. Drabe, J. Kommandeur: Rotations and electronic decay, in *Excited States*, ed. by E.C. Lim, K.K. Innes, Vol. 7 (Academic, New York 1988) pp. 107–162
2.124 H. Saigusa, A.K. Jameson, E.C. Lim: J. Chem. Phys. **79**, 5228 (1983)
2.125 S. Okajima, H. Saigusa, E.C. Lim: J. Chem. Phys. **76**, 2096 (1982)
2.126 H. Saigusa, E.C. Lim: J. Chem. Phys. **78**, 91 (1983)
H. Saigusa, E.C. Lim: J. Chem. Phys. Lett. **88**, 455 (1982)
2.127 M. Terazima, E.C. Lim: Chem. Phys. Lett. **127**, 330 (1986)
2.128 P.-N. Wang, E.C. Lim, W. Siebrand: Chem. Phys. Lett. **159**, 7 (1989)
2.129 P.M. Felker, W.R. Lambert, A.H. Zewail: Chem. Phys. Lett. **89**, 309 (1982)
2.130 P.M. Felker, A.H. Zewail: Chem. Phys. Lett. **128**, 221 (1986)
2.131 J.L. Knee, F.E. Doany, A.H. Zewail: J. Chem. Phys. **82**, 1042 (1985)
2.132 N. Ohta, M. Fujita, H. Baba: Chem. Phys. Lett. **135**, 330 (1987)

2.133 N. Ohta: Chem. Phys. Lett. **151**, 93 (1988)
2.134 G.J. Scherer, Y. Chen. R.L. Redington, J.L. Kinsey, R.W. Field: J. Chem. Phys. **85**, 6315 (1986)
2.135 G. ter Horst, D.W. Pratt, J. Kommandeur: J. Chem. Phys. **74**, 3616 (1981)
2.136 D.B. McDonald, G.R. Fleming, S.A. Rice: Chem. Phys. **60**, 335 (1981)
2.137 A.K. Jameson, S. Okajima, E.C. Lim: J. Chem. Phys. **75**, 480 (1981)
2.138 H. Baba, M. Fujita, K. Uchida: Chem. Phys. Lett. **73**, 425 (1980)
2.139 H. Baba, N. Ohta, O. Sekiguchi, M. Fujita, K. Uchida: J. Phys. Chem. **87**, 943 (1983)
O. Sekiguchi, N. Ohta, H. Baba: Chem. Phys. Lett. **106**, 387 (1984)
2.140 H. Stafast, H. Bitto, J.R. Huber: Chem. Phys. Lett. **93**, 303 (1982)
E.C. Apel, E.K.C. Lee: J. Phys. Chem. **88**, 1283 (1984)
E.C. Apel, E.K.C. Lee: J. Phys. Chem. **90**, 2837 (1986)
J.K. Frisoli, C.B. Moore: J. Phys. Chem. **92**, 5425 (1988)
2.141 H.T. Jonkman, K.E. Drabe, J. Kommandeur: Chem. Phys. Lett. **116**, 357 (1985)
2.142 P.J. de Lange, K.E. Drabe, J. Kommandeur: J. Chem. Phys. **84**, 538 (1986)
P.J. de Lange, K.E. Drabe, J. Kommandeur: J. Chem. Phys. **88**, 588 (1988)
2.143 P.J. de. Lange, B. J. van der Meer, K.E. Drabe, J. Kommandeur: J. Chem. Phys. **86**, 4004 (1987)
2.144 P.J. de Lange, J. Kommandeur: The quantum yield of non-resonant light scattering, in *Large Finite Systems*, ed. by J. Jortner, A. Pullman, B. Pullman, Jerusalem Symp. Quantum Chem. Biochem., Vol. 20 (Reidel, Dordrecht 1987) pp. 1–9
2.145 P.A.M. Uijt de Haag, W.L. Meerts: Chem. Phys. **135**, 139 (1989)
2.146 A. Lorincz, D.D. Smith, F. Novak, R. Kosloff, D.J. Tannor, S.A. Rice: J. Chem. Phys. **82**, 1067 (1985)
2.147 F. Novak, R. Kosloff, D.J. Tannor, A. Lorincz, D.D. Smith, S.A. Rice: J. Chem. Phys. **82**, 1073 (1985)
2.148 A. Amirav, J. Jortner: J. Chem. Phys. **84**, 1500 (1986)
2.149 A. Amirav: Chem. Phys. **126**, 327 (1988)
A. Amirav, Y. Oreg: Chem. Phys. **126**, 343 (1988)
A. Amirav: Chem. Phys. **126**, 365 (1988)
A. Amirav: Chem. Phys. **124**, 163 (1988)
A. Amirav: J. Chem. Phys. **86**, 4706 (1987)
A. Amirav, J. Jortner, M. Terazima, E.C. Lim: Chem. Phys. Lett. **133**, 179 (1987)
2.150 D.M. Bartels, K.G. Spears: J. Phys. Chem. **86**, 5180 (1982)
2.151 Y. Matsumoto, L.H. Spangler, D.W. Pratt: Laser Chem. **2**, 91 (1983)
Y. Matsumoto, L.H. Spangler, D.W. Pratt: Chem. Phys. Lett. **98**, 333 (1983)
Y. Matsumoto, L.H. Spangler, D.W. Pratt: Chem. Phys. Lett. **95**, 343 (1983)
Y. Matsumoto, L.H. Spangler, D.W. Pratt: J. Chem. Phys. **80**, 5539 (1984)
2.152 Y. Matsumoto, D.W. Pratt: J. Chem.Phys. **81**, 573 (1984)
2.153 M. Terazima. E.C. Lim: J. Chem. Phys. **86**, 4471 (1987)
2.154 N. Ohta, O. Sekiguchi, T. Takemura: Bull. Chem. Soc. Jpn. **62**, 2711 (1989)
N. Ohta, O. Sekiguchi, H. Baba: J. Chem. Phys. **82**, 1609 (1985)
N. Ohta, O. Sekiguchi, H. Baba: J. Chem. Phys. **88**, 68 (1988)
N. Ohta, O. Sekiguchi, H. Baba: Chem. Phys. Lett. **126**, 124 (1986)
2.155 P.-N. Wang, E.C. Lim: Chem. Phys. Lett. **142**, 389 (1987)
2.156 N. Ohta, H. Baba: Chem. Phys. Lett. **146**, 410 (1988)
2.157 N. Ohta, T. Takemura, M. Fujita, H. Baba: J. Chem. Phys. **88**, 4197 (1988)
2.158 N. Ohta, T. Takemura, H. Baba: J. Phys. Chem. **92**, 5554 (1988)
N. Ohta, T. Takemura: J. Chem. Phys. **91**, 4477 (1989)
N. Ohta, T. Takemura: J. Chem. Phys. **93**, 877 (1990)
N. Ohta, T. Takemura: J. Phys. Chem. **94**, 3466 (1990)
2.159 G.W. Loge, C.S. Parmenter: J. Chem. Phys. **74**, 29 (1981)
2.160 N.A. Borisevich, V.A. Povedailo, V.A. Tolkachev, V.V. Gavrilyuk, A.A. Samartsev: Opt. Spektrosk. **67**, 1293 (1989) [English transl.: Opt. Spectrosc. (USSR) **67**, 761 (1989)]

2.161 P. Mukherjee, H.S. Kwok: J. Chem. Phys. **87**, 128 (1987)
2.162 C.S. Parmenter: Adv. Chem. Phys. **22**, 365 (1972)
T.F. Hunter, M.G. Stock: Chem. Phys. Lett. **22**, 368 (1973)
2.163 H. Abe, Y. Ohyanagi, M. Ichijo, N. Mikami, M. Ito: J. Phys. Chem. **89**, 3512 (1985)
2.164 F. Tanaka, S. Hirayama, K. Shobatake: Chem. Phys. Lett. **164**, 335 (1989)
2.165 A. Amirav: Chem. Phys. **108**, 403 (1986)
A. Amirav: J. Chem. Phys. **88**, 2840 (1988)
A. Amirav: J. Phys. Chem. **92**, 3725 (1988)
A. Amirav: Rotational effects on intramolecular radiationless transitions – The story of pyrazine, in *Large Finite Systems*, ed. by J. Jortner, A. Pullman, B. Pullman, Jerusalem Symp. Quantum Chem. Biochem., Vol. 20 (Reidel, Dordrecht 1987) pp. 19–30
2.166 J. Kommandeur: Adv. Chem. Phys. **70**, part 1, 133–164 (1988)
2.167 J. Kommandeur, W.A. Majewski, W.L. Meerts, D.W. Pratt: Annu. Rev. Phys. Chem. **38**, 433–462 (1987)
2.168 O. Sekiguchi, N. Ohta, H. Baba: Laser Chem. **7**, 213 (1987)
2.169 W.R. Lambert, P.M. Felker, A.H. Zewail: J. Chem. Phys. **75**, 5958 (1981)
W.R. Lambert, P.M. Felker, A.H. Zewail: J. Chem. Phys. **81**, 2209 (1984)
W.R. Lambert, P.M. Felker, A.H. Zewail: J. Chem. Phys. **81**, 2217 (1984)
2.170 P.M. Felker, A.H. Zewail: Chem. Phys. Lett. **102**, 113 (1983)
P.M. Felker, A.H. Zewail: Phys. Rev. Lett. **53**, 501 (1984)
P.M. Felker, A.H. Zewail: J. Chem. Phys. **82**, 2961 (1985)
P.M. Felker, A.H. Zewail: J. Chem. Phys. **82**, 2975 (1985)
P.M. Felker, A.H. Zewail: J. Chem. Phys. **82**, 2994 (1985)
P.M. Felker, A.H. Zewail: J. Chem. Phys. **86**, 2460 (1987)
P.M. Felker, W.R. Lambert, A.H. Zewail: J. Chem. Phys. **82**, 3003 (1985)
B.W. Keelan, A.H. Zewail: J. Chem. Phys. **82**, 3011 (1985)
B.W. Keelan, A.H. Zewail: J. Phys. Chem. **89**, 4939 (1985)
2.171 P.M. Felker, A.H. Zewail: Adv. Chem. Phys. **70**, 265 (1988)
2.172 A.H. Zewail: Faraday Discuss. Chem. Soc., No. 75, 315 (1983)
A.H. Zewail: Ber. Bunsenges. Phys. Chem. **89**, 264 (1985)
2.173 N. Bloembergen, A.H. Zewail: J. Phys. Chem. **88**, 5459 (1984)
2.174 J.S. Baskin, P.M. Felker, A.H. Zewail: J. Chem. Phys. **86**, 2483 (1987)
2.175 J.S. Baskin, M. Dantus, A.H. Zewail: Chem. Phys. Lett. **130**, 473 (1986)
2.176 J.S. Baskin, T.S. Rose, A.H. Zewail: J. Chem. Phys. **88**, 1458 (1988)
2.177 G.W. Robinson, R.P. Frosch: J. Chem. Phys. **37**, 1962 (1962)
G.W. Robinson, R.P. Frosch: J. Chem. Phys. **38**, 1187 (1963)
2.178 G.W. Robinson: J. Chem. Phys. **47**, 1967 (1967)
2.179 G.W. Robinson: Radiationless transitions in gaseous benzene, in *The Triplet State*, ed. by A.B. Zahlan (Cambridge Univ. Press, London 1967) pp. 213–227
2.180 J.P. Byrn, E.F. McCoy, I.G. Ross: Austr. J. Chem. **18**, 1589 (1965)
2.181 J.P. Byrn, I.G. Ross: Austr. J. Chem. **24**, 1107 (1971)
2.182 W. Siebrand: J. Chem. Phys. **44**, 4055 (1966)
2.183 W. Siebrand, D.F. Williams: J. Chem. Phys. **46**, 403 (1967)
W. Siebrand, D.F. Williams: J. Chem. Phys. **49**, 1860 (1968)
2.184 W. Siebrand: J. Chem. Phys. **46**, 440 (1967)
W. Siebrand: J. Chem. Phys **47**, 2411 (1967)
2.185 N.A. Borisevich, G.A. Zalesskaya, A.E. Urbanovich: Opt. Spektrosk. **59**, 96 (1985) [English transl.: Opt. Spectrosc. (USSR) **59**, 56 (1985)]
N.A. Borisevich, G.A. Zalesskaya, A.E. Urbanovich: Opt. Spektrosk. **63**, 1176 (1987) [English transl.: Opt. Spectrosc. (USSR) **63**, 695 (1987)]
N.A. Borisevich: Acta Phys. Pol. **71A**, 683 (1987)
V.A. Kovarskiĭ, A.V. Belousov, É.P. Sinyavskiĭ, O. Yu. Keloglu: Zh. Eksp. Teor. Fiz. **89**, 1552 (1985) [English transl.: Sov. Phys.- JETP **62**, 900 (1985)]
2.186 A.A. Puretzky: Laser Chem. **6**, 103 (1986)

A.A. Puretzky, V.V. Tyakht: Usp. Fiz. Nauk **157**, 51 (1989) [English, transl.: Sov. Phys.- Usp. **32**, 30 (1989)]
2.187 W. Rhodes, B.R. Henry, M. Kasha: Proc. Nat. Acad. Sci. USA, **63**, 31 (1969)
2.188 W. Rhodes: J. Chem. Phys. **50**, 2885 (1969)
W. Rhodes: The role of molecular stationary states in radiationless processes, in *Transitions Non-Radiatives dans les Molécules*, J. Chim. Phys. Phys.-Chim. Biol. (Suppl.) **10**, 40 (1970)
W. Rhodes: Chem. Phys. Lett. **11**, 179 (1971)
W. Rhodes: Chem. Phys. **22**, 95 (1977)
2.189 G.W. Robinshon, C.A. Langhoff: Chem. Phys. **5**, 1 (1974)
2.190 J. M. Friedman, R.M. Hochstrasser: Chem. Phys. **6**, 155 (1975)
2.191 G.G. Konoplev, A.A. Ovchinnikov, V.G. Plotnikov: Zh. Eksp. Teor. Fiz. **66**, 1956 (1974) [English transl.: Sov. Phys.-JETP **39**, 962 (1974)]
2.192 S. Haroche: Quantum beats and time-resolved fluorescence spectroscopy, in *High-Resolution Laser Spectroscopy*, ed. by K. Shimoda, Topics Appl. Phys., Vol. 13 (Springer, Berlin, Heidelberg 1976) pp. 253–313
E.B. Alexandrov, N.I. Kaliteevskiĭ, M.P. Chaika: Usp. Fiz. Nauk **129**, 155 (1979) [English transl.: Sov. Phys.-Usp. **22**, 760 (1979)]
2.193 D.M. Burland, G. Castro, G.W. Robinson: J. Chem. Phys. **52**, 4100 (1970)
2.194 A. Penner, A. Amirav: J. Phys. Chem. **92**, 5079 (1988)
A. Penner, Y. Oreg, E. Villa, E.C. Lim, A. Amirav: Chem. Phys. Lett. **150**, 243 (1988)
2.195 K.W. Holtzclaw, L.H. Spangler, D.W. Pratt: Chem. Phys. Lett. **161**, 347 (1989)
2.196 M. Fujita, N. Ohta, T. Takemura, H. Baba: Bull. Chem. Soc. Jpn. **62**, 1787 (1988)
2.197 W.E. Howard, E.W. Schlag: Chem. Phys. **17**, 123 (1976)
W.E. Howard, E.W. Schlag: Chem. Phys. **29**, 1 (1978)
W.E. Howard, E.W. Schlag: J. Chem. Phys. **68**, 2679 (1978)
2.198 W.E. Howard, E.W. Schlag: Rotational fine structure in radiationless transitions, in *Radiationless Transitions*, ed. by S.H. Lin (Academic, New York 1980) pp. 81–134
2.199 E.K.C. Lee, G.L. Loper: Experimental measurement of electronic relaxation of isolated small polyatomic molecules, in *Radiationless Transitions*, ed. by S.H. Lin (Academic, New York 1980) pp. 1–77
2.200 M. Kac: *Probability and Related Topics in Physical Sciences* (Wiley, London 1959)
2.201 W.M. Gelbart, D.F. Heller, M.L. Elert: Chem. Phys. **7**, 116 (1975)
2.202 K.F. Freed, J. Jortner: J. Chem. Phys. **50**, 2916 (1969)
K.F. Freed, J. Jortner: J. Chem. Phys. **52**, 6272 (1970)
2.203 P.J. Robinson, K.A. Holbrook: *Unimolecular Reactions* (Wiley, London 1972)
2.204 A. Nitzan, J. Jortner: Teor. Chim. Acta **30**, 217 (1973)
2.205 E.S. Medvedev: Usp. Fiz. Nauk **161**, 31–70 (1991) [English transl.: Sov. Phys.-Usp. **34**, 16–35 (1991)]
2.206 M. Bixon, J. Jortner: J. Chem. Phys. **50**, 3284 (1969)
2.207 B.S. Neporent: Izv. Akad. Nauk, SSSR, Ser. Fiz. **54**, 380 (1990) [English transl.: Bull. Acad. Sci. SSSR, Phys. Ser. **54**, (3), 10 (1990)]
B.S. Neporent: Opt Spektrosk. **68**, 753 (1990) [English transl.: Opt. Spectrosc. (USSR) **68**, 438 (1990)]
B.S. Neporent, S.V. Kulya, A.G. Spiro, I.N. Myakisheva: Opt. Spektrosk, **68**, 1285 (1990) [English transl.: Opt. Spectrosc. (USSR) **68**, 755 (1990)]
B.S. Neporent, S.V. Kulya, A.G. Spiro: Pis'ma Zh. Eksp. Teor. Fis. **48**, 367 (1988) [English transl.: JETP Lett. **48**, 406 (1988)]
2.208 T. Yajima, H. Souma: Phys. Rev. A **17**, 309 (1978)
T. Yajima, H. Souma, Y. Ishida: Phys. Rev. A **17**, 324 (1978)
2.209 H.-D. Meyer: Chem. Phys. **82**, 199 (1983)
H-D. Meyer, H. Köppel: J. Chem. Phys. **81**, 2605 (1984)
H. Köppel, L.S. Cederbaum, W. Domcke: Chem. Phys. Lett. **110**, 469 (1984)
H. Köppel, H.-D. Meyer: Chem. Phys. Lett. **107**, 149 (1984)
2.210 F. Remacle, M. Desouter-Lecompte, J.C. Lorquet: J. Chem. Phys. **91**, 4155 (1989)
2.211 W. Domcke, H. Köppel: Chem. Phys. Lett. **140**, 133 (1987)

R. Schneider, W. Domcke: Chem. Phys. Lett. **150**, 235 (1988)
R. Schneider, W. Domcke: Chem. Phys. Lett. **159**, 61 (1989)
G. Stock, R. Schneider, W. Domcke: J. Chem. Phys. **90**, 7184 (1989)
2.212 A.I. Burstein, A.V. Storozhev: Chem. Phys. **119**, 1 (1988)
A. Boeglin, A.A. Villaeys, R. Voltz, S.H. Lin: J. Chem. Phys. **79**, 3819 (1983)
2.213 I.S. Gradstein, I.M. Ryzhik: *Tables of Integrals, Sums, Series, and Products* (Academic, New York 1980)
2.214 J.R. Christie, D.P. Craig.: Mol. Phys. **23**, 345 (1972)
2.215 C.A. Langhoff, G.W. Robinson: Mol. Phys. **26**, 249 (1973)
2.216 J.M. Delory, C. Tric: Chem. Phys. **3**, 54 (1974)
2.217 R. Lefebvre, J. Savolainen: Some exactly soluble models in the theory of excitation and decay of polyatomic molecules, in *Molecular Energy Transfer*, ed. by R.D. Levine, J. Jortner (Wiley, New York 1976) pp. 212–226
R. Lefebvre, J. Savolainen: J. Chem. Phys. **60**, 2509 (1974)
2.218 D. Chock, J. Jortner, S.A. Rice: J. Chem. Phys. **49**, 610 (1968)
2.219 V.G. Plotnikov, G.G. Konoplev: Fiz. Tverd. Tela **15**, 680 (1973) [English transl.: Sov. Phys.-Solid State **15**, 480 (1973)]
2.220 C. Tric: Chem. Phys. Lett. **21**, 83 (1973)
2.221 M. Bixon, J. Jortner, Y. Dothan: Mol. Phys. **17**, 109 (1969)
2.222 W.H. Miller, R. Hernandez, C.B. Moore: J. Chem. Phys. **93**, 5657 (1990)
2.223 J. Jortner, R.S. Berry: J. Chem. Phys. **48**, 2757 (1968)
2.224 A. Nitzan, J. Jortner, P.M. Rentzepis: Proc. R. Soc. London, Ser. A, **327**, 367 (1972)
2.225 J. Jortner, J. Kommandeur: Chem. Phys. **28**, 273 (1978)
2.226 E.D. Trifonov, V.L. Shekhtman: Fiz. Tverd. Tela **11**, 2984 (1969) [English transl.: Sov. Phys.-Solid State **11**, 2415 (1970)]
2.227 K.F. Freed, A. Nitzan: J. Chem. Phys. **73**, 4765 (1980)
2.228 A.L. Sobolewski: Chem. Phys. Lett. **139**, 528 (1987)
2.229 D.R. Demmer, J.W. Hager, G.W. Leach, S.C. Wallace: Chem. Phys. Lett. **136**, 329 (1987)
G.A. Bickel, D.R. Demmer, E.A. Outhouse, S.C. Wallace: J. Chem. Phys. **91**, 6013 (1990)
C. Lakshminarayan, J.L. Knee: J. Phys. Chem. **94**, 2637 (1990)
2.230 D.A. Dolson, K.W. Holtzclaw, D.B. Moss, C.S. Parmenter: J. Chem. Phys. **84**, 1119 (1986)
K.W. Holtzclaw, C.S. Parmenter: J. Phys. Chem. **88**, 3182 (1984)
K.W. Holtzclaw, C.S. Parmenter: J. Chem. Phys. **82**, 5283 (1985)
K.W. Holtzclaw, C.S. Parmenter: J. Chem. Phys. **84**, 1099 (1986)
2.231 G. Dujardin, S. Leach: Faraday Discuss. Chem. Soc., No. 75, 23 (1983)
G. Dujardin, S. Leach: J. Chem. Phys. **79**, 658 (1983)
2.232 S. Mukamel: J. Chem. Phys. **82**, 2867 (1985)
J. Sue, S. Mukamel: Chem. Phys. Lett. **107**, 398 (1984)
2.233 A. Amirav, J. Jortner, S. Okajima, E.C. Lim: Chem. Phys. Lett. **126**, 487 (1986)
2.234 S.H. Kable, J.W. Thoman, Jr., S. Beames, A.E.W. Knight: J. Phys. Chem. **91**, 1004 (1987)
2.235 P.A. Harmon, S.A. Asher: J. Chem. Phys. **88**, 2925 (1988)
2.236 B.R. Henry, W. Siebrand: J. Chem. Phys. **54**, 1072 (1971)
2.237 E.S. Medvedev, V.I. Osherov, V.M. Pschenichnikov: Chem. Phys. **23**, 397 (1977)
2.238 W.H. Miller: Classical-limit quantum mechanics and the theory of molecular collisions, in *Adv. Chem. Phys.*, ed. by I. Prigogine, S.A. Rice, Vol. 25 (Wiley, New York 1974) pp. 69–177
2.239 F. Duchinsky: Acta Physicochim. URSS **7**, 551 (1937)
2.240 R.J. Wolf, D.S. Bhatia, W.L. Hase: Chem. Phys. Lett. **132**, 493 (1986)
2.241 B.R. Henry, W. Siebrand: J. Chem. Phys. **49**, 5369 (1968)
2.242 J.W. Ellis: Trans. Faraday Soc. **25**, 888 (1929)
2.243 T.E. Martin, A.H. Kalantar: J. Chem. Phys. **49**, 235 (1968)
2.244 B.R. Henry: Acc. Chem. Res. **10**, 207 (1977)
B.R. Henry: Acc. Chem. Res. **20**, 429 (1987)
B.R. Henry: The local mode model, in *Vibrational Spectra and Structure*, ed. by J.R. Durig, Vol. 10 (Elsevier, Amsterdam 1981) pp. 270–319

A.C. Albrecht: Thermal lensing spectroscopy in the condensed phase, in *Advances in Laser Chemistry*, ed. by A.H. Zewail, Springer Ser. Chem. Phys., Vol. 3 (Springer, Berlin, Heidelberg 1978) pp. 235–248
M.L. Sage, J. Jortner: Adv. Chem. Phys. **47(I)**, 293 (1981)
2.245 R.J. Hayward, B.R. Henry, W. Siebrand: J. Mol. Spectrosc. **46**, 207 (1973)
R.J. Hayward, B.R. Henry: J. Mol. Spectrosc. **50**, 58 (1974)
R.J. Hayward, B.R. Henry: Chem Phys. **12**, 387 (1976)
2.246 B.R. Henry, M.A. Mohammadi: Chem. Phys. **55**, 385 (1981)
2.247 M.S. Burberry, A.C. Albrecht: J. Chem. Phys. **71**, 4631 (1979)
2.248 J.W. Perry, A.H. Zewail: J. Chem. Phys. **70**, 582 (1979)
2.249 E.J. Heller, W.M. Gelbart: J. Chem. Phys. **73**, 626 (1980)
2.250 R.G. Bray, M.J. Berry: J. Chem. Phys. **71**, 4909 (1979)
2.251 D.F. Heller, S. Mukamel: Intramolecular lineshape narrowing in the overtone spectrum of benzene, in *Picosecond Phenomena*, ed. by C.V. Shank, E.P. Ippen, S.L. Shapiro, Springer Ser. Chem. Phys., Vol. 4 (Springer, Berlin, Heidelberg 1978) pp. 51–54
2.252 R.L. Swofford, M.E. Long, A.C. Albrecht: J. Chem. Phys. **65**, 179 (1976)
2.253 C. Jaffé: J. Chem. Phys. **81**, 616 (1984)
2.254 A.L. Sobolewski, J. Prochorow: Chem. Phys. **92**, 221 (1985)
2.255 T.M. Abdul Rasheed, V.P.N. Nampoori, K. Sathianandan: Chem. Phys. **108**, 349 (1986)
2.256 A.C. Scott, P.S. Lomdahl, J.C. Eilbeck: Chem. Phys. Lett. **113**, 29 (1985)
A.C. Scott, J.C. Eilbeck: Chem. Phys. Lett. **132**, 23 (1986)
2.257 V.I. Osherov, E.S. Medvedev: Opt. Spektrosk. **67**, 97 (1989) [English transl.: Opt. Spectrosc. (USSR) **67**, 53 (1989)]
2.258 R.H. Page, Y.R. Shen, Y.T. Lee: Phys. Rev. Lett. **59**, 1293 (1987)
R.H. Page, Y.R. Shen, Y.T. Lee: J. Chem. Phys. **88**, 4621 (1988)
2.259 A. Lorincz, F.A. Novak, S.A. Rice: Chem. Phys. Lett. **111**, 322 (1984)
2.260 W. Nadler, R.A. Marcus: J. Chem. Phys. **86**, 6982 (1987)
2.261 Y. Nomura, Y. Fujimura, H. Kono: J. Chem. Phys. **88**, 1501 (1988)
2.262 Yu. T. Mazurenko, V.A. Smirnov: Opt. Spektrosk. **64**, 713 (1988) [English transl.: Opt. Spectrosc. (USSR) **64**, 427 (1988)]
2.263 B.D. Faĭnberg: Opt. Spektrosk. **65**, 1223 (1988) [English transl.: Opt. Spectrosc. (USSR) **65**, 722 (1988)]
2.264 S.O. Williams, D.G. Imre: J. Phys. Chem. **92**, 6648 (1988)
2.265 S.H. Lin, B. Fain, N. Hamer, C.Y. Yeh: Chem. Phys. Lett. **162**, 73 (1989)
2.266 R.N. Zare: J. Chem. Phys. **45**, 4510 (1965)
R.N. Zare: Acc. Chem. Res. **4**, 361 (1971)
J. Chaiken, J.D. McDonald: Adv. Laser Spectrosc. **1**, 177 (1982)
J. Mühlbach, M. Dubs, H. Bitto, J.R. Huber: Chem. Phys. Lett. **111**, 288 (1984)
J. Mühlbach, J.R. Huber: J. Chem. Phys. **84**, 3014 (1984)
2.267 O. Chesnowski, A. Amirav: Chem. Phys. Lett. **109**, 368 (1984)
J.L. Hardwick: J. Mol. Spectrosc. **109**, 85 (1985)
2.268 R. Kullmer, W. Demtröder: J. Chem. Phys. **81**, 2919 (1984)
R. Kullmer, W. Demtröder: Chem. Phys. **92**, 423 (1985)
R. Kullmer, W. Demtröder: J. Chem. Phys. **84**, 3672 (1986)
O. Snė, O. Chesnowski: Chem. Phys. Lett. **130**, 487 (1986)
2.269 J.E. Dove, H. Hippler, J. Plach, J. Troe: J. Chem. Phys. **81**, 1209 (1984)
N. Ochi, H. Watanabe, S. Tsuchiya: Chem. Phys. **113**, 271 (1987)
Y.S. Lee, D. Giblin, S.J. McNichol, S.J. Silvers: Chem. Phys. Lett. **161**, 116 (1989)
H. Abe, Y. Fukuda, H. Hayashi: J. Phys. Chem. **93**, 69 (1989)
2.270 C.X.W. Qian, H. Reisler, C. Wittig: Chem. Phys. Lett. **139**, 175 (1987)
2.271 G.D. Greenblatt, E. Nissani, E. Zaroura, Y. Haas: J. Phys. Chem. **91**, 570 (1987)
2.272 M. Ya. Ovchinnikova: Teor. Eksp. Khim. **4**, 575 (1968) [English transl.: Theor. Exp. Chem. (USSR) **4**, 367 (1968)]
2.273 R. Englman, J. Jortner: Mol. Phys. **18**, 145 (1970)

2.274 W. Siebrand: Triplet decay and intersystem crossing in aromatic hydrocarbons, in *The Triplet State*, ed. by A.B. Zahlan (Cambridge Univ. Press, London 1967) pp. 31–45
2.275 R.E. Kellogg, N.C. Wyeth: J. Chem. Phys. **45**, 3156 (1966)
R.E. Kellogg, R.P. Schwenker: J. Chem. Phys. **41**, 2860 (1964)
2.276 B.R. Henry, W. Siebrand: Spin-orbit coupling and its effect on radiative and nonradiative transitions in aromatic hydrocarbons, in *Transitions Non-Radiatives dans les Molécules*, J. Chim. Phys. Phys.-Chim. Biol. (Suppl.) **10**, 33 (1970)
2.277 S. Murata, C. Iwanaga, T. Toda, H. Kokubun: Chem. Phys. Lett. **13**, 101 (1972)
S. Murata, C. Iwanaga, T. Toda, H. Kokubun: Chem. Phys. Lett. **15**, 152 (1972)
2.278 Y. Kurabayashi, K. Kikuchi, H. Kokubun, Y. Kaizu, H. Kobayashi: J. Phys. Chem. **88**, 1308 (1984)
2.279 A. Maciejewski, A. Safarzadeh-Amiri, R.E. Verrall, R.P. Steer: Chem. Phys. **87**, 295 (1984)
2.280 K.G. Spears, S.A. Rice: J. Chem. Phys. **55**, 5561 (1971)
2.281 S.J. Formosinho, A.M. da Silva: J. Chem. Soc., Faraday Trans. 2 **72**, 2044 (1976)
2.282 G. Porter, F.J. Wright: Trans. Faraday Soc. **51**, 1205 (1951)
2.283 C.W. Ashpole, S.J. Formosinho, G. Porter: Proc. R. Soc. London Ser. A, **323**, 11 (1971)
2.284 J.L. Baptista, S.J. Formosinho, M. de Fátima Leitão: Chem. Phys. **28**, 425 (1978)
2.285 S.J. Formosinho, A.M. da Silva: Mol. Photochem. **9**, 257 (1979)
2.286 S.J. Formosinho, G. Porter, M.A. West: Chem. Phys. Lett. **6**, 7 (1970)
2.287 B. Soep, C. Michel, A. Tramer, L. Lindqvist: Chem. Phys. **2**, 293 (1973)
2.288 K.T. Chen, B.E. Forch, H. Saigusa, A.K. Jameson, E.C. Lim: Chem. Phys. Lett. **103**, 89 (1983)
S.M. Arabeĭ, K.N. Solov'yov, T.F. Kachura: Opt. Spektrosk. **65**, 354 (1988) [English transl.: Opt. Spectrosc. (USSR) **65**, 213 (1988)]
2.289 C.A. Hutchison, B.W. Mangum: J. Chem. Phys. **32**, 1261 (1960)
2.290 M.R. Wright, R.P. Frosch, G.W. Robinson: J. Chem. Phys. **33**, 934 (1960)
2.291 M.S. de Groot, J.H. van der Waals: Mol. Phys. **4**, 189 (1961)
2.292 V.L. Ermolaev: Izv. Akad. Nauk. SSSR, Ser. Fiz. **27**, 617 (1963)
2.293 T.E. Martin, A.H. Kalantar: J. Chem. Phys. **48**, 4996 (1968)
T.E. Martin, A.H. Kalantar: Phosphorescence lifetimes and radiationless transitions in the deuterobenzenes at 77 K, in *Transitions Non-Radiatives dans les Molécules*, J. Chim. Phys. Phys.-Chim. Biol. (Suppl.) **10**, 101 (1970)
2.294 S.H. Lin, R. Bersohn: J. Chem. Phys. **48**, 2732 (1968)
A. Beckett: Nature **211**, 410 (1966)
2.295 K.W. Holtzclaw, M.D. Schuh: Chem. Phys. **56**, 219 (1981)
2.296 E.B. Sveshnikova, V.P. Kondakova: Opt. Spektrosk. **50**, 870 (1981) [English transl.: Opt. Spectrosc. (USSR) **50**, 477 (1981)]
2.297 Y. Kajii, K. Obi, I. Tanaka: Chem. Phys. Lett. **111**, 347 (1984)
2.298 M.D. Schuh, S. Speiser, G.H. Atkinson: J. Phys. Chem. **88**, 2224 (1984)
W.F. Beck, M.D. Schuh, M.P. Thomas, T.J. Tront: J. Phys. Chem. **88**, 3431 (1984)
M.L. Longmire, L.A. Posey, M.D. Schuh: J. Phys. Chem. **90**, 5700 (1986)
2.299 R.B. Cundall, A.S. Davies: Trans. Faraday Soc. **62**, 1151 (1966)
2.300 H. Brühlmann, P. Russegger, J.R. Huber: Chem. Phys. Lett. **75**, 179 (1980)
2.301 E.M. Kober, J.V. Caspar, R.S. Lumpkin, T.J. Meyer: J. Phys. Chem. **90**, 3722 (1986)
2.302 B.I. Makshantsev: Chem. Phys. **49**, 397 (1980)
B.I. Makshantsev, V. N. Fleurov: Chem. Phys. **49**, 77 (1980)
2.303 K. Sushida, M. Fujita, T. Takemura, H. Baba: Chem. Phys. **88**, 221 (1984)
2.304 M. Kawasaki, K. Kasatani, H. Sato: Chem. Phys. **94**, 179 (1985)
2.305 C. Guttman, S.A. Rice: J. Chem. Phys. **61**, 651 (1974)
J.J. O'Brien, M.D. Horne, B.K. Selinger, B.R. Henry: Chem. Phys. **68**, 123 (1982)
2.306 J.O. Uy, E.C. Lim: Chem. Phys. Lett. **7**, 306 (1970)
2.307 U. Boesl, H.J. Neusser, E.W. Schlag: Chem. Phys. Lett. **31**, 1 (1975)
2.308 M. Stockburger, H. Gattermann, W. Klusmann: J. Chem. Phys. **63**, 4519 (1975)
H. Gattermann, M. Stockburger: J. Chem. Phys. **63**, 4541 (1975)

2.309 A. Amirav. M. Sonnenschein, J. Jortner: Chem. Phys. Lett. **100**, 488 (1983)
A. Amirav, C. Horwitz, J. Jortner: J. Chem. Phys. **88**, 3092 (1988)
2.310 J.L. Knee, C.E. Otis, P.M. Johnson: J. Chem. Phys. **81**, 4455 (1984)
2.311 V.L. Bogdanov, E.N. Vikktorova: Opt. Spektrosk. **61**, 211 (1986) [English transl.: Opt. Spectrosc. (USSR) **61**, 136 (1986)]
2.312 J.G. Langan, E.V. Sitzmann, K.B. Eisenthal: Chem. Phys. Lett. **110**, 521 (1984)
J.G. Langan, E.V. Sitzmann, K.B. Eisenthal: Chem. Phys. Lett. **124**, 59 (1986)
2.313 D.F. Heller, K.F. Freed, W.M. Gelbart: J. Chem. Phys. **56**, 2309 (1972)
D.F. Heller, K.F. Freed, W.M. Gelbart: Chem. Phys. Lett. **23**, 56 (1973)
2.314 W.M. Gelbart, K.G. Spears, K.F. Freed, J. Jortner, S.A. Rice: Chem. Phys. Lett. **6**, 345 (1970)
2.315 W.R. Ware, B.K. Selinger, C.S. Parmenter, M.W. Schuyler: Chem. Phys. Lett. **6**, 342 (1970)
2.316 A. Nitzan, J. Jortner: J. Chem. Phys. **55**, 1355 (1971)

Chapter 3

3.1 W. Heitler: *The Quantum Theory of Radiation*, 3rd edn. (Clarendon, Oxford 1954)
3.2 A. Nitzan, J. Jortner: Mol. Phys. **24**, 109 (1972)
3.3 A. Nitzan, J. Jortner, B.J. Berne: Mol. Phys. **26**, 281 (1973)
3.4 A.V. Lukashin, B.I. Makshantsev: Fiz. Tverd. Tela **16**, 826 (1974) [English transl.: Sov. Phys.-Solid State **16**, 533 (1974)]
3.5 W.F. Polik, C.B. Moore, W.H. Miller: J. Chem. Phys. **89**, 3584 (1988)
3.6 W.F. Polik, D.R. Guyer, C.B. Moore: J. Chem. Phys. **92**, 3453 (1990)
W.F. Polik, D.R. Guyer, W.H. Miller, C.B. Moore: J. Chem. Phys. **92**, 3471 (1990)
D.R. Guyer, W.F. Polik, C.B. Moore: J. Chem. Phys. **84**, 6519 (1986)
3.7 K. Jug, H. von Weyssenhoff: J. Chem. Phys. **56**, 517 (1972)
3.8 A. Nitzan, J. Jortner: Theor. Chim. Acta **29**, 97 (1973)
3.9 D. Neuberger, A.B.F. Dunkan: J. Chem. Phys. **22**, 1693 (1954)
3.10 K.F. Greenough, A.B.F. Dunkan: J. Am. Chem. Soc. **83**, 555 (1961)
3.11 A.E. Douglas: J. Chem. Phys. **45**, 1007 (1966)
3.12 V.M. Donnelly, F. Kaufman: J. Chem. Phys. **66**, 4100 (1977)
3.13 H. Orita, H. Morita, S. Nagakura: Chem. Phys. Lett. **81**, 33 (1981)
3.14 P. Wannier, P.M. Rentzepis, J. Jortner: Chem. Phys. Lett. **10**, 102 (1971)
P. Wannier, P.M. Rentzepis, J. Jortner: Chem. Phys. Letss. **10**, 193 (1971)
3.15 G.E. Busch, P.M. Rentzepis, J. Jortner: J. Chem. Phys. **56**, 361 (1972)
G.E. Busch, P.M. Rentzepis, J. Jortner: Chem. Phys. Lett. **11**, 437 (1971)
3.16 R.M. Hochstrasser: Acc. Chem. Res. **1**, 266 (1968)
3.17 R.M. Hochstrasser, J.E. Wessel: Chem. Phys. Lett. **19**, 156 (1973)
3.18 R. Naaman, D.M. Lubman, R.N. Zare: Time-resolved emission from benzophenon in a molecular beam, in *Advances in Laser Chemistry*, ed. by A.H. Zewail, Springer Ser. Chem. Phys., Vol 3 (Springer, Berlin, Heidelberg 1978) pp. 437–442
3.19 H.G. Küttner, H.L. Selzle, E.W. Schlag: Chem. Phys. **28**, 1 (1978)
3.20 R. Voltz: Mol. Phys. **19**, 881 (1970)
A. Tramer, R. Voltz: Time-resolved studies of excited molecules, in *Excited States*, ed. by E.C. Lim, Vol. 4 (Academic, New York 1979) pp. 281–394
3.21 M.L. Goldberger, K.M. Watson: *Collision Theory* (Wiley, New York 1964)
3.22 L. Mower: Phys. Rev. **142**, 799 (1966)
3.23 M. Berrondo, K. Weiss: J. Chem. Phys. **52**, 807 (1970)
3.24 R. Englman, M.Z. Zgierski: Chem. Phys. **94**, 187 (1985)

Chapter 4

4.1 M. Born, K. Huang: *Dynamical Theory of Crystal Lattices* (Oxford Univ. Press, London 1954)

4.2 K.K. Rebane: *Impurity Spectra of Solids* (Plenum, New York 1970)

4.3 I.B. Bersuker, V.Z. Polinger: *Vibronic Interactions in Molecules and Crystals* (Nauka, Moscow 1983) [English transl.: Springer Ser. Chem. Phys., Vol. 49 (Springer, Berlin, Heidelberg 1989)]

4.4 E.E. Nikitin, S. Ya. Umanskiĭ: *Theory of Slow Atomic Collisions*, Springer Ser. Chem. Phys., Vol. 30 (Springer, Berlin, Heidelberg 1984)
V.N. Kondrat'ev, E.E. Nikitin: *Kinetics and Mechanism of Gas-Phase Reactions* (Nauka, Moscow 1974) Chap. 3 (in Russian)

4.5 M.S. Child: *Molecular Collision Theory* (Academic, New York 1974)

4.6 E. Deumens, L. Lathouwers, P. van Leuven: Chem. Phys. Lett. **112**, 341 (1984)

4.7 S.I. Pekar: *Studies on the Electronic Theory of Crystals* (Gostekhizdat, Moscow 1951) (in Russian)

4.8 S.I. Pekar: Zh. Eksp. Teor. Fiz. **22**, 641 (1952)

4.9 V.G. Plotnikov, B.A. Dolgikh: Opt. Spektrosk. **43**, 882 (1977) [English transl.: Opt. Spectrosc. (USSR) **43**, 522 (1977)]
V.G. Plotnikov, B.A. Dolgikh, V.M. Komarov: Opt. Spektrosk. **43**, 1072 (1977) [English transl.: Opt. Spectrosc. (USSR) **43**, 634 (1977)]
V.G. Plotnikov: Opt. Spektrosk. **64**, 784 (1988) [English transl.: Opt. Spectrosc. (USSR) **64**, 468 (1988)]

4.10 G.V. Maĭer: Opt. Spektrosk. **61**, 1009 (1986) [English transl.: Opt. Spectrosc. (USSR) **61**, 631 (1986)]

4.11 D.W. Boldridge, B.L. Justus, G.W. Scott: J. Chem. Phys. **80**, 3179 (1984)
D.W. Boldridge, G.W. Scott: J. Chem. Phys. **84**, 6790 (1986)

4.12 H.C. Longuet-Higgins: Some recent developments in the theory of molecular energy levels, in *Adv. Spectrosc.*, Vol. 2 (Interscience, New York 1961) p. 429

4.13 K.F. Freed, S.H. Lin: Chem. Phys. **11**, 409 (1975)

4.14 E.S. Medvedev: Pis'ma Zh. Eksp. Teor. Fiz. **13**, 59 (1971) [English transl.: JETP Lett. **13**, 39 (1971)]
E.S. Medvedev: Cryst. Lattice Defects **5**, 79 (1974)

4.15 S.P. Karkach, V.I. Osherov: Mol. Phys. **36**, 1069 (1978)

4.16 J.A. Pople, J.W. Sidman: J. Chem. Phys. **27**, 1270 (1957)
J.N. Murrell, J.A. Pople: Proc. Phys. Soc. London, Ser. A, **69**, 245 (1956)

4.17 R. McWeeny, B.T. Sutcliffe: *Methods of Molecular Quantum Mechanics* (Academic, New York 1969)

4.18 A.R. Edmonds: *Angular Momentum in Quantum Mechanics* (Princeton Univ. Press, Princeton 1960)
M.E. Rose: *Elementary Theory of Angular Momentum* (Wiley, New York 1957)

4.19 P.A. Braun, A.A. Kiselev: *Introduction to the Theory of Molecular Spectra* (Leningrad Univ. Press, Leningrad 1983) (in Russian)

4.20 W.C. Veeman, J.H. van der Waals: Mol. Phys. **18**, 63 (1970)
S. Friedrich, F. Metz, F. Dörr: Mol. Phys. **30**, 289 (1975)

4.21 F. Metz: Chem. Phys. **9**, 121 (1975)

4.22 G.V. Maĭer, V. Ya. Artyukhov, A.V. Karynov: Opt. Spektrosk. **66**, 823 (1989) [English transl.: Opt. Spectrosc. (USSR) **66**, 482 (1989)]
G.V. Maĭer, A.I. Galeeva: Opt. Spektrosk. **67**, 1256 (1989) [English transl.: Opt. Spectrosc. (USSR) **67**, 740 (1989)]
G.V. Maĭer, O.K. Basyl', E.V. Kurshakova: Opt. Spektrosk. **64**, 704 (1988) [English transl.: Opt. Spectrosc. (USSR) **64**, 424 (1988)]
G.V. Maĭer, A.V. Karynov, O.K. Basyl', V. Ya. Artyukhov: Opt. Spektrosk. **64**, 1018 (1988) [English transl.: Opt. Spectrosc. (USSR) **64**, 606 (1988)]
A.V. Lusanov, Yu. F. Pedash: Opt. Spektrosk. **65**, 34 (1988) [English transl.: Opt. Spectrosc. (USSR) **65**, 20 (1988)]

4.23 M. Lombardi, R. Jost, C. Michel, A. Tramer: Chem. Phys. **46**, 273 (1980)
P. Dupré, R. Jost, M. Lombardi, C. Michel, A. Tramer: Chem. Phys. **82**, 25 (1983)

P. Dupré, R. Jost, M. Lombardi: Chem. Phys. **91**, 355 (1984)
E. Pebay-Peyroula, R. Jost, M. Lombardi, P. Dupré: Chem. Phys. **102**, 417 (1986)
E. Pebay-Peyroula, R. Jost, M. Lombardi, J.P. Pique: Chem. Phys. **106**, 243 (1986)
E. Pebay-Peyroula, R. Jost: J. Mol. Spectrosc. **121**, 167 (1987)
4.24 J. Schmidt, I.-Y. Chan, J.H. van der Waals: The study of radiationless decay from phosphorescent triplets by zero-field spectroscopy, in *Transitions Non-Radiatives dans les Molécules*, J. Chim. Phys. Phys.-Chim. Biol. (Suppl.) **10**, 116 (1970)
D.A. Anthenius, J. Schmidt, J.H. van der Waals: Mol. Phys. **27**, 1521 (1974)
M. Schwoerer, H. Sixl: Chem. Phys. Lett. **2**, 14 (1968)
H. Sixl, M. Schwoerer: Chem. Phys. Lett. **6**, 21 (1970)
4.25 R.H. Clarke, J.M. Hayes: Chem. Phys. Lett. **27**, 556 (1974)
4.26 R.H. Clarke, J.M. Hayes, R. H. Hofeldt: J. Magn. Res. **13**, 68 (1974)
4.27 R.H. Clarke, H.A. Franck: J. Chem. Phys. **65**, 39 (1976)
4.28 N.G. Kilmer, A.H. Kalantar: Chem. Phys. **27**, 355 (1978)
4.29 Chr. Bräuchle, H. Kabza, J. Voitländer: Chem. Phys. **32**, 63 (1978)
Chr. Bräuchle, H. Kabza, J. Voitländer: Z. Naturforsch. **34a**, 6 (1979)
Chr. Bräuchle, H. Kabza, J. Voitländer: Chem. Phys. **55**, 137 (1981)
4.30 G. Weinzierl, J. Friedrich: Chem. Phys. Lett. **80**, 55 (1981)
4.31 S. Yamauchi, N. Hirota: J. Phys. Chem. **91**, 1754 (1987)
M. Terazima, S. Yamauchi, N. Hirota: J. Chem. Phys. **84**, 3679 (1986)
M. Terazima: Chem. Phys. Lett. **140**, 6 (1987)
4.32 G. Herzberg: *Molecular Spectra and Molecular Structure*, Vol. 2, Infrared and Raman Spectra of Polyatomic Molecules (Van Nostrand, New York 1945)
G. Herzberg: *Molecular Spectra and Molecular Structure*, Vol. 3, Electronic Spectra and Electronic Structure of Polyatomic Molecules (Van Nostrand, New York 1966)
4.33 E.B. Wilson, Jr., J.C. Decius, P.C. Cross: *Molecular Vibrations. The Theory of Infrared and Raman Vibrational Spectra* (McGraw-Hill, New York 1955)
4.34 H.C. Allen, P.C. Cross: *Molecular Vib-Rotors* (Wiley, New York 1963)
4.35 G. Amat, H.H. Nielsen, G. Tarrago: *Rotation-Vibration of Polyatomic Molecules* (Marcel Dekker, New York 1971)
4.36 W.H. Kroto: *Molecular Rotation Spectra* (Wiley, London 1975)
D. Papoušek, M.R. Aliev: *Molecular Vibrational/Rotational Spectra* (Academic, Prague 1982)
L. Nemes: Sum rules for vibration-rotation interaction coefficients, in *Vibrational Spectra and Structure*, ed. by J.R. Durig, Vol. 10 (Elsevier, Amsterdam 1981) pp. 395–462
4.37 F.A. Novak, S.A. Rice: J. Chem. Phys. **71**, 4680 (1979)
F.A. Novak, S.A. Rice: J. Chem. Phys. **73**, 858 (1980)
4.38 F.A. Novak, S.A. Rice, M.D. Morse, K.F. Freed: On rotational effects in radiationless processes in polyatomic molecules, in *Radiationless Transitions*, ed. by S.H. Lin (Academic, New York 1980) pp. 135–183
4.39 A. Meenakshi, W. Chen, E.C. Lim: Chem. Phys. Lett. **148**, 507 (1988)
Y.-D. Shin, E.C. Lim: Chem. Phys. Lett. **151**, 308 (1988)
4.40 N. Ohta, H. Baba: J. Chem. Phys. **76**, 1654 (1982)
4.41 C.G. Stevens, J.C.D. Brand: J. Chem. Phys. **58**, 3324 (1973)
4.42 J.T. Hougen: Can. J. Phys. **42**, 433 (1964)
4.43 J.H. van Vleck: Rev. Mod. Phys. **23**, 213 (1951)
4.44 K.F. Freed: J. Chem. Phys. **45**, 4214 (1966)
4.45 I.M. Mills: Pure Appl. Chem. **11**, 325 (1965)
4.46 E. Riedle, H.J. Neusser, E.W. Schlag, S.H. Lin: J. Chem. Phys. **88**, 198 (1984)
4.47 E.K.C. Lee: Acc. Chem. Res. **10**, 319 (1977)
4.48 E.L. Sibert, III: J. Chem. Phys. **90**, 2672 (1989)
4.49 M. Fujii, T. Ebata, N. Mikami, M. Ito, S.H. Kable, W.D. Lawrance, T.B. Parsons, A.E.W. Knight: J. Phys. Chem. **88**, 2837 (1984)
4.50 T. Kulp, R.S. Ruoff, J.D. McDonald: J. Chem. Phys. **82**, 2175 (1985)

4.51 N.L. Garland, E.K.C. Lee: J. Chem. Phys. **84**, 28 (1986)
4.52 W. Dietz, S.F. Fischer: J. Chem. Phys. **87**, 249 (1987)
4.53 W.D. Lawrance, A.E.W. Knight: J. Phys. Chem. **92**, 5900 (1988)
4.54 J.A. Konings, W.A. Majewski, Y. Matsumoto, D.W. Pratt, W.L. Meerts: J. Chem. Phys. **89**, 1813 (1988)
4.55 L. Wunsch, H.J. Neusser, E.W. Schlag: Z. Naturforsch. **36A**, 1340 (1981)
E. Riedle, H.J. Neusser, E.W. Schlag: J. Phys. Chem. **86**, 4847 (1982)
E. Riedle, H.J. Neusser, E.W. Schlag: Faraday Discuss. Chem. Soc. No. 75, 387 (1983)
E. Riedle, H.J. Neusser: J. Chem. Phys. **80**, 4686 (1984)
4.56 E. Riedle, H. Stepp, H.J. Neusser: Chem. Phys. Lett. **110**, 452 (1984)
4.57 U. Schubert, E. Riedle, H.J. Neusser, E.W. Schlag: J. Chem. Phys. **84**, 6182 (1986)
4.58 U. Schubert, E. Riedle, H.J. Neusser: J. Chem. Phys. **84**, 5326 (1986)
U. Schubert, E. Riedle, H.J. Neusser: J. Chem. Phys. **90**, 5994 (1989)
4.59 H. Sieber, E. Riedle, H.J. Neusser: J. Chem. Phys. **89**, 4620 (1988)
Y.M. Engel, R.D. Levine: J. Chem. Phys. **89**, 4633 (1988)
4.60 E. Riedle, Th. Knittel, Th. Weber, H.J. Neusser: J. Chem. Phys. **91**, 4555 (1990)
E. Riedle: Doppler-free spectroscopy: an opportunity for state specific investigations in large molecules, in *Large Finite Systems*, ed. by J. Jortner, A. Pullman, B. Pullman, Jerusalem Symp. Quantum Chem. Biochem., Vol. 20 (Reidel, Dordrecht 1987) pp. 387–398
4.61 E. Riedle, Th. Weber, U. Schubert, H.J. Neusser, E.W. Schlag: J. Chem. Phys. **93**, 967 (1990)
4.62 A.A. Villaeys, K.F. Freed: Chem. Phys. Lett. **123**, 515 (1986)
4.63 R.L. Rosman, S.A. Rice: Chem. Phys. Lett. **132**, 351 (1986)
4.64 R.L. Rosman, A.A. Villaeys, K.F. Freed, S.A. Rice: J. Chem. Phys. **86**, 2576 (1987)
4.65 Yu.S. Demchuk, S.L. Vinokurov, S.O. Mirumyants: Opt. Spektrosk. **65**, 552 (1988) [English transl.: Opt. Spectrosc. (USSR) **65**, 329 (1988)]
4.66 S.L. Vinokurov, Yu. S. Demchuk, S.O. Mirumyants: Opt. Spektrosk. **67**, 472 (1989) [English transl.: Opt. Spectrosc. (USSR) **67**, 277 (1989)]
4.67 B.E. Forch, E.C. Lim: Chem. Phys. Lett. **110**, 593 (1984)
4.68 G.M. Nathanson, G.M. McClelland: J. Chem. Phys. **81**, 629 (1984)
G.M. Nathanson, G.M. McClelland: J. Chem. Phys. **84**, 3170 (1986)
G.M. Nathanson, G.M. McClelland: J. Chem. Phys. **85**, 4311 (1986)
G.M. Nathanson, G.M. McClelland: Chem. Phys. Lett. **114**, 441 (1985)
4.69 M. Lax: J. Chem. Phys. **20**, 1752 (1952)
4.70 J. Franck: Trans. Faraday Soc. **21**, 536 (1925)
4.71 E.U. Condon: Phys. Rev. **28**, 1182 (1926)
E.U. Condon: Phys. Rev. **32**, 858 (1928)
4.72 G. Herzberg, E. Teller: Z. Phys. Chem. (Leipzig) **B21**, 410 (1933)
4.73 V.P. Maslov: *The WKB Method in the Multidimensional Case*, Supplement to the Russian edn. of [5.10] (Mir, Moscow 1965) (in Russian)
4.74 V.P. Maslov, M.V. Fedoryuk: *Quasiclassical Approximation to Equations of Quantum Mechanics* (Nauka, Moscow 1976) (in Russian)
4.75 V.A. Kovarskii: Fiz. Tverd. Tela **4**, 1635 (1962) [English transl.: Sov. Phys. - Solid State **4**, 1200 (1963)]
4.76 V.A. Kovarskii, E.P. Smiyavskii: Fiz. Tverd. Tela **4**, 3202 (1962) [English transl.: Sov. Phys.-Solid State **4**, 2345 (1963)]
4.77 A. Nitzan, J. Jortner: J. Chem. Phys. **56**, 3360 (1972)
4.78 B. Sharf, R. Silbey: Chem. Phys. Lett. **4**, 423 (1969)
B. Sharf, R. Silbey: Chem. Phys. Lett. **4**, 561 (1970)
B. Sharf, R. Silbey: Chem. Phys. Lett. **9**, 125 (1971)
4.79 G. Orlandi, W. Siebrand: Chem. Phys. Lett. **8**, 473 (1971)
G. Orlandi, W. Siebrand: Chem. Phys. Lett. **14**, 19 (1972)
G. Orlandi, W. Siebrand: Chem. Phys. Lett. **21**, 217 (1973)
W. Siebrand, M.Z. Zgierski: Chem. Phys. Lett. **58**, 8 (1978)
W. Siebrand, M.Z. Zgierski: Chem. Phys. Lett. **72**, 411 (1980)
W. Siebrand, M.Z. Zgierski: J. Chem. Phys. **72**, 1641 (1980)

4.80 J. Fujumure, H. Kono, T. Nakajima: J. Chem. Phys. **66**, 199 (1977)
4.81 E. Gutsche: Phys. Status Solidi (b) **109**, 583 (1982)
4.82 F. Zerbetto, M.Z. Zgierski: Chem. Phys. **110**, 421 (1986)
F. Zerbetto, M.Z. Zgierski: J. Chem. Phys. **93**, 1235 (1990)
4.83 H. Hornburger: Chem. Phys. Lett. **143**, 284 (1988)
4.84 J. Tellinghuisen: J. Mol. Spectrosc. **103**, 455 (1984)
V.L. Bakhrakh, I.M. Umanskiĭ: Opt. Spektrosk. **62**, 992 (1987) [English transl.: Opt. Spectrosc. (USSR) **62**, 587 (1987)]
4.85 A.I. Voronin, V.I. Osherov: Zh. Eksp. Teor. Fiz. **66**, 135 (1974) [English transl.: Sov. Phys.-JETP **39**, 62 (1974)]
S.P. Karkach, V.I. Osherov, V.G. Ushakov: Zh. Eksp. Teor. Fiz. **68**, 493 (1975) [English transl.: Sov. Phys.-JETP **41**, 241 (1975)]
V.I. Osherov., L.V. Poluyanov, A.I. Voronin: Chem. Phys. **93**, 13 (1985)
V.I. Osherov, V.G. Ushakov, A.I. Voronin: J. Phys. B **13**, 1507 (1980)
A.I. Voronin, V.I. Osherov, L.V. Poluyanov: Khim. Fiz. **5**, 1462 (1986) [English transl.: Sov. J. Chem. Phys. (UK) **5**, 2714 (1990)]
A.I. Voronin, S.P. Karkach, V.I. Osherov, V.G. Ushakov: Zh. Eksp. Teor. Fiz. **71**, 884 (1976) [English transl.: Sov. Phys.-JETP **44**, 465 (1976)]
4.86 F.R. Gantmakher: *Theory of Matrices*, 3rd edn. (Nauka, Moscow 1967) (in Russian)
4.87 Y. Mochizuki, K. Kaya, M. Ito: Chem. Phys. **54**, 375 (1981)
4.88 J. Subbi: Chem. Phys. Lett. **109**, 1 (1984)
4.89 A.R. Bacon, J.M. Hollas: Chem. Phys. Lett. **120**, 477 (1985)
4.90 J.A. Warren, J.M. Hayes, G.J. Small: J. Chem. Phys. **80**, 1786 (1984)
J.A. Warren, J.M. Hayes, G.J. Small: Chem. Phys. **102**, 313 (1986)
4.91 J.A. Warren, E.R. Bernstein: J. Chem. Phys. **85**, 2365 (1986)
4.92 A.R. Auty, A.C. Jones, D. Phillips: Chem. Phys. **103**, 163 (1986)
4.93 C. Kryschi, H. Kupka, H.-H. Perkampus: Chem. Phys. **116**, 53 (1987)
4.94 E.V. Doktorov, I.A. Malkin, V.I. Man'ko: J. Mol. Spectrosc. **64**, 302 (1977)
4.95 T.R. Faulkner, F.S. Richardson: J. Chem. Phys. **70**, 1201 (1979)
4.96 N. Heider, S.F. Fischer: Chem. Phys. **88**, 209 (1984)
4.97 K. Krogh-Jespersen, R.P. Rava, L. Goodman: J. Phys. Chem. **88**, 5503 (1984)
4.98 H. Kupka, O.E. Polansky: J. Chem. Phys. **80**, 3153 (1984)
4.99 H. Kupka, G. Olbrich: J. Chem. Phys. **82**, 3975 (1985)
4.100 H. Kupka, P.H. Cribb: J. Chem. Phys. **85**, 1303 (1986)
4.101 D. Gruner, P. Brumer: Chem. Phys. Lett. **138**, 310 (1987)
4.102 V.I. Baranov, A.N. Solov'yov: Opt. Spektrosk. **62**, 59 (1987) [English transl.: Opt. Spectrosc. (USSR) **62**, 36 (1987)]
4.103 J. Brickmann, Y.M. Engel, R.D. Levine: Chem. Phys. Lett. **137**, 441 (1987)
4.104 J. Subbi: Chem. Phys. **122**, 157 (1988)
4.105 A.P. Chigiryov: Opt. Spektrosk. **67**, 303 (1989) [English transl.: Opt. Spectrosc. (USSR) **67**, 175 (1989)]
4.106 B.I. Makshantsev: Opt. Spektrosk. **31**, 355 (1971) [English transl.: Opt. Spectrosc. (USSR) **31**, 191 (1971)]
B.I. Makshantsev: Opt. Spektrosk. **34**, 872 (1973) [English transl.: Opt. Spectrosc. (USSR) **34**, 503 (1973)]
4.107 B.I. Makshantsev, I.P. Perstnev: Opt. Spektrosk. **30**, 371 (1971) [English transl.: Opt. Spectrosc. (USSR) **30**, 202 (1971)]
4.108 S. Fischer: Chem. Phys. Lett. **11**, 577 (1971)
4.109 W. Siebrand: Chem. Phys. Lett. **6**, 192 (1970)
4.110 V. Lawetz, G. Orlandi, W. Siebrand: J. Chem. Phys. **56**, 4058 (1972)
4.111 W. Siebrand, M.Z. Zgierski: Chem. Phys. Lett. **35**, 151 (1975)
4.112 M.A. El-Sayed, C.R. Chen: Chem. Phys. Lett. **10**, 313 (1971)
4.113 M.A. El-Sayed, W.R. Moomav, J.B. Chodak: Chem. Phys. Lett. **20**, 11 (1973)
4.114 Y. Fujimura, S. Shimakura, T. Nakajima: J. Chem. Phys. **66**, 3530 (1977)

4.115 A. Desprès, V. Lejeune, E. Migirdicyan, C. Mijoule, G. Marconi, G. Orlandi, W. Siebrand, M.Z. Zgierski: Chem. Phys. **88**, 229 (1984)
4.116 A. Desprès, V. Lejeune, E. Migirdicyan, W. Siebrand: J. Phys. Chem. **92**, 6914 (1988)
4.117 V.L. Ermolaev, E.B. Sveshnikova: Opt. Spektrosk. **21**, 134 (1966) [English transl.: Opt. Spectrosc. (USSR) **21**, 78 (1966)]
4.118 T. Suzuki, M. Sato, N. Mikami, M. Ito: Chem. Phys. Lett. **127**, 292 (1986)
4.119 A.P. Darmanyan: Chem. Phys. Lett. **102**, 329 (1983)
4.120 J. Najbar, W. Jarzeba, Z.H. Urbanek: Chem. Phys. **79**, 245 (1983)
J. Najbar, W. Jarzeba, R.M. Hochstrasser: Chem. Phys. **95**, 1 (1985)
4.121 K.F. Freed: J. Chem. Phys. **64**, 1604 (1976)
4.122 K.F. Freed, C. Tric: Chem. Phys. **33**, 249 (1978)
4.123 P.R. Harrowell, K.F. Freed: J. Chem. Phys. **83**, 6288 (1985)
4.124 S. Mukamel: Chem. Phys. Lett. **60**, 310 (1979)
4.125 K.W. Holtzclaw, D.B. Moss, C.S. Parmenter, G.W. Loge: J. Phys. Chem. **87**, 4495 (1983)
4.126 C. Michel, M. Lombardi, R. Jost: Chem. Phys. **109**, 357 (1986)
4.127 I. Oref, S. Speiser, G.H. Atkinson: J. Phys. Chem. **90**, 912 (1986)
4.128 N.A. Borisevich, L.M. Bolot'ko, A. Yu. Kosnikov, T.F. Raïchyonok: Opt. Spektrosk. **63**, 784 (1987) [English transl.: Opt. Spectrosc. (USSR) **63**, 465 (1987)]
4.129 U. Brühlmann, J.R. Huber: Chem. Phys. **94**, 273 (1985)
4.130 C.A. Thayer, J.T. Yardley: J. Chem. Phys. **61**, 2487 (1974)
C.A. Thayer, A.V. Pocius, J.T. Yardley: J. Chem. Phys. **62**, 3712 (1975)
4.131 T.H. Cheng, N. Hirota, S.W. Mao: Chem. Phys. Lett. **15**, 274 (1972)
4.132 T. Azumi: Chem. Phys. Lett. **19**, 580 (1973)
4.133 G. Orlandi, F. Barigelletti, L. Flamigni, S. Dellonte: J. Photochem. **31**, 49 (1985)
4.134 A.V. Karyakin, Yu. N. Smirnov: Zh. Prikl. Spektrosk. **49**, 774 (1988) [English transl.: J. Appl. Spectrosc. (USSR) **49**, 1144 (1988)]
4.135 M.A. Morgan, G.C. Pimentel: J. Phys. Chem. **93**, 3056 (1989)
4.136 S. Hirayama, K. Shobatake, K. Tabayashi: Chem. Phys. Lett. **121**, 228 (1985)
S. Hirayama, F. Tanaka, K. Shobatake: Chem. Phys. Lett. **140**, 447 (1987)
4.137 B. Sharf: Chem. Phys. Lett. **6**, 364 (1970)
B. Sharf: Chem. Phys. Lett. **14**, 315 (1972)
B. Sharf: Chem. Phys. Lett. **14**, 475 (1972)
B. Sharf: Chem. Phys. Lett. **19**, 35 (1973)
4.138 S.H. Lin: J. Chem. Phys. **44**, 3759 (1966)
4.139 K. Aizawa, H. Igarashi, K. Kaya: Chem. Phys. **23**, 273 (1977)
4.140 J.H. van der Waals, M.S. de Groot: In *The Triplet State*, ed. by A.B. Zahlan (Cambridge Univ. Press, London 1967) p. 101
4.141 E.C. Lim, R.J. Locke, B.T. Lim, T. Fujioka, H. Iwamura: J. Phys. Chem. **91**, 1298 (1987)
4.142 K. Aron, C. Otis, R.E. Demaray, P. Johnson: J. Chem. Phys. **73**, 4167 (1980)
4.143 Y. Takagi, M. Sumitani, N. Nakashima, D.V. O'Connor, K. Yoshihara: J. Chem. Phys. **77**, 6337 (1982)
4.144 D.V. O'Connor: Faraday Discuss. Chem. Soc. No. 75, 415 (1983)
4.145 J.H. Callomon: Faraday Discuss. Chem. Soc. No. 75, 417 (1983)
4.146 I. Yamazaki, T. Murao, T. Yamanaka, K. Yoshihara: Faraday Discuss. Chem. Soc. No. 75, 395 (1983)
I. Yamazaki, T. Murao, K. Yoshihara: Chem.Phys. Lett. **87**, 384 (1982)
4.147 D.V. O'Connor, M. Sumitani, Y. Takagi, N. Nakashima, K. Kamogawa, Y. Udagawa, K. Yoshihara: J. Phys. Chem. **87**, 4848 (1983)
D.V. O'Connor, M. Sumitani, Y. Takagi, N. Nakashima, K. Kamogawa, Y. Udagawa, K. Yoshihara: Chem. Phys. **93**, 373 (1985)
4.148 M. Sumitani, D.V. O'Connor, Y. Takagi, K. Yoshihara: Chem. Phys. Lett. **108**, 11 (1984)
4.149 Y. Takagi, M. Sumitani, D.V. O'Connor, N. Nakashima, K. Kamogawa, Y. Udagawa, K. Yoshihara: Chem. Phys. Lett. **99**, 445 (1983)

4.150 M. Sumitani, D. O'Connor, Y. Takagi, N. Nakashima, K. Kamogawa, Y. Udagawa, K. Yoshihara: Chem. Phys. Lett. **97**, 508 (1983)
M. Sumitani, D. O'Connor, Y. Takagi, N. Nakashima, K. Kamogawa, Y. Udagawa, K. Yoshihara: Chem. Phys. **93**, 359 (1985)
4.151 J.L.G. Suijker, C.A.G.O. Varma, A.H. Huizer: Chem. Phys. Lett. **107**, 496 (1984)
4.152 T.A. Stephenson, P.L. Radloff, S.A. Rice: J. Chem. Phys. **81**, 1060 (1984)
4.153 T.A. Stephenson, S.A. Rice: J. Chem. Phys. **81**, 1073 (1984)
4.154 Y. Achiba, A. Hiraya, K. Kimura: J. Chem. Phys. **80**, 6047 (1984)
4.155 D.B. Moss, C.S. Parmenter: J. Phys. Chem. **90**, 1011 (1986)
4.156 R.J. Longfellow, D.B. Moss, C.S. Parmenter: J. Phys. Chem. **92**, 5438 (1988)
4.157 E. Hiroike, Y. Fujimura: J. Chem. Phys. **86**, 2510 (1987)
E. Hiroike, Y. Fujimura: Chem. Phys. Lett. **143**, 193 (1988)
4.158 J.H. Callomon, L.W. Somers: Chem. Phys. Lett. **144**, 463 (1988)
4.159 T. Suzuki, M. Ito: J. Chem. Phys. **91**, 4564 (1990)
4.160 B.E. Forch, S. Okajima, E.C. Lim: Chem. Phys. Lett. **108**, 311 (1984)
4.161 A.K. Jameson, B.E. Forch, K.T. Chen, S. Okajima, H. Saigusa, E.C. Lim: J. Phys. Chem. **88**, 4937 (1984)
4.162 J.L. Knee, L.R. Khundkar, A.H. Zewail: J. Phys. Chem. **89**, 3201 (1985)
4.163 K. Huang, A. Rhys: Proc. R. Soc. London, Ser. A, **204**, 406 (1950)
4.164 H. Bateman, A. Erdélyi: *Higher Transcendental Functions*, Vol. 2 (McGraw-Hill, New York 1953)
4.165 D.M. Burland, G.W. Robinson: J. Chem. Phys. **51**, 4548 (1969)
4.166 P.A. Praem: Proc. Inst. Phys. Astron., Acad. Sci. Estonian SSR, Tallinn, **16**, 57 (1961) (in Russian)
4.167 E. Hutchisson: Phys. Rev. **36**, 410 (1930)
4.168 C. Manneback: Physica **17**, 1001 (1951)
4.169 B. Scharf: Chem. Phys. Lett. **68**, 242 (1979)
4.170 K.K. Rebane, O.I. Sil'd: Proc. Inst. Phys. Astron., Acad. Sci. Estonian SSR, Tallinn, **12**, 3 (1960) (in Russian)
4.171 A.A. Kovalev, B.I. Makshantsev: Opt. Spektrosk. **35**, 850 (1973) [English transl.: Opt. Spectrosc. (USSR) **35**, 494 (1973)]
A.A. Kovalev, B.I. Makshantsev: Spetrosc. Lett. **5**, 13 (1972)
4.172 B.I. Makshantsev: Spectrosc. Lett. **5**, 1 (1972)
4.173 H. Hornburger, K.-D. Reinsch, M. Melzig, S. Leach: Z. Phys. D **5**, 151 (1987)
4.174 J. Morales, L. Sandoval, A. Palma, J. Lopez-Bonilla: Chem. Phys. Lett. **135**, 149 (1987)

Chapter 5

5.1 C. Zener: Phys. Rev. **37**, 556 (1931)
C. Zener: Proc. R. Soc. London, Ser. A, **137**, 696 (1932)
5.2 K.K. Rebane: Proc. Inst. Phys. Astron., Acad. Sci. Estonian SSR, Tallinn, **16**, 49 (1961) (in Russian)
5.3 Y. Kajanuma: J. Phys. Soc. Jpn. **51**, 3526 (1982)
5.4 E.J. Heller, R.C. Brown: J. Chem. Phys. **79**, 3336 (1983)
5.5 S.J. Formosinho: J. Chem. Soc. Faraday Tran. 2 **70**, 605 (1974)
5.6 L.M. Biberman, I.T. Yakubov: Opt. Spektrosk. **8**, 294 (1960) [English transl.: Opt. Spectrosc. (USSR) **8**, 155 (1960)]
R.T. Pack, J. S. Dahler: J. Chem. Phys. **50**, 2397 (1969)
W.D. Smith, R.T. Pack: J. Chem. Phys. **52**, 1381 (1970)
M.S. Child: Mol. Phys. **29**, 1421 (1975)
5.7 L.A. Kuznetsova, N.E. Kuz'menko, Yu.Ya. Kuzyakov, Yu.A. Plastilin: *Optical Transition Probabilities for Diatomic Molecules* (Nauka, Moscow 1980) (in Russian)

N.E. Kuz'menko, L.A. Kuznetsova, Yu.Ya. Kuzyakov: *Franck-Condon Factors for Diatomic Molecules* (Moscow Univ. Press, Moscow 1984) (in Russian)

5.8 E.S. Medvedev: Khim. Fiz. **1**, 32 (1982) [English transl.: Sov. J. Chem. Phys. (UK) **1**, 59 (1983)]
E.S. Medvedev: Khim. Fiz. **2**, 188 (1982) [English transl.: Sov. J. Chem. Phys. (UK) **2** (1983)]

5.9 E.S. Medvedev: Chem. Phys. **73**, 243 (1982)

5.10 J. Heading: *An Introduction to Phase-Integral Methods*, Monogr. Phys. Subjects (Methuen, London 1962)

5.11 M.A. Lavrent'ev, B.V. Shabat: *Methods of the Theory of Functions of Complex Variable* (Nauka, Moscow 1973) (in Russian)

5.12 R. Kubo: Phys. Rev. **86**, 929 (1952)

5.13 Y. Mikami, K. Mizunoya, T. Nakajima: Chem. Phys. Lett. **30**, 373 (1975)

5.14 Y. Mikami: J. Chem. Phys. **73**, 3314 (1980)
Y. Mikami, M. Suzuki: Chem. Phys. Lett. **131**, 70 (1986)

5.15 I. Madzgalla, K.-E. Süsse, D.-G. Welsch: Ann. Phys. (Leipzig) **44**, 249 (1987)

5.16 I.H. Kühn, D.F. Heller, W.M. Gelbart: Chem. Phys. **22**, 435 (1977)

5.17 A. Sobolewski: Acta Phys. Pol. **A58**, 353 (1980)

5.18 N. Shimakura, Y. Fujimura, T. Nakajima: Theor. Chim. Acta **37**, 77 (1975)

5.19 N.S. Strokach, E.A. Gastilovich, D.N. Shigorin: Opt. Spektrosk. **35**, 238 (1973) [English transl.: Opt. Spectrosc. (USSR) **35**, 140 (1973)]

5.20 R.M. Badger: J. Chem. Phys. **2**, 128 (1934)
R.M. Badger: J. Chem. Phys. **3**, 710 (1935)

5.21 D.J. Swanton, B.R. Henry: J. Chem. Phys. **86**, 4801 (1987)

5.22 A.I. Voronin, V.I. Osherov: *Molecular Reaction Dynamics* (Nauka, Moscow 1990) (in Russian)

5.23 K.F. Freed, W.M. Gelbart: Chem. Phys. Lett. **10**, 187 (1971)

5.24 K.K. Pukhov, V.P. Sakun: Phys. Status Solidi (b) **95**, 391 (1979)
V.P. Sakun: Multiphonon relaxation of impurity centers in solids, in *Theoretical Problems of Chemical Physics*, ed. by N.M. Kuznetsov (Nauka, Moscow 1982) pp. 220–238 (in Russian)

5.25 O.V. Balagura, A.I. Ivanov: Opt. Spektrosk. **62**, 1043 (1987) [English transl.: Opt. Spectrosc. (USSR) **62**, 616 (1987)]

5.26 T. Uzer, M.S. Child: Mol. Phys. **41**, 1177 (1980)

5.27 M.J. Davis, E.J. Heller: J. Chem. Phys. **75**, 246 (1981)

5.28 S.K. Gray: Chem. Phys. **83**, 125 (1984)

5.29 D. Farrelly, T. Uzer: J. Chem. Phys. **85**, 308 (1986)

5.30 T.A. Holme, J.S. Hutchinson: J. Chem. Phys. **84**, 5455 (1986)
K.T. Marshall, J.S. Hutchinson: J. Phys. Chem. **91**, 3219 (1987)

5.31 J.S. Hutchinson: J. Chem. Phys. **85**, 7087 (1986)

5.32 E.S. Medvedev: J. Mol. Spectrosc. **114**, 1 (1985)

5.33 E.S. Medvedev: Chem. Phys. Lett. **120**, 173 (1985)

5.34 E.S. Medvedev: Spectrosc. Lett. **18**, 447 (1985)

5.35 E.S. Medvedev: Opt. Spektrosk. **58**, 1205 (1985) [English transl.: Opt. Spectrosc. (USSR) **58**, 741 (1985)]
A.Yu. Ermilov, E.S. Medvedev, A.V. Nemukhin: Opt. Spektrosk. **69**, 554 (1990) [English transl.: Opt. Spectrosc. (USSR) **69**, 331 (1990)]

5.36 E.S. Medvedev: Dokl. Akad. Nauk SSSR **284**, 841 (1985) [English transl.: Sov. Phys. -Dokl. **30**, 853 (1985)]

5.37 E.S. Medvedev: *Determination of a New Molecular Constant for Diatomic Systems* (Acad. Sci. USSR, Chernogolovka 1984)

5.38 E.N. Bodunov: Opt. Spektrosk. **62**, 770 (1987) [English transl.: Opt. Spectrosc. (USSR) **62**, 459 (1987)]

5.39 S.M. Melikova, D.N. Shchepkin: Opt. Spektrosk. **53**, 1019 (1982) [English transl.: Opt. Spectrosc. (USSR) **53**, 609 (1982)]

O.V. Bogdanova, T.D. Kolomiĭtsova, L.F. Strelkova, D.N. Shchepkin: Opt. Spektrosk. **54**, 650 (1983) [English transl.: Opt. Spectrosc. (USSR) **54**, 385 (1983)]
5.40 D.A. Chernoff, J.D. Myers, J.G. Pruett: J. Chem. Phys. **85**, 3732 (1986)
5.41 H.B. Levene, D. S. Perry: J. Chem. Phys. **80**, 1772 (1984)
D.E. Braham, D.S. Perry: Chem. Phys. Lett. **103**, 487 (1984)
A.M. de Souza, D.S. Perry: J. Phys. Chem. **90**, 4508 (1986)
5.42 C. Manzanares I, N.L.S. Yamasaki, E. Weitz: J. Phys. Chem. **90**, 3553 (1986)
5.43 O.S. Mortensen, B.R. Henry, M.A. Mohammadi: J. Chem. Phys. **75**, 4800 (1981)
O.S. Mortensen, M.K. Ahmed, B.R. Henry, A.W. Tarr: J. Chem Phys. **82**, 3903 (1985)
A.W. Tarr, B.R. Henry: Chem. Phys. Lett. **112**, 295 (1984)
A.W. Tarr, B.R. Henry: J. Chem. Phys. **84**, 1355 (1986)
M.K. Ahmed, B.R. Henry: J. Phys. Chem. **90**, 1081 (1986)
M.K. Ahmed, B.R. Henry: J. Phys. Chem. **90**, 1737 (1986)
M.K. Ahmed, B.R. Henry: J. Phys. Chem. **91**, 3741 (1987)
M.K. Ahmed, B.R. Henry: J. Phys. Chem. **91**, 5194 (1987)
M.K. Ahmed, D.J. Swanton, B.R. Henry: J. Phys. Chem. **91**, 293 (1987)
5.44 J.S. Wong, C.B. Moore: J. Chem. Phys. **77**, 603 (1982)
5.45 G.J. Scherer, K.K. Lehmann, W. Klemperer: J. Chem. Phys. **81**, 5319 (1984)
5.46 G.A. Voth, R.A. Marcus, A.H. Zewail: J. Chem. Phys. **81**, 5494 (1984)
5.47 J.E. Baggott, H.-C. Chuang, R.N. Zare, H.-R. Dübal, M. Quack: J. Chem. Phys. **82**, 1186 (1985)
5.48 J.E. Baggott, H.J. Clase, I.M. Mills: J. Chem. Phys. **84**, 4193 (1986)
5.49 J.E. Baggott, D.W. Law, P.D. Lightfoot, I.M. Mills: J. Chem. Phys. **85**, 5414 (1986)
5.50 K.M. Gough, F.L. Baudais, H.L. Casal: J. Chem. Phys. **84**, 549 (1986)
5.51 L.C. Baylor, E. Weitz, P. Hofmann: J. Chem. Phys. **90**, 615 (1989)
P. Hofmann, R.B. Gerber, M.A. Ratner, L.C. Baylor, E. Weitz: J. Chem. Phys. **88**, 7434 (1988)
5.52 A. Campargue, F. Stoeckel: J. Chem. Phys. **85**, 1220 (1986)
5.53 A. Campargue, F. Stoeckel, M. Chenevier, H.B. Kraiem: J. Chem. Phys. **87**, 5598 (1987)
H.B. Kraiem, A. Campargue, M. Chenevier, F. Stoeckel: J. Chem. Phys. **91**, 2418 (1989)
5.54 M.W. Crofton, C.G. Stevens, D. Klenerman, J.H. Gutow, R.N. Zare: J. Chem. Phys. **89**, 7100 (1988)
5.55 H.-R. Dübal, M. Lewerenz, M. Quack: J. Chem. Phys. **85**, 34 (1986)
5.56 H.-R. Dübal, M. Quack: J. Chem. Phys. **81**, 3779 (1984)
S. Peyerimhoff, M. Lewerenz, M. Quack: Chem. Phys. Lett. **109**, 563 (1984)
5.57 C. Kormann, M. Lewerenz, M. Quack: Chem. Phys. Lett. **149**, 429 (1988)
5.58 M. Lewerenz, M. Quack: Chem. Phys. Lett. **123**, 197 (1986)
M. Lewerenz, M. Quack: J. Chem. Phys. **88**, 5408 (1988)
5.59 G. Longhi, G. Zerbi, L. Ricard, S. Abbate: J. Chem. Phys. **88**, 6733 (1988)
5.60 C. Manzanares I, N.L.S. Yamasaki, E. Weitz: Chem. Phys. Lett. **144**, 43 (1988)
C. Manzanares I, N.L.S. Yamasaki, E. Weitz, J.T. Knudson: Chem. Phys. Lett. **117**, 477 (1985)
5.61 R. Nakagaki, I. Hanazaki: Spectrochim. Acta **40A**, 57 (1984)
R. Nakagaki, I. Hanazaki: Chem. Phys. Lett. **128**, 432 (1986)
5.62 J.W. Perry, A.H. Zewail: J. Chem. Phys. **80**, 5333 (1984)
5.63 J.W. Perry, D.J. Moll, A. Kuppermann, A.J. Zewail: J. Chem. Phys. **82**, 1195 (1985)
5.64 J. Segall, R.N. Zare, H.-R. Dübal, M. Lewerenz, M. Quack: J. Chem. Phys. **86**, 634 (1987)
5.65 G.A. West, R.P. Mariella, J.A. Pete, W.B. Hammond, D.F. Heller: J. Chem. Phys. **75**, 2006 (1981)
5.66 J.M. Jasinski: Chem. Phys. Lett. **123**, 121 (1986)
5.67 H.L. Fang, R.L. Swofford, M. McDevitt, A.B. Anderson: J. Phys. Chem. **89**, 225 (1985)
5.68 H.L. Fang, R.L. Swofford, D.A.C. Compton: Chem. Phys. Lett. **108**, 539 (1984)
5.69 S.L. Coy, K.K. Lehmann: J. Chem. Phys. **84**, 5239 (1986)
5.70 R.L. Swofford, M.E. Long, M.S. Burberry, A.C. Albrecht: J. Chem. Phys. **66**, 664 (1977)

5.71 J.M. Jasinski: Chem. Phys. Lett. **109**, 462 (1984)
5.72 G.O. Brink: J. Mol. Spectrosc. **103**, 295 (1984)
5.73 L.J. Butler, T.M. Ticich, M.D. Likar, F.F. Crim: J. Chem. Phys. **85**, 2331 (1986)
5.74 R.A. Bernheim, F.W. Lampe, J.F. O'Keefe, J.R. Qualley, III: Chem. Phys Lett. **100**, 45 (1983)
R.A. Bernheim, F.W. Lampe, J.F. O'Keefe, J.R. Qualley, III: J. Mol. Spectrosc. **104**, 194 (1984)
R.A. Bernheim, F.W. Lampe, J.F. O'Keefe, J.R. Qualley, III: J. Phys. Chem. **89**, 1087 (1985)
5.75 R.A. Bernheim, F.W. Lampe, J.F. O'Keefe, J.R. Qualley, III: J. Chem. Phys. **80**, 5906 (1984)
5.76 D.C. McKean, A.R. Morrisson, M.I. Kelly: Chem. Phys. Lett. **109**, 347 (1984)
5.77 A. Campargue, F. Stoeckel, M. Terrile: Chem. Phys. **110**, 145 (1986)
5.78 C.E. Brion: Comm. At. Mol. Phys. **16**, 249 (1985)
5.79 E.S. Medvedev: Khim. Fiz. **4**, 1603 (1985) [English transl.: Sov. J. Chem. Phys. (UK) **4**, 2664 (1989)]
E.S. Medvedev: Khim. Fiz. **5**, 156 (1986) [English transl.: Sov. J. Chem. Phys. (UK) **5**, 248 (1989)]
E.S. Medvedev: Teor. Mat. Fiz. **90**, 218 (1992) [English transl.: Theor. Math. Phys. **90**, 146 (1992)]
Yu.V. Kivran, E.S. Medvedev: Khim. Fiz. **12**, 1299 (1993)
E.S. Medvedev: J. Chem. Phys. **100**, 7192 (1994)
5.80 M.L. Koszykowski, D.W. Noid, R.A. Marcus: J. Phys. Chem. **86**, 2113 (1982)
5.81 J.R. Stine, D.W. Noid: J. Chem. Phys. **78**, 1876 (1983)
J.R. Stine, D.W. Noid: J. Chem. Phys. **78**, 3647 (1983)
5.82 M.V. Fedoryuk: *The Saddle-Point Method* (Nauka, Moscow 1977) (in Russian)
5.83 B. Timm, R. Mecke: Z. Phys. **98**, 363 (1936)
5.84 R. Mecke: Z. Elektrochem. **54**, 38 (1950)
5.85 A.F. Ferguson, D. Parkinson: Planet. Space Sci. **11**, 149 (1963)
5.86 M.S. Child, L. Halonen: Adv. Chem. Phys. **57**, 1 (1984)
5.87 K. Scholz: Z. Phys. **78**, 751 (1932)
L. Infeld, T.E. Hull: Rev. Mod. Phys. **23**, 21 (1952)
H.S. Heaps, G. Herzberg: Z. Phys. **133**, 48 (1952)
5.88 V.S. Vasan, R.J. Cross: J. Chem. Phys. **78**, 3869 (1983)
5.89 K.V. Reddy, D.F. Heller, M.J. Berry: J. Chem. Phys. **76**, 2814 (1982)
5.90 A.A. Radzig, B.M. Smirnov: *Reference Data on Atoms, Molecules and Ions*, Springer Ser. Chem. Phys., Vol. 31 (Springer, Berlin, Heidelberg 1985)
5.91 A. Amrein, H.-R. Dübal, M. Lewerenz, M. Quack: Chem. Phys. Lett. **112**, 387 (1984)
5.92 K.K. Lehmann, A.M. Smith: J. Chem. Phys. **93**, 6140 (1990)
E.E. Nikitin, C. Noda, R. N. Zare: J. Chem. Phys. **98**, 46 (1993)
5.93 M.S. Burberry, A.C. Albrecht: J. Chem. Phys. **70**, 147 (1979)
5.94 C.K.N. Patel, A.C. Tam, R.J. Kerl: J. Chem. Phys. **71**, 1470 (1979)
5.95 N. Yamamoto, N. Matsuo, H. Tsubomura: Chem. Phys. Lett. **71**, 463 (1980)
5.96 G. Rimpel: Z. Naturforsch. **29A**, 588 (1974)
5.97 E.S. Medvedev: Tunneling nonradiative and radiative transitions in molecules, Dissertation, Institute of Chemical Physics, Moscow (1986) (in Russian)
5.98 H.L. Fang, R.L. Swofford: J. Chem. Phys. **72**, 6382 (1980)
H.L. Fang, R.L. Swofford: J. Chem. Phys. **73**, 2607 (1980)
H.L. Fang, R.L. Swofford: Appl. Opt. **21**, 55 (1982)
5.99 M.K. Ahmed, B.R. Henry: J. Chem. Phys. **87**, 3724 (1987)
5.100 M.S. Burberry, J.A. Morrell, A.C. Albrecht, R.L. Swofford: J. Chem. Phys. **70**, 5522 (1979)
5.101 M.S. Burberry, A.C. Albrecht: J. Chem. Phys. **71**, 4768 (1979)
5.102 K.K. Lehmann: J. Chem. Phys. **91**, 2759 (1989)
5.103 A.M. Smith, U.G. Jørgensen, K.K. Lehmann: J. Chem. Phys. **87**, 5649 (1987)
5.104 A.M. Smith, K.K. Lehmann, W. Klemperer: J. Chem. Phys. **85**, 4958 (1986)
5.105 R.N. Sileo, T.A. Cool: J. Chem. Phys. **65**, 117 (1976)
5.106 V.I. Krassovsky, N.N. Shefov, V.I. Yarin: J. Atmosph. Terr. Phys. **21**, 46 (1961)

5.107 R.E. Meredith, F. G. Smith: J. Quant. Spectrosc. Radiat. Transfer **13**, 89 (1973)
5.108 A.L. Sobolewski: Chem. Phys. Lett. **124**, 187 (1986)
A.L. Sobolewski: J. Chem. Phys. **84**, 1133 (1986)
5.109 R.L. Spellicy, R.E. Meredith, F.G. Smith: J. Chem. Phys. **57**, 5119 (1972)
5.110 C. Chackerian, Jr.: J. Chem. Phys. **65**, 4228 (1976)
5.111 R.H. Tipping, J.F. Ogilvie: J. Mol. Spectrosc. **96**, 442 (1982)
5.112 P. Niay, P. Bernage, C. Coquant, R. Houdart: Can. J. Phys. **56**, 727 (1978)
5.113 R.E. Meredith: J. Quant. Spectrosc. Radiat. Transfer **12**, 485 (1972)
5.114 P. Bernage, P. Niay: J. Quant. Spectrosc. Radiat. Transfer **18**, 315 (1977)
5.115 J. Gelfand, M. Zughul, H. Rabitz, C.J. Han: J. Quant. Spectrosc. Radiat. Transfer **26**, 303 (1981)
5.116 K. Cashion: J. Mol. Spectrosc. **10**, 182 (1963)
5.117 W.S. Benedict, R. Herman, G.E. Moore, S. Silverman: J. Chem. Phys. **26**, 1671 (1957)
5.118 R.A. Toth, R.H. Hunt, E.K. Plyler: J. Mol. Spectrosc. **32**, 85 (1969)
R.A. Toth, R.H. Hunt, E.K. Plyler: J. Mol. Spectrosc. **35**, 110 (1970)
5.119 K.V. Reddy: J. Mol. Spectrosc. **82**, 127 (1980)
5.120 L.A. Young, W.J. Eachus: J. Chem. Phys. **44**, 4195 (1966)
5.121 C. Chackerian, Jr., F.P.J. Valero: J. Mol. Spectrosc. **62**, 338 (1976)
5.122 P.C. Goswami, D.J. Swanton, B.R. Henry: J. Chem. Phys. **86**, 5281 (1987)
5.123 V.I. Osherov: Chem. Phys. Lett. **1**, 539 (1968)
5.124 V.I. Osherov: Phys. Status Solidi **31**, 31 (1969)
5.125 I.S. Averbukh, V.A. Kovarskiĭ, N.F. Perel' man: Zh. Eksp. Teor. Fiz. **74**, 1230 (1978) [English transl.: Sov. Phys.-JETP **47**, 645 (1978)]
5.126 C.A. Coulson, K. Zalewskii. Proc. R. Soc. London, Ser. A, **268**, 437 (1962)
5.127 B.A. Waite, W.H. Miller: J. Chem. Phys. **73**, 3713 (1980)
B.A. Waite, W.H. Miller: J. Chem. Phys. **74**, 3910 (1981)
5.128 B.K. Holmer, M.S. Child: Faraday Discuss. Chem. Soc. No. 75, 131 (1983)
5.129 V.K. Bykhovskiĭ, E.E. Nikitin, M.Ya. Ovchinnikova: Zh. Eksp. Teor. Fiz. **47**, 750 (1964) [English transl.: Sov. Phys.-JETP **20**, 500 (1965)]
5.130 A. Bárány: J. Phys. B **11**, L399 (1978)
A. Bárány: J. Phys. B **12**, 2841 (1979)
5.131 E.C.G. Stueckelberg: Helv. Phys. Acta **5**, 369 (1932)
5.132 V.B. Nagaev, V.I. Osherov: Opt. Spektrosk. **25**, 810 (1968) [English transl.: Opt. Spectrosc. (USSR) **25**, 453 (1968)]
5.133 V.I. Osherov: Chem. Phys. Lett. **1**, 182 (1967)
5.134 V.I. Osherov: Teor. Eksp. Khim. **4**, 299 (1968)
5.135 M. Matsuzawa: J. Phys. Soc. Jpn. **23**, 1383 (1967)
5.136 A. Sobolewski, I. Deperasińska, J. Prochorow: Acta Phys. Pol. **56A**, 535 (1979)
5.137 A. Warshel, M. Karplus: Chem. Phys. Lett. **32**, 11 (1975)
5.138 A. Warshel: Nature **260**, 679 (1976)
A. Warshel: J. Phys. Chem. **86**, 2218 (1982)
5.139 J.-K. Hwang, A. Warshel: Chem. Phys. Lett. **115**, 281 (1985)
5.140 A. Warshel, J.-K. Hwang: J. Chem. Phys. **82**, 1756 (1985)
A. Warshel, J.-K. Hwang: J. Chem. Phys. **84**, 4938 (1986)
5.141 E.E. Nikitin, M.Ya. Ovchinnikova: Chem. Phys. **138**, 45 (1989)
5.142 R. Feynman, A. Hibbs: *Quantum Mechanics and Path Integrals* (McGraw-Hill, New York, 1965)
5.143 P. Pechucas: Phys. Rev. **181**, 166 (1969)
5.144 S.P. Karkach: Mol. Phys. **57**, 473 (1986)
S.P. Karkach: Khim. Fiz. **7**, 1598 (1988)
5.145 R.D. Coalson, J.L. Kinsey: J. Chem. Phys. **85**, 4322 (1986)
R.D. Coalson: J. Chem. Phys. **86**, 995 (1987)

Chapter 6

6.1 E.B. Sveshnikova, V.L. Ermolaev: Opt. Spektrosk. **30**, 379 (1971) [English transl.: Opt. Spectrosc. (USSR) **30**, 208 (1971)]
E.B. Sveshnikova, V.L. Ermolaev: Izv. Akad. Nauk SSSR, Ser. Fiz. **35**, 1481 (1971)
E.B. Sveshnikova, V.L. Ermolaev: Izv. Akad. Nauk SSSR, Ser. Fiz. **46**, 226 (1982) [English transl.: Bull. Acad. Sci. USSR, Phys. Ser. **46**, 10 (1982)]
6.2 E.B. Sveshnikova, B.S. Neporent: Izv. Akad. Nauk SSSR, Ser. Fiz. **36**, 1088 (1972)
E.B. Sveshnikova, B.S. Neporent: Izv. Akad. Nauk SSSR, Ser. Fiz. **37**, 378 (1973)
6.3 E.B. Sveshnikova, N.T. Timofeev, V.M. Zolotaryov: Izv. Akad. Nauk SSSR, Ser. Fiz. **44**, 722 (1980) [English transl.: Bull. Acad. Sci. USSR, Phys. Ser. **44**, 31 (1980)]
6.4 E.B. Sveshnikova, A.A. Stroganov: Izv. Akad. Nauk SSSR, Ser. Fiz. **52**, 725 (1988) [English transl.: Bull. Acad. Sci. USSR, Phys. Ser. **52**, 91 (1988)]
6.5 V.L. Ermolaev, E.B. Sveshnikova: Chem. Phys. Lett. **23**, 349 (1973)
6.6 V.L. Ermolaev, E.B. Sveshnikova: J. Lumin. **20**, 387 (1979)
6.7 V.P. Gapontsev, M.R. Sirtlanov, A.K. Gromov, A.A. Isineev: Proc. Int'l Conf. Lasers 81, USA (1981) p. 763
6.8 V.P. Gapontsev, A.A. Isineev, Yu.E. Sverchkov, M.R. Sirtlanov: Kvantovaya Elektronika **8**, 1824 (1981) [Engl. transl.: Sov. J. Quant. Electron. **11**, 1101 (1981)]
6.9 V.P. Gapontsev, M.R. Sirtlanov, W. Yen: J. Lumin. **31&32**, 201 (1984)
6.10 E.S. Medvedev, V.I. Osherov: Abstracts of papers, 5th Int'l Congr. Quant. Chem. Montréal, Canada (1985) Abstr. No. 45D
6.11 M.E. Paige, C.B. Harris: J. Chem. Phys. **93**, 1481 (1990)
6.12 J. Stone, M.F. Goodman: J. Phys. Chem. **89**, 1250 (1985)
6.13 K. Morokuma, K.F. Freed: J. Chem. Phys. **61**, 4342 (1974)
6.14 I.H. Kühn, F. Metz: Chem. Phys. **33**, 137 (1978)
6.15 I. Madzgalla, H. Paerschke, K.-E. Süsse, D.-G. Welsch: Ann. Phys. (Leipzig) **38**, 434 (1981)
6.16 M.A. Krivoglaz: Zh. Eksp. Teor. Fiz. **25**, 191 (1953)
6.17 R. Kubo, Y. Toyozawa: Progr. Theor. Phys. **13**, 160 (1955)
6.18 J.V. Caspar, B.P. Sullivan, E.M. Kober, J. Meyer: Chem. Phys. Lett. **91**, 91 (1982)
6.19 M.P. Tsvirko, G.F. Stelmakh, V.E. Pyatosin, K.N. Solovyov, T.F. Kachura, A.S. Piskarskas, R.A. Gadonas: Chem. Phys. **106**, 467 (1986)
6.20 T.A. Shakhverdov, V.N. Kalinin, R. Ergashev: Opt. Spektrosk. **61**, 518 (1986) [Engl. transl.: Opt. Spectrosc. (USSR) **61**, 325 (1986)]
6.21 M.F.H. Schuurmans, J.M.F. van Dijk: Physica B **123**, 131 (1984)
6.22 L.E. Brus, V.E. Bondybey: Spectroscopic and time-resolved studies of small molecule relaxation in the condensed phase, in *Radiationless Transitions*, ed. by S.H. Lin (Academic, New York 1980) pp. 259–288
6.23 E.B. Sveshnikova, N.T. Timofeev, A.O. Ivanov, S.G. Lunter, Yu.K. Fyodorov: Opt. Spektrosk. **54**, 259 (1983) [English transl.: Opt. Spectrosc. (USSR) **54**, 153 (1983)]
N.T. Timofeev: Nonradiative transitions trends in rare-earth ions from the point of view of the inductive-resonant theory, Candidate's Thesis, State Optical Institute, Leningrad (1984)
6.24 O.S. Mortensen, W. Siebrand, A.W. Tarr: Chem. Phys. **125**, 231 (1988)
6.25 S.I. Pekar, M.A. Krivoglaz: Proc. Inst. Phys. Acad. Sci. Ukranian SSR **4**, 37 (1953)
6.26 A.S. Davydov: Zh. Eksp. Teor. Fiz. **24**, 397 (1953)
6.27 S. Fischer: J. Chem. Phys. **53**, 3195 (1970)
6.28 M. Pagitsas, K.F. Freed: Chem. Phys. **23**, 387 (1977)
6.29 R.C. O'Rourke: Phys. Rev. **91**, 265 (1953)
6.30 S. Fischer, S. Schneider: Chem. Phys. Lett. **10**, 392 (1971)
A.L. Stanford, S.F. Fischer: Chem. Phys. **1**, 99 (1973)
6.31 R.R. Dogonadze, A.M. Kuznetsov, M.A. Vorotyntsev: Phys. Status Solidi (b) **54**, 125 (1972)
6.32 S.H. Lin: J. Chem. Phys. **58**, 5760 (1973)
6.33 E.S. Medvedev, V.I. Osherov, V.M. Pschenichnikov: Acad. Sci. USSR, Chernogolovka (1978) (unpublished)

6.34 E.C. Lim: J. Phys. Chem. **90**, 6770 (1986)
E.C. Lim: Acta Phys. Pol. **71A**, 695 (1987)
6.35 M. Terazima, S. Yamauchi, N. Hirota: J. Phys. Chem. **90**, 4294 (1986)
S. Yamauchi, K. Mibu, Y. Komada, N. Hirota: J. Phys. Chem. **91**, 6173 (1987)
6.36 U. Brühlmann, M. Nonella, P. Russegger, J.R. Huber: Chem. Phys. **81**, 439 (1983)
6.37 H. Hornburger, J. Brand: Chem. Phys. Lett. **88**, 153 (1982)
H. Hornburger, H. Schröder, J. Brand: J. Chem. Phys. **80**, 3197 (1984)
H. Hornburger, H. Kono, S.H. Lin: J. Chem. Phys. **81**, 3554 (1984)
H. Hornburger, C.M. Sharp, S. Leach: Chem. Phys. **101**, 67 (1986)
6.38 T. Suzuki, M. Ito: J. Phys. Chem. **91**, 3537 (1987)
6.39 H. Kono, S.H. Lin, E.W. Schlag: Chem. Phys. Lett. **145**, 280 (1988)
6.40 A.L. Sobolewski: Chem. Phys. **115**, 469 (1987)
A.L. Sobolewski, R. Czermiński: Chem. Phys. **130**, 123 (1989)
6.41 P. Russegger, J.R. Huber: Chem. Phys. **61**, 205 (1981)
P. Russegger, J.R. Huber: Chem. Phys. Lett. **92**, 38 (1982)
H. Stafast, J. Opitz, J.R. Huber: Chem. Phys. **56**, 63 (1981)
6.42 J.L. Knee, P.M. Johnson: J. Chem. Phys. **80**, 13 (1984)
J.L. Knee, P.M. Johnson: J. Phys. Chem. **89**, 948 (1985)
6.43 T.G. Dietz, M.A. Duncan, R.E. Smalley: J. Chem. Phys. **76**, 1227 (1982)
T.G. Dietz, M.A. Duncan, A.C. Puiu, R.E. Smalley: J. Phys. Chem. **86**, 4026 (1982)
M.D. Morse, A.C. Puiu, R.E. Smalley: J. Chem. Phys. **78**, 3435 (1983)
6.44 J.A. Syage, P.M. Felker, D.H. Semmes, F. Al Adel, A.H. Zewail: J. Chem. Phys. **82**, 2896 (1985)
L.R. Khundkar, A.H. Zewail: J. Chem. Phys. **84**, 1302 (1986)
6.45 G.A. Zalesskaya, V.T. Pavlova: Zh. Prikl. Spektrosk. **45**, 501 (1986)
N.A. Borisevich, G.A. Zalesskaya: Spectrosc. Lett. **19**, 113 (1986)
6.46 H.G. Löhmannsröben, K. Luther, M. Stuke: J. Phys. Chem. **91**, 3499 (1987)
6.47 O. Sneh, D. Dünn-Kittenplon, O. Chesnovsky: J. Chem. Phys. **91**, 7331 (1990)
6.48 S. Kamei, K. Okuyama, H. Abe, N. Mikami, M. Ito: J. Phys. Chem. **90**, 93 (1986)
S. Kamei, N. Mikami, M. Ito: J. Phys. Chem. **90**, 2321 (1986)
6.49 E. Villa, A. Amirav, E.C. Lim: J. Phys. Chem. **92**, 5393 (1988)
6.50 D.H. Semmes, J.S. Baskin, A.H. Zewail: J. Chem. Phys. **92**, 3359 (1990)
6.51 V.I. Osherov: Phys. Status Solidi **13**, 239 (1966)
6.52 F. Metz: Chem. Phys. **18**, 385 (1976)
6.53 D.-H. Lu, W.L. Hase: Chem. Phys. Lett. **142**, 187 (1987)
D.-H. Lu, W.L. Hase: J. Phys. Chem. **92**, 3217 (1988)
D.-H. Lu, W.L. Hase: J. Chem. Phys. **89**, 6723 (1988)
6.54 C.E. Keffer, C.P. Conner, W.H. Smith: Chem. Phys. Lett. **104**, 475 (1984)
6.55 V.S. Letokhov: *Nonlinear Selective Photoprocesses in Atoms and Molecules* (Nauka, Moscow 1983) (in Russian)
E.S. Medvedev: Chem. Phys. **41**, 103 (1979)
S.S. Mitra, S.S. Bhattacharyya: J. Phys. B **23**, 1453 (1990)
6.56 H.S. Taylor: Acc. Chem. Res. **22**, 263 (1989)
J.M. Gomez Llorente, H.S. Taylor: Phys. Rev. A **41**, 697 (1990)
6.57 J. Tennyson, S.C. Farantos: Chem. Phys. **93**, 237 (1985)
J. Tennyson, G. Brocks, S.C. Farantos: Chem. Phys. **104**, 399 (1986)
6.58 W. Rhodes, A. Velenik: Hamiltonian versus non-Hamiltonian dynamics in molecular processes: an adiabatic subsystem approach, in *Mathematics and Computational Concepts in Chemistry*, ed. by N. Trinajstić (Wiley, New York 1986) Chap. 24
6.59 J.P. Pique, Y. Chen. R.M. Field, J.L. Kinsey: Phys. Rev. Lett. **58**, 475 (1987)
6.60 K. Evans, D. Heller, S.A. Rice, R. Scheps: J. Chem Soc. Faraday Trans. 2 **69**, 856 (1973)
6.61 G.D. Gillespie, E.C. Lim: Chem. Phys. Lett. **34**, 513 (1975)
6.62 C.E. Otis, J.L. Knee, P.M. Johnson: J. Chem. Phys. **78**, 2091 (1983)
C.E. Otis, J.L. Knee, P.M. Johnson: J. Phys. Chem. **87**, 2232 (1983)

6.63 P.M. Borell, H.-G. Löhmannsröben, K. Luther: Chem. Phys. Lett. **136**, 371 (1987)
6.64 H. von Weyssenhoff, F. Kraus: J. Chem. Phys. **54**, 2387 (1971)
6.65 A. Amirav, U. Even, J. Jortner: J. Chem. Phys. **71**, 2319 (1979)
6.66 S. Hirayama: J. Chem. Phys. **85**, 6867 (1986)
6.67 A.D. Brailsford, T.Y. Chang: J. Chem. Phys. **53**, 3108 (1970)
6.68 K.F. Freed, D.F. Heller: J. Chem. Phys. **61**, 3942 (1974)
6.69 B.K. Selinger, W.R. Ware: J. Chem. Phys. **53**, 3160 (1970)
6.70 C.S. Parmenter, M.W. Schuyler: Chem. Phys. Lett. **6**, 339 (1970)
6.71 A.S. Abramson, K.G. Spears, S.A. Rice: J. Chem. Phys. **56**, 2291 (1972)
6.72 J.M. Blondeau, M. Stockburger: Chem. Phys. Lett. **8**, 436 (1971)
6.73 E.W. Schlag, S. Schneider, D.W. Chandler: Chem. Phys. Lett. **11**, 474 (1971)
L. Wunsch, H.J. Neusser, E.W. Schlag: Chem. Phys. Lett. **32**, 210 (1975)
6.74 F.M. Behlen, S.A. Rice: J. Chem. Phys. **75**, 5672 (1981)
F.M. Behlen, D.B. McDonald, V. Sethuraman, S.A. Rice: J. Chem. Phys. **75**, 5685 (1981)
6.75 C. Kolmeder, W. Zinth, W. Kaiser: Chem. Phys. Lett. **91**, 323 (1982)
6.76 M. Noble, E.K.C. Lee: J. Phys. Chem. **87**, 4360 (1983)
M. Noble, E.K.C. Lee: J. Chem. Phys. **80**, 134 (1984)
6.77 Y.J. Xie, H.S. Kwok: J. Chem. Phys. **89**, 6771 (1988)
6.78 H.-P. Lutz, H. Bitto, J.R. Huber: Chem. Phys. lett. **155**, 251 (1989)
6.79 S.F. Fischer: Chem. Phys. Lett. **4**, 333 (1969)
S.F. Fischer, E.W. Schlag: Chem. Phys. Lett. **4**, 393 (1969)
S.F. Fischer, E.W. Schlag, S. Schneider: Chem. Phys. Lett. **11**, 583 (1971)
6.80 T.J. Bevilacqua, B.K. Andrews, J.E. Stout, R.B. Weisman: J. Chem. Phys. **92**, 4627 (1990)
6.81 V.I. Osherov, E.S. Medvedev: Chem. Phys. Lett. **8**, 550 (1971)
6.82 E.N. Victorova, V.V. Zelinskii, D.A. Kozlovskii: Opt. Spektrosk. **23**, 820 (1967) [English transl.: Opt. Spectrosc. (USSR) **23**, 445 (1967)]
H.-P. Dorn, A. Müller: Chem. Phys. Lett. **130**, 426 (1986)
P. Dao, A.J. Twarowski: J. Chem. Phys. **85**, 6823 (1986)
6.83 M. Terazima, T. Azumi: Chem. Phys. Lett. **153**, 27 (1988)
6.84 S.A. Payne, R.M. Hochstrasser: J. Phys. Chem. **90**, 2068 (1986)
6.85 J.S. Langer: Ann. Phys. (New York) **41**, 108 (1967)
6.86 R. Kubo: A stochastic theory of line shape and relaxation, in *Fluctuation, Relaxation and Resonance in Magnetic Systems*, ed. by D. Ter Haar (Oliver and Boyd, Edinburgh 1962) pp. 23–68
R. Kubo: J. Phys. Soc. Jpn. **12**, 570 (1957)
6.87 E.W. Schlag, H. von Weyssenhoff: J. Chem. Phys. **51**, 2508 (1969)
6.88 J.O. Uy, E.C. Lim: J. Chem. Phys. **56**, 3374 (1972)
6.89 S.H. Lin: J. Chem. Phys. **56**, 2648 (1972)
6.90 I.-F. Hung, B.R. Henry: Chem. Phys. **27**, 263 (1978)
6.91 G.F. Hatch, M.D. Erlitz, G.C. Nieman: Environmental effects on phosphorescent lifetimes of benzene, in *Molecular Luminescence*, ed. by E.C. Lim (Benjamin, New York 1969) pp. 21–38
G.F. Hatch, M.D. Erlitz, G.C. Nieman: J. Chem. Phys. **49**, 3723 (1968)
6.92 S.G. Hadley, H.E. Rast, R.A. Keller: J. Chem. Phys. **39**, 705 (1963)
6.93 P.F. Jones, A.R. Calloway: Temperature and matrix effects on the decay of excited electronic states of molecules. III. Intersystem crossing in Naphthalene, in *Transitions Non-Radiatives dans les Molécules*, J. Chim. Phys. Phys.-Chim. Biol. (Suppl.) **10**, 110 (1970)
6.94 S. Okuyama, D.W. Oxtoby: J. Chem. Phys. **88**, 2405 (1988)
6.95 M. Kaschke, J. Kleinschmidt, M. Riegler: Chem. Phys. **107**, 89 (1986)
6.96 S.E. Stein, B.S. Rabinovitch: J. Chem. Phys. **58**, 2438 (1973)
6.97 J. Franck, H. Sponer: J. Chem. Phys. **25**, 172 (1956)
6.98 M. Gouterman: J. Chem. Phys. **36**, 2846 (1962)
6.99 Th. Förster: Ann. Phys. (Leipzig) **2**, 55 (1948)
6.100 D.L. Dexter: J. Chem. Phys. **21**, 836 (1953)
6.101 M.D. Galanin, I.M. Frank: Zh. Eksp. Teor. Fiz. **21**, 114 (1951)

M.D. Galanin: Zh. Eksp. Teor. Fiz. **21**, 126 (1951)
V.M. Agranovich, M.D. Galanin: *Electronic Excitation Energy Transfer in Condensed Matter* (Nauka, Moscow 1978) (in Russian)
6.102 E.B. Sveshnikova, N.T. Timofeev: Opt. Spektrosk. **51**, 833 (1981) [English transl.: Opt. Spectrosc. (USSR) **51**, 462 (1981)]
N.T. Timofeev, E.B. Sveshnikova: Opt. Spektrosk. **54**, 1005 (1983) [English transl.: Opt. Spectrosc. (USSR) **54**, 595 (1983)]
6.103 E.B. Sveshnikova, S.P. Naumov: Opt. Spektrosk. **44**, 127 (1978) [English transl.: Opt. Spectrosc. (USSR) **44**, 72 (1978)]
S.P. Naumov, E.B. Sveshnikova: Opt. Spektrosk. **45**, 903 (1978) [English transl.: Opt. Spectrosc. (USSR) **45**, 761 (1978)]
6.104 V.L. Ermolaev, E.B. Sveshnikova, V.S. Tachin: Opt. Spektrosk. **41**, 343 (1976) [English transl.: Opt. Spectrosc. (USSR) **41**, 198 (1976)]
6.105 E.N. Bodunov, E.B. Sveshnikova: Opt. Spektrosk. **36**, 340 (1974) [English transl.: Opt. Spectrosc. (USSR) **36**, 196 (1974)]
E.N. Bodunov: Opt. Spektrosk. **40**, 942 (1976) [English transl.: Opt. Spectrosc. (USSR) **40**, 537 (1976)]
6.106 V.L. Ermolaev, E.B. Sveshnikova: Opt. Spektrosk. **68**, 780 (1990) [English transl.: Opt. Spectrosc. (USSR) **68**, 455 (1990)]
6.107 I.B. Neporent, E.B. Sveshnikova, A.P. Serov: Izv. Akad. Nauk SSSR, Ser. Fiz. **39**, 1959 (1975)
6.108 E.B. Sveshnikova, A.P. Serov, V.P. Kondakova: Opt. Spektrosk. **39**, 285 (1975) [English transl.: Opt. Spectrosc. (USSR) **39**, 162 (1975)]

Chapter 7

7.1 A. Nitzan, J. Jortner, P.M. Rentzepis: Chem. Phys. Lett. **8**, 445 (1971)
7.2 C.A. Langhoff, G.W. Robinson: Chem. Phys. **6**, 34 (1974)
7.3 J.O. Berg, C.A. Langhoff, G.W. Robinson: Chem. Phys. Lett. **29**, 305 (1974)
7.4 J.O. Hirschfelder, Ch. F. Curtiss, R.B. Bird: *Molecular Theory of Gases and Liquids* (Wiley, New York 1954)
7.5 N.N. Korst, V.I. Osherov: Zh. Eksp. Teor. Fiz. **51**, 825 (1966) [English transl.: Sov. Phys. - JETP **24**, 550 (1967)]
7.6 E.S. Medvedev, V.I. Osherov: Dokl. Akad. Nauk SSSR **180**, 905 (1968) [English transl.: Dokl. Phys. Chem. **180**, 410 (1968)]
7.7 E.S. Medvedev, V.I. Osherov: Fiz. Tverd. Tela **10**, 2509 (1968) [English transl.: Sov. Phys. - Solid State **10**, 1968 (1969)]
7.8 V.I. Osherov, E.S. Medvedev: Phys. Status Solidi **36**, 69 (1969)
7.9 V.I. Osherov: J. Chem. Phys. **47**, 3885 (1966)
7.10 J.H. Callomon, T.M. Dunn, I.M. Mills: Philos. Trans. R. Soc. London, Ser. A, **259**, 499 (1966)
7.11 J.M. Hollas, S.N. Thakur: Mol. Phys. **22**, 203 (1971)
7.12 K.Y. Tang, P.W. Fairchild, E.K.C. Lee: J. Chem. Phys. **66**, 3303 (1977)
7.13 R. Kubo, K. Tomita: J. Phys. Soc. Jpn. **9**, 888 (1954)
7.14 P.S. Hubbard: Rev. Mod. Phys. **33**, 249 (1961)
7.15 V.I. Osherov: Zh. Eksp. Teor. Fiz. **55**, 539 (1968) [English transl.: Sov. Phys. -JETP **28**, 282 (1969)]
7.16 A. Nitzan, J. Jortner: J. Chem. Phys. **58**, 2412 (1973)

Subject Index

MIX
Papier aus verantwortungsvollen Quellen
Paper from responsible sources
FSC® C105338

If you have any concerns about our products,
you can contact us on
ProductSafety@springernature.com

In case Publisher is established outside the EU,
the EU authorized representative is:
Springer Nature Customer Service Center GmbH
Europaplatz 3, 69115 Heidelberg, Germany

Printed by Libri Plureos GmbH
in Hamburg, Germany